Communications in Computer and Information Science 1441

More information about this series at http://www.springer.com/series/7899

Mayank Singh · Vipin Tyagi ·
P. K. Gupta · Jan Flusser ·
Tuncer Ören · V. R. Sonawane (Eds.)

Advances in Computing and Data Sciences

5th International Conference, ICACDS 2021
Nashik, India, April 23–24, 2021
Revised Selected Papers, Part II

Editors
Mayank Singh
Consilio Research Lab
Tallinn, Estonia

P. K. Gupta
Jaypee University of Information Technology
Waknaghat, Himachal Pradesh, India

Tuncer Ören
University of Ottawa
Ottawa, ON, Canada

Vipin Tyagi
Jaypee University of Engineering and Technology
Guna, Madhya Pradesh, India

Jan Flusser
Institute of Information Theory and Automation
Prague, Czech Republic

V. R. Sonawane
MVPS's Karmaveer Adv. Baburao Ganpatrao Thakare College of Engineering
Nashik, Maharashtra, India

ISSN 1865-0929 ISSN 1865-0937 (electronic)
Communications in Computer and Information Science
ISBN 978-3-030-88243-3 ISBN 978-3-030-88244-0 (eBook)
https://doi.org/10.1007/978-3-030-88244-0

This Springer imprint is published by the registered company Springer Nature Switzerland AG
The registered company address is: Gewerbestrasse 11, 6330 Cham, Switzerland

Preface

Computing techniques like big data, cloud computing, machine learning, and the Internet of Things (IoT) etc. are playing a key role in the processing of data and retrieving of advanced information. Several state-of-art techniques and computing paradigms have been proposed based on these techniques. This volume contains papers presented at 5th International Conference on Advances in Computing and Data Sciences (ICACDS 2021) held during April 23–24, 2021, by MVPS'S Karmaveer Adv. Baburao Ganpatrao Thakare College of Engineering, Nashik, Maharashtra, India. Due to the COVID-19 pandemic, ICACDS 2021 was organized virtually. The conference was organized specifically to help bring together researchers, academicians, scientists, and industry experts and to derive benefits from the advances of next generation computing technologies in the areas of advanced computing and data sciences.

The Program Committee of ICACDS 2021 is extremely grateful to the authors who showed an overwhelming response to the call for papers, with over 781 papers submitted in the two tracks of Advanced Computing and Data Sciences. All submitted papers went through a double-blind peer-review process, and finally 103 papers were accepted for publication in the Springer CCIS series. We are thankful to the reviewers for their efforts in finalizing the high-quality papers.

The conference featured many distinguished personalities like Vineet Kansal, Pro-Vice Chancellor, A. P. J. Abdul Kalam Technical University, India; Komal Bhatia, J. C. Bose University of Science and Technology, India; Tuncer Ören, University of Ottawa, Canada; Aparna Pandey, Gulf Medical University, UAE; Shailendra Mishra, Majmaah University, Saudi Arabia; Robert Siegfried, Aditerna GmbH, Germany; Anup Girdhar, Sedulity Group, India; Arun Sharma, Indira Gandhi Delhi Technical University for Women, India; Sarika Sharma, Symbiosis International (Deemed University), India; Shashi Kant Dargar, University of KwaZulu-Natal, South Africa; and Prathmesh Churi, NMIMS University, India, among many others. We are very grateful for the participation of all speakers in making this conference a memorable event.

The Organizing Committee of ICACDS 2021 is indebted to Smt. Neelimatai V. Pawar, Sarchitnis, MVPS, India, for the confidence that she gave to us during organization of this international conference, and all faculty members and staff of MVPS'S Karmaveer Adv. Baburao Ganpatrao Thakare College of Engineering (KBTCOE), India, for their support in organizing the conference and for making it a grand success.

We would also like to thank N. S. Patil, N. B. Desale, and S. P. Jadhav, KBTCoE, India; Sameer Kumar Jasra, University of Malta, Malta; Hemant Gupta, Carleton University, Canada; Nishant Gupta and Archana Sar, MGM CoET, India; Arun Agarwal, University of Delhi, India; Kunj Bihari Meena, Neelesh Jain, Nilesh Patel, and Kriti Tyagi, JUET Guna, India; Vibhash Yadav, REC Banda, India; Sandhya Tarar, Gautam Buddha University, India; Vimal Dwivedi, Abhishek Dixit, and Vipin

Deval, Tallinn University of Technology, Estonia; Sumit Chaudhary, Indrashil University, India; Supraja P., SRM Institute of Science and Technology, India; Lavanya Sharma, Amity University, Noida, India; Sheshang Degadwala, Sigma Institute of Engineering, India; Poonam Tanwar and Rashmi Agarwal, MRIIS, India; Rohit Kapoor, SK Info Techies, Noida, India; and Akshay Chaudhary and Tarun Pathak, Consilio Intelligence Research Lab, India, for their support.

Our sincere thanks to Consilio Intelligence Research Lab, India; the GISR Foundation, India; SK Info Techies, India; and Print Canvas, India, for sponsoring the event.

August 2021

Mayank Singh
Vipin Tyagi
P. K. Gupta
Jan Flusser
Tuncer Ören
V. R. Sonawane

Organization

Steering Committee

Alexandre Carlos Brandão Ramos	UNIFEI, Brazil
Mohit Singh	Georgia Institute of Technology, USA
H. M. Pandey	Edge Hill University, UK
M. N. Hooda	BVICAM, India
S. K. Singh	IIT BHU, India
Jyotsna Kumar Mandal	University of Kalyani, India
Ram Bilas Pachori	IIT Indore, India
Alex Norta	Tallinn University of Technology, Estonia

Chief Patron

Neelimatai V. Pawar	Sarchitnis, Maratha Vidya Prasarak Samaj, India

Patrons

Tushar R. Shewale (President)	Maratha Vidya Prasarak Samaj, India
Manikrao M. Boraste Sabhapati	Maratha Vidya Prasarak Samaj, India

Honorary Chairs

N. S. Patil (Education Officer and Principal)	MVPS's KBTCOE, India
N. B. Desale (Vice-principal)	MVPS's KBTCOE, India

General Chairs

Jan Flusser	Institute of Information Theory and Automation, Czech Republic
Mayank Singh	Consilio Research Lab, Estonia

Advisory Board Chairs

Shailendra Mishra	Majmaah University, Saudi Arabia
P. K. Gupta	JUIT Solan, India
Vipin Tyagi	JUET Guna, India

Technical Program Committee Chairs

Tuncer Ören	University of Ottawa, Canada
Viranjay M. Srivastava	University of KwaZulu-Natal, South Africa
Ling Tok Wang	National University of Singapore, Singapore
Ulrich Klauck	Aalen University, Germany
Anup Girdhar	Sedulity Group, Delhi, India
Arun Sharma	Indira Gandhi Delhi Technical University for Women, India

Conference Chair

V. R. Sonawane	MVPS's KBTCOE, India

Conference Co-chair

S. P. Jadhav	MVPS's KBTCOE, India

Conveners

Sameer Kumar Jasra	University of Malta, Malta
Hemant Gupta	Carleton University, Canada

Co-conveners

V. C. Shewale	MVPS's KBTCOE, India
Ghanshyam Raghuwanshi	Manipal University, India
Prathamesh Churi	NMIMS, Mumbai, India
Lavanya Sharma	Amity University, Noida, India

Organizing Chairs

Shashi Kant Dargar	University of KwaZulu-Natal, South Africa
V. S. Pawar	MVPS's KBTCOE, India

Organizing Co-chairs

Abhishek Dixit	Tallinn University of Technology, Estonia
Vibhash Yadav	REC Banda, India
Nishant Gupta	MGM CoET, India

Organizing Secretaries

Akshay Kumar	CIRL, India
Rohit Kapoor	SKIT, India
M. P. Kadam	MVPS's KBTCOE, India

Creative Head

Tarun Pathak	Consilio Intelligence Research Lab, India

Program Committee

A. K. Nayak	Computer Society of India, India
A. J. Nor'aini	Universiti Teknologi MARA, Malaysia
Aaradhana Deshmukh	Alaborg University, Denmark
Abdel Badeeh Salem	Ain Shams University, Egypt
Abdelhalim Zekry	Ain Shams University, Egypt
Abdul Jalil Manshad Khalaf	University of Kufa, Iraq
Abhhishek Verma	Indian Institute of Information Technology and Management, Gwalior, India
Abhinav Vishnu	Pacific Northwest National Laboratory, USA
Abhishek Gangwar	Center for Development of Advanced Computing, India
Aditi Gangopadhyay	IIT Roorkee, India
Adrian Munguia	AI MEXICO, USA
Amit K. Awasthi	Gautam Buddha University, India
Antonina Dattolo	University of Udine, Italy
Arshin Rezazadeh	University of Western Ontario, Canada
Arun Chandrasekaran	National Institute of Technology Karnataka, India
Arun Kumar Yadav	National Institute of Technology Hamirpur, India
Asma H. Sbeih	Palestine Ahliya University, Palestine
Brahim Lejdel	University of El-Oued, Algeria
Chandrabhan Sharma	University of the West Indies, West Indies
Ching-Min Lee	I-Shou University, Taiwan
Deepanwita Das	National Institute of Technology Durgapur, India
Devpriya Soni	Jaypee Institute of Information Technology, India
Donghyun Kim	Georgia State University, Georgia
Eloi Pereira	University of California, Berkeley, USA
Felix J. Garcia Clemente	Universidad de Murcia, Spain
Gangadhar Reddy Ramireddy	RajaRajeswari College of Engineering, India
Hadi Erfani	Islamic Azad University, Iran
Harpreet Singh	Alberta Emergency Management Agency, Canada
Hussain Saleem	University of Karachi, Pakistan
Jai Gopal Pandey	CSIR-Central Electronics Engineering Research Institute, Pilani, India
Joshua Booth	University of Alabama in Huntsville, Alabama
Khattab Ali	University of Anbar, Iraq
Lokesh Jain	Delhi Technological University, India
Manuel Filipe Santos	University of Minho, Portugal
Mario José Diván	National University of La Pampa, Argentina
Megat Farez Azril Zuhairi	Universiti Kuala Lumpur, Malaysia

Mitsunori Makino	Chuo University, Japan
Moulay Akhloufi	Université de Moncton, Canada
Naveen Aggarwal	Panjab University, India
Nawaz Mohamudally	University of Technology, Mauritius
Nileshkumar R. Patel	Jaypee University of Engineering and Technology, India
Nirmalya Kar	National Institute of Technology Agartala, India
Nitish Kumar Ojha	Indian Institute of Technology Allahabad, India
Paolo Crippa	Università Politecnica delle Marche, Italy
Parameshachari B. D.	GSSS Institute of Engineering and Technology for Women, India
Patrick Perrot	Gendarmerie Nationale, France
Prathamesh Chur	SVKM's NMIMS Mukesh Patel School of Technology Management and Engineering, India
Pritee Khanna	Indian Institute of Information Technology, Design and Manufacturing Jabalpur, India
Purnendu Shekhar Pandey	Indian Institute of Technology (Indian School of Mines) Dhanbad, India
Quoc-Tuan Vien	Middlesex University, UK
Rubina Parveen	Canadian All Care College, Canada
Saber Abd-Allah	Beni-Suef University, Egypt
Sahadeo Padhye	Motilal Nehru National Institute of Technology, India
Sarhan M. Musa	Prairie View A&M University, Texas, USA
Shamimul Qamar	King Khalid University, Saudi Arabia
Shashi Poddar	University at Buffalo, USA
Shefali Singhal	Madhuben & Bhanubhai Patel Institute of Technology, India
Siddeeq Ameen	University of Mosul, Iraq
Sotiris Kotsiantis	University of Patras, Greece
Subhasish Mazumdar	New Mexico Tech, New Mexico
Sudhanshu Gonge	Symbiosis International University, India
Tomasz Rak	Rzeszow University of Technology, Poland
Vigneshwar Manoharan	Bharath Corporate, India
Xiangguo Li	Henan University of Technology, China
Youssef Ouassit	Hassan II University, Morocco

Sponsor

Consilio Intelligence Research Lab, India

Co-sponsors

GISR Foundation, India
Print Canvas, India
SK Info Techies, India

Contents – Part II

Predicting Seasonal Vaccines and H1N1 Vaccines Using Machine Learning Techniques 1
Sourav P. Adi, Keshav V. Bharadwaj, and Vivek Bettadapura Adishesha

Handling Class Imbalance in Electroencephalography Data Using Synthetic Minority Oversampling Technique 12
Vibha Patel, Jaishree Tailor, and Amit Ganatra

Dissemination of Firm's Market Information: Application of Kermack-Mckendrick SIR Model 22
Renji George Amballoor and Shankar B. Naik

Improving Image-Based Dialog by Reducing Modality Biases 33
Jay Gala, Hrishikesh Shenai, Pranjal Chitale, Kaustubh Kekre, and Pratik Kanani

Dependency Parser for Hindi Using Integer Linear Programming 42
R. Sai Kesav, B. Premjith, and K. P. Soman

Medical Records Management Using Distributed Ledger and Storage 52
Samia Anjum, R. Ramaguru, and M. Sethumadhavan

Detection of Depression and Suicidal Ideation on Social Media: An Intrinsic Review 63
Sanat Madkar, Tanay Maheshwari, Mann Merani, Rahil Merchant, and Pankti Doshi

Frequency Based Feature Extraction Technique for Text Documents in Tamil Language 76
M. Mercy Evangeline, K. Shyamala, L. Barathi, and R. Sandhya

An Approach of Devanagari License Plate Detection and Recognition Using Deep Learning 85
Pankaj Raj Dawadi, Manish Pokharel, and Bal Krishna Bal

COMBINE: A Pipeline for SQL Generation from Natural Language 97
Youssef Mellah, Abdelkader Rhouati, El Hassane Ettifouri, Toumi Bouchentouf, and Mohammed Ghaouth Belkasmi

Recognition of Isolated Gestures for Indian Sign Language Using Transfer Learning 107
Kinjal Mistree, Devendra Thakor, and Brijesh Bhatt

A Study of Five Models Based on Non-clinical Data for the Prediction of Diabetes Onset in Medically Under-Served Populations 116
Rohit Srivastava, Sandeep Kumar, Vivudh Fore, and Ravi Tomar

Representation and Visualization of Students' Progress Data Through Learning Dashboard 125
Anagha Vaidya and Sarika Sharma

Denoising of Computed Tomography Images for Improved Performance of Medical Devices in Biomedical Engineering 136
Harjinder Kaur, Deepti Gupta, and Mamta Juneja

Image Dehazing Through Dark Channel Prior and Color Attenuation Prior. . . 147
Jacob John and Prabu Sevugan

Predicting the Death of Road Accidents in Bangladesh Using Machine Learning Algorithms 160
Md. Abu Bakkar Siddik, Md. Shohel Arman, Afia Hasan, Mahmuda Rawnak Jahan, Majharul Islam, and Khalid Been Badruzzaman Biplob

Numerical Computation of Finite Quaternion Mellin Transform Using a New Algorithm. 172
Khinal Parmar and V. R. Lakshmi Gorty

Predictive Modeling of Tandem Silicon Solar Cell for Calculating Efficiency. 183
S. V. Katkar, K. G. Kharade, N. S. Patil, V. R. Sonawane, S. K. Kharade, and R. K. Kamat

Text Summarization of an Article Extracted from Wikipedia Using NLTK Library 195
K. G. Kharade, S. V. Katkar, N. S. Patil, V. R. Sonawane, S. K. Kharade, T. S. Pawar, and R. K. Kamat

Grapheme to Phoneme Mapping for Tamil Language 208
M. Geerthana Anusha, D. Govind, and Vijay Krishna Menon

Comparative Study of Physiological Signals from Empatica E4 Wristband for Stress Classification 218
Varun Chandra, Ankit Priyarup, and Divyashikha Sethia

An E-Commerce Prototype for Predicting the Product Return Phenomenon Using Optimization and Regression Techniques 230
Vidya Rajasekaran and R. Priyadarshini

Crop Yield Prediction for India Using Regression Algorithms. 241
Devansh Hiren Timbadia, Sughosh Sudhanvan, Parin Jigishu Shah, and Supriya Agrawal

A Novel Framework for Multimodal Twitter Sentiment Analysis Using Feature Learning 252
Jamuna S. Murthy, Amulya C. Shekar, Drishti Bhattacharya, R. Namratha, and D. Sripriya

An Iterative Approach Based Reversible Data Hiding with Weight Update for Dual Stego Images 262
C. Shaji and I. Shatheesh Sam

Lower and Upper Bounds for 'Useful' Renyi Information Rate. 271
Pankaj Prasad Dwivedi and D. K. Sharma

Sign Language Recognition Using Convolutional Neural Network 281
Mihir Gandhi, Priyam Shah, Devansh Solanki, and Prasanna Shete

Prediction of Stock Price for Indian Stock Market: A Comparative Study Using LSTM and GRU 292
Shwetha Salimath, Triparna Chatterjee, Titty Mathai, Pooja Kamble, and Megha Kolhekar

Early Prediction of Cardiovascular Disease Among Young Adults Through Coronary Artery Calcium Score Technique. 303
Anurag Bhatt, Sanjay Kumar Dubey, and Ashutosh Kumar Bhatt

Confidentiality Leakage Analysis of Database-Driven Applications 313
Angshuman Jana and Anwesha Kashyap

Comparative Learning of on Request Direction-Finding Procedures in WSNs 324
Rakesh Kumar Saini, Mayank Singh, and Nishant Gupta

ECG Based Stress Detection in Automobile Drivers Using Long Short-Term Memory (LSTM) Network 333
Ramyashri B. Ramteke and Vijaya R. Thool

Evaluation of Soil Moisture for Estimation of Irrigation Pattern by Using Machine Learning Methods 343
Abhishek Khanna and Sanmeet Kaur

Sentiment Analysis in Online Learning Environment: A Systematic Review 353
Sarika Sharma, Vipin Tyagi, and Anagha Vaidya

Image Splicing Forgery Detection Techniques: A Review 364
Kunj Bihari Meena and Vipin Tyagi

A Predictive Model for Classification of Breast Cancer Data Sets 389
S. Venkata Achuta Rao and Pamarthi Rama Koteswara Rao

Impact of COVID-19 on the Health of Elderly Person 404
Ravindra Kumar

The Determinants of Visit Frequency and Buying Intention at Shopping Centers in Vietnam 412
Dam Tri Cuong and Nguyen Thanh Long

Correction to: Frequency Based Feature Extraction Technique for Text Documents in Tamil Language C1
M. Mercy Evangeline, K. Shyamala, L. Barathi, and R. Sandhya

Author Index 423

Contents – Part I

An Energy-Efficient Hybrid Hierarchical Clustering Algorithm for Wireless Sensor Devices in IoT 1
Nitesh Chouhan and S. C. Jain

Fund Utilization Under Parliament Local Development Scheme: Machine Learning Base Approach 15
Arun Sharma and Deepa Paliwal

Implementing Automatic Ontology Generation for the New Zealand Open Government Data: An Evaluative Approach 26
Paramjeet Kaur and Parma Nand

Blockchain Based Framework to Maintain Chain of Custody (CoC) in a Forensic Investigation 37
Sarishma, Abhishek Gupta, and Preeti Mishra

Parameters Extraction of the Double Diode Model for the Polycrystalline Silicon Solar Cells 47
T. Suganya, V. Rajendran, and P. Mangaiyarkarasi

A Light SRGAN for Up-Scaling of Low Resolution and High Latency Images. 56
Archan Ghosh, Kalporoop Goswami, Riju Chatterjee, and Paramita Sarkar

Energy Efficient Clustering Routing Protocol and ACO Algorithm in WSN 68
Shalini Subramani, M. Selvi, S. V. N. Santhosh Kumar, and A. Kannan

Efficient Social Distancing Detection Using Object Detection and Triangle Similarity 81
Vidya Zope, Nikhil Joshi, Srivatsan Iyengar, Krish Mahadevan, and Meher Singh

Explaining a Black-Box Sentiment Analysis Model with Local Interpretable Model Diagnostics Explanation (LIME) 90
Kounteyo Roy Chowdhury, Arpan Sil, and Sharvari Rahul Shukla

Spelling Checking and Error Corrector System for Marathi Language Text Using Minimum Edit Distance Algorithm 102
Kavita. T. Patil, R. P. Bhavsar, and B. V. Pawar

A Study on Morphological Analyser for Indian Languages: A Literature Perspective 112
Jayashree Nair, L. S. Aiswarya, and P. R. Sruthy

Cyber Safety Against Social Media Abusing 124
Yuvraj Anil Jadhav, Sakshi Jitendra Jain, Bhushan Sanjay More, Mayur Sunil Jadhav, and Bhushan Chaudhari

Predictive Rood Pattern Search for Efficient Video Compression 137
Hussain Ahmed Choudhury

An Effective Approach for Classifying Acute Lymphoblastic Leukemia Using Hybrid Hierarchical Classifiers 151
Sharath Sunil, P. Sonu, S. Sarath, R. Rahul Nath, and Vivek Viswan

Abnormal Blood Vessels Segmentation for Proliferative Diabetic Retinopathy Screening Using Convolutional Neural Network 162
Vasavi Agarwal, Ridhi Sipani, and P. Saranya

Predictive Programmatic Classification Model to Improve Ad-Campaign Click Through Rate 171
Nisheel Saseendran and C. Sneha

Live Stream Processing Techniques to Assist Unmanned, Regulated Railway Crossings 181
Jacob John, Mariam Varkey, and M. Selvi

Most Significant Bit-Plane Based Local Ternary Pattern for Biomedical Image Retrieval 193
Nilima Mohite, Manisha Patil, Anil Gonde, and Laxman Waghmare

Facial Monitoring Using Gradient Based Approach 204
Arush Jain, Mani Sachdeva, and Paramita De

Overlapped Circular Convolution Based Feature Extraction Algorithm for Classification of High Dimensional Datasets 214
Rupali Tajanpure and Akkalakshmi Muddana

Binary Decision Tree Based Packet Queuing Schema for Next Generation Firewall 224
Manthan Patel and P. P. Amritha

Automatic Tabla Stroke Source Separation Using Machine Learning 234
Shambhavi Shete and Saurabh Deshmukh

Classification of Immunity Booster Medicinal Plants Using CNN: A Deep Learning Approach 244
Md. Musa, Md. Shohel Arman, Md. Ekram Hossain, Ashraful Hossen Thusar, Nahid Kawsar Nisat, and Arni Islam

Machine Learning Model Interpretability in NLP and Computer Vision Applications 255
Navoneel Chakrabarty

Optimal Sizing and Siting of Multiple Dispersed Generation System Using Metaheuristic Algorithm 268
Lokesh Kumar Yadav, Mitresh Kumar Verma, and Puneet Joshi

Design of a Fused Triple Convolutional Neural Network for Malware Detection: A Visual Classification Approach 279
Santosh K. Smmarwar, Govind P. Gupta, and Sanjay Kumar

Mobile Agent Security Using Lagrange Interpolation with Multilayer Perception Neural Network 290
Pradeep Kumar, Niraj Singhal, Mohammad Asim, Ajay Kumar, and Mahboob Alam

Performance Analysis of Channel Coding Techniques for 5G Networks 303
Mrinmayi Patil, Sanjay Pawar, and Zia Saquib

An Ensemble Learning Approach for Software Defect Prediction in Developing Quality Software Product 317
Yakub Kayode Saheed, Olumide Longe, Usman Ahmad Baba, Sandip Rakshit, and Narasimha Rao Vajjhala

A Study on Energy-Aware Virtual Machine Consolidation Policies in Cloud Data Centers Using Cloudsim Toolkit 327
Dipak Dabhi and Devendra Thakor

Predicting Insomnia Using Multilayer Stacked Ensemble Model 338
Md. Sabab Zulfiker, Nasrin Kabir, Al Amin Biswas, and Partha Chakraborty

A Novel Encryption Scheme Based on Fully Homomorphic Encryption and RR-AES Along with Privacy Preservation for Vehicular Networks 351
Righa Tandon and P.K. Gupta

Key-Based Decoding for Coded Modulation Schemes in the Presence of ISI 361
Vanaja Shivakumar

Optimizing the Performance of KNN Classifier for Human Activity Recognition 373
Ali Al-Taei, Mohammed Fadhil Ibrahim, and Nada Jasim Habeeb

Face Recognition with Disguise and Makeup Variations Using Image Processing and Machine Learning 386
Farah Jawad Al-ghanim and Ali mohsin Al-juboori

Attention-Based Deep Fusion Network for Retinal Lesion Segmentation in Fundus Image 401
A. Mary Dayana and W. R. Sam Emmanuel

Visibility Improvement in Hazy Conditions via a Deep Learning Based Image Fusion Approach 410
Satbir Singh, Asifa Mehraj Baba, Md. Imtiyaz Anwar, Ayaz Hussain Moon, and Arun Khosla

Performance of Reinforcement Learning Simulation: x86 v/s ARM 420
Sameer Pawanekar and Geetanjali Udgirkar

A Performance Study of Probabilistic Possibilistic Fuzzy C-Means Clustering Algorithm 431
J. Vijaya and Hussian Syed

Optimized Random Forest Algorithm with Parameter Tuning for Predicting Heart Disease 443
Ajil D. S. Vins and W. R. Sam Emmanuel

Machine Learning Based Techniques for Detection of Renal Calculi in Ultrasound Images 452
Harsha Herle and K. V. Padmaja

Unsupervised Change Detection in Remote Sensing Images Using CNN Based Transfer Learning 463
Josephina Paul, B. Uma Shankar, Balaram Bhattacharyya, and Alak Kumar Datta

Biological Sequence Embedding Based Classification for MERS and SARS 475
Shamika Ganesan, S. Sachin Kumar, and K. P. Soman

Supply Path Optimization in Video Advertising Landscape 488
Ujwala Musku and Prakhar Yadav

Stack-Based CNN Approach to Covid-19 Detection 500
V. S. Suryaa and Z. Sayf Hussain

Performance Analysis of Various Classifiers for Social Intimidating Activities Detection 512
Mansi Mahendru and Sanjay Kumar Dubey

Technique for Enhancing the Efficiency and Security of Lightweight IoT Devices 528
Santosh P. Jadhav, Georgi Balabanov, and Vladimir Poulkov

Performance Improvement in Deep Learning Architecture for Phonocardiogram Signal Classification Using Spectrogram 538
R. Sai Kesav, M. Bhanu Prakash, Krishanth Kumar, V. Sowmya, and K. P. Soman

Performance Analysis of Machine Learning Techniques in Device Free Localization in Indoor Environment 550
K. S. Anusha, R. Ramanathan, and M. Jayakumar

D-Leach: An Energy Optimized Deterministic Sub-clustering and Multi-hop Routing Protocol for Wireless Sensor Networks 561
Subhash Chandra Gupta and Mohammad Amjad

Robust Image Watermarking Using Support Vector Machine and Multi-objective Particle Swarm Optimization 571
Kapil Jain and Parmalik Kumar

Generalized Intuitionistic Fuzzy Entropy on IF-MARCOS Technique in Multi-criteria Decision Making 592
Rishikesh Chaurasiya and Divya Jain

Feature Selection in Machine Learning by Hybrid Sine Cosine Metaheuristics 604
Nebojsa Bacanin, Aleksandar Petrovic, Miodrag Zivkovic, Timea Bezdan, and Milos Antonijevic

Rainfall Prediction Using Logistic Regression and Support Vector Regression Algorithms 617
Srikantaiah K. C and Meenaxi M. Sanadi

Collaborative Recommender System (CRS) Using Optimized SGD - ALS . . . 627
Gopal Behera and Neeta Nain

Violence Detection from CCTV Footage Using Optical Flow and Deep Learning in Inconsistent Weather and Lighting Conditions 638
R. Madhavan, Utkarsh, and J. V. Vidhya

Speech Based Multiple Emotion Classification Model Using Deep Learning 648
Shakti Swaroop Patneedi and Nandini Kumari

A Legal-Relationship Establishment in Smart Contracts: Ontological Semantics for Programming-Language Development 660
Vimal Dwivedi and Alex Norta

Aspect Based Sentiment Analysis – An Incremental Model Learning Approach Using LSTM-RNN 677
Alka Londhe and P. V. R. D. Prasada Rao

Cloud Based Exon Prediction Using Maximum Error Normalized Logarithmic Algorithms 690
Md. Zia Ur Rahman, Annabathuni Chandra Haneesh, Bhimireddy Shanmukha Sai Reddy, Sala Surekha, and Putluri Srinivasareddy

Unsupervised Learning of Visual Representations via Rotation and Future Frame Prediction for Video Retrieval. 701
Vidit Kumar, Vikas Tripathi, and Bhaskar Pant

Application of Deep Learning in Classification of Encrypted Images 711
Geetansh Saxena, Girish Mishra, and Noopur Shrotriya

Ear Recognition Using Pretrained Convolutional Neural Networks 720
K. R. Resmi and G. Raju

An Adaptive Service Placement Framework in Fog Computing Environment. 729
Pankaj Sharma and P. K. Gupta

Efficient Ink Mismatch Detection Using Supervised Approach 739
Garima Jaiswal, Arun Sharma, and Sumit Kumar Yadav

Author Index 747

Predicting Seasonal Vaccines and H1N1 Vaccines Using Machine Learning Techniques

Sourav P. Adi[1(✉)], Keshav V. Bharadwaj[2], and Vivek Bettadapura Adishesha[3]

[1] Accenture, Bengaluru, Karnataka, India
[2] Mad Street Den, Bengaluru, Karnataka, India
[3] Strukton-BHEL, Bengaluru, Karnataka, India

Abstract. A swine flu virus dubbed as H1N1 hit the planet in 2009, impacting millions of people. Swine flu appears to be an infection of the respiratory tract, meaning it damages the respiratory apparatus. This life-threatening virus proved to be a nightmare as it affected millions of people which mainly included children and the middle-aged. It became incredibly important to get vaccinated against this infection as it could save lives. A Couple of vaccines initially proved to be effective in various parts of the world and also among different aged and health conditioned people. In this research, a machine learning model is developed to help in estimating the probability of a person receiving seasonal and H1N1 vaccines. The data for this research is provided by the National 2009 H1N1 Flu Survey (NHFS) team, which conducted a study via telephone calls in the United States. The data consists of a total of 36 attributes. A Gradient boosting classifier is developed to address the problem. The model was evaluated on the separate test data given by the NHFS and the performance metric was the area under Receiver Operating Characteristic (ROC) curve. With the model developed using necessary parameter tuning we were able to achieve the best score of 0.8368.

Keywords: Swine flu · Machine learning · Gradient boosting classifier · Roc curve

1 Introduction

Swine flu [1,5–7] is a respiratory infection caused by influenza strain. It all started in the year 1919 (Spanish Flu) [7–9] and even today it circulates as the seasonal flu virus. Pain in the muscle areas, dry cough, gastrointestinal problems like vomiting, diarrhea, fatigue, breathing problems were some of the common symptoms observed. The people who were more prone to get infected were the children, pregnant women, and the elderly. Analgesics, Antiviral medications were used for the treatment. Later in April 2009 the virus made a comeback and combining with the Eurasian pig flu virus it was called Swine Flu. Researchers estimated the actual number of infected people to be around 700 million to 1.4

M. Singh et al. (Eds.): ICACDS 2021, CCIS 1441, pp. 1–11, 2021.
https://doi.org/10.1007/978-3-030-88244-0_1

billion (11%–21% of the globe). It is estimated that the flu pandemic has caused about 284,000 deaths and WHO reports say about 250,000–500,000 people die every year from seasonal flu. The first case of H1N1 was detected in California in the year 2009. The vaccines [20,21] for the virus were developed by growing virus in the chicken eggs and finally delivered in November 2009. Two vaccines [19–21] that proved effective were TIV (TrivalentInactivated Vaccine) and LAIV (Live Attenuated Influenza Vaccine) [16,22]. In early 2010, the United States of America conducted a survey in which respondents were asked if they had received H1N1 and Seasonal Flu vaccines. The survey covered various topics about the respondents like their financial status, demographic background, the effectiveness of vaccines, etc. Several researchers have used various methodologies to understand the virus, forecast [17] the behavior, etc. In this research, we use machine learning models to predict the probability of an individual receiving the H1N1 vaccine and seasonal vaccine i,e. two probabilities. To achieve the specified task, we have used the gradient boosting classifier [15,23] and OneVsRest Classifier. The performance of the built models was measured using different test data and the performance parameter used was the area under the receiver's operating characteristic curve (ROC Curve) [10–12].

2 Literature Survey

Geeta Chhabra et al. [2] presented an overview of different imputation techniques for manipulation of missing data. The authors discuss about the traditional techniques like mean imputation, hot-deck imputation etc. and Modern and Hybrid methodologies like Regression imputation, KNN and clustering their advantages and limits. The paper concludes as a reference to various imputation methods so as to suit the problem statement.

Q. Song and Martin Shepperd [3] discuss about missing data imputation techniques to the KDD and ML communities. The paper gives an insight about three missing data techniques taxonomy. Single imputation techniques like PCA, Sequential imputation, mean imputation etc. and multiple imputation methods which proves as a rectifier to disadvantages of single imputation techniques. Concluding the paper authors suggest a standard procedure to follow while imputation is being used.

Aizaz Chaudhry et al. [4] propose a novel approach to convert a single variable to multivariate form. Proposed method is later tested on LTE spectrum data. Proposed methodology is later tested against Kalman filtering method for multivariate imputation proving that the method outperforms Kalman filtering.

Mujoriya Rajesh Z, David M. Morens, Anthony S. Fauci [9] conducted a research to answer various questions about the 1918 influenza pandemic. Authors try to partially answer the questions linking to origin, epidemiologic features, cause for mortality etc. The authors talk about pathogenesis leading to deaths among various aged people. The authors do mention in the paper that the mortality rate was high among the youngsters than aged with necessary graphs. There is also a mention about the pandemic in 3 different waves, predicting the pandemics, preventing mortality. The research concludes that if a pandemic as that of 1918 reappears, with the help of aggressive medical interference, 1.9 million fatalities could be prevented.

Samithamby Senthilnathan [13] makes an attempt to explain correlation analysis between two variables which helps to identify level of multicollinearity in the model. Karl Pearson's coefficient of correlation has been used as a scope of research. Coefficient of Determination (R^2) and its significance in explaining variance of dependent and independent variable is mentioned. An insight about Data based and structure-based multicollinearity can also be observed. VIF (Variance Inflation Factor) and its analysis to measure multicollinearity is clearly mentioned.

Nashreen Sultana et al. [19] discuss various time series forecasting models in order to forecast future cases of the virus. Box-Cox transformation, Neural Nets, Exponential Smoothing, Seasonal Naive are the models used. Applied forecasting models are calculated with various metrics like Root Mean Squared Error (RMSE), Mean Absolute Error (MAE), Auto Correlation Function (ACF) etc. The researchers claim that the neural network forecasting model outperforms all others with 98.4% accuracy.

Dr. Colin R Simpson et al. [20] conducted a research to analyse the effectiveness of the vaccine for a sample of Scottish people. Retrospective cohort design methodology which involved linking data of patient, virological swabs, death certification was used. Their outcome was that the effectiveness of vaccination was 19.5% in case of prevention of emergency hospitalization and 77% in case of prevention of laboratory confirmed influenza.

William W. Busse et al. [21] conducted a research to analyse vaccine effectiveness in asthma patients. 390 people were involved in the study and within each severity group, people were injected 15–30 μg of H1N1 vaccine intramuscularly twice 21 d apart. The outcome was that the 21 d seroprotection was observed for both 15 μg and 30 μg inmoderate asthmaticpatients while for severe asthmatic patients, seroprotection was 77.9% and 94.1% for15 μg and 30 μg doses respectively. The authors conclude that the monovalent inactivated 2009 H1N1 pandemic influenza vaccine was safe and assured seroprotection and also for people older than 60 years should be diagnosed with 30 μg dose.

3 Data Analysis

3.1 Dataset

The dataset for this research is provided by the National Centre for Health Statistics, US Department of Health and Human Services (DHHS) [22]. The dataset consists of 36 columns (attributes) of which respondent_id is unique and 35 other columns out of which most of them were binary type and others were of ordinal type. The goal was to predict the probability of a person receiving seasonal and H1N1 vaccines. The predicted values were later evaluated based on the ROC [10–12] curve. The problem falls under multilabel problem [14] not a multiclass problem.

3.2 Data Cleaning and Pre-processing

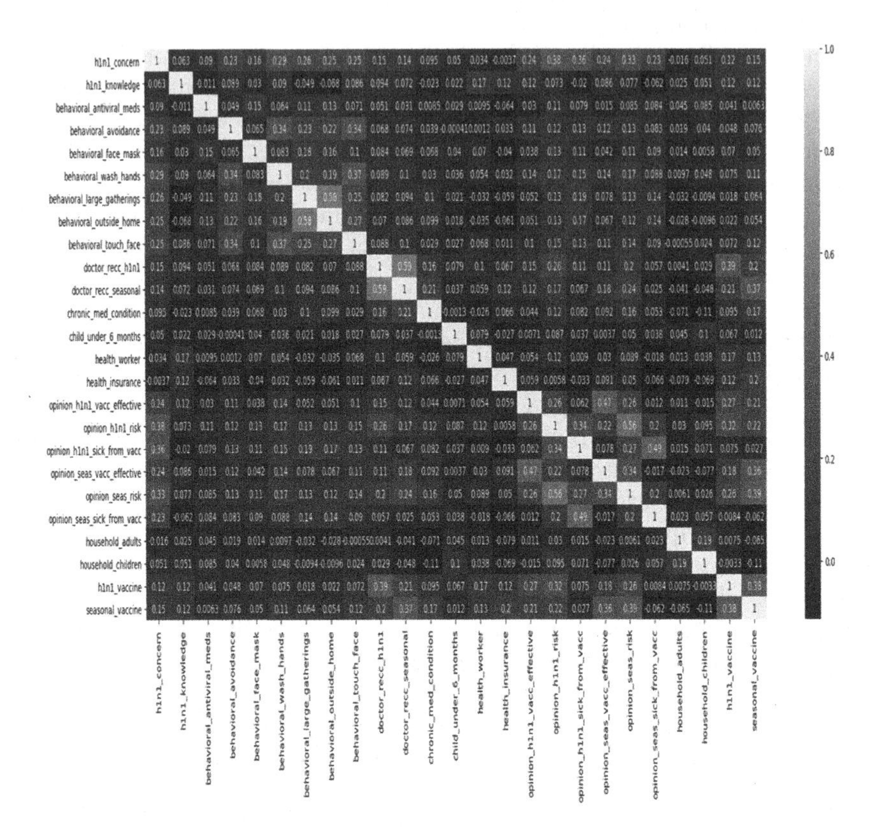

Fig. 1. Correlation heatmap for the dataset

The first step in the data analysis was to check if the data has any duplicate and missing values, duplicates were not observed, but the dataset had many missing values in different attributes ranging from 19 to 13470. So, our first concern was to clean the data. Since the attributes employment_concern and employment_occupation had the highest missing data (13470 values) we decided to drop the same along with respondent id as it was of no use. The next step involved checking of correlation [13] between the attributes. This was done using the seaborn library's correlation heatmap Fig. 1. It was noted that two attributes named doctor_recc_H1N1_vaccine and doctor_recc_seasonal_vaccine attributes were positively correlated by 60%. Hence, only one attribute has been retained. In the next step, the missing data were handled. Two different imputation methods are implemented in this project, the first one being mean imputation and the second one forward filling imputation [2–4]. The last stage involved converting all object type attributes to categorical type to achieve one-hot encoding. After the data is cleaned and pre-processed, graphical visualizations are generated.

3.3 Data Visualization

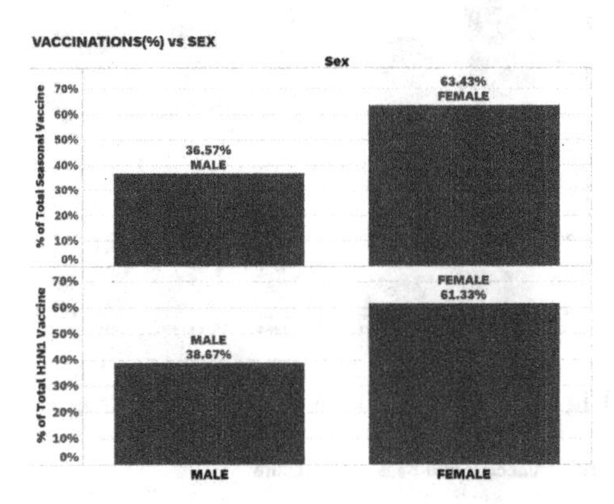

Fig. 2. Vaccination for male and female

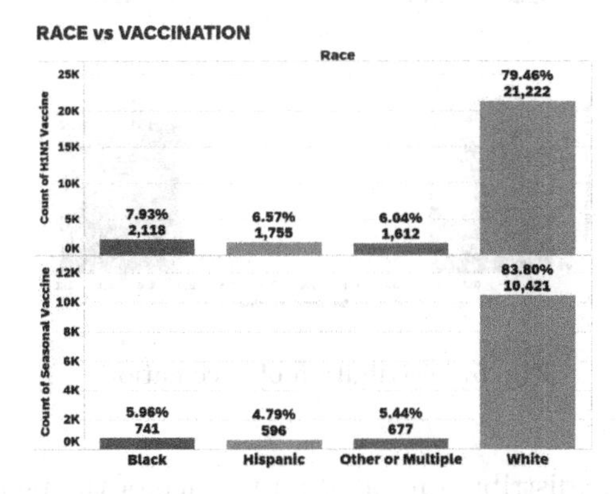

Fig. 3. Vaccination for different race

Different attributes are plotted to check how the data is distributed.

Firstly, it is observed that out of 5674 people who received the H1N1 vaccines 61.33% were female and only 38.67% were male. The same case was also observed with seasonal medicines through which one can conclude that women are more prone to get affected than men Fig. 2.

Next, the seasonal and H1N1 vaccine attributes were plotted. It was observed that about 12,435 (46.561%) of individuals received the seasonal vaccine while 14,272 (53.439%) did not. On the other hand, 21,033 (78.75%) of individuals did not receive the H1N1 vaccine while only 5674 (21.25%) received the same Fig. 5.

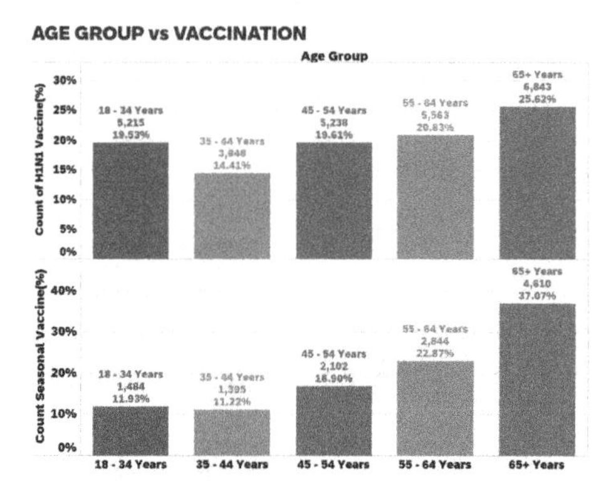

Fig. 4. Vaccination for different age groups

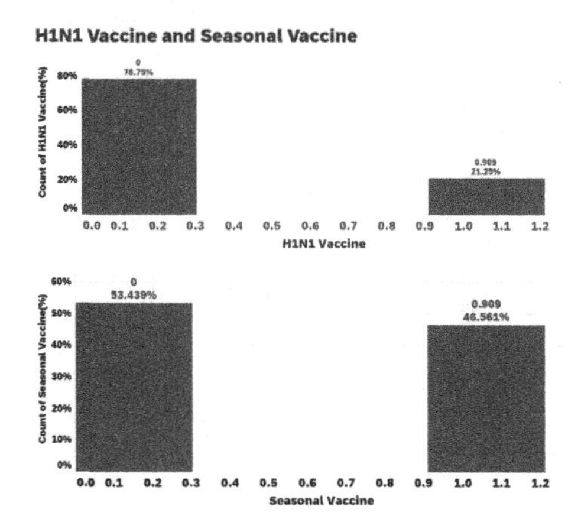

Fig. 5. Distribution of vaccination

Next, the vaccine distribution based on the age of the people is checked. It was observed that in both cases people above 65 years received more vaccines than any other aged individuals Fig. 4.

A graph to conclude white people received the highest vaccination than any other race is depicted in Fig. 3.

Finally, Based on the income it was observed that people with an income level of less than or equal to $75,000 (Above the poverty line) received vaccination than any others Fig. 6.

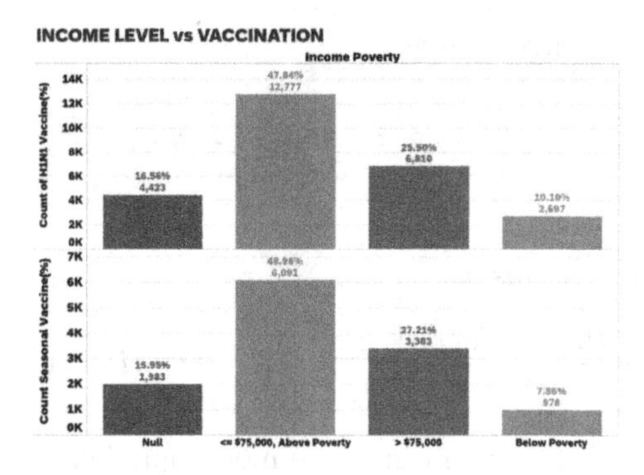

Fig. 6. Vaccination received based on income level

4 Training and Testing the Model

Before the algorithm is used, the data is broken down into training and testing sets. Data is divided as 85% training and 15% testing. Before adding one-hot encoding, some of the attributes are converted to a categorical form. One Hot Encoding was accomplished by the Pandas get_dummies function.

Two different approaches are utilized to achieve the task assigned. The first one was to use OnevsRest Classifier and the second one was to split the problem into two individual problems and develop two different classifiers and predict the output.

5 Methodologies

5.1 Method 1 - OneVsRest

The first method involved developing a single gradient boosting classifier, passing it to the OneVsRest classifier, and fitting it to the training data. The method was used along with forward fill imputation and was able to achieve descent accuracy and AUC score. 83.82% accuracy was obtained on H1N1 vaccine predictions and 77.91% of accuracy was obtained on Seasonal vaccines. Overall AUC score was 0.746. Figure 7a and Fig. 7b.

5.2 Gradient Boosting

Algorithm

- Initialize training set $(x_i, y_i)_{i=1}^{n}$, a differentiable loss function $\mathrm{L}\left(\mathrm{y}, \mathrm{F}\left(\mathrm{x}\right)\right)$ and number of Iterations to M

- Initialize model with fixed value:

$$F_0(x) = \underset{\gamma}{\operatorname{argmin}} \sum_{i=1}^{n} L(y_i, \gamma) \tag{1}$$

- For m = 1 to M
  Compute pseudo residuals as

$$r_{im} = -\frac{\partial L(y_i, F(x_i))}{\partial F(x_i)}\bigg|_{F(x)=F_{m-1}(x)} \quad for\ (i) = 1, 2, 3, ..., n \tag{2}$$

  where i is the row number and m is the model number.
- Fit a base learner $h_m(x)$ to the pseudo-residuals
- Compute γ_m using

$$\gamma_m = \underset{\gamma}{\operatorname{argmin}} \sum_{i=1}^{n} L\left(y_i, F_{m-1}\left(x_i\right) + \gamma h_m\left(x_i\right)\right) \tag{3}$$

  where γ_m is the predicted value by the model in m^{th} iteration.
- Update the model as

$$F_m(x) = F_{m-1}(x) + \gamma h_m(x) \tag{4}$$

- Finally output $F_M(x)$

The term L in (1, 2, and 3) is the Loss Function which can be Mean Squared Error, Mean Absolute Error, Root mean squared error for regression, and for classification it can be Log-loss, hinge loss, etc. The important point to consider while choosing the loss function is that the loss function must be differentiable. For our problem, we use Binomial Deviance as loss function which is calculated as

$$L = \log\left(1 + 2e^{-(2y\hat{f})}\right) \tag{5}$$

Working of Gradient Boosting

- Calculate the initial prediction as

$$Predicted\,Value = \frac{e^{\log\left(\frac{Positives}{Negatives}\right)}}{1 + e^{\log\left(\frac{Positives}{Negatives}\right)}} \tag{6}$$

  Equation (6) is the same formula as that of logistic regression.
- Calculate the residuals as

$$Residual = Actual\,Value - Predicted\,Value \tag{7}$$

  where,
  Actual Value = 1 and 0 for Positive and Negative examples respectively
  Predicted Value = Log odds calculated using Eq. (7)

- Build the tree using all other attributes as input and Residuals calculated in previous step as output.
- Apply the transformation formula to calculate the output value for all leaf nodes using

$$\frac{\sum Residual_i}{\sum [Previous\ Probability_i * (1 - Previous\ Probability_i)]} \tag{8}$$

- For each entity in the dataset the log(odds) predicted value will be

$$\begin{aligned} \log{(odds)}_{Prediction} = & \ Prediction_{Initial} \\ & + (Learning\_Rate * Output_{Leafnode}) \end{aligned} \tag{9}$$

and the probability using

$$Probability = \frac{e^{\log(odds)}}{1 + e^{\log(odds)}} \tag{10}$$

Method 2 - Gradient Boosting Classifier. In the second method, two different gradient boosting classifiers are used to solve the challenge. One classifier predicting seasonal vaccine and other H1N1 vaccines. The parameters were optimally tuned to achieve an AUC score of ~0.74 on the test set and .8366 on the validation set. Good accuracy of 86% on the prediction of H1N1 vaccines and 79.96% on Seasonal Vaccines are achieved. ROC curve was plotted to understand how the model was performing Fig. 8a, 8b. Finally, the developed classifiers were applied on validation data supplied by the driven data platform which resulted in an AUC score of 0.8366.

6 Results and Evaluation

The model results were evaluated based on the Area under the ROC curve. ROC curve is a plot between True Positive Rate(TPR)/sensitivity and False Positive Rate(FPR)/(1-specificity) on the Y and X-axis respectively. ROC curve depicts the overall performance of the model. The diagonal line in Fig. 8a, 8b represents the case of not using the model. The line above the diagonal line captures how sensitivity and precision shift as the odds change. A model with a higher AUC is always preferred. In our case, an AUC score of 0.8368 for the OneVsRest classifier and 0.8366 for the gradient boosting method were reported. The AUC scores indicate that both the models are fairly good.

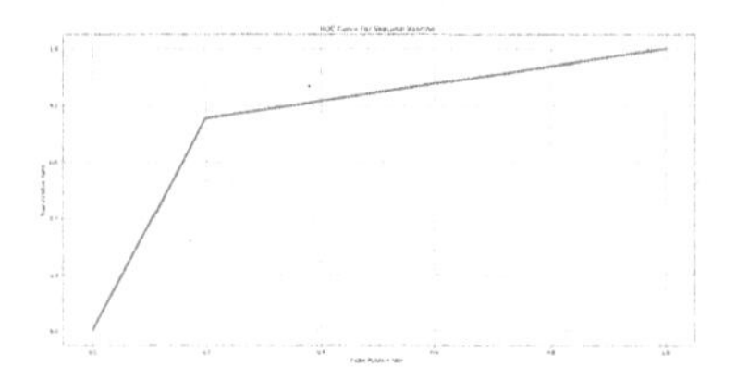

(a) ROC Curve for Seasonal Vaccine

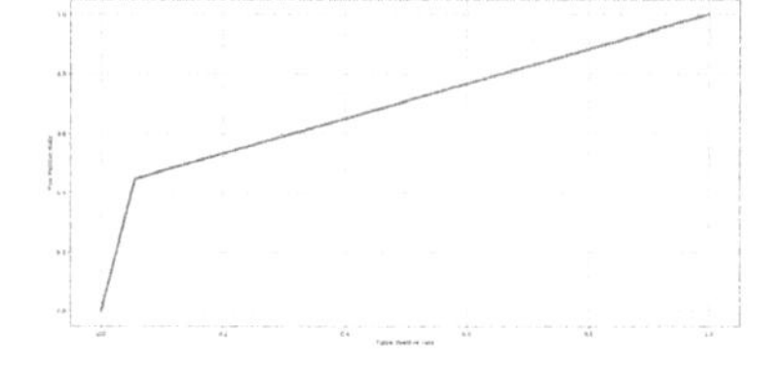

(b) ROC Curve for H1N1 Vaccine

Fig. 7. ROC curves plotted using forward fill imputation

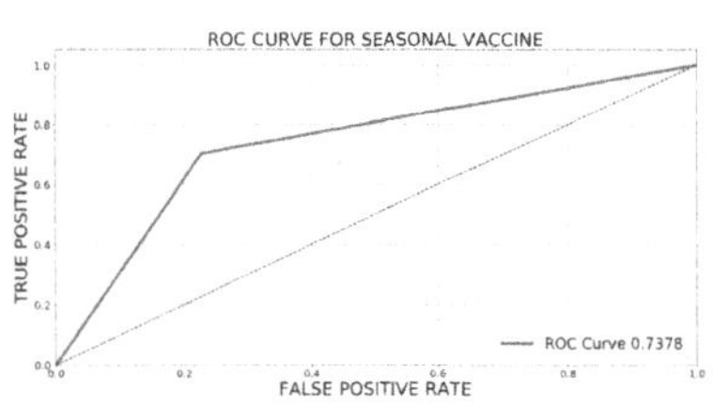

(a) ROC Curve for Seasonal Vaccine

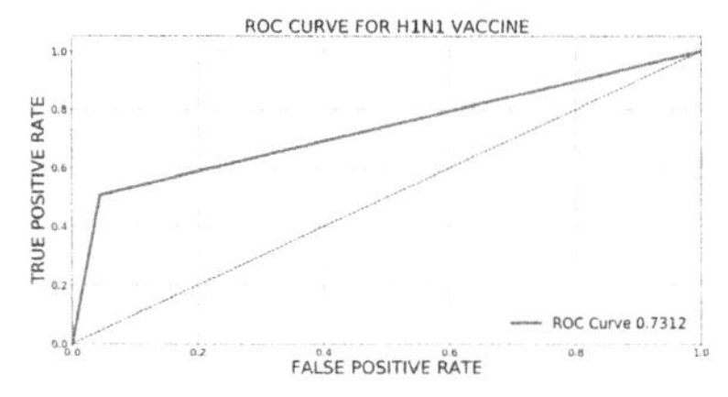

(b) ROC Curve for H1N1 Vaccine

Fig. 8. ROC curves plotted using mean imputation

7 Conclusion

To conclude, a fairly good model has been built to predict probabilities of Seasonal and H1N1 vaccinations received by individuals. In the research, the power of gradient boosting classifier was utilized to achieve an AUC score of 0.8366 and 0.8368 respectively. As a future scope of this project, different algorithms can be tested with the application of more profound mathematics to the data, which may improve the AUC score. An enhanced effort in data cleaning and visualization can be carried out to come up with more accurate conclusions about the data. The drawback of using a gradient boosting classifier is its increased complexity and time-consumption with more estimators and depth. Hence building better models to reach the goal in an improvised time constraint is preferred.

References

1. Baria, H.G., et al.: International journal community medicine. Public Health **4**(10), 3668–3672 (2017)
2. Chhabra, G., Vashisht, V., Ranjan, J.: A review on missing data value estimation using imputation algorithm. J. Adv. Res. Dyn. Control Syst. **11**, 312–318 (2019)
3. Song, Q., Shepperd, M.: Missing data imputation techniques. IJBIDM **2**, 261–291 (2007). https://doi.org/10.1504/IJBIDM.2007.015485
4. Chaudhry, A., Li, W., Basri, A., Patenaude, F.: A method for improving imputation and prediction accuracy of highly seasonal univariate data with large periods of missingness. Wirel. Commun. Mob. Comput. **2019**, 1–13 (2019). https://doi.org/10.1155/2019/4039758

5. Sinha, M.: Swine flu. J. Infect. Public Health **2**, 157–166 (2009). https://doi.org/10.1016/j.jiph.2009.08.006
6. Mujariya, R., Kishore, D., Bodla, R.: A review on study of swine flu. Indo-Global Res. J. Pharm. Sci. **1**, 47–51 (2011)
7. Tsoucalas, G., Kousoulis, A., Sgantzos, M.: The 1918 Spanish Flu Pandemic, the Origins of the H1N1-virus Strain, a Glance in History. Euro. J. Clin. Biomed. Sci. **2** (2016)
8. Taubenberger, J.K.: The origin and virulence of the 1918 "Spanish" influenza virus. Proc. Am. Philos. Soc. **150**(1), 86–112 (2006). PMID: 17526158; PMCID: PMC2720273
9. Morens, D.M., Fauci, A.S.: The 1918 influenza pandemic: insights for the 21st century. J. Infect. Dis. **195**(7), 1018–1028 (2007)
10. Hajian-Tilaki, K.: Receiver Operating Characteristic (ROC) curve analysis for medical diagnostic test evaluation. Caspian J Intern Med. **4**(2), 627–635 (2013). PMID: 24009950; PMCID: PMC3755824
11. Park, S.H., Goo, J.M., Jo, C.H.: Receiver Operating Characteristic (ROC) curve: practical review for radiologists. Korean J. Radiol. **5**(1), 11–18 (2004)
12. Bradley, A.P.: The use of the area under the ROC curve in the evaluation of machine learning algorithms. Pattern Recogn. **30**(7), 1145–1159 (1997). ISSN 0031–3203
13. Senthilnathan, S.: Usefulness of correlation analysis. SSRN Electron. J. (2019). https://doi.org/10.2139/ssrn.3416918
14. Li, R., Liu, W., Lin, Y., Zhao, H., Zhang, C.: An ensemble multilabel classification for disease risk prediction. J. Healthcare Eng. **2017**, 1–10 (2017). https://doi.org/10.1155/2017/8051673
15. Natekin, A., Knoll, A.: Gradient boosting machines. A Tutorial. Front. Neurorobot. **7**, 21 (2013). https://doi.org/10.3389/fnbot.2013.00021
16. TIV and LAIV Influenza Vaccines for 2011–2012. http://www.epi.alaska.gov/bulletins/docs/b2011_25.pdf
17. Sultana, N., Sharma, N.: Statistical Models for Predicting Swine F1u Incidences in India. In: 2018 First International Conference on Secure Cyber Computing and Communication (ICSCCC), pp. 134–138. Jalandhar, India (2018). https://doi.org/10.1109/ICSCCC.2018.8703300
18. Simpson, C.R., et al.: Effectiveness of H1N1 vaccine for the prevention of pandemic influenza in Scotland, UK: a retrospective observational cohort study. Lancet Infect. Dis. **12**(9), 696–702 (2012)
19. Busse, W.W., et al.: Vaccination of patients with mild and severe asthma with a 2009 pandemic H1N1 influenza virus vaccine. J. Allergy Clin. Immunol. **127**(1), 130–137.e3 (2011)
20. Wu, J., et al.: Safety and Effectiveness of a 2009 H1N1 Vaccine in Beijing. N. Engl. J. Med. **363**, 2416–23 (2010). https://doi.org/10.1056/NEJMoa1006736
21. Ambrose, C.S., Levin, M.J., Belshe, R.B.: The relative efficacy of trivalent live attenuated and inactivated influenza vaccines in children and adults. Influenza Other Respir. Viruses **5**, 67–75 (2011). https://doi.org/10.1111/j.1750-2659.2010.00183.x
22. Flu Shot Learning: Predict H1N1 and Seasonal Flu Vaccines. https://www.drivendata.org/competitions/66/flu-shot-learning/
23. Gradient Boosting In Classification: Not a Black Box Anymore! https://blog.paperspace.com/gradient-boosting-for-classification/

Handling Class Imbalance in Electroencephalography Data Using Synthetic Minority Oversampling Technique

Vibha Patel[1(✉)], Jaishree Tailor[2], and Amit Ganatra[3]

[1] Department of Computer Engineering, Chhotubhai Gopalbhai Patel Institute of Technology, Uka Tarsadia University, Bardoli, Gujarat, India
vibha.patel@utu.ac.in

[2] Shrimad Rajchandra Institute of Management and Computer Application, Uka Tarsadia University, Bardoli, Gujarat, India

[3] Department of Computer Engineering, DEPSTAR, Charotar University of Science and Technology, Anand, Gujarat, India

Abstract. The issue of class imbalance is very common in medical diagnosis applications. The problem of class imbalance arises when the distribution of samples across the known classes is skewed. It affects the performance of robust machine learning algorithms. The performance evaluation parameters used to measure the machine learning models are also very important in the case of class imbalanced datasets. This paper critically evaluates the performance of random forest classifier for the application of seizure state recognition using electroencephalography (EEG) data. Synthetic Minority Oversampling Technique (SMOTE) is used to balance the EEG dataset. Parameters such as accuracy, sensitivity, specificity, precision, false positive rate (FPR), f1-score, and area under the receiver operating characteristic curve (AUC) are considered for evaluation and analysis of the performance of actual class imbalanced dataset and class balanced dataset using SMOTE. The results indicate that the effect of balancing class does not affect the accuracy of the model much, however, it improves the sensitivity and AUC parameters significantly; which are the important parameters in case of measuring the predictive performance of class imbalanced dataset.

Keywords: Machine learning · Class imbalance · Synthetic Minority Oversampling Technique (SMOTE) · Random forest · Electroencephalography (EEG)

1 Introduction

The real-world applications are vastly incorporating machine learning techniques. The real-world data tend to be imperfect because of many reasons which include missing values, inconsistency, noise, uncertainty, etc. Therefore, the machine learning algorithms need to be robust to extract the important patterns from the imperfect data. It is reported that some robust machine learning algorithms are also showing poor results for real-world applications [1]. One of the reasons is the class imbalance problem. The class

M. Singh et al. (Eds.): ICACDS 2021, CCIS 1441, pp. 12–21, 2021.
https://doi.org/10.1007/978-3-030-88244-0_2

imbalance problem arises when there is a vast difference in the ratio of each class [2, 3], i.e. one set of classes, called majority class, dominates the other set of classes, called minority classes. In the case of binary classification, one class dominates the other. It implies machine learning models to be biased towards majority classes. The problem of class imbalance may appear in the applications like disease detection, earthquake prediction, intrusion detection, and spam email filtering.

The class imbalance problem handling mechanism is an important aspect to elevate the predictive performance of the classifier while working with highly imbalanced datasets [1, 4, 5]. There are various approaches to deal with the class imbalance problem [6–9]. The first approach is Data Preprocessing, which includes resampling, and feature extraction and selection techniques. The second approach is cost-sensitive learning, which includes methods for training data modification, changing learning process or learning objective to build a cost-sensitive classifier, and methods based on Bayes theory.

Medical diagnosis datasets are generally suffering from class imbalance problem [1, 10]. The work presented here critically evaluates the performance of applying oversampling technique Synthetic Minority Oversampling Technique (SMOTE) [6], on electroencephalography (EEG) dataset for the application of epileptic seizure state recognition. Epilepsy is a chronic brain disorder in which a person with epilepsy suffers from recurrent seizures [11, 12]. EEG signal recordings are the most common technique to collect brain signals through invasive or non-invasive methods [11]. EEG signals are widely used for seizure detection and seizure prediction [13]. The EEG dataset is highly imbalanced due to the vast difference in the presence of normal and abnormal signals [12, 14]. Machine learning algorithms that use these datasets generally suffer from poor performance due to the problem of class imbalance [15]. Even if the machine learning model exhibits very high accuracy, it is possible that the model is not able to find a single instance of the seizure in case of seizure state recognition problem [16]. So, the use of appropriate performance evaluation parameters is also very crucial in machine learning prediction problems with imbalanced datasets. The work by Khaled Mohamad Almustafa [17] observes that the random forest classifier exhibits better performance than all other machine learning classifiers for the application of seizure state recognition. In this work, random forest algorithm has been used as the classifier.

The organization of the rest of the paper is as follows: the following section describes the dataset used and preprocessing required for the task of EEG based epileptic seizure state recognition. Then the details of the epileptic seizure state recognition model with SMOTE and random forest algorithm are given, which is followed by the discussion on important parameters to consider for class imbalance problem, results of the experiments, and conclusion.

2 Dataset Information and Preprocessing

The Children's Hospital Boston and the Massachusetts Institute of Technology (CHB-MIT) dataset [18] is used in this work to perform the seizure state recognition task. The dataset of Children's Hospital Boston (CHB) and the Massachusetts Institute of Technology (MIT), known as CHB-MIT is considered for the task of epileptic seizure state recognition. This dataset is open source and downloaded from its official web-page: https://physionet.org/content/chbmit/1.0.0/. The dataset is consisting of the electroencephalography (EEG) recording of 22 subjects, grouped into 23 cases. The patients are 5 males with ages between 3 to 22 years, and 17 females with ages between 1.5 to 19 years. The dataset comes with a subject-info text file that contains the gender and age of each subject. The detailed information is available on the official web page: physionet.org. The dataset contains variations in channel numbers and types used amongst patients. The duration of the EEG recordings also varies. Preprocessing is very crucial in using the dataset of CHB-MIT. The dataset does not contain labels of seizure states. Seizure events are explicitly mentioned in the annotation file with seizure start and end times. So, preprocessing also includes seizure label generation. This study included subject numbers Chb01, Chb02, Chb03, Chb05, Chb06, Chb07, Chb08, Chb10, and Chb23 because these cases contain uniform 23 channels. The summary file of each subject is parsed to generate the labels, 'Class 0' for the negative tuple and 'Class 1' for the positive tuple. Table 1 contains statistics of the dataset in which 'Class 0' indicates normal signal and 'Class 1' indicates abnormal signal. As the abnormal signals are very few into the recording, the dataset tends to be highly imbalanced.

3 Epileptic Seizure State Recognition Model

There are two approaches used in this study to evaluate the effect of balancing class on prediction results: first uses original class imbalanced data to train and test the model, and second uses SMOTE to balance the classes in the training set. Statistics of the training set and testing set samples for both the cases are given in Table 1. Both these approaches are evaluated using the random forest classifier as mentioned in the introduction section. Following is the description of SMOTE and random forest classifier used in literature for the task of epileptic seizure state recognition.

3.1 Synthetic Minority Oversampling Technique (SMOTE)

SMOTE is an oversampling technique that is different from traditional replacement based oversampling. In the highly skewed dataset, SMOTE performs oversampling of the class with fewer samples, called minority class, by generating synthetic examples [6]. SMOTE uses feature space instead of data space to generate synthetic examples. The minority class is oversampled by taking each minority class sample and introducing a synthetic sample along the line segment joining all or any of the k minority class nearest neighbor. This technique is used successfully in the literature to handle the class imbalance problem of EEG data [14, 19–21]. The work presented here uses the imbalanced-learn library of Scikit-learn, a machine learning library for Python.

3.2 Random Forest Classifier

The model used in this study does not use explicit feature selection as random forest classifier performs implicit feature selection using a small subset of strong variables [22]. This classifier outperforms other traditional machine learning classifiers for the task of epileptic seizure state recognition [14, 17, 23–27].

Table 1. Dataset preprocessing statistics of CHB-MIT dataset

Subject	Total	Original		Train/Test split		Train classes ratio		Test classes ratio		Train SMOTE	After SMOTE	
		Class 0	Class 1	Train samples	Test samples	Class 0	Class 1	Class 0	Class 1		Class 0	Class 1
Chb01	37372928	37257984	114944	25039861	12333067	24962767	77094	12295217	37850	49925534	24962767	24962767
Chb02	32501504	32456704	44800	21776007	10725497	21745961	30046	10710743	14754	43491922	21745961	21745961
Chb03	35022336	34917632	104704	23464965	11557371	23395038	69927	11522594	34777	46790076	23395038	23395038
Chb05	35944960	35800832	144128	24083123	11861837	23986545	96578	11814287	47550	47973090	23986545	23986545
Chb06	61502976	61461248	41728	41206993	20295983	41178992	28001	20282256	13727	82357984	41178992	41178992
Chb07	61795328	61711360	83968	41402869	20392459	41346779	56090	20364581	27878	82693558	41346779	41346779
Chb08	18437888	18201344	236544	12353384	6084504	12194752	158632	6006592	77912	24389504	12194752	12194752
Chb10	46101504	45985280	116224	30888007	15213497	30810095	77912	15175185	38312	61620190	30810095	30810095
Chb23	24476160	24365824	110336	16399027	8077133	16325020	74007	8040804	36329	32650040	16325020	16325020

4 Performance Evaluation Parameters

The robust machine learning algorithms which use imbalanced data as a training set produces models with high predictive accuracy and specificity, i.e. majority class samples predicted correctly. However, it shows low sensitivity, i.e. minority class samples predicted correctly. Thus, it is recommended to use compatible evaluation parameters for class imbalanced data [1, 28]. Past publications on seizure detection application focus majorly on the accuracy measure. However, accuracy cannot be considered as a sufficient parameter for performance measure as the high accuracy might be achieved by the dominating class, and the model is not capable of detecting a single case of a positive sample. So, this study uses the confusion matrix and its derivations: sensitivity, specificity, precision, accuracy, and false positive rate (FPR) as the evaluation parameters. F-measure, known as f1-score, which is the weighted harmonic mean of precision and recall [8, 14], and area under receiver operating curve (AUC) [1] is also considered. Equations 1 to 7 are used in this work to derive the results of the aforementioned parameters. Where TP is true positive, TN is true negative, FP is false positive, and FN is false negative. TP and TN indicate the correct numbers of positive and negative predictions respectively. FP and FN indicate the number of incorrect predictions for negative and positive cases respectively. It is important to note that, sensitivity is also known as recall or true positive rate (TPR), and specificity is called true negative rate (TNR). Precision is also called positive predictive value (PPV).

$$Accuracy = \frac{TP + TN}{TP + TN + FP + FN} = \frac{TP + TN}{P + N} \tag{1}$$

$$Sensitivity = \frac{TP}{TP + FN} = \frac{TP}{P} \tag{2}$$

$$Specificity\ (SP) = \frac{TN}{TN + FP} = \frac{TN}{N} \tag{3}$$

$$Precision = \frac{TP}{TP + FP} \tag{4}$$

$$Recall = \frac{TP}{TP + FN} \tag{5}$$

$$False\ Positive\ Rate\ (FPR) = \frac{FP}{TN + FP} = 1 - SP \tag{6}$$

$$F1 = 2 \times \left(\frac{Precision \times Recall}{Precision + Recall}\right) \tag{7}$$

5 Results and Discussion

AUC is considered an effective way of analyzing the predictive performance of the classifier. AUC is evaluated on the value scale 0 to 1, where 0 indicates a perfectly inaccurate result whereas 1 indicates a perfectly accurate result. More specifically, the AUC value of 0.5 suggests that the classifier is not able to discriminate the positive and negative samples, 0.7 to 0.8 can discriminate the positive and negative samples with limited perfection, and 0.9 to 1 can discriminate the positive and negative samples with outstanding perfection [29]. For the parameters of accuracy, sensitivity, specificity, precision, and f1-score the best performance value is 1, whereas the worst performance value is 0. For the parameter of false positive rate, which is a test result that wrongly indicates a positive sample, the best value is 0 and the worst value is 1.

Table 2 shows the predictive performance of epileptic seizure state recognition with class balancing using SMOTE. Whereas Table 3 shows the results of the same with the class imbalance dataset. It can be derived from the results that there is not much difference in the accuracy of the model in both the approaches as the average accuracy obtained is 0.99 without applying SMOTE and 0.98 with applying SMOTE. However, the AUC and sensitivity show significant differences in both approaches. Figure 1 shows the sensitivity results and Fig. 2 shows the AUC results of both the approaches used in the study. The model which uses class imbalanced training data can achieve average sensitivity of 0.34 and an average AUC of 0.67. In comparison to these, the model which uses SMOTE as an oversampling method for balancing the classes has achieved an average sensitivity of 0.62 and an average AUC of 0.80. An increased sensitivity and AUC values indicate that the model is generalized for both minority and majority classes. The parameters of specificity, precision, and f1-score do not show much difference in the predictive performance of class balanced dataset and class imbalanced dataset. However, the performance is slightly degraded in terms of false positive rate. When the number of samples of the positive class is extremely less as compared to the negative class, i.e. class imbalance issue, the average false positive rate of the classifier is 0.002. After resolving the issue of class imbalance using SMOTE, i.e. when the number of samples of the positive class and negative class is equal, the false positive rate increases. The average false positive rate of random forest classifier with class balanced data obtained is 0.013.

After analyzing the results, it can be observed that accuracy cannot be considered as the only parameter for evaluating the predictive performance of models which use a highly imbalanced dataset. Specifically, for medical diagnosis, the parameters of AUC, sensitivity, and false positive rate carry more weightage than the other parameters under consideration in this study. Also, higher performance is obtained in terms of AUC and sensitivity for the random forest classifier which uses SMOTE for resolving class imbalance issues, which can be visualized in Fig. 1 and Fig. 2.

Table 2. Performance of seizure state recognition with SMOTE.

Evaluation parameter	Measures with applying SMOTE								
	Chb01	Chb02	Chb03	Chb05	Chb06	Chb07	Chb08	Chb10	Chb23
AUC	0.8379	0.8008	0.8323	0.8372	0.6368	0.8205	0.8535	0.8492	0.7991
Accuracy	0.9892	0.9934	0.9854	0.9813	0.9934	0.9928	0.9667	0.9914	0.9782
Sensitivity	0.6855	0.6077	0.6783	0.6921	0.2798	0.6477	0.7375	0.7064	0.6185
Specificity	0.9902	0.9939	0.9863	0.9824	0.9939	0.9933	0.9696	0.9921	0.9798
Precision	0.177	0.1209	0.1299	0.1368	0.03	0.1167	0.2395	0.1849	0.1214
FPR	0.0098	0.0061	0.0137	0.0176	0.0061	0.0067	0.0304	0.0079	0.0202
F1-score	0.2813	0.2016	0.2181	0.2285	0.0541	0.1978	0.3616	0.2931	0.203

Table 3. Performance of seizure state recognition without applying SMOTE.

Evaluation parameter	Measures without applying SMOTE								
	Chb01	Chb02	Chb03	Chb05	Chb06	Chb07	Chb08	Chb10	Chb23
AUC	0.7072	0.6406	0.6742	0.7297	0.5195	0.6882	0.7263	0.7113	0.653
Accuracy	0.9964	0.998	0.996	0.9956	0.9986	0.9983	0.9858	0.9971	0.9937
Sensitivity	0.4163	0.2824	0.3506	0.4618	0.0399	0.3774	0.46	0.4242	0.3094
Specificity	0.9982	0.999	0.998	0.9977	0.9992	0.9991	0.9927	0.9985	0.9967
Precision	0.4116	0.274	0.3392	0.4474	0.0332	0.3735	0.4494	0.4198	0.2986
FPR	0.0018	0.001	0.002	0.0023	0.0008	0.0009	0.0073	0.0015	0.0033
F1-score	0.4139	0.2782	0.3448	0.4545	0.0363	0.3755	0.4546	0.422	0.3039

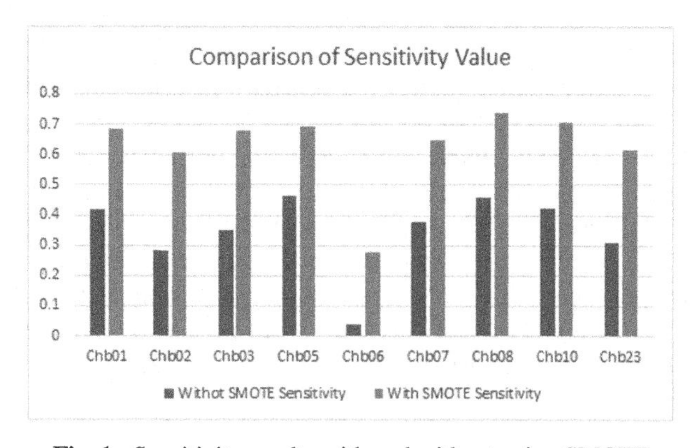

Fig. 1. Sensitivity results with and without using SMOTE

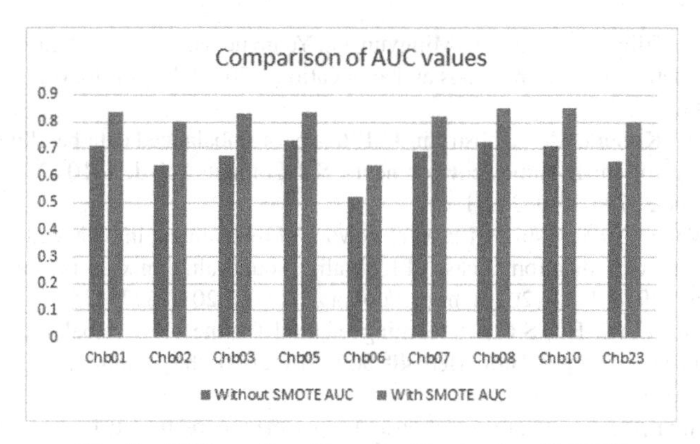

Fig. 2. AUC results with and without using SMOTE

6 Conclusion and Future Work

Medical diagnosis datasets exhibit the issue of class imbalance, which affects the predictive performance of the machine learning models. Robust algorithms that perform exceptionally well also decay if exposed to such class imbalanced data. A random forest classifier based epileptic seizure state recognition model is critically evaluated in this work to address the class imbalance issue. Oversampling technique SMOTE is used to balance the positive and negative classes of the EEG dataset. The results indicate that the model which uses balanced class training data exhibits higher sensitivity and AUC values compared to the model which uses class imbalanced data. It can be derived that the SMOTE is an effective approach to handle the issue of class imbalance. Experiments of class balancing using SMOTE on the different imbalanced datasets can be done as future work to affirm the derivation.

References

1. Mena, L., Gonzalez, J.A.: Symbolic one-class learning from imbalanced datasets: application in medical diagnosis. Int. J. Artif. Intell. Tools **18**, 273–309 (2009). https://doi.org/10.1142/S0218213009000135
2. Ling, C.X., Sheng, V.S.: Class Imbalance Problem. In: Sammut, C., Webb, G.I. (eds.) Encyclopedia of Machine Learning and Data Mining. Springer, Boston (2017). https://doi.org/10.1007/978-1-4899-7687-1_110
3. Japkowicz, N., Stephen, S.: The class imbalance problem: a systematic study. Intell. Data Anal. **6**, 429–449 (2002). https://doi.org/10.3233/ida-2002-6504
4. Chawla, N.V., Japkowicz, N., Drive, P.: Editorial: special issue on learning from imbalanced data sets Aleksander Kolcz. ACM SIGKDD Explor. Newsl. **6**, 1–6 (2004)
5. Provost, F.: Machine learning from imbalanced data sets 101. In: Proceedings of AAAI 2000 Workshop on Imbalanced Data Sets (2000)
6. Chawla, N.V., Bowyer, K.W., Hall, L.O., Kegelmeyer, W.P.: SMOTE: synthetic minority over-sampling technique. J. Artif. Intell. Res. **16**, 321–357 (2002). https://doi.org/10.1613/jair.953

7. Haixiang, G., Yijing, L., Shang, J., Mingyun, G., Yuanyue, H., Bing, G.: Learning from class-imbalanced data: review of methods and applications (2017). https://doi.org/10.1016/j.eswa.2016.12.035
8. Hamad, R.A., Kimura, M., Lundström, J.: Efficacy of imbalanced data handling methods on deep learning for smart homes environments. SN Comput. Sci. **1**, 1–10 (2020). https://doi.org/10.1007/s42979-020-00211-1
9. Zhao, Y., Wong, Z.S.Y., Tsui, K.L.: A framework of rebalancing imbalanced healthcare data for rare events' classification: a case of look-alike sound-alike mix-up incident detection. J. Healthc. Eng. **2018**, 1–11 (2018). https://doi.org/10.1155/2018/6275435
10. Li, D.C., Liu, C.W., Hu, S.C.: A learning method for the class imbalance problem with medical data sets. Comput. Biol. Med. **40**, 509–518 (2010). https://doi.org/10.1016/j.compbiomed.2010.03.005
11. Kim, T., et al.: Epileptic seizure detection and experimental treatment: a review (2020). https://doi.org/10.3389/fneur.2020.00701
12. Roy, Y., Banville, H., Albuquerque, I., Gramfort, A., Falk, T.H., Faubert, J.: Deep learning-based electroencephalography analysis: a systematic review (2019). https://doi.org/10.1088/1741-2552/ab260c
13. Patel, V., Buch, S., Ganatra, A.: A review on EEG based epileptic seizure prediction using machine learning techniques. In: Pandian, A.P., Ntalianis, K., Palanisamy, R. (eds.) ICICCS 2019. AISC, vol. 1039, pp. 384–391. Springer, Cham (2020). https://doi.org/10.1007/978-3-030-30465-2_43
14. Siddiqui, M.K., Morales-Menendez, R., Huang, X., Hussain, N.: A review of epileptic seizure detection using machine learning classifiers (2020). https://doi.org/10.1186/s40708-020-00105-1
15. Yuan, Q., et al.: Epileptic seizure detection based on imbalanced classification and wavelet packet transform. Seizure **50**, 99–108 (2017). https://doi.org/10.1016/j.seizure.2017.05.018
16. Fernández, A., García, S., Herrera, F., Chawla, N.V.: SMOTE for learning from imbalanced data: progress and challenges, marking the 15-year anniversary (2018). https://doi.org/10.1613/jair.1.11192
17. Almustafa, K.M.: Classification of epileptic seizure dataset using different machine learning algorithms. Inform. Med. Unlocked **21** (2020). https://doi.org/10.1016/j.imu.2020.100444
18. Shoeb, A.: Application of machine learning to epileptic seizure onset detection and treatment (2009)
19. Stojanović, O., Kuhlmann, L., Pipa, G.: Predicting epileptic seizures using nonnegative matrix factorization. PLoS ONE **15**, e0228025 (2020). https://doi.org/10.1371/journal.pone.0228025
20. Akter, M.S., et al.: Multiband entropy-based feature-extraction method for automatic identification of epileptic focus based on high-frequency components in interictal iEEG. Sci. Rep. **10**, 1–17 (2020). https://doi.org/10.1038/s41598-020-62967-z
21. Fergus, P., Hussain, A., Hignett, D., Al-jumeily, D.: A machine learning system for automated whole-brain seizure detection. Appl. Comput. Inform. **12**, 70–89 (2016). https://doi.org/10.1016/j.aci.2015.01.001
22. Menze, B.H., et al.: A comparison of random forest and its Gini importance with standard chemometric methods for the feature selection and classification of spectral data. BMC Bioinform. **10**, 1–16 (2009). https://doi.org/10.1186/1471-2105-10-213
23. Singh, K., Malhotra, J.: IoT and cloud computing based automatic epileptic seizure detection using HOS features based random forest classification. J. Ambient Intell. Humaniz. Comput. (2019). https://doi.org/10.1007/s12652-019-01613-7
24. Donos, C., Dümpelmann, M., Schulze-Bonhage, A.: Early seizure detection algorithm based on intracranial EEG and random forest classification. Int. J. Neural Syst. **25** (2015). https://doi.org/10.1142/S0129065715500239

25. Wang, X., Gong, G., Li, N., Qiu, S.: Detection analysis of epileptic EEG using a novel random forest model combined with grid search optimization. Front. Hum. Neurosci. **13**, 52 (2019). https://doi.org/10.3389/fnhum.2019.00052
26. Mursalin, M., Zhang, Y., Chen, Y., Chawla, N.V.: Automated epileptic seizure detection using improved correlation-based feature selection with random forest classifier. Neurocomputing **241**, 204–214 (2017). https://doi.org/10.1016/j.neucom.2017.02.053
27. Wang, Y., Cao, J., Lai, X., Hu, D.: Epileptic state classification for seizure prediction with wavelet packet features and random forest. In: Proceedings of the 31st Chinese Control and Decision Conference, CCDC 2019 (2019). https://doi.org/10.1109/CCDC.2019.8833249
28. Johnson, J.M., Khoshgoftaar, T.M.: Survey on deep learning with class imbalance. J. Big Data **6**, 1–54 (2019). https://doi.org/10.1186/s40537-019-0192-5
29. Mandrekar, J.N.: Receiver operating characteristic curve in diagnostic test assessment. J. Thorac. Oncol. **5**, 1315–1316 (2010). https://doi.org/10.1097/JTO.0b013e3181ec173d

Dissemination of Firm's Market Information: Application of Kermack-Mckendrick SIR Model

Renji George Amballoor(✉) and Shankar B. Naik

Higher Education, Government of Goa, Goa, India

Abstract. In a market economy with high degree of competition, the dissemination of Firm's market information assumes lot of significance. In an information driven economy, any information dissemination is an important social process which creates huge amount of big data. Many times due to the lack of a healthy partnership between the practitioner of computer science, and social sciences, optimal utilization of the hidden patterns and meanings from the big data is missing. The information dissemination is similar to the spread of infectious disease through person to person transmission and spread within given population. Kermack- Mckendrick SIR (Susceptible-Infectious-Recovered) model explains the process of the viral spread in epidemiology. The model is adapted and used in various non-epidemic studies especially to understand the effect of product launch or the product itself, on the potential buyers over time. The model considers the impact of various information dissemination techniques in terms of transmission and spread. The model has been analysed by varying the transmission rate and recovery rate parameters to understand the firm's information dissemination. The present model is good because it can capture the market dynamics and consumer behaviour on a real time basis more effectively.

Keywords: Business intelligence · Computational economics · Complexity economics · Marketing · Kermack-Mckendrick SIR model · Data mining · Information dissemination

1 Introduction

The spread of disruptive digital technologies along with Internet of Things has ushered a world of digital convergence [1] resulting in the emergence of new types of real-time data sets generated at a break-neck speed from the interactions of people, machine and things, which when analysed can create value for the society and economy through data-driven discovery and decision making. In the new era of big data, data sets are being produced from the most unusual sources referred to as digital breadcrumbs [2] or digital exhaust [3] like online transactions, email exchanges, videos, mobile tower locations, sensors, etc. in addition to those created by satellite images, medical devices, sensors, people and machines interactions,

M. Singh et al. (Eds.): ICACDS 2021, CCIS 1441, pp. 22–32, 2021.
https://doi.org/10.1007/978-3-030-88244-0_3

digital modes, simulations to understand social and technological process, etc. The usage of big data sets has gone beyond the monopoly of technology companies like Google and Facebook and its impact is felt across the economy [4].

Laney [5] proposed the 3V model to showcase the novelty of big data. The 3V's consisted of the size of the data represented by Volume, the speed of the data collection and transfer by Velocity and the assorted data sources and type by Variety. The big data sets are diverse in matter relating to levels of complexity, scope, granularity, reliability, etc.

According to Chang et al. [1], it is a big challenge to tap the probable benefits of big data. The data mining techniques available under the Knowledge Discovery in Databases (KDD) process through its structure of collecting-cleaning-formatting-analysing is unable to generate all possible meaningful patterns. In order to extract value from big data, Business Intelligence (BI) and Business Analytics (BN) have emerged to provide methodological and technological capabilities for data analysis [6,7].

The epidemic growth of social big data sets has raised the trillion-dollar question among social scientists and researchers about how to use the computational reactor [8] to optimize the value from it [9]. This naturally occurring social data[1] which is both structured and unstructured provides a lot of insights into human behaviour, expressions, interactions, networks, preferences, beliefs, etc.

The challenges of theoretical complexity of social science, difficulty levels in securing the required observational data and the hardships of experimentally deriving large-scale socially relevant data has led to the emergence of Computational Social Science (CSS) [10]. Computational Social Science provides quantitative and qualitative models for understanding the complexities of real-world economy and society. One of the very popular arsenals with Computational Social Science is social stimulation through Agent-Based Modelling (ABM) used for building generative models. All social theories, problems and phenomena revolve around people who are the agents. The new alliance will radically change the social science discipline in 4 ways; study real behaviour rather than self-reported attitudes, experiments will move from labs to actual social environment, include larger sample size and more trans-disciplinary collaborative research [11].

The difficulty in coming together of Computer & Social Scientists is very well captured by Hopkins and King [12]. Computer Scientists are focused on drawing prediction from data while social scientists are concerned with the feasible explanations for the observed data.

This paper is an attempt to apply the Kermack-Mckendrick SIR Model to study the dissemination of Firm's market information thereby bridging the gap between the two research disciplines. Huge amount of data is generated during the product launch or the product itself, on the potential buyers over time. At present Economics and other Social Science is unable to analyse huge data sets and derive hidden patterns. The huge data sets on its on does not provide utility

[1] The traditional social science data collection methods like surveys experiments involve the intervention of the researcher on the sample population.

to Social sciences. Hence, Economics - Computer Science partnership is used to study the dynamics of information dissemination.

The SIR model which is generally used in epidemiology studies is used in the paper to understand firm's market based information diffusion. The Python codes are used to stimulate different scenarios with different sets of S, I and R. Further the collaboration between Economics and Computer Science for analysing big data also adds to the novelty.

The objective of the study are the following:

1. Application of Kermack-Mckendrick SIR model to understand the transmission rate of information about goods and services
2. To develop a model for information dissemination by factoring in the impact of sources of information.
3. To understand the rate at which the susceptible individuals get infected and recovered when the advertisement assumes an epidemic form.

2 Related Work

To do away with the practice of introducing restrictive assumptions for analysing complex economic models, researchers in Economics and Computer science came together. They applied computing techniques to economic problems and economic models to computing resulting in the baptism of a new discipline called Computational Economics. The role of Computational Economics assumes importance because of the complex economy with heterogeneous agents. The interacting agents keep on changing their actions, reactions and strategies in response to the impact they jointly create [13]. According to Complexity Economics [14], the economy is in a state of non-equilibrium i.e. the economy is continuously changing and evolving to create a state of flux because of uncertainty and technological disruptions. The bounded rationality [15] and cognitive limitation makes the economic models more complex necessitating the need for Computational Economics.

Information dissemination or diffusion involves a process whereby rumour, information, disease, etc. disperses from person to person [16].

According to Murray [17], Kermack & McKendrick's paper in 1927 developed the mathematical epidemiological model for a better understanding of infectious diseases. It is an SIR (Susceptible-Infected-Recovered) model or Compartmental model where the population is divided into S the number of susceptible to disease, I the proportion of infected persons and R the number of those recovered. The total population is constant and given as $N = S + I + R$ [19].

The Kermack-Mckendrick SIR Model has been applied in many fields where the concept of spread and diffusion is similar to an epidemic [20] in areas like marketing, advertising, share market, political campaigns, studies on demonstration effect, sociology, etc. Researchers have used the models for epidemics to explain the diffusion of knowledge especially the diffusion of opinion and ideas on different web forums because of the similar patterns observed in the spreading of epidemics & social contagion process [21].

Dongliang [22] has used the classical infectious disease model, the Kermack-Mckendrick SIR framework to understand the dissemination of information in micro-blogging platforms. They have modified the traditional model by introducing internet marketer for reducing the spread of rumours. Kandhway e al. [23], in their paper studies campaign information can be disseminated at a high spread rate in a cost-effective framework. Two models were proposed- the information spread through SIS (Susceptible-Infected-Susceptible) process and the second through the SIR. They also highlight how optimal control strategy can improve the campaign effectiveness.

The topic diffusion in Online forums is explained using the epidemiology model [21]. The optimal control strategy is used in the modified SIR model to bring about a reduction in the spread of false information through posts, videos and images in social media [24].

2.1 The Kermack-Mckendrick SIR Model

The Kermack-Mckendrick SIR Model is a communicable disease model which involves differential equations. According to the model, the population under study is divided into three partitions namely, S, I, and R. S is the susceptible class consisting of individuals who are not as yet infected and have the chance of being infected later in time. I is the set of individuals who are infected and can pass on the infection to the individuals in S, while R is the set of individuals who are recovered and cannot pass on the infection to other individuals. The SIR model assumes that the individuals recover with immunity.

The formulation of the Kermack-Mckendrick SIR model is

$$\frac{dS}{dt} = -\beta SI \tag{1}$$

$$\frac{dI}{dt} = \beta SI - \nu I \tag{2}$$

$$\frac{dR}{dt} = \nu I \tag{3}$$

where, β is the transmission rate and ν is the recovery rate.

3 Methodology

The study proposes to use the Kermack-Mckendrick SIR model to understand the dissemination of firm's market information on a real time basis.

Python codes were used to integrate various equations and generate diverse situations with different sets of S, I, and R.

3.1 Product Information Dissemination Model

Intending to maximize the sale of their products or services, reaching out to maximum section of people becomes important for the firms. Hence, major focus is on adopting and following the best practices of advertising with aim on maximum information dissemination among the potential buyers.

Let S denote the number of people, called as susceptible, who are not yet aware of the product. Let I be the people, called as the infected, who are aware about the product, impressed by it and have the ability to pass on the product information to others through various means. The set I is divided into three sets I_p, I_w and I_o. Set I_p represents those infected who have purchased the product influence others with the display and use of the product. Whereas, the set I_w represents those infected who have not purchased the product and transmit product information by word of mouth while set I_o the once infected having not purchased the product and transmit product information through online tools such as messengers.

Let R be the set of people, called as recovered, who no longer transmit any information about the product.

The model for product information dissemination in its simplest form is expressed using the Eqs. 1, 2 and 3.

Since set I consists of three types of infected people, each type of infected person will transmit information at different rates. Let β_p,β_w and β_o be the transmission rates of I_p, I_w and I_o, respectively. Considering this scenario, Eq. 1 can be rewritten as

$$\frac{dS}{dt} = -\beta_p S I_p - \beta_w S I_w - \beta_o S I_o \tag{4}$$

Equation 4 assumes that information is transmitted to a susceptible person from another infected person only, which need not be true. Firms, in order to influence their potential customers, adopt to use different techniques of advertisements such as using celebrities or influential personalities speak for their products, display banners or hoardings, online campaigning and print media. These techniques do not belong to the population and are factors external which can be called as the original cause of converting susceptible to infected ones.

Let $\{e_1, e_2, e_3, ...\}$ be the set of numbers of such external factors which when exposed to susceptible persons, converts them to infected ones. Considering the transmission effect of these external factors, Eq. 4 can be rewritten as

$$\frac{dS}{dt} = (-\beta_p S I_p - \beta_w S I_w - \beta_o S I_o) - \sum_{i=1}^{n} S \beta_i e_i \tag{5}$$

where β_i is the transmission rate of the external factor i and n is the number of external factors.

The recovery rate is different for different sets of I. Persons who purchased the product will tend to transmit information as long as they use and display the product which is much longer than the transmission time of those who spread information by word of mouth or online means.

Let ν_p, ν_w and ν_o be the recovery rates of I_p, I_w and I_o, respectively. The model for product information dissemination is

$$\frac{dS}{dt} = (-\beta_p S I_p - \beta_w S I_w - \beta_o S I_o) - \sum_{i=1}^{n} S \beta_i e_i \tag{6}$$

$$\frac{dI}{dt} = \beta_p S I_p + \beta_w S I_w + \beta_o S I_o + \sum_{i=1}^{n} S \beta_i e_i - (\nu_p I_p + \nu_w I_w + \nu_o I_o) \tag{7}$$

$$\frac{dR}{dt} = \nu_p I_p + \nu_w I_w + \nu_o I_o \tag{8}$$

3.2 State Changes in Product Information Transmission

Let S_t, I_t and R_t be the number of susceptible, infected and recovered persons, respectively, at time t. An individual from set S can move to either of the sets I_p, I_w or I_o. Not only that, individuals from sets I_w and I_o can also move to set I_p, which is possible when individuals who are either speaking about the product or passing information using online tools themselves later purchase the product. In this case, an infected person remains infected. The entry of this individual in R will be delayed. Thus, this phenomenon does not effect S_t, but affects I_t and R_t.

Let β_{wp} and β_{op} denote the rate at which an individual shifts to set I_p from the sets I_w and I_o. Accordingly, we can conclude

$$\frac{dI_p}{dt} = \beta_{wp} I_w + \beta_{op} I_o \tag{9}$$

The effect of 9 on the set R_t is expressed as

$$\frac{dR}{dt} = \nu_p I_p + \nu_w I_w + \nu_o I_o - \frac{dI_p}{dt} = \nu_p I_p + \nu_w I_w + \nu_o I_o - \beta_{wpexpression} I_w - \beta_{op} I_o \tag{10}$$

Considering the scenario represented in the Eq. 4, the model for product information dissemination is

$$\frac{dS}{dt} = -\beta_p S I_p - \beta_w S I_w - \beta_o S I_o) - \sum_{i=1}^{n} S \beta_i e_i \tag{11}$$

$$\frac{dI}{dt} = \beta_p S I_p + \beta_w S I_w + \beta_o S I_o + \sum_{i=1}^{n} S \beta_i e_i - (\nu_p I_p + \nu_w I_w + \nu_o I_o - \beta_{wp} I_w - \beta_{op} I_o) \tag{12}$$

$$\frac{dR}{dt} = \nu_p I_p + \nu_w I_w + \nu_o I_o - \beta_{wp} I_w - \beta_{op} I_o \tag{13}$$

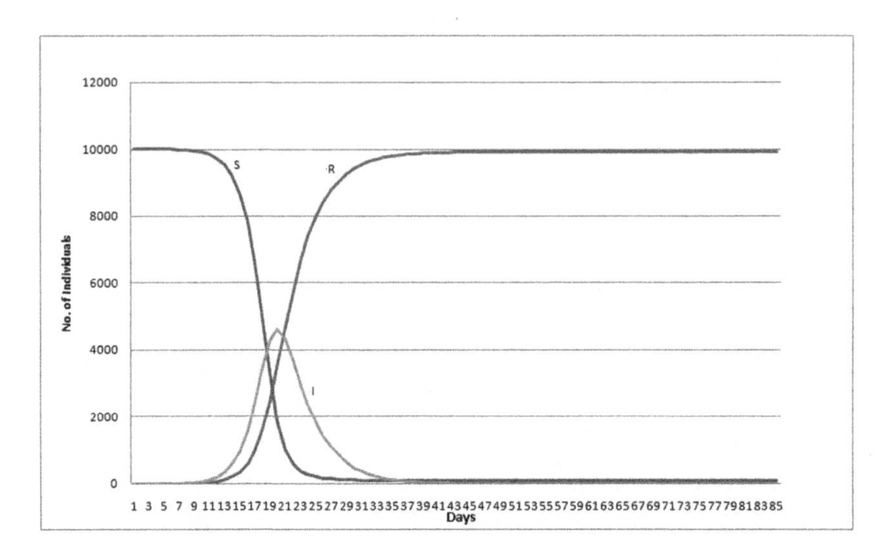

Fig. 1. Rate of change in S, I and R for $\beta = 0.0001$ and $\nu = 0.25$

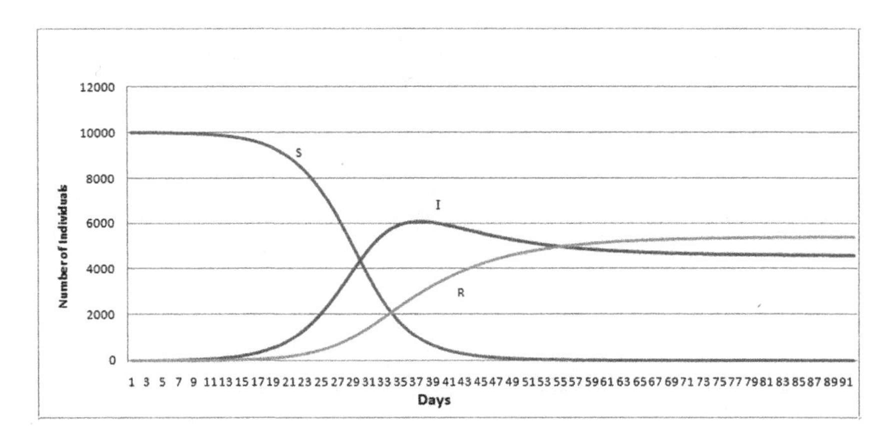

Fig. 2. Rate of change in S, I and R for $\nu_w = 0.1$

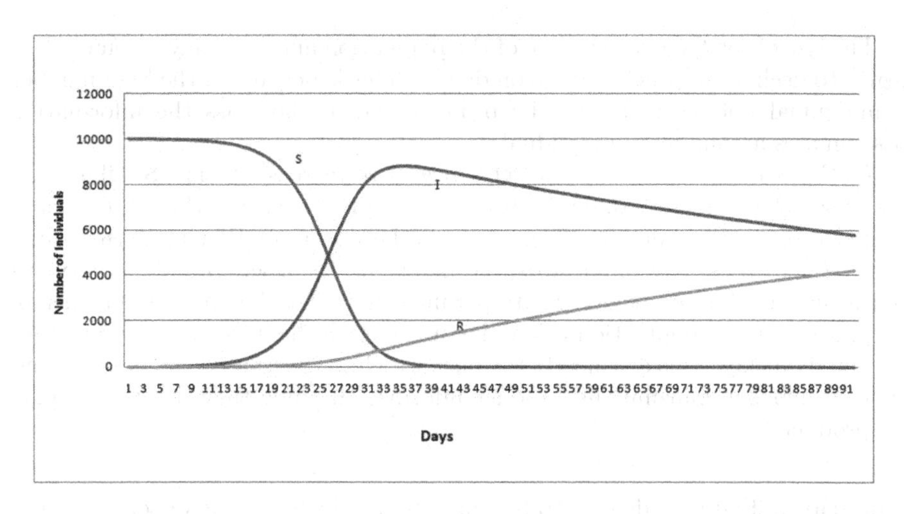

Fig. 3. Rate of change in S, I and R for $\nu_w = 0.01$

4 Experiments

Experiments were done to analyse the proposed model to understand the trends in changes in the sizes of set S, I and R over time by varying the transmission and recovery rate parameters. The analysis was done using Python programming language.

4.1 For Eqs. 1, 2 and 3

Figure 1 demonstrates the rate at which the size of sets S, I and R change for the total population size of $10K$, $\beta = 0.0001$ and $\nu = 0.25$. These basic values are arbitrarily chosen.

Initially, the number of susceptible individuals is the same as the total size of the population. The number of individuals aware about the product is zero. A sudden decline is seen in the number of susceptible individual from day 13. While at the same time the number of individuals receiving product information increases. As the number of recovered individuals increase the number of infected ones begin to decrease from day 21.

For $\beta = 0.0001$ and $\nu = 0.25$, the size of infected individual curve has a normal distribution with $\mu = 21$. This is because $\nu > \beta$.

4.2 For Eqs. 11, 12 and 13

Scenario 1. Figure 2 demonstrates the rate at which the size of sets S, I and R change for the total population size of $10K$, $\beta_p = 0.00003$, $\beta_w = 0.00004$,$\beta_o = 0.00003$,$\beta_1 = 0.0001$, $\beta_2 = 0.0001$, $\beta_{wp} = 0$, $\beta_{op} = 0$, $e_1 = 1$, $e_2 = 1$, $\nu_p = 0.0001$, $\nu_w = 0.1$ and $\nu_o = 0.1$.

The size of set S equals the size of the population and from day 15 onwards it begins to declines to touch to *zero* on day 47. This is because of the high number of individual not aware about the product and as day pass the information disseminates among more individuals.

As the size of set S decreases, set I begins to increase in size. Similarly the size of set R also increases. The rate of increase in the size of the set I is more than that of set R till day 39. Thereafter, as the size of set R increase the size of the set I decreases. Day 55 onwards, the change in size of sets R and I is very low indicating that for the recovery rate parameters values chosen, the influence of the product on the population is long-lasting. Not all the infected individuals get recovered quickly. This are mostly the ones who have purchased the product and are continuously transmitting product information while they use and display the product.

Scenario 2. Figure 3 demonstrates the rate at which the size of sets S, I and R change for the total population size of $10K$, $\beta_p = 0.00003$, $\beta_w = 0.00004$,$\beta_o = 0.00003$,$\beta_1 = 0.0001$, $\beta_2 = 0.0001$, $\beta_{wp} = 0$, $\beta_{op} = 0$, $e_1 = 1$, $e_2 = 1$, $\nu_p = 0.0001$, $\nu_w = 0.01$ and $\nu_o = 0.1$.

In this case, the recovery rate individuals in set I_w is reduced from 0.1 and set as $\nu_w = 0.01$. This implies that infected individuals who are spreading product information continue to do so for ten times longer than in the case of scenario 1. As an effect, the rate at which the size of set R increases is much lower than that of set I.

This is possible when the launch product or the product itself is highly impressive and it takes more time for individuals to stop discussing or stop using the product.

5 Conclusion and Future Work

The Kermack- McKendrick SIR model which is used in the epidemiological studies is used in this paper to study the dissemination/ diffusion of firm's market information on products and services among potential buyers. Information dissemination and pandemic exhibits similar tendencies as far transmission and spread is concerned.

A new model has been proposed which describes the impact, of the launch of the product or the product itself, on the potential buyers over time. The model not only considers the phenomenon of information transmission between the individuals of the population but also considers the impact of the various advertising techniques used on the duration of the impact. The paper also presents the analysis of the model by changing the transmission rate β and recovery rate ν parameters.

The experiments have shown that in case of the Kermack-McKendrick SIR model, the infected individual curve has a normal distribution and the recovered individual curve has a sharp increase around the mean of the infected individual curve. Whereas, in case of the proposed model, the infected individual curve decreases slightly after reaching a maximam and then remains almost constant and the recovered individual curve increase around the the maxima of the infected individual curve and then remains almost constant. This is due to the high influences of the techniques used in the form of advertisements by the companies so that the individuals are constantly kept under the influence of the product and services.

The limitation of the study is that the values of N, β, ν and other parameters are arbitrarily chosen as the model has not been applied to real-time scenario. In future, we propose to apply this model on a real-time scenario.

The key players in the dissemination of the firm's market information are the transmission rate β and the recovery rate ν. Higher values of β implies that the techniques used by the companies in creating awareness about their product are highly effective while lower values of ν imply that the product has a high impact on individuals. Hence, companies should opt and implement strategies in choosing the product, quality enhancement and marketing mix which guarantee high transmission rate and very low recovery rates.

References

1. Chang, R.M., et al.: Understanding the paradigm shift to computational social science in the presence of big data. Decision support systems (2013). https://doi.org/10.1016/j.dss.2013.08.008
2. Lyseggen, J.: Outside Insight: Navigating a World Drowning in Data. United States. Ideapress Publishing, Virginia (2019)
3. Manyika, J., et al.: Big Data: The Next Frontier for Innovation. Competition & Productivity, MCKinsey Global Institute Report (2011)
4. Huberty, M.: Awaiting the second big data revolution: from digital noise to value creation. J. Ind. Compet. Trade **15**(1), 35–47 (2015). https://doi.org/10.1007/s10842-014-0190-4
5. Laney, D.: 3D data management: controlling data volume, velocity & variety. application delivery status (2001). http://blogs.gartner.com/doug-laney/files/2012/01/ad949-3D-Data-Management-Controlling-Data-Volume-Velocity-andVariety.pdf
6. Acito, F., Vijay, K.K.: Business analytics: why now and what next? Bus. Horizons **57**, 565–570 (2014)
7. Llave, M.R.: Data Lakes in Business Intelligence: Reporting from the Trenches. Procedia Comput. Sci. **138**, 516–524 (2018)
8. Evans, J., Jacob, G.F.: Computation & the sociological imagination. Contexts **18**(4) (2019). https://doi.org/10.1177/1536504219883850
9. Shah, D.V., et al.: Big Data, Digital Media, and Computational Social Science: Possibilities and Peril, p. 659. The Annals of the American Academy (2015)
10. Watts, D.J.: Computational social science: exciting progress and grand challenges. In: Proceedings of the 22nd ACM SIGKDD International Conference on Knowledge Discovery & Data Mining (2016)
11. Wouter, N.A., Tai-Quan, P.: When communication meets computation: opportunities, challenges, and pitfalls in computational communication science. Commun. Methods Measur **12**, 2–3 (2018). https://doi.org/10.1080/19312458.2018.1458084

12. Hopkins, D.J., King, G.A.: Method of Automated Non-parametric Content Analysis for Social Science. Am. J. Polit. Sci. **54**(1), 229–247 (2010)
13. Arthur, W.B.: Complexity and the economy. Science **284**(5411), 107–109 (1999)
14. Arthur, W.B.: Complexity Economics: A Different Framework for Economic Thought, Oxford University Press Oxford (2013)
15. Simon, H.A.: Administrative Behavior: A Study of Decision-Making Processes. Administrative Organization. Macmillan, New York (1957)
16. Cliff, A., Haggett, P.: Modeling diffusion processes. In: Kempf Leonard, K. (ed) Encyclopedia of Social Measurement, pp 709–72 . Academic Press, London (2005)
17. Mathematical Biology. IAM, vol. 17. Springer, New York (2002). https://doi.org/10.1007/978-0-387-22437-4_9
18. Kermack, W.O., McKendrick, A.G.: A contribution to the mathematical theory of epidemics. Proc. R. Soc. **115**(772), 367 (1927)
19. Shiller, R.J.: Narrative Economics. Cowles Foundation Discussion Paper-2069. Yale University (2017)
20. Rodrigues, H.S.: Application of SIR epidemiological model: new trends. Int. J. Appl. Math. Inform. **10** (2016)
21. Woo, J., Hsinchun, C.: Epidemic Model for information diffusion in web forums: experiments in marketing exchange and political dialog. Springerplus **5**(66) (2016). https://doi.org/10.1186/s40064-016-1675-x
22. Dongliang, X., et al.: Information dissemination model of micro-blogging with internet marketers. J. Inf. Process. Syst. **15**(4) (2019). https://doi.org/10.3745/JIPS.04.0126
23. Kandhway, K., Joy, K.: How to Run a campaign: optimal control of SIS and SIR information epidemics. Appl. Math. Comput. **231** (2014). https://doi.org/10.1016/j.amc.2013.12.164
24. Hamza, B., Sara, B., Omar, Z., Mostafa, R.: A new simple epidemic discrete-time model describing the dissemination of information with optimal control strategy. Discrete Dyn. Nat. Soc. **2020**, 7465761, 11p (2020). https://doi.org/10.1155/2020/7465761

Improving Image-Based Dialog by Reducing Modality Biases

Jay Gala, Hrishikesh Shenai(✉), Pranjal Chitale, Kaustubh Kekre, and Pratik Kanani

Department of Computer Engineering, Dwarkadas J. Sanghvi College of Engineering, University of Mumbai, Mumbai, India

Abstract. Machines cannot outperform human intelligence yet; however, an image-based dialog can enable machines to perceive cues from different modalities and process information in a more human-like manner. The proposed solution is an AI-based agent that can have engaging conversations with humans by considering an image and answering questions about its visual content, taking into account both visual and textual context. Deep learning-based techniques like Recurrent Neural Networks (RNNs) with self-attention mechanisms have been employed. Responses generated by such models, in some cases, are more biased towards dialog history and are not very relevant to the actual question asked. The proposed work focuses on reducing the modality biases without compromising dialog history and improving the visual context through the use of dense captions to describe various entities in the image and generate relevant answers.

Keywords: Attention · Computer Vision · Deep learning · DenseCap · Dense captions · Faster Region-based Convolutional Neural Network (Faster R-CNN) · Long Short Term Memory (LSTMs) · Modality bias · Natural language processing (NLP) · RNNs

1 Introduction

Artificial Intelligence (AI) has seen tremendous progress through the years in performing a variety of tasks ranging from classifying images, detecting objects to complex AI tasks such as learning to play video games, summarizing videos, and even answering questions about objects in images and videos. The advancements at the intersection of Computer Vision (CV) and AI are also noteworthy. The ability to have engaging conversations with humans about visual content is something that the 'next-gen' Visual AI systems will need to possess. In recent times, various sectors have started incorporating intelligent agents in their existing systems. These systems will gain prominent benefits for various applications with the aim of reducing human intervention. Few examples of such systems are as follows - Helping visually impaired users to understand their surroundings and help them to navigate ("How much time for the signal to turn green?" 30 s), ("How many steps to climb?" 9 steps); Aiding users to analyze huge quantities of surveillance data ("How many people left the building between 5 pm and 6 pm" 102 people), ("Was there a person with a green backpack"); Enabling users to have a dialog with intelligent

M. Singh et al. (Eds.): ICACDS 2021, CCIS 1441, pp. 33–41, 2021.
https://doi.org/10.1007/978-3-030-88244-0_4

assistants ("Is the front door open?" Yes, "Did any person enter?"); Rescue operations where robots/drones send visual data without human intervention ("Is there a fire in the room?" Yes, "Are there any people in the room?").

A number of techniques have been produced over the years, which use techniques such as object detection, feature extraction from an image, along with Neural Networks to analyze the images and obtain data from them. Image Captioning is a technique that generates a textual description of an image. Generally, Convolutional Neural Networks (CNNs) and RNNs are used in an Encoder-Decoder type architecture model for captioning. LSTMs, a type of RNNs, accomplish the goal of tracking long-term dependencies by keeping the contextual information for longer spans, thereby allowing the information to flow from the current state to the next state via the input, forget, and output gates.

Proceeding further, Visual Question Answering (VQA) is another image-based technique that attempts to answer questions about the image asked in natural language; however, without storing the history of questions and answers. It also performs image recognition (understanding patterns and extracting features) using CNNs and uses RNNs for natural language processing, which are combined to form an intelligent system. As an improvement to VQA, Visual Dialog [1–4] is a recently developed technique that attempts to answer questions asked in natural language, not only based on the image but also on a history of dialog (Questions and Answers) about the image, producing better and consistent results.

2 Literature Survey

The system proposed by Abhishek Das et al. [1] is an AI-based agent that holds dialog with humans about visual content in the image provided. This model also allows for follow-up questions based on previously given answers. The dataset collected for this implementation overcomes the 'visual priming bias.' The image is free of longer answers, and the questioner is not exposed to them. An encoder-decoder architecture is used. The encoder creates image and dialog embeddings, whereas the decoder generates an answer on a word-by-word basis depending on the probability distribution of the overall words in the corpus/vocabulary. In both the encoder and decoder, embeddings are generated and given as inputs to the LSTM based architecture.

Stefan Lee et al. [2] have presented a cooperative guessing game based on images between two agents, an A-BOT and a Q-BOT. These bots are under constant communication in the dialog of considered language so that Q-BOT can choose a yet unseen image from the dataset. Training and modeling of these agents is done using reinforcement learning. Dialogs are treated as static supervised learning problems in question/answer systems and image/video captioning systems. They are not considered interactive agent learning problems.

The system implemented by Satwik Kottur et al. [3] is a neural network architecture based on modules for tackling the problem which occurs at a word level. The system makes use of two modules, Exclude and Refer. These modules perform explicit resolution at a finer word level. A question encoder and a multi-layer LSTM are used. For any given question, the new entity detected is grounded first, and then the other entities are resolved using the reference pool. This system works at a word level compared to a traditional encoding that works on a sentence level.

Sungjin Park et al. [4] have proposed a multi-view attention-based network for the application of visual dialog. Objects appearing in an image are represented using a bottom-up attention-based mechanism. The question and the semantic intent are considered simultaneously, which enables the effective alignment of the visual and textual information using different alignment processes. Topic aggregation is performed using the initially used word embeddings to extract meaning from the sentences. The answer candidates are represented using uni-directional LSTM as their sequence lengths are shorter as compared to the dialog history and questions.

The system presented by Vishvak Murahari et al. [5] intends to lead to an even better and enhanced dialog, i.e., a less repetitive dialog while still being comparatively relevant. This system devises a basic objective that gives incentives to Q-BOT when it asks diverse questions, thus helping the A-BOT in exploring a greater space during reinforcement learning. This also exposes the system to more vision-based topics to converse about.

Hao Tan et al. [6] have implemented a multimodal-based approach for visual dialog. Two separate models are maintained, an image-based model and an image-dialog history joint model. A better model is obtained by combining the complementary abilities of the above-stated models. The models are integrated using consensus dropout fusion and ensemble.

In some cases, Image-based Dialog models tend to focus much more on the dialog history. Thus, the answers generated for the current question of the user are not very relevant in all the cases. There have been approaches to reduce modality biases towards dialog history in the generated responses that tend to reduce attention over dialog history. This paper reduces the dialog history bias by generating local, object-level captions that describe image-level features. By attending to these captions, the model generates answers that are more relevant to the current question. Moreover, another advantage of this approach is that it also reduces the bias towards prominent image-level features, as the dense captions describe each entity in the image, thus improving the performance of the model and enabling it to answer more subtle questions about the entities present in the image and generate more relevant answers.

3 Methodology

3.1 System Workflow

The proposed system works in two phases, as shown in Fig. 1. Phase 1 consists of detecting objects and extracting information from the images so as to generate a caption. This process of feature extraction is carried out using a Faster R-CNN [10] based model. The images are passed through the object detection model, which generates a set of encoded representations of different parts of the image. It also stores the object names and the location of their frames, which will be required in Phase 2. Dense captions are generated by the Dense Captioning module and are required for Phase 2. The encoded representations are passed through an attention network that generates a caption. Phase 2 accepts an input query from the user about the image. Objects the user is referring to are identified and mapped to a standard term of generalization using NLP. The attention mechanism is applied by the User Query Resolver for context-based answer generation. The Region of Interest (ROI) is obtained from objects and the frames stored in phase 1.

The system analyses the ROI while considering the conversation history as the context to resolve the query and stores the current conversation as well.

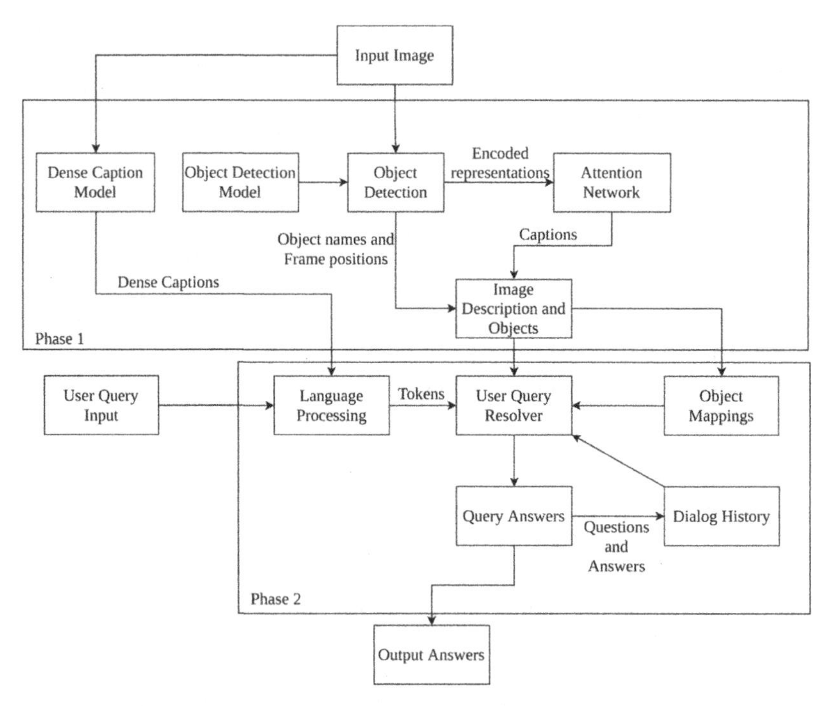

Fig. 1. System workflow

3.2 Dataset Used

The Visual Dialog [1] dataset contains images and rounds of dialog based on the image. The images are obtained from the Microsoft COCO (Common Objects in Context) [7] dataset. The training set contains 123K images and 10 rounds of dialog on each image. The validation set has 2K images and 10 rounds of dialog, and the test set has 8K images with 1 round of dialog. Microsoft COCO captioning dataset (2014) consists of over 100K image-caption pairs. Out of these, 82,783 pairs make the training set, and the remaining are split equally into test and validation sets. Dense captions generated from the DenseCap model by Justin Johnson et al. [8], pre-trained on the Visual Genome dataset [9]. The base model was trained on 120K images. For the proposed model, due to hardware constraints (RAM constraints due to in-memory processing), a subset consisting of randomly sampled 85K images from the Visual Dialog dataset was used.

3.3 Feature Extraction and Dense Caption Generation

A Faster R-CNN model, pre-trained on the Visual Genome dataset, is used to extract visual features from the input image. Faster R-CNN unifies the Region Proposal Networks (RPN) with the Fast R-CNN [11]. The Faster R-CNN algorithm comprises two modules. The regions are proposed by the RPN, which is the first module, and then used by the Fast R-CNN detector, which is the second module.

The Dense Caption Generation task refers to localizing and describing all the entities present in the image. Dense captions provide more visual context, as they contain all entities present in the image and carry significant information about their properties. These dense captions provide valuable information to the model and help in answering questions about subtleties present in the image. Perhaps, because of the natural language form of these dense captions, they prove to be much more beneficial than other visual features [12].

The system uses a CNN-based model, DenseCap by Justin Johnson et al. [8], which has been pre-trained on the Visual Genome dataset [9]. DenseCap provides a grounded natural language description of various entities. The peculiarity of the DenseCap model is that the set of Dense captions generated provides a detailed description of the visual features that also include object properties like shape, color, and even relations between various entities. The dense captions generated are separately encoded using an LSTM model and are passed to the next stage.

3.4 Attention Mechanism

The motivation behind the attention mechanism, proposed by Bahdanau et al. [13], is human attention. Human attention is how humans are able to correlate words in a sentence or pay attention to separate regions of an image. The way human visual attention works is that it allows humans to focus on certain regions having high resolution. So, the attention mechanism in deep learning can be intuitively considered as a vector having precedence weights. This vector is used to infer how strongly an image element correlates with other elements present. During the inference, the values of all other elements (considering the weights in the attention vector) are used so as to approximate the value of the target element.

3.5 Network Architecture

The network consists of an encoder-decoder stack with a self-attention mechanism, which is shown in Fig. 2.

Late Fusion Encoder. The encoder encodes the given question, dense captions from the Dense Caption Generation Module, and dialog history (concatenated as a lengthy string) using three different LSTMs and a set of individual inputs (i.e., images, dialog history, given question) linearly transformed to an encoded vector of joint representation.

Generative Decoder. The encoded vector of joint representation generated from the encoder works as the starting state of the LSTM model. The model tries to maximize the log-likelihood of the answer sequence word by word based on corresponding encoded vector representation and ranks the candidate answers.

Discriminative Decoder. The decoder computes the similarity between the encoded vector of joint representation and the LSTM encoded representation of the options

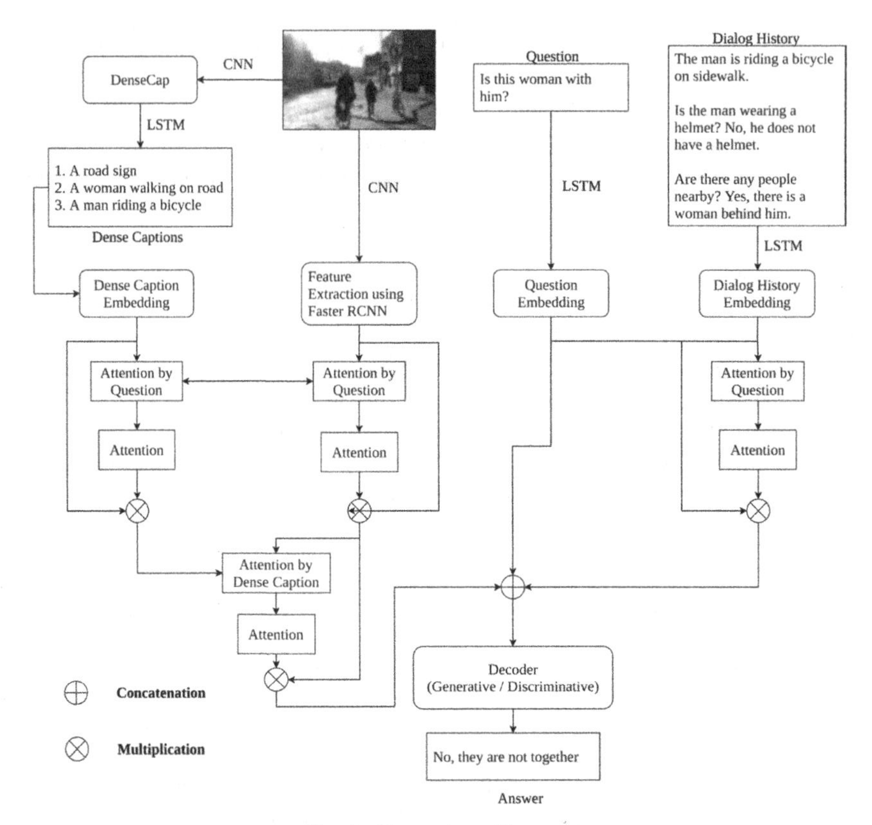

Fig. 2. Network architecture

(answer) chosen using dot product attention, which are then passed to a softmax function so as to compute the posterior probability over the answer choices. Here, the log-likelihood of the correct answer is maximized by the model.

4 Experimentation and Results

The training was carried out on Nvidia's Tesla V100 GPU with 24 GB of memory. The model was trained on a subset of the Visual Dialog dataset consisting of 85K images for multiple epochs. During the training, metrics such as Normalized Discounted Cumulative Gain (NDCG) and Mean Reciprocal Rank (MRR) are used to analyze the model performance. NDCG is the extent to which a ranking generated by the model is conforming with the ideal ranking, taking into account the significance of each element in the list of things to rank. It is the ratio of the model's DCG score over the ideal ranking's DCG score. MRR is a measure to evaluate a ranked list of answers. The rank denotes the position of the highest-ranked answer. MRR is the mean of all the reciprocal ranks.

Table 1. Model metrics

Model variants	NDCG	MRR
Base model discriminative decoder (Trained on 120K images)	0.5162	0.6041
Base model generative decoder (Trained on 120K images)	0.5421	0.4657
Discriminative decoder with semantic attention (Trained on 85K images)	0.5216	0.6019
Generative decoder with semantic attention (Trained on 85K images)	0.5593	0.4720

Table 1 compares the metrics of the base model and the proposed model. Results of both generative and discriminative decoder-based models are mentioned.

As it is seen from the table, Discriminative models tend to outperform the Generative models on the MRR metric, while the Generative models outperform their Discriminative counterparts on the NDCG metric. In practical use-case scenarios, Discriminative models cannot be deployed where a set of dialog responses are not available. Figures 3, 4, and 5 show the image-based dialog output obtained from the proposed model, along with the generated Dense Captions.

Test Case 1. In the input image in Fig. 3, the fire hydrant is predominantly yellow, and only the top of the hydrant is red. The base model only attends to the image features and dialog history; thus, it is observed that the model tends to attend more towards the prominent features, and hence when asked about the color of the top of the fire hydrant, the base model outputs "Yellow" as the answer. In the case of the proposed model, DenseCap is used to generate object-level captions, and therefore even the less prominent features are attended and described in the dense captions. Attention on the dense captions ensures that questions about the less prominent regions of the image are also answered appropriately. Therefore, the proposed model correctly answers the same question about the top of the fire hydrant as "Red."

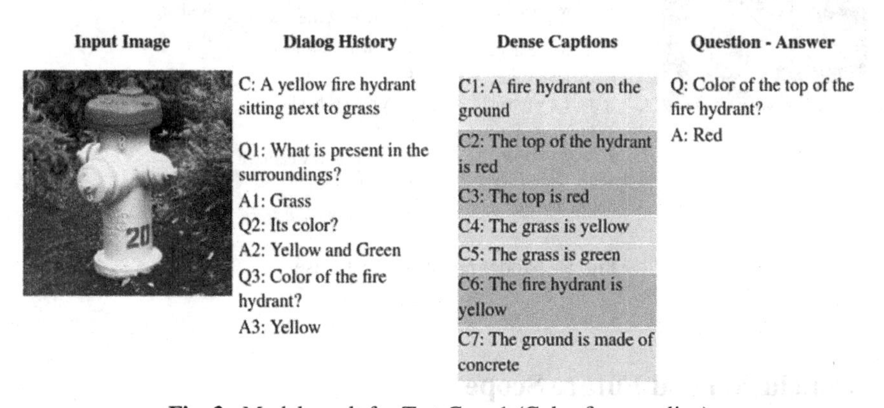

Fig. 3. Model result for Test Case 1 (Color figure online)

Test Case 2. In the input image in Fig. 4, the color of the truck is green, and the wheels are black and yellow. As explained in Test Case 1, the base model tends to attend more towards the prominent features and hence answers the color of wheels as "Green." However, with the use of generated object-level captions, the proposed model answers the color of wheels as "Black and Yellow," thus ensuring that the less prominent features are attended and relevant answers are generated.

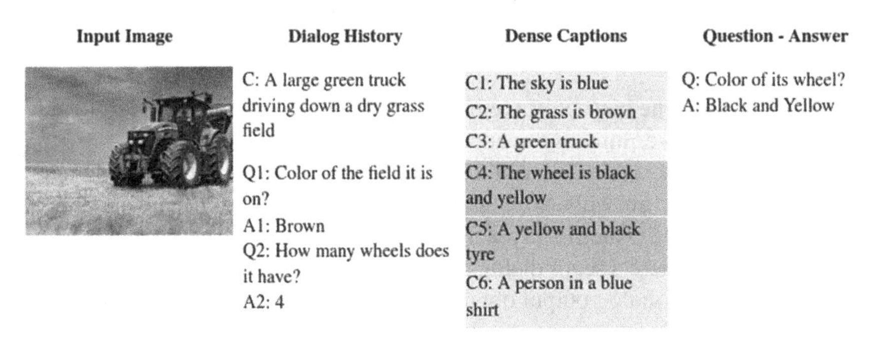

Fig. 4. Model result for Test Case 2 (Color figure online)

Test Case 3. In the input image in Fig. 5, there are 4 people walking; however, one person is only partially visible. Dense captions are generated by DenseCap by attending to object-level features of the image, but in this case, as the features are only partially visible, it is unable to detect and identify some objects and thereby draw inferences. Hence, the proposed model answers the number of people walking as "3."

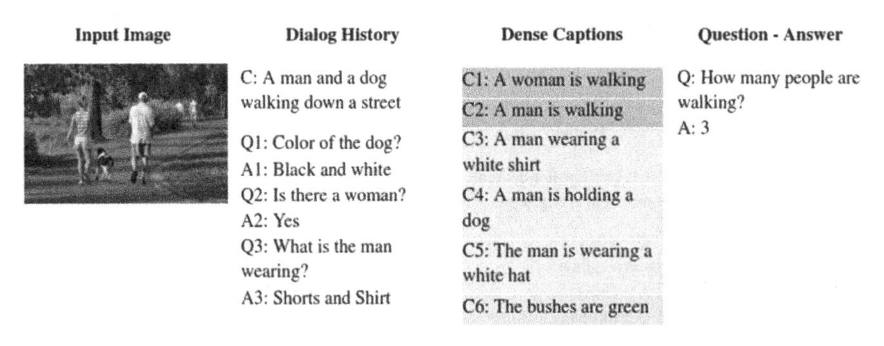

Fig. 5. Model result for Test Case 3

5 Conclusion and Future Scope

The proposed model is capable of having human-like conversations about an image and has been trained on the Visual Dialog dataset. The proposed system shows an approach to reduce the bias towards the dialog history and the prominent image-level features by attending to the generated object-level dense captions. Even though a smaller subset is considered in the case of the proposed model, the proposed model is able to generate comparable results. Thus, the performance gets boosted with the introduction of enhanced visual context using dense captions. In some cases, it can be seen that too much attention over the dialog history makes the model biased towards historical context rather than generating the answer of the current question based on visual information. So, this makes the task of modality balancing extremely important as we move towards more human-like dialogue systems.

Currently, recognizing complex human activities from a single image is difficult. Techniques on incorporating more visual context from a sequence of images should be exploited to understand human activities and enable systems to generate context-based answers based on multiple frames, thus leading to a more human-like conversation and can have applications in various fields like automation and information retrieval in surveillance.

References

1. Das, A., et al.: Visual dialog. In: IEEE Conference on Computer Vision and Pattern Recognition (CVPR), pp. 1080–1089 (2017). https://doi.org/10.1109/CVPR.2017.121
2. Lee, S., Das, A., Kottur, S., Moura, J., Batra, D.: Learning cooperative visual dialog agents with deep reinforcement learning. In: IEEE International Conference on Computer Vision (ICCV), pp. 2970–2979 (2017). https://doi.org/10.1109/ICCV.2017.321
3. Kottur, S., Moura, J.M.F., Parikh, D., Batra, D., Rohrbach, M.: Visual coreference resolution in visual dialog using neural module networks. In: Ferrari, V., Hebert, M., Sminchisescu, C., Weiss, Y. (eds.) ECCV 2018. LNCS, vol. 11219, pp. 160–178. Springer, Cham (2018). https://doi.org/10.1007/978-3-030-01267-0_10
4. Park, S., Whang, T., Yoon, Y., Lim, H.: Multi-view attention network for visual dialog. arXiv Preprint arXiv:2004.14025 (2020)
5. Murahari, V., Chattopadhyay, P., Batra, D., Parikh, D., Das, A.: Improving generative visual dialog by answering diverse questions. arXiv Preprint arXiv:1909.10470 (2019)
6. Kim, H., Tan, H., Bansal, M.: Modality-balanced models for visual dialogue. arXiv Preprint arXiv:2001.06354v1 (2020)
7. Lin, T.-Y., et al.: Microsoft COCO: common objects in context. In: Fleet, D., Pajdla, T., Schiele, B., Tuytelaars, T. (eds.) ECCV 2014. LNCS, vol. 8693, pp. 740–755. Springer, Cham (2014). https://doi.org/10.1007/978-3-319-10602-1_48
8. Johnson, J., Karpathy, A., Fei-Fei, L.: DenseCap: fully convolutional localization networks for dense captioning. arXiv Preprint arXiv:1511.07571v1 (2015)
9. Krishna, R., Zhu, Y., et al.: Visual genome: connecting language and vision using crowdsourced dense image annotations. arXiv Preprint arXiv:1602.07332v1 (2016)
10. Ren, S., He, K., Girshick, R., Sun, J.: Faster R-CNN: towards real-time object detection with region proposal networks. IEEE Trans. Pattern Anal. Mach. Intell. **39**(6), 1137–1149 (2017). https://doi.org/10.1109/TPAMI.2016.2577031
11. Girshick, R.: Fast R-CNN. In: IEEE International Conference on Computer Vision (ICCV), pp. 1440–1448 (2015). https://doi.org/10.1109/ICCV.2015.169
12. Tyagi, V.: Understanding Digital Image Processing. CRC Press, Boca Raton (2018). https://doi.org/10.1201/9781315123905
13. Bahdanau, D., Cho, K., Bengio, Y.: Neural machine translation by jointly learning to align and translate. arXiv Preprint arXiv:1409.0473 (2014)

Dependency Parser for Hindi Using Integer Linear Programming

R. Sai Kesav, B. Premjith(✉), and K. P. Soman

Centre for Computational Engineering and Networks (CEN),
Amrita School of Engineering, Amrita Vishwa Vidyapeetham,
Coimbatore, Tamil Nadu, India
b_premjith@cb.amrita.edu, kp_soman@amrita.edu

Abstract. This paper is an attempt to show a dependency parser for Hindi Language using Integer Linear Programming. A new approach is described for developing the constraint parser by convexifying the integer values taken as input, depicting the source groups (nouns) and demand groups (verbs) in a sentence. The convexified input is further used to find the Integer dependency values, which validates the word's dependency relations in any given Hindi Language sentence. Based on the Karaka constraint conditions, Karaka charts are computed by referencing the Karaka word values obtained from the constraint formulations, which solve the inequalities generated in a sentence from its source groups and demand groups based on the concepts of Paninian Grammar.

Keywords: Integer linear programming · Dependency parsing · Convex optimization

1 Introduction

Parsing is a significant task in understanding the natural language and is also utilised in several natural language applications. Dependency Parsing is the task of uncovering and analysing the grammatical structure, which helps to identify the relations between the words to understand the semantic relations of those words better. An ever-increasing number of languages are computationally utilised for various purposes, which prompted the advancement of varied kinds of grammar-driven, data-driven, and hybrid parsers. Languages like (Hindi, Telugu, Malayalam, Sanskrit, etc.) are morphologically rich and exhibit free-word order (MoR-FWO) [9,10]. A dependency parsing framework for these languages provides features for various Natural Language Processing (NLP) tasks like Machine Translation (MT). The dependency-based sentence parsing techniques like Malt-Parser [5] have been analysed and used in various languages for various NLP applications.

Several methods have been proposed to address dependency parsing in Hindi [3]. Integer Linear Programming (ILP) is one such method [4]. The ILP formulations help learn the dependency among the words in a sentence by analysing the

M. Singh et al. (Eds.): ICACDS 2021, CCIS 1441, pp. 42–51, 2021.
https://doi.org/10.1007/978-3-030-88244-0_5

semantic relations of the terms, which can be used for various tasks like Machine Translation, Question Answering, etc. [4]. Linear Programming is a tool for solving optimization problems that aim to maximise and minimise a linear function concerning its set of linear constraints. In Integer Linear Programming, all the decision variables are taken as integers [7].

This work describes the grammar-driven approach where the sentence is parsed based on its karaka relations Karta, Karma, and Karana karakas to solve the dependency relations among the words in a Hindi sentence on the Paninian Grammar. The proposed work uses a formulated framework for solving the constraints among the local word groupings in a Hindi sentence. The algorithm translates the problem and the corresponding constraints into the Integer Linear Programming (ILP) framework. The proposed algorithm solves the problem by convexifying the constraints to obtain a Hindi sentence's and to solve it's parse structure. This method is language-independent and can be used to solve the dependency parsing problem in other related languages such as Sanskrit, Malayalam, and Tamil.

2 Literature Survey

A significant amount of works have been done to solve the dependency parsing problem in Indian languages [13]. In [5], Nivre et al. proposed a method for addressing the multi-lingual dependency parsing. The initial attempt at building the parsing system was using the constraint-based parsing approach, which works with a dependency-based representation of semantic structures for the Indian languages with free-word-order [3]. A sentence hailing from any particular language also exhibits syntactic and semantic relations, and their analysis is used for finding the dependency between the words in a sentence [15]. The introduction of syntactically annotated tree-banks paved ways for statistical methods to solve the dependency parsing in Indian languages using statistical approaches, which require less effort to build parsers. Their results are entirely dependent on the size of the tree-banks being used.

In general, the Integer Linear Programming formulation's development enables a quick way of extracting features from the constraints compared to traditional algorithms [4]. The Integer Linear Programming approach focuses more on modelling problems and feasible resolvability than algorithm design. The General methodology was suggested by [6] (Reidel, Clarke et al.), which mentions the applicability of Integer Linear Programming to a more complex set of problems by adding all the necessary constraints to the decision variables and capturing the correct formulation. In the reign of exploring the possibilities of formulating the parsers for Paninian literature and morphologically rich and free-word-order sentences, the semantic relations are analysed for each sentence which is done by using the karaka relations. The semantic relations are a part of the Paninian grammar explained in the detailed format along with the formulations of the constraint parser and the conditions required to solve the equalities and inequalities generated by implementing of the integer programming formulations in the Paninian framework.

Karaka Relations are formulated as per the Integer Linear Programming by using the Karaka constraint conditions. These constraint conditions enable taking the source groups and demand groups with a specific integral value for its respective karaka relation solving the karaka conditions, as mentioned in Sect. 3.1. In most cases, there are similar Parts of Speech words occurring in a sentence which are solved by Local word grouping by giving specific integral value to the source and demand groups [1–3].

3 Paninian Grammar

For many centuries now, Paninian Grammar is used as a base for understanding the grammatical contexts in Hindi by extracting the karaka relations, syntactic-semantic relations from a sentence, and here we have used the same Paninian approach for making a parser. This framework is developed in a few millennia and has initially been designed and originated from Sanskrit. Still, as other languages evolved from Sanskrit, the application of Paninian grammar grew to the languages which exhibit free word order, and vibhakti [14]. Indian languages display a higher inflection rate where diversity is expressed in a word with attribute and grammatical meaning. For example, eat, eating, ate display only function in English. Still, in the Indian languages, the same word display different meanings and different tense as per the context they are used in, and for understanding these deflections in terms, Paninian grammar is used [14]. In Paninian grammar, a dependency is used in the notion of karaka relations and semantic relations between a verb and other grammatical constituents in a sentence describing its activity, which consists of the karaka dependency and the result expressing the intention of the uttered sentence.

The Constraint conditions were first explained in Paninian Grammar. These constraint conditions deal with the equalities and inequalities generated while selecting the demand and source groups and classifying them into Karta, Karma, and Karana as per their Parts of Speech in the language they are to be parsed through the Grammar Dependencies. Every Language has different Grammar constraints, which enable in deciding the needed Parts of Speech for the chosen word in the sentence [3].

3.1 Karaka Relations

The syntactic relation between nouns and verbs in a sentence is defined as karaka, and a literal sentence meaning is made from these karakas. They are denoted in a sentence using Vibhaktis (cases). The three Karta, Karma, and Karana karakas are used to formulate the dependencies between the words in a sentence. "ϕ" represents no presence of vibhakti with the noun or the auxiliary verb in a sentence.

Table 1. Basic Karaka Chart [5]

Karaka	Vibhakti	Presence
Karta	ϕ	Mandatory
Karma	ko or ϕ	Mandatory
Karana	se or dwaraa	Optional

The vibhaktis play a crucial role in determining the mandatory and optional presence of karakas in a sentence, as mentioned in Table 1, which specifies the mapping of the karakas in the computational utilisation of the Paninian grammar. The nouns and verbs in a sentence are the source and demand groups. A karaka is determined in a sentence when a noun group satisfies the karaka restrictions based on the karaka chart of the verbs in a sentence. Hence the verb groups are determined as demand groups, and the nouns are determined as the source groups as they satisfy the demands.

A parse is a sub-graph for the constraint graph, which contains all the nodes of a constraint graph and satisfies the following conditions [2].

- Each mandatory karaka from the karaka chart for each source group is expressed exactly only once.
- Each optional karaka from the karaka chart for each demand group is expressed exactly only once.
- Each source group in the sentence will be connected to the demand groups through the karakas or to the chosen root word.

A sentence becomes ambiguous if there are many sub-graphs of a constraint graph that satisfies the above conditions. A sentence becomes indeterminate if there are no sub-graphs that meet the constraint conditions, then that sentence ought to be not having a parse and is ill-formed.

4 Methodology

The input taken for the formulation of the words from the demand and the source groups are in taken as per their arrangement in the sentence were in different matrices as per the length of the sentence, which made the input semi-definite and symmetric for the whole sentence, which is solved by convexifying and adding a diagonal matrix to the input which makes it positive and definite and compensates by subtracting the same using a linear part of the objective function, and let A representing the input and the following steps are:

- Making A symmetric by:

$$A := 0.5(A + A^T) \tag{1}$$

It gives us the same solution, but with this real eigenvalues are calculated.

– Calculating the smallest eigenvalue:

$$\lambda 1 := min_i \lambda i \tag{2}$$

– The $\lambda 1 < 0$ is replaced by A then

$$A := A + (|\lambda 1| + tolI) \tag{3}$$

and also a linear part is introduced in the objective:

$$minx^T Ax - (|\lambda 1| + tol) \sum_i xi \tag{4}$$

if $\lambda 1 \geq 0$ solves the problem, then we use a tolerance value such that the smallest eigenvalue is $\lambda 1 \geq tol$. From Eq. 4, x represents the column vector $(n \times 1)$ of the non-convexified and raw input taken from the chosen integers 0, 1, 2, 3, and 4 representing the words in a sentence.

For the sentence "Bachaa haath se Kela khaata hein", we give it integer values ranging from 0 to 4, including the root, which provides us with non-symmetric and indefinite values, which are convexified from the Eqs. (1) (2) (3) (4) and are chosen as an input. This input is used to solve the equalities and inequalities generated by the source groups (nouns) and demand groups (verbs) and the karaka relation between them by using the Integer Programming constraint formulations for the Karaka Relations [1–4].

In Paninian grammar, there is more than one possible way to produce the local word groups and karakas for the source groups and the demand groups. For the same sentence "Bachaa haath se Kela khaata hein" segmented into 4 parts, Bachaa = *l1*; haath se = *l2*; kela = *l3*; khaata hein = *l4*. The sentence is segmented into *l1*, *l2*, *l3*, *l4* ranging from *l1* to *l4*.

Here, depending upon the sentence we choose for solving, there would be two words *l1* and *l3* which are in a local word group belonging to the source groups or be nouns, and the remaining two words *l2* and *14* which are of the same local group verbs and belong to the demand groups.

Based on the formulations of karaka constraints which solve the equalities and inequalities between source groups and demand groups with their respective Karaka connections, we get two sets of values when we interchange the karaka connections with respect to the unsuitable words zero value is found. Depending on the generated values are found and based on the non-zero values and comparing it with the Constraint conditions of the Paninian grammar [2] [3]. The Methodology followed is briefly explained in the Fig. 1. Algorithm 4 is the psuedocode of the methodology used, where in the three constraint conditions i represents source groups, j represents demand groups, k represents the karakas through which the words are connected to each other in the sentence. $B_{i,k}$ represents the number of equalities holding for each of its demand group and respective karaka in the sentence. $C_{j,k}$ represents the number of inequalities holding for each demand group for the optional or desirable karakas. D_j represents the number of equalities holding for the source groups in the sentence [3].

Algorithm 1. Pseudo-code

Input: Semi-definite & Symmetric Integer values;
Output1: Integer Dependency Values;
Output2: Karaka Word Values;
def (Constraint conditions):
$B_{i,k} : \sum_j x_{i,k,j} = 1;$
$C_{j,k} : \sum_j x_{i,k,j} \leq 1;$
$D_j : \sum_{i,k} x_{i,k,j} = 1;$
`return` $\sum i, j, k$

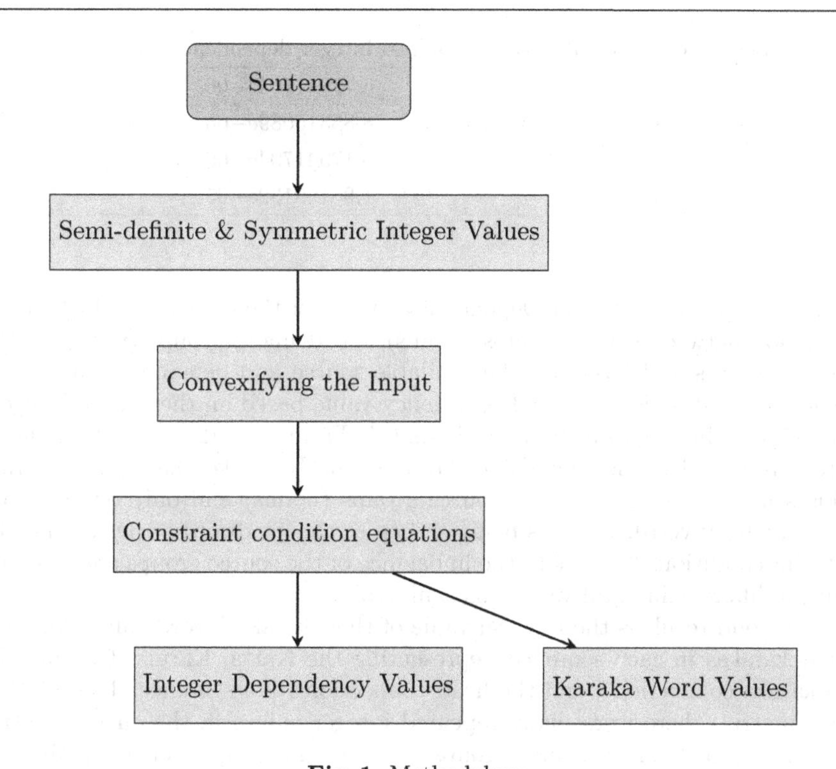

Fig. 1. Methodology.

5 Results and Discussions

After convexifying the input and the input data is solved through the given constraint conditions, two results were achieved, one being the integral values which explain the karaka dependency as per the constraint conditions based on the default karaka chart Table 1, and the same value was attained which show the dependency being, for sentences with a single karaka chart, the arrangement of the words in a sentence is different when compared to the structure of the words for a sentence which comprises for two karaka charts. The integer outputs representing the word's dependency for one and two karaka charts are plotted in Table 2. Few sentences used for testing the model are:

1. Bachaa haath se kela khaata hein
2. Ram ne Ravan ko teer se maara
3. Ped se Baaga mein patta Gira
4. Ramesh ne kiran ko chaaku se maara

Table 2. Integer Dependency values of the sentences

Sentence	Formulated karaka charts	Integer dependency values
1	01	6.17941704e−06
2	02	8.85315089e−06
3	01	6.17941704e−06
4	01	1.96199132e−05

Table 2 depicts the Integer Dependency values for the sentences, which define the relation between source groups (nouns) and demand groups (verbs) in the chosen sentences and based on the available source and demand groups. Each sentence has a specified Integer Dependency value based on the chosen integers as the chosen integers are 0, 1, 2, 3, and 4. From Table 2, sentence 4 has a different Integer Dependency Value. "Ramesh ne Kiran ko chaaku se maara", In this sentence, there are three source groups (nouns) and only one demand group (auxiliary-verb), which is being chosen as the headword as per the karaka constraint conditions [2]. Due to the imbalance of the source groups and demand groups, a different integral value was achieved.

The second result is the integral value of their karakas' word values for each of their karakas in each sentence, representing the Karta, Karma, Karana, the three karaka connections, and the head connections of a sentence. There in the output matrix, there were many repeated integer values in the output matrix due to the words being as source groups and demand groups, being explained in Table 3, for the sentence "Bachaa haath se Kela khaata hein".

Table 3. Word values of their karakas in the sentence "Bachaa haath se kela khaata hein"

Karta Karaka	Karma karaka	Karana karaka	Head Word	Words
−1.98346301	−2.38689688	−0.79337811	−0.79337811	Bachaa
−2.38683688	−4.36585312	−0.56760995	0.0948927	Kela
−0.79337811	−0.56760995	−1.98346301	0.50000005	Haath se
−0.79337811	0.0948927	0.50000005	−1.98346301	Khaata Hein

The result obtained from the formulations depicting the words in the sentence was then used to plot the dependencies by hand utilising the Karaka relations. The integers ranging from 0 to 4 were used in numerating the position of the words in the sentence, and the approach of utilising the mixed-integer quadratic equations solved under the Linear Programming approach using convex optimization (CVXPY) [12] in a python based environment with an ECOS-BB integer solver, which is explicitly used for solving mixed-integer problems.

The results obtained from Table 3 depict each value obtained for its respective karaka relation where Karta karaka only have the words which are nouns and subjects and Karma Karaka have just the words which are objects and nouns, but the third karaka, Karana karaka, only takes in words which is a noun-verb and an adjoining adjective as mentioned in Table 1. After being solved through the ILP formulations, the dependency graph is based on the Paninian grammar constraint conditions. Out of the ten randomly chosen sentences, the two sentences "Ram Lanka se vaapas aaya" and "Geeta ne Seeta se mithai liya" show an ambiguity. For, the sentence "Ram Lanka se vaapas aaya", where both words Ram and Lanka are nouns and does not have any vibhakti (adjective) to support the word for its existence as Karma karaka and "vaapas" is backed by vibhakti (se) making it as the Karana karaka and "aaya" being an auxiliary verb is the Karana karaka. In the same way "Geeta ne seeta se mithai liya", both the words Seeta and Geeta are nouns and can be simultaneously Karta and Karma. In this case, "Mithayi" and "liya" both are auxiliary verbs where "Mithayi" is supported with a vibhakti (se), making it suitable for Karana karaka and "liya" as the headword. Despite the word values, as shown in Table 3, are the same for all the chosen sentences, the sentences were parsed based on the constraint conditions Karta, karna, and Karana karakas physically. In that case, from the generated values, the dependency graph with non-zero values was when charted gives the correct dependency graph as per the constraint conditions based on the Paninian Grammar [2,3].

After sentences being parsed using the ILP approach, the sentences were tested with the sentences' parsed structure using Malt-Parser. After comparison, when the Hindi sentence is being parsed using the traditional Malt-Parser mostly the words are misinterpreted from their original Parts of Speech. But, through this approach where the karakas - Karta, Karma, and Karana are depicted through integers ranging from 0 to 4. The existence of dependence between the words is established through the formulation after being normalized and convexified. This methodology enabled the parsing of the sentences in the Hindi language as per their grammar norms of Karta, Karma, and Karana.

For a few sentences, the main word's Part of Speech identified was similar to that of the dependency by Malt-Parser. From the chosen 10 sentences for the implementation of the ILP approach only 5 sentences were such that, when tested against the parsed structure of Malt-parser the dependency parsing was found to be similar.

6 Conclusion and Future Scope

In this work, a new approach was implemented to solve the constraint parsers for Hindi using multiple integers and to convexify the input for optimal results. This parser is wholly based on translating the grammatical constraints in Hindi of a sentence to an integer programming constraints approach. The linear Programming approach enabled a maximum optimization of the dependency values and the karaka constraint dependencies between the words, which helped in structuring the dependency parsing of the sentences. The dependency parser method and approach explained in this work is in the preliminary stage and is found to be working for non-ambiguous and simple Hindi sentences, which are morphologically rich and follow the free-word-order.

The formulated parse structures can be further expanded by using a Deep Learning approach for future works to solve the ambiguous and complicated sentences with even lesser semantic relations and an explicit free-word-order.

References

1. Aparnna, T., Raji, P.G., Soman, K.P.: Integer linear programming approach to dependency parsing for MALAYALAM. In: 2010 International Conference on Recent Trends in Information, Telecommunication and Computing, Kochi, Kerala, pp. 324–326 (2010). https://doi.org/10.1109/ITC.2010.97
2. Bharati, A., Sangal, R., Reddy, T.P.: A constraint based parser using integer programming. In: Language Technologies Research Centre, Inernational Institute of Information Technology, Hyderabad 500019, Andhra Pradesh, India
3. Bharati, A., Sangal, R.: A Karaka Based Approach to Parsing of Indian Languages, pp. 25–29 (1990). https://doi.org/10.3115/991146.991151
4. Martins, A., Smith, N., Xing, E.: Concise Integer Linear Programming Formulations for Dependency Parsing, pp. 342–350 (2009). https://doi.org/10.3115/1687878.1687928
5. Nivre, J., Hall, J., Nilsson, J.: MaltParser: a data-driven parser-generator for dependency parsing. In: Proceedings of LREC (2006)
6. Bharati, A., Chaitanya, V., Sangal, R.: Natural Language Processing, a Paninian Perspective. Department of Computer Science and Engineering, Indian Institute of Technology, Kanpur
7. Riedel, S., Clarke, J.: Incremental integer linear programming for non-projective dependency parsing. In: EMNLP (2006)
8. Nagaraju, G., Mangathayaru, N., Padmaja Rani, B.: Dependency parser for Telugu language. In: Proceedings of the Second International Conference on Information and Communication Technology for Competitive Strategies (ICTCS 2016). Association for Computing Machinery, New York, NY, USA, Article 138, pp. 1–5 (2016). https://doi.org/10.1145/2905055.2905354
9. Ratnam, D.J., Kumar, M.A., Premjith, B., Soman, K.P., Rajendran, S.: Sense disambiguation of English simple prepositions in the context of English–Hindi machine translation system. In: Margret Anouncia, S., Wiil, U.K. (eds.) Knowledge Computing and Its Applications, pp. 245–268. Springer, Singapore (2018). https://doi.org/10.1007/978-981-10-6680-1_13

10. Premjith, B., Soman, K.P., Kumar, M.: A deep learning approach for Malayalam morphological analysis at character level. Procedia Comput. Sci. **132**, 47–54 (2018). https://doi.org/10.1016/j.procs.2018.05.058
11. Dhanalakshmi, V., Kumar, M.A., Murugesan, C.: Dependency Parser for Tamil classical literature-Kurunthokai (2012)
12. Diamond, S., Boyd, S.: CVXPY: a python-embedded modeling language for convex optimization. J. Mach. Learn. Res. JMLR **17**(1), 2909–2913 (2016)
13. Makwana, M., Vegda, D.: Survey: Natural Language Parsing For Indian Languages (2015)
14. Das, A., Halder, T., Saha, D.: Automatic extraction of Bengali root verbs using Paninian grammar. In: 2nd IEEE International Conference on Recent Trends in Electronics, Information and Communication Technology (RTEICT), Bangalore, India, pp. 953–956 (2017). https://doi.org/10.1109/RTEICT.2017.8256739
15. Zhou, J., Li, Z., Zhao, H.: Parsing All: Syntax and Semantics, Dependencies and Spans (2019)

Medical Records Management Using Distributed Ledger and Storage

Samia Anjum(✉), R. Ramaguru(✉), and M. Sethumadhavan

TIFAC-CORE in Cyber Security, Amrita School of Engineering,
Amrita Vishwa Vidyapeetham, Coimbatore, India
cb.en.p2cys19014@cb.students.amrita.edu,
{r_ramaguru,m_sethu}@cb.amrita.edu

Abstract. In the last decade, blockchain technology has seen adoption to infinite domains, health care sector is one of the major domains where there are a greater opportunity and advantage to leverage the benefits of distributed ledgers in storing and securing patient medical records. The Government of India (GoI) is also very keen on digitization in addition to their wider adoption of blockchain technology to serve the citizens by ensuring their privacy and security of personally sensitive . In this paper, we propose a framework for the secure management of patient medical records (PMR) based on the Ethereum blockchain. The PMR is tokenized using the ERC-721 standard which is a Non-Fungible Token (NFT) that can be uniquely mapped to the individual patient. To ensure a higher level of security and non-redundancy, the medical records are stored and maintained in distributed storage like InterPlanetary File System (IPFS); to ensure privacy, the patient's sensitive information is anonymized through a privacy-preserving scheme. To provide complete control of the medical record to individual patients, we have employed a secret sharing scheme with essential share that is owned by the patient. The proposed framework aims to provide correctness and consistency in managing the PMR over its life cycle.

Keywords: Patient medical record · Electronic health record · IPFS · ERC-721 · PACS

1 Introduction

A Patient Medical Record (PMR) is the historical document that maintains the patient's health details like patient's health history, clinical findings, diagnostic test results, pre-and post-operative care, patient's health progress, and prescribed medication, etc. [1]. The medical practitioner can use the information provided in a medical record to understand the patient's present condition and to provide informed care. These records are created, updated, and maintained either by the medical practitioners or by the health care providers. It is the most important part of dispensing medical care to the patient. The need

M. Singh et al. (Eds.): ICACDS 2021, CCIS 1441, pp. 52–62, 2021.
https://doi.org/10.1007/978-3-030-88244-0_6

for maintaining a good medical record includes monitoring the patient, medical research and audit, statistical study, and insurance cases, criminal cases. PMR is important to understand the health condition of the patient at the same time it contains a lot of personal sensitive information that needs to be kept secure and confidential.

The challenges in managing medical records are storage, access permissions, privacy, and security. Hospitals with manual and paper-based medical record systems face numerous challenges like the revision of forms, management of active and inactive records, destruction of records [1]. The Digital India initiative of the MeitY has developed the e-hospital, the e-blood bank, and Online Registration System (ORS) which is a cloud-based application to provide healthcare services to all the citizens across the country [2]. This outlines the importance and the need for digitizing medical records & health care services and the absence of such a system among healthcare providers. Generally, the record is the responsibility of the health care provider, but a copy of the record should be given to the patient when requested. Ownership and maintenance of PMR vary from country to country. The U.S. Federal Law, to protect sensitive patient health information called Health Insurance Portability and Accountability Act (HIPAA) of 1996 outlines that the data within the medical record owned by the patient and the physical form belongs to the organization responsible for maintaining the record. If a patient finds any error in the record, the patient can petition the health care provider to update the correct data. According to Indian Medical Council (IMC) Regulations, 2002, the medical records of the patients should be maintained by medical practitioners for at least a period of three years. Also, they are supposed to keep the patient's data confidential even the details about the patient's personal and domestic life. If the patient or authorized legal personnel make any request for medical records, the medical practitioner is supposed to make such records available within 72 h. If the medical practitioner is found guilty of performing professional fraud, dismissal of the name from the record of the licensed practitioner entirely or for a specified time. Under European Union law, the General Data Protection Regulation (GDPR) is a data protection and privacy regulation. There is a right for patients to access their health records through a subject access request (SAR) according to the GDPR and Data Processing Agreement (DPA), 2018, which can be submitted by the patient or patient-authorized third party. It should be made mandatory to secure the medical record of every patient so that it can't be misused in any manner. If any third party can get the medical records of patients, it's a breach of patient's confidentiality and privacy. The lack of interoperability among various systems, i.e. the patient data held in the systems of different hospitals, is probably the biggest issue with Electronic Health Records. Different systems must be able to interact with each other to provide a clear and complete understanding of the medical history of the patient.

Blockchain Ledger is a data structure similar to a linked list, every block is a container linked to each other cryptographically. Blockchain Technology is a decentralized computation and distributed ledger platform to immutably store transactions in a verifiable manner efficiently, through a rational decision-making process among multiple parties in an open and public system [3]. Blockchain which is the major technology underlying cryptocurrencies has found its effective use in various domains like Digital Rights Management (DRM) [4], IoT [5], Supply Chain Management, Identity Management, including the health care sector for insurance coverage and securely storing medical records. By adopting blockchain to healthcare record maintenance could help the PMR information be distributed, and prevents intentional and unintentional alteration to the PMR. As in distributed healthcare blockchain network, the PMR can be accessed and updated in real-time with the patient's consent and knowledge.

The rest of the paper is divided as follows. Section 2 outlines the related works of blockchain in health care. In Sect. 3, we have detailed our proposed work with the system architecture and its use-cases. We have concluded and discussed the scope for future work in Sect. 4.

2 Related Works

In this section, we refer and provide an outline to the related works on electronic health records or medical records maintenance through blockchain technology.

Most of the work defines the accessibility of patient's data and interoperability of medical records among hospitals. An electronic medical record storage system using Ethereum blockchain, named MedRec [6] which gives the patient and medical practitioner easy access to the patient's medical records. It incentivizes medical researchers and health care providers by releasing access to aggregate and anonymized medical data. Medchain [7] proposed a similar incentive model based on Proof of Authority (PoA) which incentivizes health care providers to create, validate, and append new blocks with timed-based smart contracts to govern transactions and to control access to the medical records. OmniPHR, a publish-subscribe model using a chord algorithm ensures the balanced, scalable, and elastic service for integrating Personal Health Records (PHRs). OpenEHR standard connects with other health data standards, such as HL7, LOINC, SNOMED-CT, and DICOM, which is the prescribed standard for this model. [8].

Anclie [9], an Ethereum-based solution for privacy-preserving and interoperable healthcare records uses smart contracts along with proxy re-encryption to provide improved access control, data obfuscation, and advanced security. Similarly, a work based on smart contract to provide an access control mechanism and use a privacy-preserving scheme framework called MedBloc [10] using permissioned blockchain. Madine et al. [11] proposed a system to give patients control on the PHR through Ethereum blockchain-based smart contracts. Additionally, it employs InterPlanetary File Systems (IPFS) to securely store and share patients' medical records through trusted reputation-based re-encryption oracles.

3 Proposed Solution

In this section, we present our proposed blockchain-based framework for patient medical records maintenance.

3.1 System Architecture

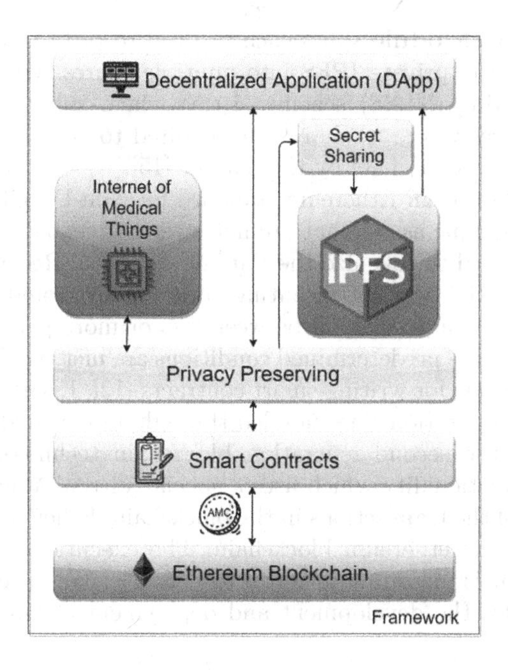

Fig. 1. System architecture

The proposed framework uses the below components as seen in Fig. 1.

Decentralized Application (DApp) is an application that integrates the front-end with smart contracts[12]. The hospital staff and the doctors would use this to register patients and record their medical conditions, and the patients can use them to view their health reports. DApp is created using Ethereum JavaScript API called Web3.js.

Internet of Medical Things (IoMT) also known as Healthcare IoT, consists of medical devices and applications that generate and collect medical information and uploads the data to the healthcare provider's network. Examples of IoMT devices are wearable devices, smart pills with a sensor, etc. [13]. The data from IoMT devices of the patient can be stored in the blockchain using smart contracts by referring to the unique Patient-ID.

InterPlanetary File System (IPFS) is a content-addressing distributed file system for storing and sharing data. It uniquely identifies each file stored by a cryptographic hash value [14]. The medical records that are large like lab and diagnostic reports are securely stored in private IPFS and referenced in the blockchain using their hash value.

Privacy Preserving Module helps in maintaining the privacy of the PMR details that are stored in the private IPFS. We are using the anonymization method in addition to the NLSS Scheme.

Secret Sharing Module splits each medical record (X-Ray, Scan reports, etc.,) that is stored in private IPFS into multiple shares. We are using a Non-Linear Secret Sharing (NLSS) scheme (1,t,n) which splits the given medical record into 'n' shares where 't' shares are required to get back the record with one essential share (secret) held by the patient [15].

ERC-721 Token is an Ethereum standard for Non-Fungible Token (NFT), used to represent unique assets that are not interchangeable [16]. We have used the ERC-721 standard to tokenize the Patient's Medical Record (PMR).

Smart Contracts is a self-executing code deployed and executed on top of blockchain which is agreed upon between two or more parties. They execute automatically when the predetermined conditions are met. We have used solidity programming language for writing smart contracts [12]. Each stakeholder is represented and their operations are handled through a dedicated smart contract.

Ethereum is the second-generation blockchain technology that provides smart contracts functionality which uses a special Virtual Machine called EVM for the execution of the transactions in the blockchain. Ether (ETH) is the native cryptocurrency of the ethereum blockchain. The execution of every operation requires a cost that is measured in terms of Gas [17]. We have used Ethereum-Ropsten Testnet for the development and deployment of our proposed framework.

3.2 Usecases

Hospital and Doctor Registration play an important role in ensuring the authenticity and verification of service of the medical service provider and the medical practitioner. The hospital needs to register itself by providing information regarding their *Government Registration ID, Name, Location, and their Specialization, and Contact Details* as shown in Fig. 2. These details are stored in the blockchain after verification by regulatory bodies like medical associations, which generates a unique *Hospital-ID.*

Once the hospital is registered on the blockchain, the medical practitioners can be on-boarded to the blockchain by the hospital by providing the details like *Registration Number, Name, Specialization, and Contact Details.* The *Doctor-ID* is generated by the smart contract which is then used to uniquely identify the medical practitioner and the actions taken by the corresponding stakeholder. Both these IDs are in turn referenced in the Patient Medical Record.

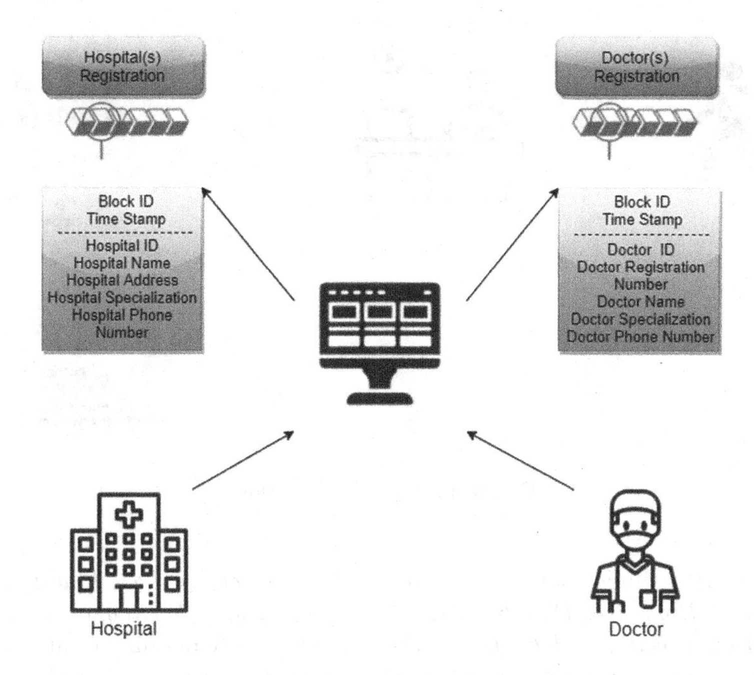

Fig. 2. Hospital and doctor registration

Patient Registration is required the first time when a patient visits the hospital, and for further consultancy, the same *Patient-ID* could be referred. The patient gets registered by the hospital's registration desk. The details captured does not only contains the patient's information but also the attendant's information which would be helpful in case of medical fraud or negligence. The details requested from the patient could include *Name, Age, Gender, Weight, Height, Contact Details* are stored in blockchain along with the details of the attendant (if any).

The attendant details requested to store could include the *Attendant's Name, Phone Number, and Relationship with the patient* as shown in Fig. 3. The smart contract generates a *Patient-ID* that uniquely identifies the patient and is referenced in the Patient Medical Record.

Patient Consulting Medical Practitioner. The patient along with the attendant can visit the medical practitioner for detailed diagnosis and consultancy as shown in Fig. 4. The medical practitioner creates the PMR of the patient based on the information received from the patient's past illness & diagnosis performed and this PMR is mapped to the *Patient-ID*. The PMR of the patient is created based on the date-of-visit and is identified by a unique *PMR-ID* which is the *Patient-ID* appended with the date in the format DDMMYYYY. (Say, a Patient with *Patient-ID* 85437 visits the medical practitioner on 26th January 2021, then the *PMR-ID* generated will be 8543726012021).

Fig. 3. Patient's registration

The PMR includes details like *Present Illness Details, Past Illness Details, Provisional Diagnosis Details, Treatment Summary, Lab Diagnostics Report, IOMT Device Data, and Insurance Details.* The PMR details are anonymized through the privacy-preserving scheme to ensure the privacy of sensitive information. These details are tokenized using the ERC-721 standard which creates a unique, non-fungible token with *PMR-ID* as *TOKEN-ID.*

Lab Diagnostics is recommended by the medical practitioner for a patient based on the preliminary diagnosis. The diagnosis can be X-Ray Imaging, MRI Scan, CT Scan, Ultrasound, Mammography, etc. As shown in Fig. 5, let us say a patient is advised to go for a digital X-Ray. In general, most health care providers use advanced PACS Systems (Picture Archiving and Communication System) which provide economical storage, retrieval, management, distribution, and presentation of medical images [18]. These medical images are normally stored and transferred in DICOM (Digital Imaging and Communications in Medicine) format. To improve the storage efficiency and to provide additional security, these lab reports are anonymized through our privacy-preserving module and securely split into multiple shares with one essential share that is given to the patient. These shares are stored into private IPFS which is uniquely identified by a *hash-value*, which in turn is referenced in the PMR. The details of the Lab report like *Patient-ID, Date of Capture, Reason, and Observations, IPFS-Hash* are stored in the blockchain.

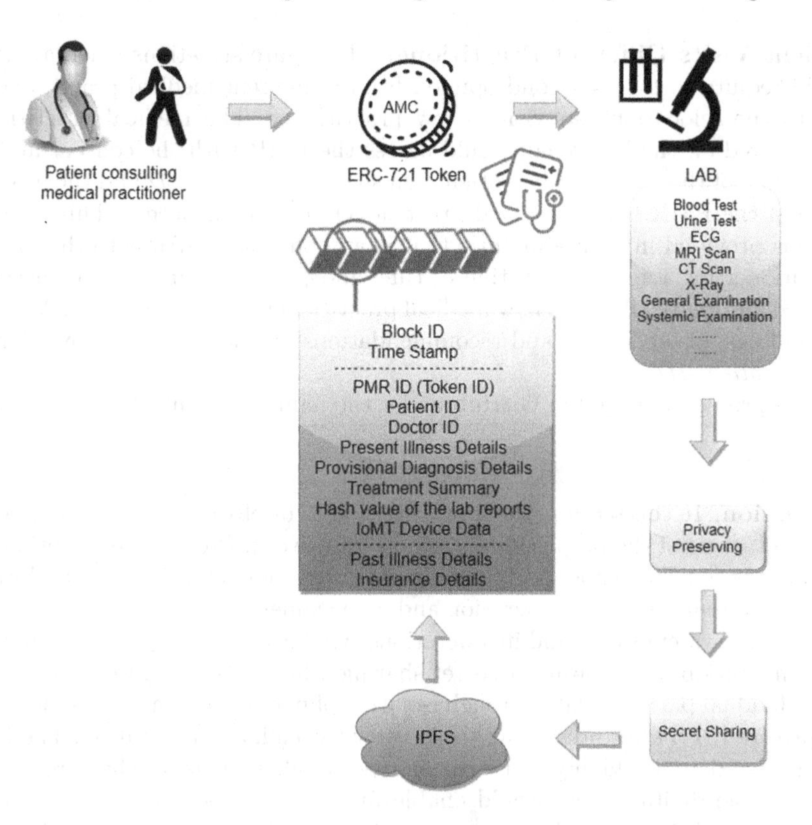

Fig. 4. Patient consulting medical practitioner

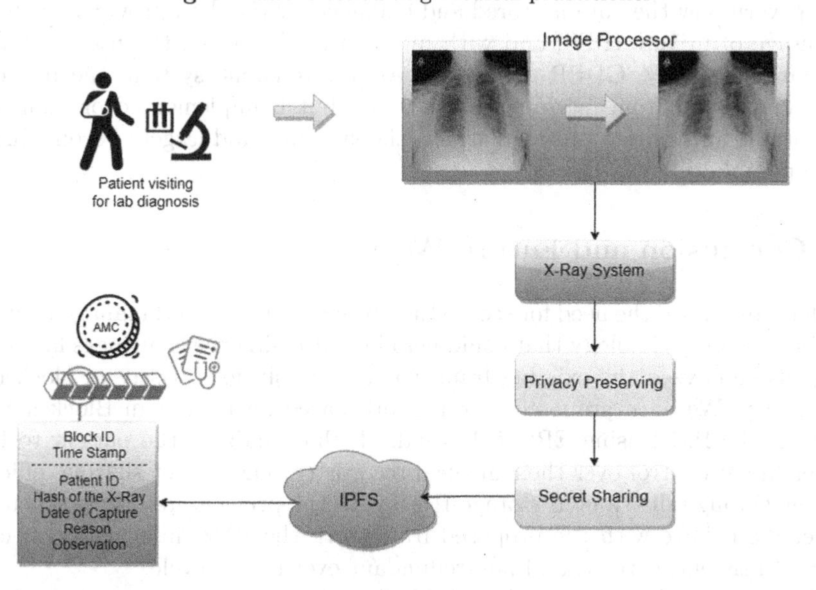

Fig. 5. Lab diagnostics

Patient Visits Different Practitioner. There are situations when a patient would require to take a second opinion from a different medical practitioner or due to the relocation can visit a new practitioner. The medical practitioner, if registered on the blockchain could access the PMR with the consent and the knowledge of the patient. The patient can share the *Patient-ID* and the essential share of the PMR to the medical practitioner to provide access. This essential share is provided in such a manner that it will only be available to the medical practitioner for a specific duration of time and post the duration, the access is revoked automatically. This new medical practitioner can also create a PMR with the findings, observations, and recommendations on the blockchain by referring to the *Patient-ID*.

The project is hosted in GitHub under the name "Patient-Medical-Records" [19].

Discussion. In this section, we draw attention to briefly describe the advantages and limitations of the proposed blockchain framework. Proxy Re-encryption and its variants like Threshold and Time-based proxy re-encryption are used in the existing systems. The key generation and management can incur additional overhead to the system's cost and its operations. We have addressed this through the implementation of a non-linear secret sharing scheme. Besides, we have secured the individual patient's data through isolation of personally sensitive information via private IPFS after preserving the privacy through anonymization. The PMR is considered to be highly sensitive and personal, tokenizing these real-world non-fungible digital assets would enable for easy, efficient, and effective management which is new and unique in our framework. Data Privacy and related laws govern how the data is stored and managed by the data providers, through the usage of private IPFS, and with minimal modification, the proposed work can be made 100% GDPR compliant. As a standalone system, the proposed work may suffer from interoperability issues but if implemented or mandated by suitable authorities and regulations, the system could largely become highly interoperable.

4 Conclusion and Future Work

We have discussed the need for secure Electronic Health Record management and the blockchain technology that could help in addressing the challenges faced. We have also analyzed the existing frameworks and solutions based on blockchain technology. We have proposed a framework based on Ethereum Blockchain to tokenize the PMR using ERC-721 standard, thus enabling the patient to have ownership & control over their medical records through secret sharing, efficient storage through distributed storage IPFS with the privacy-preserving scheme in place. We believe with the proposed framework the PMR information is complete, consistent, correct, and non-redundant over its life cycle.

The proposed blockchain-based Medical Records Management could be extended to a Hospital Management System. The proposed system currently uses the ERC-721 standard to tokenize the medical records, for efficient usage, this can be replaced with the new Multi-token standard ERC-1155. Additionally, the Machine Learning layer could be added to the proposed framework to perform real-time analysis of the patient record and provide useful insights and can even help in the early detection of abnormal conditions [20]. The new policy proposed by the Medical Council of India to issue a Unique Permanent Registration Number (UPRN) to medical practitioners to maintain one record for one doctor can be implemented in our proposed framework.

References

1. Bali, A., Bali, D., Iyer, N., et al.: Management of medical records: facts and figures for surgeons. J. Maxillofac. Oral Surg. **10**(3), 199–202 (2011). https://doi.org/10.1007/s12663-011-0219-8
2. Digital India Initiative - e-hospital. Available. https://ehospital.gov.in/ehospitalsso/. Accessed on 25 Dec 2020
3. Ramaguru, R., Minu, M.: Blockchain terminologies. NamChain Open Initiative Research Lab, (2021). https://github.com/NamChain-Open-Initiative-Research-Lab/Blockchain-Terminologies. Accessed 25 Dec 2020
4. Kripa, M., Nidhin Mahesh, A., Ramaguru, R., Amritha, P.P.: Blockchain framework for social media DRM based on secret sharing. In: Senjyu, T., Mahalle, P.N., Perumal, T., Joshi, A. (eds.) Information and Communication Technology for Intelligent Systems (ICTIS 2020). Smart Innovation, Systems and Technologies, vol 195. pp. 451–458. Springer (2020). https://doi.org/10.1007/978-981-15-7078-0_43
5. Samal, D., Arul, R.: A novel privacy preservation scheme for internet of things using blockchain strategy. In: Bindhu, V., Chen, J., Tavares, J. (eds.) International Conference on Communication, Computing and Electronics Systems, vol 637, pp. 695–705. Springer, Singapore (2020) . https://doi.org/10.1007/978-981-15-2612-1_66
6. Azaria, A., Ekblaw, A., Vieira, T., Lippman, A.: MedRec: using blockchain for medical data access and permission management. In: 2nd International Conference on Open and Big Data (OBD), Vienna, 2016, pp. 25–30 (2016). https://doi.org/10.1109/OBD.2016.11
7. Daraghmi, E., Daraghmi, Y., Yuan, S.: MedChain: a design of blockchain-based system for medical records access and permissions management. IEEE Access **7**, 164595–164613 (2019). https://doi.org/10.1109/ACCESS.2019.2952942
8. Roehrs, A., André da Costa, C., da Rosa Righi, R.: OmniPHR: a distributed architecture model to integrate personal health records. J. Biomed. Inform. vol. 71, pp. 70–81 (2017). https://doi.org/10.1016/j.jbi.2017.05.012
9. Dagher, C.G., Mohler, J., Milojkovic, M., Babu, A., Marella, A.: Privacy-preserving framework for access control and interoperability of electronic health records using blockchain technology. Sustain. Cities Soc. **39**, 283–297 (2018). https://doi.org/10.1016/j.scs.2018.02.014
10. Huang, J., Qi, Y.W., Asghar, M.R., Meads, A., Tu, Y.: MedBloc: a blockchain-based secure EHR system for sharing and accessing medical data. In:18th IEEE International Conference On Trust, Security And Privacy in Computing And Communications/13th IEEE International Conference On Big Data Science And Engineering (TrustCom/BigDataSE), Rotorua, New Zealand,, pp. 594–601 (2019). https://doi.org/10.1109/TrustCom/BigDataSE.2019.00085

11. Madine, M.M., Battah, A.A., Yaqoob, I., Salah, K., Jayaraman, R., Al-Hammadi, Y., Pesic, S., Ellahham, S.: Blockchain for giving patients control over their medical records. IEEE Access **8**, 193102–193115 (2020). https://doi.org/10.1109/ACCESS.2020.3032553
12. Vitalik. B.: A next-generation smart contract and decentralized application platform. White paper 3.37(2014). https://cryptorating.eu/whitepapers/Ethereum/Ethereum_white_paper.pdf
13. Internet of Medical Things: https://healthtechmagazine.net/article/2020/01/how-internet-medical-things-impacting-healthcare-perfcon. Accessed 25 Dec 2020
14. IPFS - Content Addressed, Versioned, P2P File System. https://github.com/ipfs/papers/raw/master/ipfs-cap2pfs/ipfs-p2p-file-system.pdf. Accessed 08 Dec 2020
15. Aishwarya Nandakumar, P.P. Amritha, K.V. Lakshmy, V.S.T.: Non linear secret sharing for gray scale images. Procedia Eng. **30**, 945–952 (2012). https://doi.org/10.1016/j.proeng.2012.01.949
16. ERC-721 Token Standard. http://erc721.org/. Accessed 08 Dec 2020
17. Wood, G.: Ethereum: a secure decentralised generalised transaction ledger. Ethereum project Yellow Paper. 151.2014 (2014). https://ethereum.github.io/yellowpaper/paper.pdf
18. PACS Systems: Everything You Need to Know About them. https://blog.peekmed.com/pacs-systems/. Accessed 08 Dec 2020
19. Patient Medical Records GitHub Page. https://amrita-tifac-cyber-blockchain.github.io/Patient-Medical-Records/
20. Misha Abraham, A.M., Vyshnavi, H., Srinivasan, C., P.K. Namboori, K.: Healthcare security using blockchain for pharmacogenomics. J. Int. Pharm. Res. **46**(1), 529–533 (2019)

Detection of Depression and Suicidal Ideation on Social Media: An Intrinsic Review

Sanat Madkar(✉), Tanay Maheshwari, Mann Merani, Rahil Merchant, and Pankti Doshi

Department of Computer Engineering, Mukesh Patel School of Technology Management and Engineering, NMIMS University, Mumbai, India
{sanat.madkar56,tanay.maheshwari59,mann.merani68, rahil.merchant69}@nmims.edu.in, pankti.doshi@nmims.edu

Abstract. Depression is one of the most prevalent mental health disorders. The advent of social media has only escalated its impact on the modern generation. It has an impact on the language usage reflected in the written text. Despite its alarming growth, detection of depression has always been a challenging task, especially over social media. Research on mental health disorders has also been advanced in light of the ever-growing popularity of social media platforms and their increased impact on society. Unlike traditional observational cohort studies conducted through questionnaires and self-reported surveys, a reliable alternative would be the use of machine learning algorithms to help detect depression levels of a user. The key objective of our study is to review such machine learning algorithms proposed by previous authors and discuss the various types of dataset characteristics, feature extraction and modelling techniques to classify whether a given text is of depressive nature or not.

Keywords: Natural Language Processing · Depression · Social media · Preprocessing · Feature extraction · Topic modeling · Neural networks · Deep learning

1 Introduction

One in four people in the world will be affected by mental or neurological illness at some point in their lives as per a report by the World Health Organization [13]. Despite the increasing success in the identification of these mental illnesses, many instances do remain very difficult to identify. Automated methods are becoming increasingly capable of detecting symptoms commonly associated with mental illnesses. These symptoms may be recognized in content generated by users on social media websites and discussion forums. Patterns in language and online activity of users can be used to screen those that may potentially be suffering from mental health issues. Traditional methods utilized surveys in order

M. Singh et al. (Eds.): ICACDS 2021, CCIS 1441, pp. 63–75, 2021.
https://doi.org/10.1007/978-3-030-88244-0_7

to diagnose depression but it is apparent that automated detection methods ought to be incorporated to further aid screening procedures.

In recent times, people have begun to share their experiences and challenges with mental health issues through online forums, micro-blogs, or tweets [1,8]. Their online activities have encouraged many researchers to introduce new types of health care solutions and early depression diagnostic procedures. If the automated procedure can determine high levels of stress for the user, appropriate support and treatment can be provided to the person before it gets out of hand.

Natural Language Processing (NLP) [6,21] defines the sentiment expression of a specific subject, and classifies the polarity of the sentiment lexicons. NLP can identify fragments of a text that may contribute to sentiment identification, instead of classifying the sentiment of the whole text based on the specific subject. Feature extraction algorithm is one of the NLP techniques. It can be used to extract subject-specific features, extract sentiment of each sentiment-bearing lexicon, and associate the extracted sentiment with specific subject.

Depression detection on social media is based on automated analysis of the content posted by the users. This is achieved by performing predictive analysis using various models, and these models employ 'features' that have been extracted from social media data. These features are then treated as independent variables in an algorithm to predict the dependent variable of an outcome of interest (e.g. users' mental health). Predictive models are trained, using an algorithm, on part of the data (the training set) and then are evaluated on the other part (the test set) to avoid overfitting – a process called cross-validation. The prediction performances are then reported as one of several possible metrics.

Apart from this, instead of using features from social media data, there have been a few lexicon-based approaches [14,22]. In general, these approaches use a lexicon-dictionary based method for assigning an overall depression score to subjects. However, lexicon-based approaches suffer from low recall and are highly dependent on the quality of the created lexicon.

The outline of NLP is given by the workflow of gathering data, preprocessing, extracting features and topics for modelling and finally the various models used to classify text. The specific domain of depression detection and suicide ideation requires extensive linguistic preprocessing and feature extraction to highlight emotional undertones and specific language and jargon used by users affected by depression or suicidal thoughts.

The objective is to bridge the gap in research for depression and suicidal intent over social media and to employ modern intuitive approaches that enable fast detection and help in early prevention of suicides amongst people online. With the help of NLP, a lot of these tasks have been expedited and with newer approaches bred from the inception of Deep Learning, the process of detecting depression online can only get more facile.

2 Datasource and Dataset Characteristics

An extensive dataset is required for thorough analysis and preprocessing of language and sentiment for depression. Various sources can be utilized for procurement ranging from social media sites using APIs to readily available datasets for analysis. Various papers have implemented datasets aggregated from different online sources as given in Table 1.

Reddit is an online discussion forum consisting of multiple 'subreddits' or communities of various specific topics. The content on these subreddits are structured by posts, each of which has a title, some content and comments by users as well as a rating system for likes and dislikes known as 'upvotes' or 'downvotes' respectively. Similarly, Twitter is another social networking service where users can post public messages known as 'tweets'. A tweet may also be accompanied with some username tags to denote mention of other twitter users or # tags (hashtags) which denote the mention of a certain topic along with the post. By filtering messages into depressive subreddits or hashtags, it is easy to analyze the nature of depressive content on these sites.

Some of the aforementioned sources include searching public Tweets for keywords to identify users who have shared their mental health diagnosis, user language on mental illness related forums, collecting public Tweets that mention mental illness keywords for annotation, or through searching on depression specific chat boards for posts and comments by users and obtaining all posts by a user who has posted before in that chat board. The approaches using public data have the advantage that much larger samples can, in principle, be collected

Table 1. Papers and their Datasets

Ref No.	Paper	Year	Dataset name	Platform	User characteristic
[38]	Yazdavar et al.	2017	Custom Twitter Search API generated	Twitter	Self-Declared
[29]	Trotzek et al.	2018	CLEF eRisk 2017	Reddit	Self-Declared
[11]	Orabi et al.	2018	CLPsych 2015, Bell's Let's Talk	Twitter	Self-Declared
[33]	Wang et al.	2018	CLEF eRisk 2018	Reddit	Self-Declared
[35]	Wongkoblap et al.	2019	myPersonality project	Facebook	Survey (CES-D)
[34]	Wolohan et al.	2018	Custom Python Reddit API generated	Reddit	Self-Declared
[3]	Aldarwish et al.	2017	Custom gathered dataset	Livejournal, Twitter, Facebook	Self-Declared
[30]	Tsugawa S et al.	2015	Twitter API and CES-D survey of Japanese users	Twitter	Survey (CES-D)
[2]	Aladağ et al.	2016	Publicly available dataset	Reddit	Self-Declared
[25]	Sawhney et al.	2018	Custom Twitter Rest API generated	Twitter	Self-Declared
[27]	Tadesse et al.	2018	Reddit dataset by Pirina, Coltekin [24]	Reddit	Self-Declared
[9]	Du et al.	2017	Custom Twitter Streaming API generated	Twitter	Self-Declared
[4]	Almeida et al.	2017	CLEF eRisk 2017	Reddit	Self-Declared
[10]	Gkotsis et al.	2016	Full Reddit Submission Corpus (2006 thru August 2015) from r/datasets	Reddit	Self-Declared

The survey dataset [CES-D] includes questions such as how often they experienced symptoms over the past week over certain situations and asking their users to rate symptoms on a fixed scale.

faster and more cheaply than through the administration of surveys, though survey-based assessment generally provides a higher degree of validity.

Table 1 shows the source, platforms, and characteristics of the datasets used by the papers reviewed.

The purpose of data annotation is to make sense of the data, by adding relevant metadata labels, in order to train supervised models for better understanding of new unannotated data for language detection. Annotation can either be done manually by the authors or by a group of specialized annotators[4,29,38]. The scale rating can be from 0 to 1 [2,4,9,25] or it can be categorized into different levels of depression, as many as upto 9 [38] (Patient Health Questionnaire-9 (PHQ-9) questionnaire) [17]. Specific datasets such as eRisk [4,29,33] and Center for Epidemiological Studies-Depression (CES-D) [30,35] have been annotated by the dataset authors themselves [19]. Certain lexicons such as Valence Aware Dictionary and sEntiment Reasoning (VADER) sentiment lexicon used in [29] are also pre-annotated by authors [12].

3 Methodology/Assessment Criteria for Modeling

3.1 Preprocessing

Preprocessing is an essential part of Machine Learning and especially Natural Language Processing. Any given piece of text, to make sense of it needs to be preprocessed effectively to reduce ambiguity and noise which may be present in the dataset.

The papers using Reddit datasets have taken some common steps to preprocess these datasets, which include removing user or subreddit mentions either by simply removing them or replacing or generalizing them to some normal form [29]. The same could be said for Twitter datasets wherein tweets containing retweets [11], URLs [4,11,25] and username links [11] were removed or simply replaced by their respective keywords [38]. URLs or other links have also been removed in order to retain normalized text [9,11,25,27,38].

The common preprocessing pipeline is displayed in the following diagram. The first stage of the pipeline is to remove noisy entities such as punctuations, non-alphanumeric, non-Unicode, non-ASCII, space delimiters and non-English characters. This stage also includes filtering text with suicidal keywords and collecting text with more frequent use of first person pronouns. The next step is tokenization; tokens can be directly implemented by Recurrent Neural Networks (RNN), Gated Recurrent Units (GRU) and other deep learning models. Tokens can be entire words or individual characters of the word itself and can be used to distinguish topics in case of topic modeling. Stopword removal gets rid of words that are insignificant for modeling such as articles and prepositions which do not usually depict sentiment. The final step involves stemming and lemmatization - both processes used to convert a word to its root form. Stemming involves the use of an algorithm to decompose a given word, which may or may not be an actual word whereas lemmatization uses a corpus and part of speech to compare words, therefore always giving a valid decomposed form of the original word.

Most of the papers have implemented steps or different alterations of the given stages of the pipeline suited best to remove features specific to their research (Fig. 1).

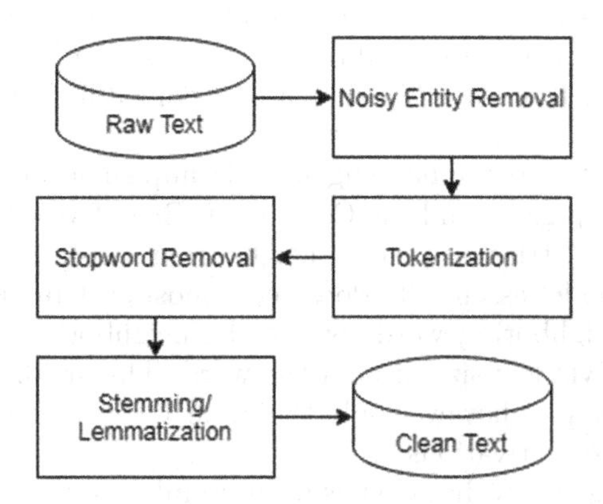

Fig. 1. Preprocessing pipeline

In the case of survey responses, users who completed the survey before the given amount of time could be considered as bogus responses, a condition handled by the authors by discarding such responses [30]. Furthermore, a specific limit can be imposed on the number of words [25,30] or posts are taken into consideration, in case of selecting the most frequently used words [10,33], a minimum limit of words considered for evaluation [35], or the most recent authored posts [34].

3.2 Feature Extraction

After preprocessing, the text obtained is still in a raw form. Because the machine learning algorithms cannot process the raw text directly, instead relying on matrices and vectors, another step is needed to convert raw data to usable form. Feature extraction converts this raw text into a matrix or a vector of 'features' which can be used by the machine learning algorithms for learning. A feature value can be obtained by a number of suitable processes, whether from a given corpus, does a given sentence contain specific words or a binary vectorization of text with the words obtained from the corpus. For depression analysis, features are important so as to classify the occurrence of common vocabulary among texts closely associated with depressive intent.

Single set features such as n-grams and bag of words may be used individually or along with other feature extraction or topic modeling methods [27]. The bag of words method does not consider the order of words and n-grams can be used to group 'n' words together for analysis. Sentiment analysis to denote polarity of words may also be used for a document [38].

Deep Learning architectures are unable to process raw plaintext and strings, they require the text to be converted to processable structure in the form of numbers, which is why word embedding is used. Word embedding involves mapping of individual words to vectors of numbers. It essentially maps words uniquely as a dictionary, where each word has a different vector representation. There are different approaches to word embeddings; they may be based on frequency of words, or be a prediction based in the case of embeddings utilizing the neural network approach.

Two examples of word embedding models implementing a neural network approach are the Skip-gram and the Continuous Bag of Words (CBoW) models. The skip-gram model tries to predict neighboring words in the given window while the CBoW model essentially does the opposite; it predicts a target word based on all the neighboring words where the neighboring words selected are determined by the window size around the word. The architecture remains the same for both techniques however only the input layer dimensions change along with the hidden layer activations.

Transfer learning as a technique has been useful for deep learning approaches in the usage of extensively pre-trained models for tasks on computer vision and NLP. The concept of transfer learning is to be able to reuse a previously developed model and use it as a basis for another model on a different task while being relatively time effective and improving model performance. It forms the basis for a lot of NLP tasks including the recent research of transformers. Pretrained word embeddings are extensively designed on large datasets to capture the word definitions over a larger context. The lack of training data and a large number of training parameters makes it difficult to design custom word embeddings hence pre-trained embeddings are preferred during processing.

The word2vec word embedding by Google [20] is made up of two neural network architectures: the Continuous Bag of Words (CBoW) and the Continuous Skip-gram (SG) architecture. The Global Vectors for Word Representation (GloVe) embedding published by Stanford NLP [23] combines the context-based learning and matrix factorization models like the Latent Semantic Analysis (LSA). Trotzek et al. [29] trained a 50-dimensional GloVe model on a Wikipedia and a Gigaword 5 news corpus along with a 300 dimension fastText model on a similar corpus. They also utilize a custom fastText model trained on Reddit comments and the implementation in training a 300 dimension CBoW model. A word2vec model is also trained for comparison with the fastText and GloVe models. Along with using two word2vec models, using skip-gram and CBoW representation, Orabi et al. [11] expanded their word embedding system by using labels for improving estimations by using the multi-task deep learning (MTL) [7] using Rectified Linear Unit for activation.

The Linguistic Inquiry and Word Count (LIWC) tool [28] developed by Pennebaker, Booth and Francis, is used to classify words in a given text representing different emotions, temporal orientation and informal language, chiefly dividing them into psychological groups. It calculates four variables that denote emotion, authenticity, confidence and analytical thinking, giving a measure as to which psychological category, each word falls in. The LIWC tool has been utilized by authors in [2,25,27,29,34].

The binary firefly algorithm is a discrete space based modification proposed to the original firefly algorithm [26] utilized by [25]. It aims to minimize the number of features in order to reduce high number of dimensions and redundancies. A feedback mechanism to the classifier is used to fine tune the features. The input is in the form of a binary vector of 0 s and 1 s.

The martingale framework [32] used for change point detection is useful to detect emotion changes. The authors in [31] use a strangeness measure to distinguish a tweet from previous tweets, which are then ranked to create a family of martingales to detect the mobility of the tweets. The change is measured as a Euclidean distance from the mean of other data points.

3.3 Topic Modeling

Topic modeling constitutes a part of feature extraction that involves statistical modeling for distinguishing various 'topics' as part of a document. Grouping words provides better context and distinction between different words in a text as user generated text may be ambiguous and unclear to understand purely from a machine point of view, especially categorizing amongst various symptoms of depressions such as hyperactivity, self-harm and sleep disorders.

An example of topic modeling is Latent Dirichlet Allocation (LDA) which is used to classify certain texts in a document to these various topics. The aim is to build models that aggregate the words per topic and subsequently the topics per document modeled as Dirichlet distribution. LDA involves setting a parameter for n which is the number of topics and parameters alpha, and beta for composites-versus-topics and parts-versus-topic respectively.

The number of topics taken can vary distantly from 10 [10] to as many as 70 [27] and can be used additionally with other sampling methods [30]. The number of topics parameter forms the basis for the complexity of categorizing topics for the document.

Another example of topic modeling is the Term Frequency - Inverse Document Frequency (TF-IDF) measure. TF denotes the frequency of a term in a document while the inverse document frequency measures the rarity of a word in the entire set of documents. The TF-IDF vectorizer converts each word to a unique integer vector form ranging from low [25,33] to high values. It can also be used on multiple matrices [2], used with combinations like n-grams and LIWC features [34] or base models ranging such as SVMs [11].

3.4 Model Approaches

Numerous machine learning and neural network algorithmic approaches have been considered, all depending on the prior approaches taken for preprocessing and extracting features or topics. A range of convolutional and recurrent neural network architectures have been used in recent researches while regular machine learning algorithms have also proven to be useful for the task. The advantages that neural networks hold over conventional machine learning algorithms is that they provide better learning and predictions due to the numerous layers involved, the recursivity of the algorithms and the fact that learning is unsupervised.

Convolutional neural networks (CNN) can read words as vector representations for NLP tasks. The number of filters, characteristics of the layers, pooling, and softmax functions can be added as per input or output requirements. Trotzek et al. [29] modified a one-layer CNN architecture proposed by Zhang and Wallace [39] with a total of 100 filters, with height as 2 and a width of given vector dimensions. Concatenated Rectified Linear Units (CReLU) activation gives a 200-dimensional vector output which is then propagated with applied dropout and softmax, resulting in a output matrix corresponding to the word embedding. Wang et al. [33] take the top 300 Reddit posts, concatenate the title and text, and process this with a CNN-based sentence encoder. The classifier predicts risk for the user if more than a certain number of posts are labeled as positive for depressive content and a non-risk if more than a certain number are negative. In the last chunk, the risk is predicted for more than a given number of posts or comments classified as depressive.

Du et al. [9] use a short text classification CNN architecture proposed by Kim et al. [15] to build a binary classifier. Max pooling and dropout are added to the pooling layer and applied to a softmax output. Orabi et al. [11] utilize three different CNN models. CNNwithMax has a 1-dimensional convolution operator with 250 filters and kernel size 3 and a global max pooling strategy. The MultiChannelCNN has 3 convolutions with 128 features each, for filter lengths 3, 4, and 5. One dimensional operators and max-pooling layer process a single output of features. MultiChannelPooling CNN uses the same strategy as MultiChannelCNN with two different max-pooling sizes of 2 and 5.

Gkotsis et al. [10] use filter size of 3 for their CNN model along with a feedforward architecture. In both architectures, embedding layers of dimension 16 are followed by a dense structure of 64 neurons for Feed Forward and 256 neurons for CNN as feature space and dropout of 0.25. The CNN has an extra convolutional layer for length 5 and max-pooling of length 2. Rectified Linear Units (ReLU) is used as activation and categorical cross-entropy is used for the objective function.

Wongkoblap et al. [35] work on the multiple instance learning network (MIL-NET) approach [5,16] using Long Short Term Memory (LSTM) layers, as an alternative to a CNN post encoding previously developed by them [36]. The word embedding vectors are encoded using bidirectional GRUs. A softmax classifier and another layer of bidirectional GRU is encoded subsequently. A one-layer multilayer perceptron (MLP) is used for the attention vector and user vectors are calculated from these measures. Sawhney et al. [25] analyze a combination of LSTM and CNN called CNN-LSTM along with regular LSTM models for comparison with supervised models such as XGBoost and Random Forest classifiers used in tandem with the Binary Firefly Algorithm.

Orabi et al. [11] along with CNN models, use an RNN Bidrectional Long Short Term Memory (BiLSTM) architecture with attention. It captures temporal and abstract sequence information bidirectionally. Context-aware attention mechanism [37] is used as a classification feature vector in the form of a weighted summation of all words.

Yazdavar et al. [38] propose the semi-supervised topic modeling over time (ssToT), working on the principle of Latent Dirichlet Allocation (LDA), which is more effective in categorizing and labeling terms than the LDA. It seeds words for each topic, the most relevant words to a symptom. Common themes and symptoms of depression form the various topics which would otherwise be unlabelled under LDA.

Amongst the different activation functions used, the softmax function provides a probability distribution value which gives the probability of the output class being depressive or not while the more effective ReLU implements threshold operation to rectify output values into respective classes providing faster computability. Both aforementioned activation functions work better than a regular sigmoid function, which simply works on feedforward networks and gives poor results and gradient irregularities for backpropagation methods.

Almeida et al. [4] work on an information retrieval based approach merged with conventional supervised learning-based systems. A decision function calculates if a user belongs to the depression class with individual indicator variables for each learning system. The supervised learning models were constructed through three classification algorithms, a logistic model tree [18] and Random Forests.

Support Vector Machines are common as a part of traditional classifier models based on modeling vector spaces. SVMs are used by [2,3,34] (two SVMs with TF-IDF and LIWC features) and [30]. Additionally, Aldarwish et al. [3] use the Naive Bayes classifier and Aladağ et al. [2] use the logistic regression classifier for comparison. Tadesse et al. [27] also use a wide range of classifiers including the SVM, Logistic Regression, Random Forest, Ada Boost, and the machine layer perceptron artificial neural network with two hidden layers along with different features and combination of features.

4 Discussion

Upon comparison of the aforementioned techniques, the approaches can be roughly divided into two categories: the traditional classifier models such as SVMs, Logistic Regression, Random Forest and the neural network architectures such as CNNs and RNNs. Over traditional classifier models, SVM was used extensively and compared with other classifiers, usually with topic modelling done beforehand with several experiments tested. In almost all experimentations [2,27,30], the SVM yielded the most favourable results except Aldarwish et al. [3] where the Naive Bayes Classifier prevailed. The CNN architecture tends to yield higher accuracies than both conventional supervised learning approaches and RNN models as evident by experimentations [9–11], subsequently outperforming the traditional classifier models. The CLEF eRisk dataset [19] calculates a parameter called the Early Risk Detection Error (ERDE) defined by the authors themselves. It calculates a penalty variable based on how early a true positive alert is noticed and rewards them, conversely penalizing late alerts. Experiments conducted by [4,29,33], show that some models have a higher ERDE score but a low F1, precision or recall scores and vice versa. There is a tradeoff between the different goals. The model with a higher F-score usually suffers from poor ERDE score. Independent experiments conducted by [26,34,35,38] also show better results compared to their respective baseline models tested.

5 Conclusion

The review of the literature on depression detection is a very exhaustive task, however, in the long run, shall prove to be a task of important sorts.

The studies reviewed here suggest that depression, mental illnesses and suicidal ideation are detectable on several online environments, but the generalizability of these studies to larger samples and gold standard clinical criteria has not been established.

It is observed that proposed approaches, ensemble models tend to perform better than regular modeling methods. This is due to consideration taken with respect to dataset and specificity in feature extraction.

6 Future Work

Limited pre-processing done in papers was evident in the fact that the noise gathered at the output compromised with the accuracy. Also, social media content can be ambiguous for classifiers, as they are not advanced enough to deal with sarcasm or context of some of the discussions, while a manual annotator may notice this.

Apart from this, transformer models such as Bidirectional Encoder Representations from Transformers (BERT) and A Lite BERT for Self-Supervised Learning of Language Representations (ALBERT) have hardly been implemented by researchers for the purpose of this study, which have shown great potential in the fields of language translation and sentimental analysis.

References

1. Aalbers, G., McNally, R., Heeren, A., de Wit, S., Fried, E.: Social media and depression symptoms: a network perspective. J. Exp, Psychol. Gen. (2018, in press). https://doi.org/10.1037/xge0000528
2. Aladağ, A.E., Muderrisoglu, S., Akbas, N., Zahmacioglu, O., Bingol, H.: Detecting suicidal ideation on forums and blogs: proof-of-concept study. J. Med. Internet Res. **20**, e215 (2018). https://doi.org/10.2196/jmir.9840
3. Aldarwish, M.M., Ahmad, H.F.: Predicting depression levels using social media posts. In: 2017 IEEE 13th international Symposium on Autonomous decentralized system (ISADS), pp. 277–280. IEEE (2017)

4. Almeida, H., Briand, A., Meurs, M.: Detecting early risk of depression from social media user-generated content. In: CLEF (2017)
5. Angelidis, S., Lapata, M.: Multiple instance learning networks for fine-grained sentiment analysis. CoRR abs/1711.09645 (2017)
6. Chowdhury, G.G.: Natural language processing. Ann. Rev. Inf. Sci. Technol. **37**(1), 51–89 (2003). https://doi.org/10.1002/aris.1440370103
7. Collobert, R., Weston, J.: A unified architecture for natural language processing: deep neural networks with multitask learning. In: ICML 2008 (2008)
8. De Choudhury, M., Counts, S., Horvitz, E.: Social media as a measurement tool of depression in populations. In: Proceedings of the 5th Annual ACM Web Science Conference (WebSci 2013), pp. 47–56. Association for Computing Machinery, New York (2013). https://doi.org/10.1145/2464464.2464480
9. Du, J., et al.: Extracting psychiatric stressors for suicide from social media using deep learning. In: BMC Medical Informatics and Decision Making 2018 (2018)
10. Gkotsis, G., et al.: Characterisation of mental health conditions in social media using informed deep learning. Sci. Rep. **7**, 45141 (2017). https://doi.org/10.1038/srep45141
11. Husseini Orabi, A., Buddhitha, P., Husseini Orabi, M., Inkpen, D.: Deep learning for depression detection of twitter users. In: Proceedings of the Fifth Workshop on Computational Linguistics and Clinical Psychology: From Keyboard to Clinicpp, pp. 88–97, January 2018. https://doi.org/10.18653/v1/W18-0609
12. Hutto, C., Gilbert, E.: Vader: A parsimonious rule-based model for sentiment analysis of social media text. In: ICWSM (2014)
13. Härtl, G.: The world health report 2001: Mental disorders affect one in four people, September 2001, https://www.who.int/news/item/28-09-2001-the-world-health-report-2001-mental-disorders-affect-one-in-four-people
14. Karmen, C., Hsiung, R., Wetter, T.: Screening internet forum participants for depression symptoms by assembling and enhancing multiple NIP methods. Comput. Methods Prog. Biomed. **120** (2015). https://doi.org/10.1016/j.cmpb.2015.03.008
15. Kim, Y.: Convolutional neural networks for sentence classification. In: Proceedings of the 2014 Conference on Empirical Methods in Natural Language Processing (EMNLP)(2014)
16. Kotzias, D., Denil, M., de Freitas, N., Smyth, P.: From group to individual labels using deep features. In: Proceedings of the 21th ACM SIGKDD International Conference on Knowledge Discovery and Data Mining (KDD 2015), pp. 597–606. Association for Computing Machinery, New York (2015). https://doi.org/10.1145/2783258.2783380
17. Kroenke, K., Spitzer, R.: The phq-9: a new depression diagnostic and severity measure. Psychiatric Ann. **32**, 509–515 (2002)
18. Landwehr, N., Hall, M., Frank, E.: Logistic model trees. Mach. Learn. **59**, 161–205 (2005). https://doi.org/10.1007/s10994-005-0466-3
19. Losada, D.E., Crestani, F.: A test collection for research on depression and language use. In: Fuhr, N., et al. (eds.) CLEF 2016. LNCS, vol. 9822, pp. 28–39. Springer, Cham (2016). https://doi.org/10.1007/978-3-319-44564-9_3
20. Mikolov, T., Chen, K., Corrado, G., Dean, J.: Efficient estimation of word representations in vector space. In: Proceedings of Workshop at ICLR 2013, January 2013
21. Nadkarni, P., Ohno-Machado, L., Chapman, W.: Natural language processing: an introduction. J. Am. Med. Inf. Assoc. JAMIA **18**(5), 544–51 (2011)

22. Neuman, Y., Cohen, Y., Assaf, D., Kedma, G.: Proactive screening for depression through metaphorical and automatic text analysis. Artif. Intell. Med. **56**(1), 19–25 (2012). https://doi.org/10.1016/j.artmed.2012.06.001
23. Pennington, J., Socher, R., Manning, C.D.: Glove: Global vectors for word representation. In: EMNLP (2014)
24. Pirina, I., Çagri Çöltekin: Identifying depression on reddit: the effect of training data. In: EMNLP 2018 (2018)
25. Sawhney, R., Shah, R.R., Bhatia, V., Lin, C., Aggarwal, S., Prasad, M.: Exploring the impact of evolutionary computing based feature selection in suicidal ideation detection. In: 2019 IEEE International Conference on Fuzzy Systems (FUZZ-IEEE), pp. 1–6 (2019). https://doi.org/10.1109/FUZZ-IEEE.2019.8858989
26. Sawhney, R., Mathur, P., Shankar, R.: A firefly algorithm based wrapper-penalty feature selection method for cancer diagnosis, pp. 438–449, July 2018. https://doi.org/10.1007/978-3-319-95162-1_30
27. Tadesse, M.M., Lin, H., Xu, B., Yang, L.: Detection of depression-related posts in reddit social media forum. IEEE Access **7**, 44883–44893 (2019). https://doi.org/10.1109/ACCESS.2019.2909180
28. Tausczik, Y., Pennebaker, J.: The psychological meaning of words: Liwc and computerized text analysis methods. J. Lang. Soc. Psychol. **29**, 24–54 (2010)
29. Trotzek, M., Koitka, S., Friedrich, C.: Utilizing neural networks and linguistic metadata for early detection of depression indications in text sequences. IEEE Trans. Knowl. Data Eng. **32**, 588–601 (2018). https://doi.org/10.1109/TKDE.2018.2885515
30. Tsugawa, S., Kikuchi, Y., Kishino, F., Nakajima, K., Itoh, Y., Ohsaki, H.: Recognizing depression from twitter activity. In: Proceedings of the 33rd Annual ACM Conference on Human Factors in Computing Systems, pp. 3187–3196, April 2015. https://doi.org/10.1145/2702123.2702280
31. Vioulès, M.J., Moulahi, B., Azé, J., Bringay, S.: Detection of suicide-related posts in twitter data streams. IBM J. Res. Dev. **62**, 7 (2018)
32. Vovk, V., Nouretdinov, I., Gammerman, A.: Testing exchangeability on-line. In: Proceedings of the 29th International Conference on Machine Learning, pp. 768–775, January 2003
33. Wang, Y.T., Huang, H.H., Chen, H.: A neural network approach to early risk detection of depression and anorexia on social media text. In: CLEF (2018)
34. Wolohan, J., Hiraga, M., Mukherjee, A., Sayyed, Z., Millard, M.: Detecting linguistic traces of depression in topic-restricted text: attending to self-stigmatized depression with NLP (2018)
35. Wongkoblap, A., Vadillo, M.A., Curcin, V.: Predicting social network users with depression from simulated temporal data. In: IEEE EUROCON 2019-18th International Conference on Smart Technologies, pp. 1–6 (2019). https://doi.org/10.1109/EUROCON.2019.8861514
36. Wongkoblap, A., Vadillo Nistal, M., Curcin, V.: Modeling depression symptoms from social network data through multiple instance learning (2019)
37. Yang, Z., Yang, D., Dyer, C., He, X., Smola, A., Hovy, E.: Hierarchical attention networks for document classification. In: Proceedings of the 2016 Conference of the North American Chapter of the Association for Computational Linguistics: Human Language Technologies, pp. 1480–1489. Association for Computational Linguistics, San Diego, California, June 2016. https://doi.org/10.18653/v1/N16-1174

38. Yazdavar, A., et al.: Semi-supervised approach to monitoring clinical depressive symptoms in social media. In: Proceedings of the IEEE/ACM International Conference on Advances in Social Network Analysis and Mining, vol. 2017, August 2017. https://doi.org/10.1145/3110025.3123028
39. Zhang, Y., Wallace, B.: A sensitivity analysis of (and practitioners' guide to) convolutional neural networks for sentence classification. In: Proceedings of the The 8th International Joint Conference on Natural Language Processing (2016)

Frequency Based Feature Extraction Technique for Text Documents in Tamil Language

M. Mercy Evangeline[1(✉)], K. Shyamala[1], L. Barathi[2], and R. Sandhya[2]

[1] Dr Ambedkar Govt Arts College, Vyasarpadi, Chennai, Tamilnadu, India
[2] Women's Christian College, Chennai, Tamilnadu, India

Abstract. In Natural Language Processing, text classification involves the process of categorizing documents into organized groups. Classifiers are trained to analyze the documents and categorize them into groups based on pre-defined tags. In this paper, a framework has been devised to extract features from the text documents. The documents are preprocessed, tagged and then the features are extracted. Preprocessing includes removal of stop words, punctuations. Preprocessed Words were tagged under two categories, Noun and Verb. Tagging of words was based on morphophonemic rules pertaining to Tamil Language. Feature extraction is based on frequency of the occurrences of words tagged as Noun. Extraction of features through the proposed framework enhances dimensionality reduction of data considered for classification process.

Keywords: Information retrieval · Text classification · Knowledge management · Feature extraction · Feature selection · Part-of-speech tagging

1 Introduction

In our everyday life, we come across a huge volume of data in the form of text, number, diagrams, pictures, videos etc. Information may be structured or unstructured form. Acquiring the necessary information from a large amount of data is challenging in this digital world. Unstructured data is available in different forms like of text files, Emails, Social Media data, log files etc. Information Extraction normally deals with the data semantically. For categorization process, the keyword has to be identified and the data has to be gathered under a particular label. In document classification, keywords or features are used for grouping the documents under a category.

1.1 Feature Extraction and Feature Selection

Analyzing huge data collection becomes a cumbersome work, as the information is more and may be redundant. It is tough to go through all the documents, organize them and find the data of interest. The process of Feature Extraction and Feature Selection is embraced to reduce the excess usage of computing resources. It becomes important to extract the needed information from the unstructured text dataset [11]. Figure 1 gives an overall representation of data analysis and interpretation. Feature Extraction (FE)

The original version of this chapter was revised: The author's name has been corrected as "K. Shyamala". The correction to this chapter is available at
https://doi.org/10.1007/978-3-030-88244-0_39

M. Singh et al. (Eds.): ICACDS 2021, CCIS 1441, pp. 76–84, 2021.
https://doi.org/10.1007/978-3-030-88244-0_8

is basically defined as the process of extracting useful features from initial data set. It is different from Feature selection, where a subset of features will be chosen from the original features. Both these methods are used to remove redundant and irrelevant information, which in turn enhances the process of Classification of new data to the best of accuracy.

As the dimension of the data increases, it becomes necessary to reduce the number of features considered for analyzing the data. This will reduce the computational cost and time involved in data analysis and interpretation.

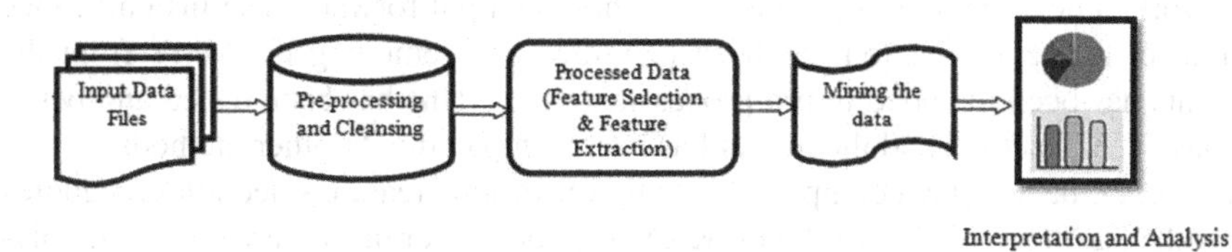

Fig. 1. Feature extraction and data analysis

Feature Extraction. Feature Extraction involves extraction of useful features from raw input data by transforming, reformatting and changing the data. The transformation of data is carried out till the dataset is suitable for the Machine Learning Model considered for analyzing the data. At the end, the dimension of the data will be reduced.

Identification of important features is considered as an important phase in Classification process which increases the performance of any classification process. Using an effective algorithm for feature extraction will increase the classification accuracy with decrease in the processing time.

Feature Extraction considers only relevant and useful information for analysis process. Choosing useful information involves expertise with the data being considered. Features Extraction applications includes Image processing, Medical Industry, Oil Industry, Diagnosis system, Machine Learning, Natural Language processing and many other emerging applications.

Feature Selection. Feature Selection (FS) is another way of reducing the dimensionality of input data considered for analysis. It is quite similar to Feature Extraction but with a subtle difference. Given a set of input raw data, Feature Selection is aimed in selecting some of the features and discarding the rest. This is based on the redundancy of the data or its irrelevance. In Feature Selection, a subset of original data is chosen and the Machine Learning model is applied. The rest of the data is not considered, thus reducing the volume of the data. It doesn't involve in transformation of data as Feature Extraction.

Different ways adopted for Feature Selection are the Wrappers, Filters and Embedded method. The Wrapper method involves evaluation of the best suited model for the dataset and sequentially choosing different subsets of features, finding the subset which gives the best result. It makes use of learning technique to find which features are useful and it proves to be of high cost.

The filter method doesn't apply any learning technique, just highlights the features which are of high potential with the problem being considered and it chooses the features listed on the top. The third method is the embedded way which considers the selection of features from the training dataset and then applying them on the test dataset.

2 Literature Survey

E. Kiliç et al. has proposed two new methods for document weighting based on Term Frequency-Inverse Document Frequency (TF-IDF). The two methods adopted in this paper are TESDF and SADF. These two methods have been used along with TF [1]. In this work, a new method of preprocessing has been put forward. The insignificance of Verbs for text classification has been identified and removing the Verbs from the documents has been included in pre-processing phase. This has been tested and better result has been observed and the output has been compared with other methods.

Zia T et al. have done a comparative study on various feature selection procedures along with classifiers like Naïve Bayes, KNN, support vector machine with linear, polynomial and radial basis kernels and decision tree [2]. The corpus used for the study included two data set, one from EMILLE corpus and other to be naïve collection of documents. For small sized EMILLE data set, Naïve Bayes classifier proved to be good with any of the feature selection methods. For a modest size of data like naïve collection, SVM with any of feature selection methods has performed better.

Khanam M H and Sravani have proposed a text summarization technique for given text document in Telugu language. This work has given a Frequency based approach where the important sentences were identified [3]. These sentences give a meaningful understanding of the document. The input document which is large in size is first summarized by the process of tokenization and stop word removal. Then words with highest frequency are identified and the corresponding sentences containing these high frequency words are selected. An abstract of useful sentences are generated which precisely shows the richness of the information leading to inflation of information retrieved.

Shirbhate, A. G., and Deshmukh, S. N. has proposed an approach for automatic classification of Twitters tweets into three main categories [4]. The categories included were positive, negative and neutral sentiment state. The classification was done for English language and it showed a better result compared to previously used methods. The tweets collected were preprocessed and tagged using Tree tagger method. After tagging process, the Naïve Bayes classifier with Mutual Information Future Selection algorithm is used for classification. This classifier combination has showed a better accuracy compared to normal Naïve Bayes classifier.

R. Dzisevič and D. Šešok have examined three different feature extraction procedures for text documents [5]. The main classification was carried out on short sentences and phrases. This classification procedure was carried along with neural network, for finding the highest accuracy classifier in representing the text features. The procedures considered for the classification included plain TF-IDF method and two of its modification - the Latent Semantic Analysis and Linear Discriminant Analysis. The outcome shows that the simple TF-IDF allows the classifier for higher accuracy in case of larger dataset and the TF-IDF with LSA for smaller dataset.

Saxena, Dixa et al. have presented a survey of feature extraction for both supervised and unsupervised text dataset [9]. They have discussed the pros and cons of adopting different techniques with previous methods. An insight has been given that application of neural network for text dataset provides a good scope in the future.

Vidhya, Asir D. and Jebamalar E. proposed an algorithm where Term Frequency method was combined with stemmer-based feature extraction [10]. This has been tested along with various classifiers and the results have shown that the proposed method an out performed compared to other methods.

Resham N. Waykole and Anuradha D. Thakare have worked towards applying the common Feature Extraction techniques and assessed their effectiveness with the two algorithms – Vanilla Logistic Regression and Basic Random Forest Classifier [12]. They have concluded that the wor2vec feature extraction proved to be a good method with random forest classifier.

Srinivasan, R., and C. N. Subalalitha have proposed Named Entity Recognition (NER) model where Feature extraction was carried out using Regex, Morphological extraction and Context extraction methods [16]. Extracted featured based on these three methods were further classified into classes and was used for prediction of Entity class using Naïve Bayes algorithm.

Rajkumar, N., et al. has proposed a model for feature selection from a large set of key words based on normalized weight [17]. Feature selection procedure reduces the key word list size which enhances the efficiency of text classification algorithms.

Rajkumar, N., et al. has proposed a model to extract features based on Term Frequency – Inverse Document Frequency (TF-IDF) method and Chi-square test for choosing optimal feature set [18]. Using the features subset different algorithms were implemented for the document corpus and outcome was measured in terms of accuracy, precision, recall and FI score.

3 Feature Extraction Techniques

Feature extraction technique is used to learn from initial data set and extract features. The algorithm learns the nature of the data using the training dataset and applies the knowledge gained for analyzing test data. Many algorithms are available for Feature Extraction process with different levels of accuracy with classifiers considered [13]. Multilevel methods for dimensionality reduction can also be implemented for having a superior performance compared to single level method [15].

The most popular methods of feature extraction include Bag-of-words (BOW) method, BOW with TF-IDF method and Word Embedding. The BOW method is one of the fundamental methods used for extracting features from a document. In this method, each word in the document is considered as a feature for training the classifier. This is the simplest, common method used among the different methods available of Feature Extraction. The main disadvantage of this method is that the word with higher frequency will become dominant compared to low frequency words which can be of importance to the dataset being considered for classification.

After pre-processing, a vocabulary of unique words is created from the given corpus. A matrix is created with each unique word as a column and the row entry indicates the presence and absence of the particular word. The main disadvantage in this method is that the order of occurrence of the word is lost. This can be avoided by taking N-gram of words instead of unigram. This may result in a huge sparse matrix which may give rise to complexity in the analysis process. The high frequency N-grams are usually articles, determiners etc. which occur more in any document. They can be removed from the document. N-grams with low frequency can also be removed from document as they appear very rarely in the document. Some words of importance which seldom appear in the document will be removed by this consideration. In order to avoid such a phase, another variation in BOW model known as TF-IDF vectorizer can be used.

Term Frequency-Inverse Document Frequency method has two parts the Term Frequency and Inverse Document Frequency. This method highlights rare words which does not occur too frequently but is of more importance. This method can be improved for a better accuracy in classification process [14]. The Term Frequency identifies the number of times a particular word has occurred in the document. The Inverse Document frequency gives the frequency of the word across all the documents. Both the term refer to the scoring of weight for the word in the corpus. It is a statistical measure for finding the importance of a word in the document collection. There are many variations available with TF-IDF. This method is mostly used in Information Retrieval and Text Mining.

Word Embedding is another method available for Feature Extraction. In this method, semantics of the word and the context in which it appears is computed. By using this technique, words which are similar in meaning will be identified more accurately and it will be used for Feature Extraction.

4 Frequency-Based Feature Extraction Architecture

The architecture represented in Fig. 2 identifies the words of higher importance and with higher frequencies in the given dataset. The input is given in the form of Text data file which is in UTF-8 format. The data file mainly consists of stories which are available as online resource. They have been saved as text file in UTF-8 format. The input file undergoes various transitions through each module and the features in the form of words of importance, with higher frequencies are identified and are further used in the classification process.

Initially, the input file goes through the Tokenization module. Here the words are tokenized, punctuations removed and removal of stop word takes place. Stop words are mainly words which come under the category of prepositions, determiners and pronouns. The output generated has an array of words. For the removal of stop words, the Dictionary Based Stop Word Removal Algorithm (DBSWRA) proposed earlier is used [6]. This array of words then goes through the next module for tagging. Mainly two kinds of tagging takes place, NOUN and VERB tagging. Tagging of words is based on identifying words which have undergone transition in accordance to morphophonemic rules and applying reverse splitting procedure.

Verb Identification Using Morphophonemic Rules (VIMR). Verb Identification module checks for transition with every word from the array. If there is a transition, they are tagged as VERB and remaining words are tagged as UNIDENTIFIED collection. Every word is considered for pattern matching, if they fall within a particular matching pattern, then they are tagged as VERB [8].

In the proposed module, Verb features are considered as insignificant for classification process. So they are not considered for Feature Extraction procedure. So the verbs in the given document are identified and removed from the word collection. The next level involves the tagging of words as NOUN and for this phase the words which are tagged as UNIDENTIFIED are considered.

Noun Identification Using Morphophonemic Rules (NIMR). For all the documents from the training data set, both manual identification and transition identification of Nouns is performed. Initially, for every document, nouns are manually identified and added to data file, which serves as a Dictionary of Nouns. The array of words from the previous phase (after VIMR) goes through two levels of identification. The first level includes identification of Nouns which have not undergone transition according to morphophonemic rules. The Nouns which are available in its root form are identified using the Dictionary of Nouns. In the second level, it goes through the NIMR module for identifying the nouns which have undergone transition according the morphophonemic rules [7].

In the NIMR procedure, a word identified with a matching pattern goes through the stemming process. In the stemming process, reverse splitting method is adopted for finding the root word. The output of the stemming process consists of an array of words. This is compared with the Dictionary of nouns and the NOUN is identified. If there are any new Noun words, they are added into the dictionary. Thus a dictionary of Nouns is created for facilitating identification procedure for future purpose.

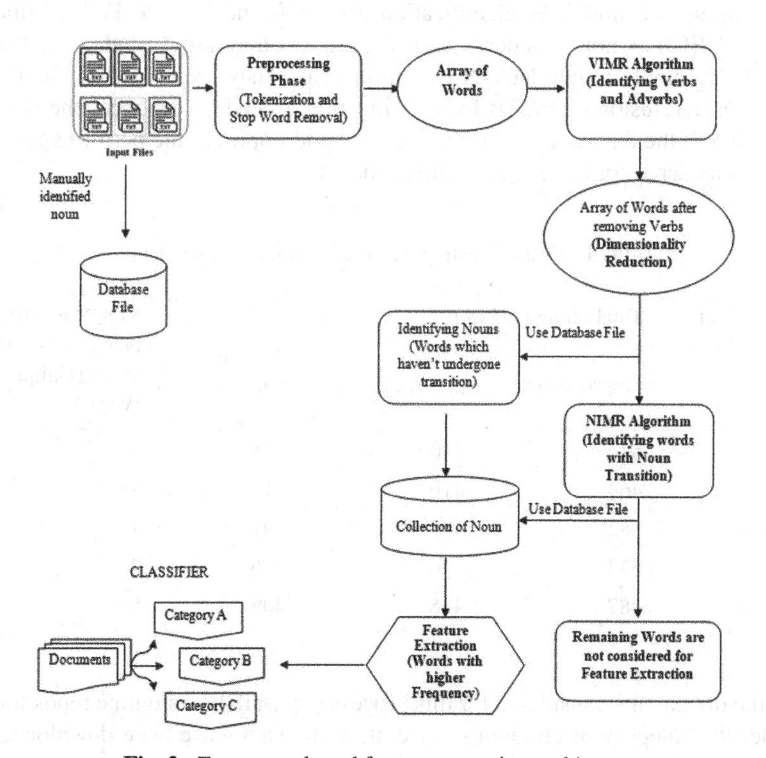

Fig. 2. Frequency-based feature extraction architecture

The Noun identification process generates a list of words. Words identified as NOUN are used as features extracted from the given document. The frequency of words is identified by finding the number of occurrences of the word within the list. The words with higher frequency are termed as features for the particular document. For training documents, the same procedure is adopted and features are extracted. Features for documents which are similar are pooled together and are used for the classification procedure.

5 Results and Discussion

In this proposed model, VERB and NOUN are identified for the given document. In this procedure, the VERB is considered as insignificant and nouns are considered significant in the classification procedure. So VERB is not considered and NOUN are considered for the classification procedure. So NOUN is considered as features from the documents pertaining to the training dataset. The frequency count of the Noun words was examined.

Table 1 gives an assessment of the files considered and its outcome after the different phases of the Feature Extraction process. The input file goes through three main modules. First they are pre-processed, then VERB is identified and finally the NOUN is identified. In the pre-processing stage, the main consideration is given for removing stop word which include articles and pronouns. After the pre-processing module the number of words reduced is very least which results in a negligible amount of reduction in the dimension of the data. So the next level of tagging is considered further.

The subsequent module is identification of VERB and NOUN. For classification procedure, VERB is not considered as a feature. So they are excluded for the next module. In the final module NOUN is identified and they are considered as features for classifier. Exclusion of VERB for classification procedure enhances the reduction in dimension of the dataset to a considerate level and improves the performance of any classifier considered for the categorization process.

Table 1. Features extracted from the training document

Input file (Text File in UTF – 8 Format)	Total number of words			NOUN identified as Features excluding Verbs (Unique Noun Words)
	Before preprocessing	After preprocessing	After verb removal	
File 1	570	530	475	79
File 2	464	416	416	57
File 3	582	497	497	75
File 4	573	496	496	74
File 5	487	456	409	59

All the documents considered for this process are available in online repository and fall under the category of children story collection. They have been downloaded and saved in UTF-8 format. The number of words for the input file has minimum of 450 words. After the identification of NOUN, the frequency count was calculated and the words with frequency range of 10 and above was considered for features extraction. These features are collected from all the training dataset documents considered for the classification process. The features extracted are represented in the form of vector matrix and is given as input to the classifier model.

Figure 3 gives a graphical representation for the data considered for the proposed algorithm and their output. It shows a comparison of total number of words in the document and number of words considered as features for the classifier. For the proposed algorithm, the files considered have a word count ranging from 450 words to 600 words before preprocessing. After going through the preprocessing phase and Verb identification removal, the word count is reduced to a subsequent level.

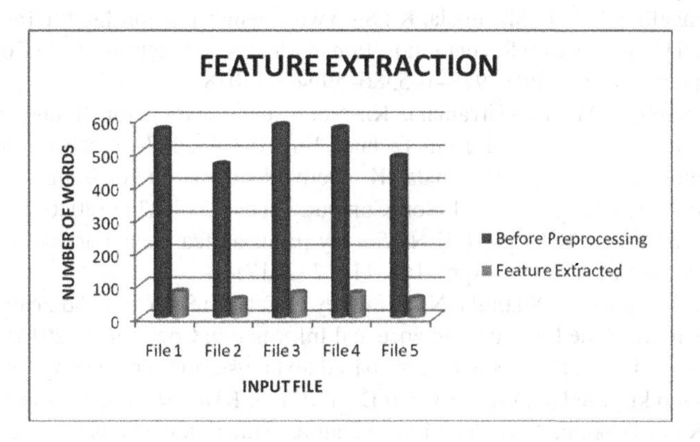

Fig. 3. Graphical representation of number of words before and after processing

6 Conclusion

Feature Extraction procedure has been applied for five documents which belonged to Children Stories of one particular interest. The Nouns have been identified from these documents by applying morphophonemic rules. The reverse splitting process has been adopted for the stemming procedure. The grammatical rules are used for identifying the nouns correctly and then are stemmed. This facilitates the Noun Identification process with a good accuracy. From the nouns identified, the frequency of the words is identified. Features are extracted based on the frequency of the words.

As future work, number of documents for training the procedure has to be increased with voluminous amount of words. Text document from different areas are to be collected and the algorithm is to be trained for the Feature Extraction procedure. Classification algorithm has to be applied for test data with the features extracted from the training dataset and their performance has to be studied.

References

1. Kiliç, E., Ateş, N., Karakaya, A., Şahin, D.Ö.: Two new feature extraction methods for text classification: TESDF and SADF. In: 23nd Signal Processing and Communications Applications Conference (SIU), pp. 475–478. Malatya (2015)
2. Zia, T., Akhter, M.P., Abbas, Q.: Comparative study of feature selection approaches for Urdu text categorization. Malays. J. Comput. Sci. **28**(2), 93–109 (2015)
3. Khanam, M.H., Sravani, S.: Text summarization for Telugu document. IOSR J. Comput. Eng. **18**(6), 25–28 (2016)
4. Shirbhate, A.G., Deshmukh, S.N.: Feature extraction for sentiment classification on twitter data. Int. J. Sci. Res. **5**(2), 2183–2189 (2016)
5. Dzisevič, R., Šešok, D.: Text Classification using Different Feature Extraction Approaches, Open Conference of Electrical, Electronic and Information Sciences (eStream). Vilnius, Lithuania, pp. 1–4 (2019)
6. Mercy Evangeline, M., Dr. Shyamala, K.: Stop word removal algorithm for Tamil Language, International Conference on Recent Innovation in Electrical, Electronics and Communication Engineering, CFP18P8 – PRT: 978–1–5386–5994–6 (2018)
7. Mercy Evangeline, M., Dr. Shyamala, K.: Noun identification for Tamil language using morphophonemic rules. Int. J. Recent Technol. Eng. ISSN: 2277–3878 **8**(4) (2019)
8. Mercy Evangeline, M., Dr. Shyamala, K.: Verb identification for Tamil language using morphophonemic rules. ICTACT J. Soft Comput. **11**(1), 2237–2243 (2020)
9. Saxena, D., Saritha, S.K., Prasad, K.N.: Survey paper on feature extraction methods in text categorization. Int. J. Comput. Appl. **166**, 11–17 (2017)
10. Samina, K., Tehmina, K., Shamila, N.: A Survey of Feature Selection and Feature Extraction Techniques in Machine Learning, Science and Information Conference (2014)
11. Beil, F., Ester, M., Xu, X.: Frequent term-based text clustering, Proceedings of International Conference on knowledge Discovery and Data Mining KDD '02, pp. 436–442 (2002)
12. Waykole, R.N., Thakare, A.D.: A review of feature extraction methods for text classification. Int. J. Adv. Eng. Res. Dev. **5**(04) (2018)
13. Elavarasan, N., Dr. Mani, K.: A survey on feature extraction techniques. Int. J. Innov. Res. Comput. Commun. Eng. **3**(1), 52–55 (2015)
14. Patil, L.H., Mohammed, A.: A novel approach for feature selection method TF-IDF in document clustering. In: 3rd IEEE International Advance Computing Conference (IACC), pp. 858–862 (2013)
15. Veerabhadrappa, L.R.: Multilevel dimensionality reduction methods using feature selection and feature extraction. Int. J. Artif. Intell. Appl. **1**(4), 54–58 (2010)
16. Srinivasan, R., Subalalitha, C.N.: Automated named entity recognition from Tamil documents. In: IEEE 1st International Conference on Energy, Systems and Information Processing (ICESIP), pp. 1–5. IEEE (2019)
17. Rajkumar, N., Subashini, T.S., Rajan, K., Ramalingam, V.: Feature selection using normalized weight method for Tamil text classification. Int. J. Recent Technol. Eng. **9**(1), 9–14 (2020)
18. Rajkumar, N., Subashini, T.S., Rajan, K., Ramalingam, V.: An efficient feature extraction with subset selection model using machine learning techniques for Tamil documents classification. Int. J. Adv. Res. Eng. Technol. **11**(10), 66–81 (2020)

An Approach of Devanagari License Plate Detection and Recognition Using Deep Learning

Pankaj Raj Dawadi(✉), Manish Pokharel, and Bal Krishna Bal

Department of Computer Science and Engineering, Kathmandu University, Dhulikhel, Kavre, Nepal
{pdawadi,manish,bal}@ku.edu.np

Abstract. This paper proposes an automatic license plate recognition system for the Devanagari license plate (LP) in a static environment. The LP region is detected and localized using YOLOv3. The RGB mask of localized LP region is converted into HSV color space; and saturation mask of HSV color is processed using the CLAHE algorithm to segment unwanted regions from LP. The noise-free LP mask is acquired using different image preprocessing techniques. Skew correction is performed in the subsequent stage, followed by segmentation of LP characters using horizontal and vertical projection profiles. The characters are learned and predicted using a Convolution Neural Network (CNN). The CNN model is trained on the 19663 self-created Nepali LP character dataset. The proposed system has 99.2% accuracy on individual character recognition of LP characters and 95.5% accuracy on recognizing whole LP characters. The system is tested on both frontal and rear LPs of private two-wheelers and four-wheelers of Bagmati Zone.

Keywords: Automatic license plate recognition (ALPR) · CLAHE · Vertical segmentation · Horizontal segmentation · YOLOv3 · Convolution Neural Network (CNN)

1 Introduction

Automatic License Plate Recognition (ALPR), which is a mass surveillance strategy of reading a vehicle LP without human intervention, possess a challenge in intelligent transportation systems (ITS). It is a surveillance technique to detect and recognize vehicle LP and can be used for automatic traffic control, electronic toll collection, vehicle tracking, monitoring, border crossing, and security. Computer vision, pattern recognition along with machine learning and deep learning approaches, are widely used in image-based detection, segmentation, and recognition. Many problems are encountered while implementing the ALPR system, such as poor resolution, poor illumination conditions, blurry inputs, plate occlusion, different font sizes, and a variety of plate structures. The typical ALPR system consists of four stages: LP detection, LP localization, character segmentation, and character recognition [1].

The purpose of this study is to develop a system that detects and localizes, and segments and recognizes the Devanagari (Nepalese) LP in the static image using a deep learning approach. The system is implemented for both frontal and rear view LPs on private vehicles (Red LP).

M. Singh et al. (Eds.): ICACDS 2021, CCIS 1441, pp. 85–96, 2021.
https://doi.org/10.1007/978-3-030-88244-0_9

1.1 Nepali License Plate (NLP)

The ALPR technology deployed in Nepal varies because of various laws and orders related to LP colors and regional characters. Nepalese LP characters have pre-define order in the LP area and are selected from the group of Devanagari characters in a particular order. Various characteristics of the number plates such as zonal category, vehicle type, and lot number are demonstrated by the order of characters. NLPs vary in LP background and foreground color because of three reasons, namely load type, ownership, and category. For instance, a public bus has a black color background as LP color, and a white color foreground characters as a vehicle identifiers. Contrarily, a private bus has a red color background as LP color and a white color foreground characters as vehicle identifier (Broad details on vehicle plate characteristics is covered in another paper [2]). Likewise, the characters are written in a pre-defined order and three LP structures are used to denote vehicle characters: 1-line LP, 2-line LP, and 3-line LP, respectively. The third category is recently introduced by the Nepal government and this study does not include these plates due to the unavailability of training characters. A typical LP structure for a 2-line and a 1-line is shown in Fig. 1. In both LP structures, the first character is a zonal code, followed by a 1 to 2 digits lot number, followed by a character that designates the vehicle load type. In 2-line LP, the second row, and in 1-line LP, the remaining characters point out vehicle identification numbers, where at least one character or at most four characters are used to represent the vehicle ID. Every vehicle has LP attached on both rear and front sides with two different-size rectangular plates. The width to height ratio of LP is usually 4: 1 for a 1-line LP and 4:3 for a 2-line LP. Both front and rear side of a vehicle may have 1-line, 2-line or mixed LP structure. Further, the character has no pre-defined size; and each character is written in different format and style.

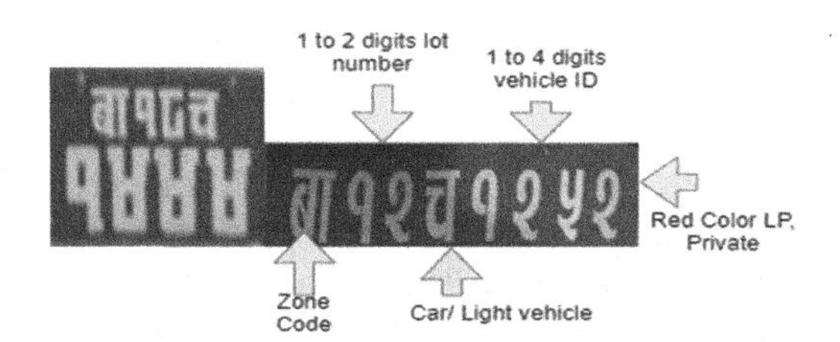

Fig. 1. Segmented LP mask, 2-line LP (left) and 1-line LP (right)

The Devanagari sample character sets for private vehicle is shown in Fig. 2. Two styles of writing four (4) are shown in e1 and e2. Similarly, two styles of writing nine (9) are shown in j1 and j2. It should be noted that other Devanagari characters are also written in different styles. For instance, 1, 2 5, 7, and 8 are found written differently.

The rest of this paper is organized as follows: Sect. 2 discusses related works; Sect. 3 presents the proposed method; Sect. 4 explains the datasets, experiments, and results; and Sect. 5 presents the conclusion and future directions.

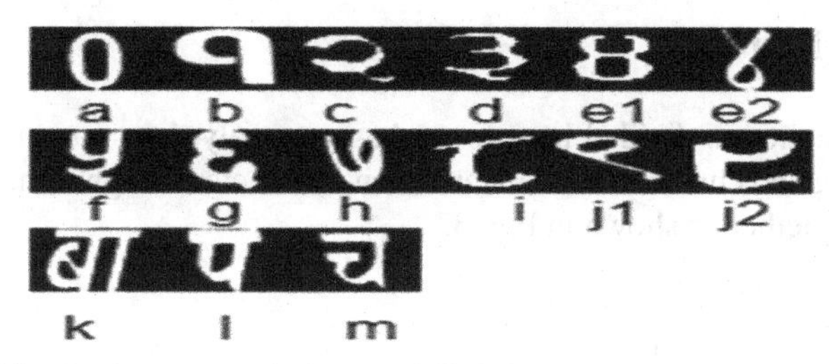

Fig. 2. Sample binarized Devanagari characters for Private LP (a). 0 (b). 1 (c). 2 (d). 3 (e1, e2). 4 (f). 5 (g). 6 (h). 7 (i). 8 (j1, j2). 9 (k). BA (l). PA (m). CHA

2 Related Works

Vehicle Identification work has been carried out by the computer vision community to address the problem of automatic license plate detection and recognition [1]. This task is divided into three categories: the first category has applied region-based approaches; the second has implemented pixel-to-pixel based [3] approaches; and the third has implemented color based approaches [4]. An input image is segmented into smaller regions in region-based approaches, where some pre-defined attributes of the license plate are located in successive regions. Morphological, high pass filtering, and others approach are devised to effectively perform detection tasks [5–7]. In the pixel-to-pixel based approach, every pixel is evaluated in the image among its neighboring pixels to form a coarse rectangular box, and the whole image is scanned pixel-by-pixel using a detection window. Subsequently, a novel color-based approach is developed by Azad et al. [8], by converting the RGB color image to HSV-color space.

Over the last few years, deep convolution neural networks (CNN) and its variants are extensively used worldwide to detect, segment, and recognize objects. CNNs have already exhibited remarkable performance for text and optical character recognition [9–11]. Under the high demand for robustness, some alternative methods employ the features extracted by CNN instead of hand-crafted features. In [12], the authors enrich the character dataset with certain hierarchical data augmentation strategies to improve the character recognition rate.

In Nepal, Pant et al. [2] propose an ALPR system based on HSV color space wherein HoG descriptor works as a feature extractor, and Support Vector Machines trains and predicts LP characters. The character recognition accuracy of their system falls par below due to the limited dataset for training, and the test dataset is also limited. In [13], the authors propose vehicle and LP detection using two passes on the same CNN, and then recognize the English LP characters using a second CNN for Brazilian and European LPs. For Non-English LP, [14] proposes image processing based LP detection and segmentation approach followed by CNN based character recognition system.

3 Proposed Method

The proposed method is divided into three phases namely detection and localization, image processing and character segmentation, and character recognition. The flowchart of the proposed method is shown in Fig. 3.

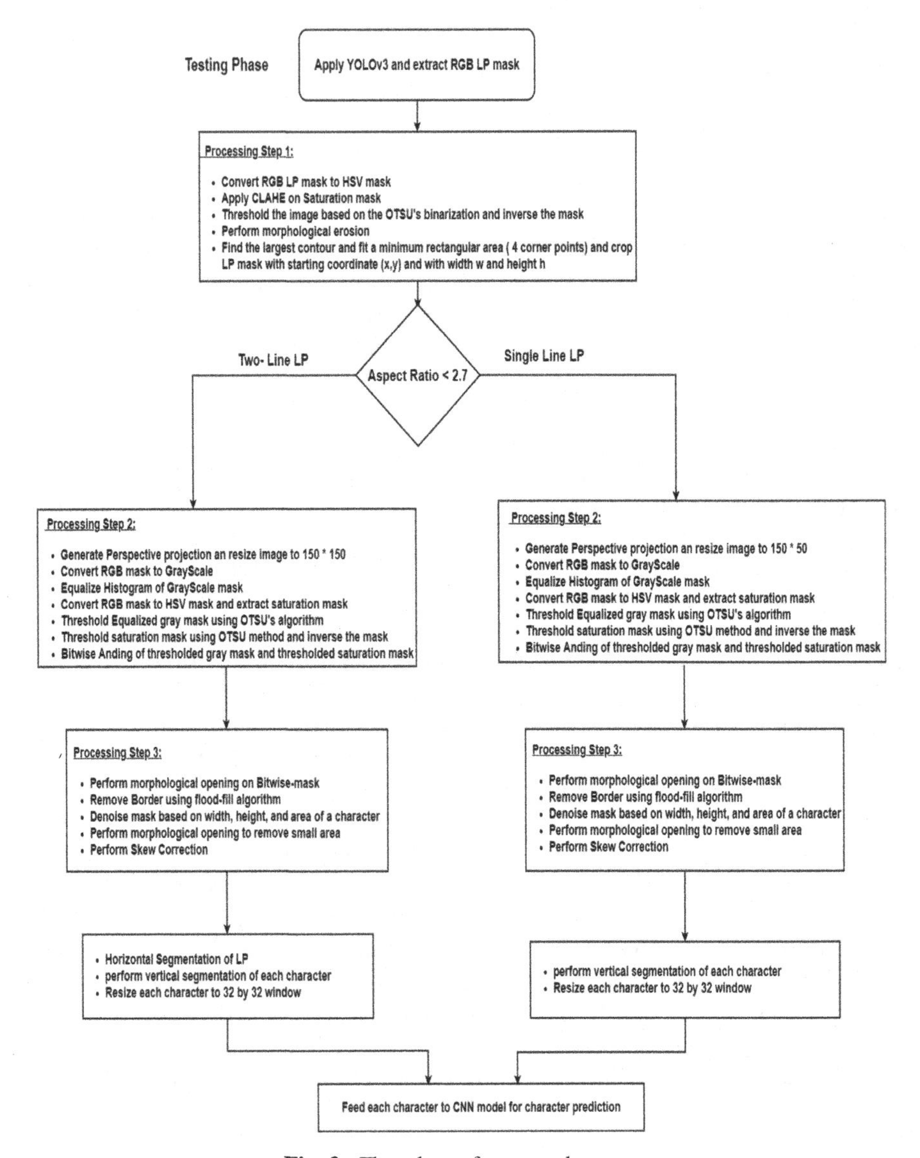

Fig. 3. Flowchart of proposed system

3.1 Detection and Localization

We used YOLOv3 [15] state-of-the-art technique for the detection and localization of LPs due to variations in the LPs in the Nepalese context. YOLOv3 takes care of different formats of LP irrespective of color information, orientation, and quality of the image. YOLOv3 is a unified detection framework based on CNN, which regards detection as a regression problem [13].

For the experiment purpose, we collected primary dataset of vehicles. We captured 1700 images of vehicles with LP in a road environment; some of them were captured in real-time and rest were taken statically around Kathmandu University premises and around parking lots. The captured images are from different private vehicles at different orientations and lighting conditions. The vehicle dataset was annotated and trained to capture the LP mask using the modified version of darknet [16]. Given the constraint in resources, Google Colaboratory was used to train the model since most of the libraries are pre-configured there. During the training, the average validation loss occurred which is shown in Fig. 4. With the average validation loss of around 2% during 2000 epochs, the system was trained to detect and localize the LP mask. Detection and localization step returns a RGB LP mask with a bounding box x, y, w, h where x and y are starting coordinates and w and h are width and height of LP, respectively.

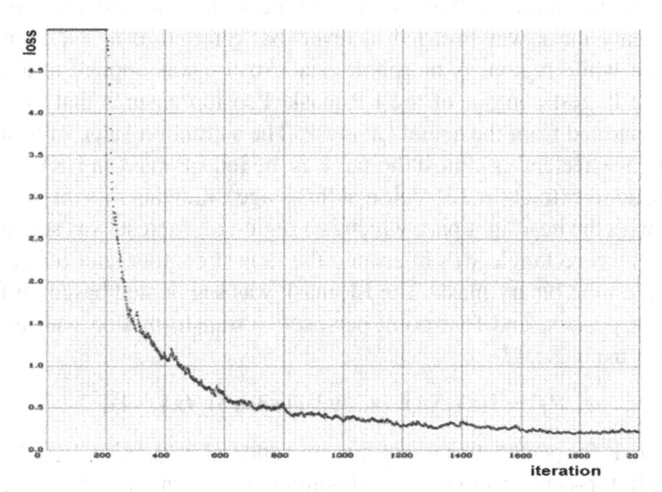

Fig. 4. Average validation loss on LP data

3.2 Image Processing and Segmentation of LP Characters

The extracted RGB LP mask was either a two-line LP or a one-line LP. The aspect ratio of the LP mask was calculated using the Eq. 1.

$$Aspect\ ratio = \frac{width\ of\ mask}{height\ of\ mask} \tag{1}$$

From the experiment, it was found that any LP having aspect ratio more than 2.7 was 2-line LP. The largest resolution needed more processing time and thus LP mask was resized to 150×150 for a 2-line plate and 150×50 for 1-line plate where the former represents the width and latter is the height of LP mask.

Adaptive histogram equalization (AHE) is a technique that distributes contrast in an image. The adaptive method calculates several histograms, each corresponding to a distinct section of the image, and uses them to redistribute the lightness values of the image. AHE is suitable for improving the local contrast and enhancing the definitions of edges in each region of an image, though AHE tends to over amplify noise in relatively homogeneous regions of an image. Contrast Limited Adaptive Histogram Equalization (CLAHE) [17, 18] prevents this by limiting the amplification. In the proposed method, we used adaptive histogram equalization using CLAHE to extract the actual LP mask; it helped improve the contrast distribution in the images. The RGB LP mask was split into HSV color space. The saturation of LP was calculated using Eq. (2).

$$Saturation(HSV) = \frac{\max(\mathrm{R,G,B}) - \min(\mathrm{R,G,B})}{\max(\mathrm{R,G,B})} \tag{2}$$

The saturation mask (background) of LP was dominantly white, whereas characters (foreground) along with other surrounding areas within LP area were dominantly black. The CLAHE algorithm was applied to threshold contrast of image based on OTSU's binarization method [19]. OTSU's binarization automatically picks the thresholding point in a deep valley based on the histogram of the saturation mask and returns a binary mask. The binary mask was inverted to revert background as a black region and the characters as a white region. A morphological erosion was applied to a binary mask to separate the largest contour of the LP mask. Erosion ensures that other unwanted regions are separated from the actual LP mask. The saturation value of headlight at the rear side of two-wheelers' is almost the same as the red LP mask in HSV color space; it was problematic to extract the LP region at this stage following this method. So, the LP mask, along with the headlight part, which is a region of interest, was segmented at this stage. A minimum rectangle was fit around the largest contour, and four corner points were detected on the binary mask. The LP mask was segmented based on four corners point in a binary mask; and four-point perspective transformation was applied on the RGB LP mask using Eq. (3).

$$PT = [(\mathrm{x1, y1}), (\mathrm{x2, y2}), (\mathrm{x3, y3}), (\mathrm{x4, y4})] \tag{3}$$

Where, PT stands for perspective transformation and (x1, y1), (x2, y2), (x3, y3) and (x4, y4) are top-left, top-right, bottom-left and bottom-right corner points of largest contour (minimum area rectangle), respectively.

The segmented RGB mask was split into HSV color space and the CLAHE filter was applied again on a saturation plane. The image was thresholded using OTSU's binarization, inverted, and a binary mask of the LP was acquired in this stage. In the context of the Nepali license plate, bolts might be attached very near to the character window which might contaminate the positional character area, to prevent this RGB mask was converted into a gray-scale image using Eq. (4).

$$\mathrm{Gray(i, j)} = 0.33 * \mathrm{R(i, j)} + 0.34 * \mathrm{G(i, j)} + 0.33 * \mathrm{B(i, j)} \tag{4}$$

Where i and j are any pixels inside the gray-scale image, R (i, j), G (i, j), B (i, j) are the three channels of RGB mask for the color image.

Histogram equalization was executed on a gray-scale mask as the next step. The CLAHE filter was applied to thus equalized histogram gray-scale mask and was thresholded using OTSU's binarization method. This step ensures areas acquired by bolts is minimum and can be filtered out after applying some image processing techniques. The binarized saturation mask and binarized gray-scale mask were bit-wise Anded to generate the final LP mask. Morphological opening and closing were performed to suppress small noise such as surface text, dust, and distortions.

Image denoising was performed in the next stage to clean the region of interest by suppressing noise from unclean LP mask. Larger noises were reduced using denoising process which suppresses the noise based on width, height, and pixel areas reserved by each character on a LP. Two aspects–the ratio of each character to the total LP area, and 8-connected components of each character from its center – were also considered. It should be noted that these criteria vary for a 1-line LP and a 2-line LP.

Projection profile based de-skewing algorithm [20] was applied in the next step to correct the angle of a mask to de-skew LP mask, either in clock-wise or anti-clockwise direction. De-skewing calculates an amount of skew and corrects a slope by rotating an image in the opposite direction. This results in a horizontally and vertically aligned image where the characters run horizontally across the localized plate, rather than at an angle. The method was applied to improve the accuracy and performance of the system before feeding the LP mask to the projection profile.

The skew-corrected binarized mask was processed using a horizontal and vertical projection profile. While scanning LP vertically, a 2-line LP is found to have a deep valley at initial space, indicating local minima followed by a peak showing local maxima, and again a deep valley that is followed by another peak and finally a deep valley which indicates the bottommost part of LP region. Horizontal projection uses this concept and splits the two local maxima as the upper window and lower window of a LP mask. With the 1-line LP, only a maxima was observed in the middle of the LP; so, the whole window was considered as a horizontal segment of the LP. Similarly, vertical profile projection split each peak as a character window while traversing in the horizontal direction for both 2-line and 1-line LP. Figure 5 shows the histogram of the horizontal profile and vertical profile of a two-line LP.

Fig. 5. 2-line LP, vertical profile (left), original mask (middle), and horizontal profile (right)

3.3 Feature Extraction and Character Training

The vertical projection profile extracts each character as a feature window in the plate. The extracted characters were considered for Optical Character Recognition (OCR) and were fed into the feature extraction subsystem. In the proposed method, Convolution Neural Network (CNN) was used to train the character dataset; the task was completed on Intel i7–8750, 2.2 GHz processor. The proposed CNN model was trained for 30 epochs for 19663 data separated into 13 classes.

The first two convolutions layers extracted the meaningful features from 32 × 32 character window with 60 filters. Each layer of the network received the output from its immediate previous layer as input and passes the current output as input to the next layer. The max-pooling operation down-sampled the output samples received from the convolution layer by the factor of 2. The operation chooses maximum value among four adjacent pixel values from the input samples. The higher convolutions layer is responsible to capture high-level features that are again down-sampled by the max-pooling layer. The output from higher convolution layers is flattened; and a dropout [21] operation is introduced to handle the problem of over-fitting. Dropout can help to reduce the complexity of the model for several reasons, including the co-adaptation of neurons, and the non-reliance of neurons on the presence of another set of neurons. Owing to these reasons, it can prevent over-fitting the model. However, the major disadvantage of the dropout operations is that it may take more iteration than usual to reach the global minima. We used 50% drop-out after performing max-pooling in the second stage to prevent the proposed model from over-fitting. The output was again converted to 1 × 500 one-dimensional output as a dense layer, and another 50% drop-out was used again to prevent over-fitting. The classification layer calculates the class probability, and errors are calculated between desired and expected outputs, and errors are back-propagated to update weights. The final dense layer outputs were converted into 13 classes; and the Softmax activation function was used to classify and predict the characters' classes. The details of proposed CNN model is shown in Table 1.

Table 1. Proposed CNN model

Layer type	Parameters
Input	32 × 32
Convolution + ReLU	28 × 28, #filters:60, k = 3 × 3
Convolution + ReLU	24 × 24, #filters:60, k = 3 × 3
Max Pooling	k: 2 × 2, s: 1
Convolution + ReLU	10 × 10, #filters:30, k = 3 × 3
Convolution + ReLU	8 × 8, #filters:30, k = 3 × 3, p1
Max Pooling	k: 2 × 2, s: 1
Flattened	#neurons: 480
Dropout	0.5

(continued)

Table 1. (*continued*)

Layer type	Parameters
Dense	# neuron: 500
Dropout	0.5
Fully connected + Softmax	# neuron: 13

4 Experiments and Results

The vehicle test dataset comprised of 200 RGB images with each image having 800 × 800 resolution. The test dataset was further partitioned into two folds: 1-line LP of 100 plate counts, and 2-line LP of 100 plate counts. The LP mask was detected, localized, and extracted successfully in all vehicles. The recognition accuracy for each segmented character and entire characters in a LP are considered to measure the accuracy of the proposed system.

4.1 Dataset

1) Vehicle Dataset: The vehicle dataset comprised of 1700 vehicle images. The images were captured from the road, around University premises, and parking lots. They were captured at a different orientation, pose, and lighting conditions. The vehicle dataset was used to train the system using YOLOv3, which detects, localizes and returns extracted LP mask of the vehicles.

2) Character Dataset: The Character dataset contained 19663 samples which were further categorized into 13 classes. The initial dataset was cropped from various LPs using thresholding and morphological operations such as dilation and erosion. Additionally, some segmented characters from the system were selected and included in the dataset. The dataset was augmented; and each label of training data was multiplied in at least four folds with the help of Keras' image augmentation tool. Each character was rotated, zoomed, sheared, and shifted in width and height. We ensured to incorporate at least 1000 samples for each character class. Each character was rescaled to 32 × 32 and stored as a portable network graphics dataset. The dataset is available at the link [22]. The details on the number of samples for each class are shown in Fig. 6. The highest number of letters (BA) was approximately 3500, and the lowest number of letters (7) was around 1050.

The characters were trained using the proposed CNN model. In the proposed system, we used 80% character dataset as training labels and the remaining as test labels.

4.2 LP Detection, Localization and Segmentation Results

The proposed system was tested on 200 randomly selected LP images of private vehicles. These images were resized to 800 × 800 mask to increase processing efficiency. All LP masks were correctly detected and localized by the object detection algorithm. The system was able to generate a perspective transformation of all LPs, which were captured

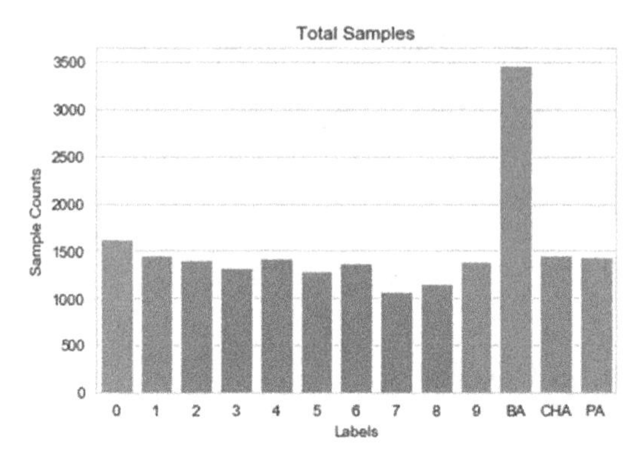

Fig. 6. Devanagari LP character set

in different orientations and angles. Character segmentation from the LP was found to be a challenging task in some LPs since they were highly influenced by the rust, mud, dirt around character windows. Segmentation accuracy also largely depended upon the spacing between the characters. The proposed system segments all characters from 200 LPs. Some LP characters were not segmented because of internal noise within a plate, that is, two consecutive characters appeared as a single one. In all the LPs, horizontal profile projection plays an important role while segmenting two rows from two-line LP; and vertical profile projection plays a prominent role while segmenting each character.

4.3 Character Prediction Results

In LPs written in Devanagari script, there are a minimum of 4 valid characters (for instance: BA 1 PA 9) and a maximum of 8 valid characters (for instance: BA 98 CHA 9871) in a LP. The proposed system recognized 745 characters out of 750 characters in 100 1-line LP and 840 characters out of 847 characters in 100 2-line LP, respectively. The overall character recognition was 99.2%. Similarly, of the total 200 LPs, 191 LPs were correctly recognized with LP recognition accuracy of 95.5%; only 12 characters were incorrectly predicted in 9 LPs. The accuracy of the proposed system in recognizing LP characters is shown in Table 2.

Table 2. Character recognition result

Plate structure	Plate counts	Total actual characters	Total characters correctly recognized	Total LP correctly recognized
1-line LP	100	750	745	96
2-line LP	100	847	840	95
Total	200	1597	1585	191
		Recognition rate	**99.2%**	**95.5%**

5 Conclusion

This study proposes a deep-learning based LP recognition system using YOLOv3 as an object detector and localizer, applying CLAHE algorithm along with various image processing techniques to clean the LP, and segmenting characters from the LP. The segmented characters are further trained and predicted using CNN model. The system was tested for both frontal and rear views LP in private vehicles of the Bagmati Zone. The proposed system was experimented with our own Devanagari character dataset. The overall character recognition was 99.2% and LP recognition accuracy was 95.5%.

Compared with the available studies [2] for NLPs, this study is more advanced in a number of ways. Firstly, the study sample (plate counts) is higher (8 compared to 200); secondly, the character recognition accuracy of the proposed system is higher (93.2% compared to 99.2%). Thirdly, the whole LP character recognition is also higher (75% compared to 96.5%) in the proposed method.

The experiment results demonstrated that the proposed method could be used efficiently for static application; its applicability for dynamic conditions needs further study. The finding of this study can be extended to recognize number plate in real-time situations. In the future, we intend to deploy the proposed method to detect, localize and recognize all types of Devanagari LP of Nepal under dynamic situation.

Acknowledgment. The authors would like to thank Khagendra Acharya for his English editing.

References

1. Du, S., Ibrahim, M., Shehata, M., Badawy, W.: Automatic License Plate Recognition (ALPR): a state-of-the-art review. IEEE Trans. Circuits Syst. Video Technol. **23**(2), 311–325 (2013)
2. Pant, A.K., Gyawali, P.K., Acharya, S.: Automatic nepali number plate recognition with support vector machines. In: Proceedings of the 9th International Conference on Software, Knowledge, Information Management and Applications (SKIMA), pp. 92–99 (2015)
3. Muhammad, J., Altun, H.: Improved license plate detection using HOG-based features and genetic algorithm. In: 2016 24th Signal Processing and Communication Application Conference (SIU), pp. 1269–1272 (2016). https://doi.org/10.1109/SIU.2016.7495978
4. Ahmad Yousef, K., Al-Tabanjah, M., Hudaib, E., Ikrai, M.: SIFT based automatic number plate recognition. In: International Conference on Information and Communication Systems. ICICS 2015, pp. 124–129 (2015). https://doi.org/10.1109/IACS.2015.7103214
5. Bai, H., Liu, C.: A hybrid license plate extraction method based on edge statistics and morphology. In: Proceedings of the 17th International Conference on Pattern Recognition, 2004. ICPR 2004, vol. 2, pp. 831–834 (2004). https://doi.org/10.1109/ICPR.2004.1334387.
6. Rabee, A., Barhumi, I.: License plate detection and recognition in complex scenes using mathematical morphology and support vector machines. In: IWSSIP 2014 Proceedings, pp. 59–62 (2014)
7. Martín-Rodríguez, F., García, M., Alba-Castro, J.L.: NEW METHODS FOR AUTOMATIC READING OF VLP's (2002)
8. Azad, R., Davami, F., Azad, B.: A novel and robust method for automatic license plate recognition system based on pattern recognition. Adv. Comput. Sci. an Int. J. **2**(3), 64–70 (2013)

9. Jaderberg, M., Simonyan, K., Vedaldi, A., Zisserman, A.: Reading text in the wild with convolutional neural networks. Int. J. Comput. Vision **116**(1), 1–20 (2015). https://doi.org/10.1007/s11263-015-0823-z
10. Wang, T., Wu, D.J., Coates, A., Ng, A.Y.: End-to-end text recognition with convolutional neural networks. In: Proceedings of the 21st International Conference on Pattern Recognition (ICPR2012), pp. 3304–3308 (2012)
11. Radzi, S.A., Khalil-Hani, M.: Character Recognition of License Plate Number Using Convolutional Neural Network. In: Zaman, H.B., et al. (eds.) Visual Informatics: Sustaining Research and Innovations, pp. 45–55. Springer Berlin Heidelberg, Berlin, Heidelberg (2011). https://doi.org/10.1007/978-3-642-25191-7_6
12. Wang, Q., Gao, J., Yuan, Y.: A joint convolutional neural networks and context transfer for street scenes labeling. IEEE Trans. Intell. Transp. Syst. **19**(5), 1457–1470 (2018). https://doi.org/10.1109/TITS.2017.2726546
13. Silva, S.M., Jung, C.R.: Real-time license plate detection and recognition using deep convolutional neural networks. J. Vis. Commun. Image Represent. **71**, 102773 (2020). https://doi.org/10.1016/j.jvcir.2020.102773
14. Rabbani, G., Islam, M.A., Azim, M.A., Islam, M.K., Rahman, M.M.: bangladeshi license plate detection and recognition with morphological operation and convolution neural network. 2018 21st International Conference of Computer and Information Technology ICCIT 2018, pp. 1–5 (2019). https://doi.org/10.1109/ICCITECHN.2018.8631937
15. Redmon, J., Farhadi, A.: YOLOv3: An Incremental Improvement [Online] (2018). http://arxiv.org/abs/1804.02767
16. GitHub - AlexeyAB/darknet: YOLOv4 (v3/v2) - Windows and Linux version of Darknet Neural Networks for object detection. https://github.com/AlexeyAB/darknet. Accessed on 05 May 2020
17. Zuiderveld, K.: Contrast Limited Adaptive Histogram Equalization. In: Graphics Gems IV, pp. 474–485. Academic Press Professional Inc, USA (1994)
18. Pizer, S., et al.: Adaptive histogram equalization and its variations. Comput. Vis. Graph. Image Process. **39**(3), 355–368 (1987). https://doi.org/10.1016/S0734-189X(87)80186-X
19. Otsu, N.: A Threshold Selection Method from Gray-Level Histogram. Automatica **11**, 285–296 (1975)
20. Ninno, E., Nicchiotti, G., Ottaviani, E.: A general and flexible deskewing method based on generalized projection. In: Bimbo, A. (ed.) ICIAP 1997. LNCS, vol. 1311, pp. 632–638. Springer, Heidelberg (1997). https://doi.org/10.1007/3-540-63508-4_177
21. Hinton, G.E., Srivastava, N., Krizhevsky, A., Sutskever, I., Salakhutdinov, R.R.: Improving neural networks by preventing co-adaptation of feature detectors (2012)
22. Character_dataset. https://www.kaggle.com/pankajdawadi/nepali-lp-dataset. Accessed on 18 Mar 2021

COMBINE: A Pipeline for SQL Generation from Natural Language

Youssef Mellah[1,2](✉), Abdelkader Rhouati[1], El Hassane Ettifouri[1], Toumi Bouchentouf[2], and Mohammed Ghaouth Belkasmi[2]

[1] Novelis Lab, Paris, France
{ymellah,arhouati,eettifouri}@novelis.io
[2] LARSA/ENSAO Laboratory, Mohammed First University, Oujda, Morocco
{t.bouchentouf,m.belkasmi}@ump.ac.ma

Abstract. Accessing data stored in relational databases requires an understanding of the database schema and mainly a query language such as SQL, which, while powerful, is difficult to master. In this sense, recent researches try to approach systems to facilitate this task, in particular by making Text-to-SQL models that attempt to map a question in Natural Language (NL) to the corresponding SQL query. In this paper, we present COMBINE, a pipeline for SQL generation from NL, in which we combine two existing models, RATSQL (We used the version RAT-SQL v3+BERT; paper's url: arxiv.org/abs/1911.04942.) and BRIDGE (We used the version BRIDGE v1+BERT; paper's url: aclweb.org/anthology/2020.findings-emnlp.438/.), that are based on recent advances in Deep Learning (DL) for Natural Language Processing (NLP). Our model is evaluated on the Spider challenge, using Exact Matching Accuracy (EMA) and Execution Accuracy (EA) metrics. Our experimental evaluation demonstrates that COMBINE outperforms the two used models in the same challenge, and at the time of writing, achieving the state of the art in EA with 70%, and competitive result in EMA with 71.4%, on Spider Dev Set.

Keywords: SQL · Text-to-SQL · COMBINE · RATSQL · BRIDGE · DL · NLP · Spider · EMA · EA

1 Introduction

Semantic Parsing (SP) is among the fundamental tasks in NLP, it consists of understanding the meaning of text in NL and mapping them to formal meaning representations [3–6], often to machine-executable programs, for many tasks such as question/answering [7], robotic control [8] and intelligent tutoring systems [9]. And as a sub-area of SP, we address the problem of mapping text written in NL to executable relational DB queries, which is known to be difficult due to the ambiguity and the flexibility in NL, as well as the complexity of relational DB.

We are interested in converting NL questions to executable SQL queries, due to the popularity of SQL as the domain-specific language used to query and manage data

M. Singh et al. (Eds.): ICACDS 2021, CCIS 1441, pp. 97–106, 2021.
https://doi.org/10.1007/978-3-030-88244-0_10

stored in most available relational DB [10]. SQL is hard to master although powerful, and thus out of reach for many users who need to query data. Despite the importance of the task, researches have recently appeared to approach advanced neural networks (NN) approaches synthesizing SQL queries in an end-to-end manner and achieve good results on most Text-to-SQL benchmarks such as GeoQuery, ATIS and WikiSQL [11–16], however, these challenges have flaws in either addressing DataBases (DB) from a single table or dealing with simple SQL queries (like the case in WikiSQL), or assuming the same DB in the trainSet and the testSet (like the case in GeoQuery and ATIS).

To overcome this limitation, Spider [17] is appeared as a new cross-domain complex Text-to-SQL Dataset, which contains multiple tables in different DBs, as well as a large number of complex queries. Spider necessarily requires a generalization model to address invisible DB schemas, as it uses different DBs during training and testing/evaluation tasks.

In this work, we propose COMBINE, a pipeline for generating SQL queries from NL utterances, which is based on the two models: RATSQL and BRIDGE. Our experimental evaluation on the Spider Dev Set demonstrates that our pipeline outperforms the two models, and reaching competitive results with the State-Of-The-Art (SOTA) in both metrics, EMA and EA.

The rest of this paper is organized as follows: in the second part, we shows the literature. In the third part, we formulate the Text-to-SQL task and describe the Spider DataSet in more detail. In fourth part, we present our COMBINE pipeline and we show and discuss the obtained results in the fifth part, then finally we draw the conclusion and future work.

2 Related Work

Building NL interfaces for databases (NLIDBs) has been a significant challenge in the SP area. Old works [18–20] focused on rule-based techniques, then later using simple keywords [21–23].

The next step forward is to enable the processing of more complex NL questions by applying approaches based on pattern [20, 24].

Additionally, to improve the accuracy of NLIDBs, approaches based on grammar have been introduced using certain predefined grammatical rules [25, 26]. Reference [27] describes these types of systems. Most of these approaches perform well for specific DBs (with a small set of NL models and few keywords), yet they are often not competitive with complex user questions, in a cross-domain context.

More new models use advanced NN architectures to generate SQL queries given a user question, precisely the use of a classical Encoder/Decoder (ED) architecture based on Recurrent Neural Network (RNN) with Long Short-Term Memory (LSTM) networks like in the work [28], then Seq2SQL [29] adds a reinforcement learning technique to learn query generation policies, then execute queries against the DB.

Then most new works use the Transformer encoder [30] based architecture, the first attempt to use it is SQLova [31], which tried to solve the WikiSQL challenge, got very good results.

And now, we focus on new works that use Spider for the evaluation, as a complex and cross-domain Dataset, Spider is currently the most considered dataset for evaluating NLIDBs, precisely text-to-SQL models based on DL approaches.

The approach in reference [32] present a novel technique for generating training data by inverting the data annotation process. The power of this proposition is to cover a wider range of queries and to generate training data more quickly.

IRNet [33] for example uses a transformer ED based on an RNN with LSTM cells. An Intermediate Representation (IR) is further introduced based on an Abstract Syntax Tree (AST) as an alternative to directly synthesizing SQL.

Recent architectures proposed for this problem joining representation of textual-tabular data and Pre-training. In this sense, two LMs for jointly representing textual and tabular data pre-trained over millions of web tables appeared: TABERT [34] and TaPas [35]. Both focus on contextualizing text with a single table. TaBERT was applied to Spider by encoding each table individually and modeling cross-table correlation through hierarchical attention, by adopting the "content snapshot" mechanism which looks for the DB entries most similar to the question and code them together with the table columns (header).

Similarly, BRIDGE uses BERT for encoding the hybrid sequence, with minimal subsequent layers, and achieves text contextualization with DB via deep attention honed in BERT. The question and the DB schema are represented as the input sequence of that model, plus values taken from the question. Combined with a pointer generator decoder with schema consistency that led to search space pruning, BRIDGE also predicts SQL query values and achieves competitive results on the Spider challenge EA metric.

RAT-SQL, is another text-to-SQL approach that has achieved, at the time of writing, the top results on the Spider Challenge using the EMA metric. This model provides a new BERT-based relationship-aware self-attention mechanism with promising results on non-trivial DB schemas, to solve the problem of schema encoding and linking.

3 Text-To-SQL Task

In this section, we describe and explain the Text-to-SQL task as a sub-area of SP tasks, as well as the DataSet used for evaluating our work.

3.1 Task Definition

We treat in this work the specific SM task, which consists of the mapping of a NL question to an SQL query, to find the answer to the original question by executing the SQL query generated on a given real DB. In particular, we use the currently cross-domain largest NL questions to SQL dataset, Spider (described with more details in the next Section) to evaluate our pipeline.

3.2 DataSet

We operate on Spider[1], a large, complex, multi-domain Text-to-SQL DataSet built by Yale university students. It covert 138 different domains treated in multiple tables, with 10181 questions, 200 DBs (unique) and 5693 SQL queries (most of them are complex).

DBs and SQL queries are different in the TrainSet and the TestSet of this DataSet, so systems must therefore be generalized to new DB schemas as well as to new SQL queries.

It is called "Spider" because it looks like a spider that traverses several complex nests (see the example in Fig. 1). It differs from previous semantic analysis tasks:

- **Academic, ATIS, and Geo:** they contain a limited number of SQL queries and a single DB, and have the same SQL queries in trainSet and testSet.
- **WikiSQL:** even if the number of SQL queries and tables is significantly high, it contains simple SQL queries, and each DB is just a simple table with no foreign key.

These features make Spider the first cross-domain and complex semantic and text-to-SQL analysis dataset.

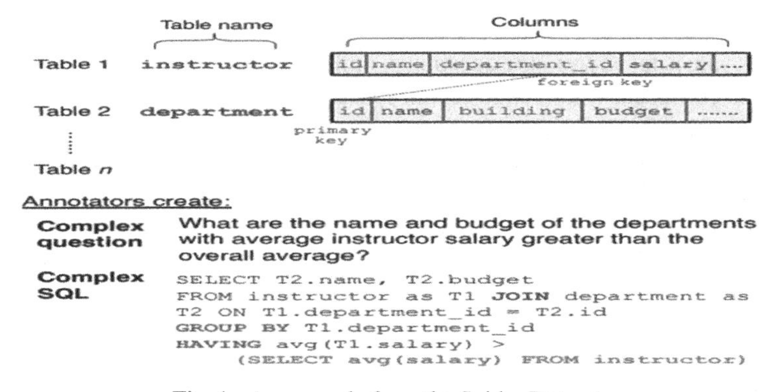

Fig. 1. An example from the Spider Dataset.

4 COMBINE: Our Pipeline for SQL Generation

In this part, we present our pipeline, inspired from the Ensemble Learning approach and which use 2 existing models.

4.1 Ensemble Learning (EL)

EL is the process by which several models are combined to solve a particular intelligence problem, it is mainly used to reduce the probability of a bad prediction of one model

[1] https://drive.google.com/uc?export=download&id=1_AckYkinAnhqmRQtGsQgUKAnTH xxX5J0.

or to improve the performance of another. There are several techniques for making 'Ensembling', such as Max Voting, taking the output predicted by the majority of models, and Averaging, taking the model predictions, then calculating the mean in order to come out with a final prediction. As there are other advanced techniques like Stacking, Blending, Bagging, and Boosting.

With our COMBINE pipeline, we propose another idea of "Ensembling", that can be used for the Text-to-SQL task. This time we make up a "Conditional Soft Voting", based on the final performance of the two models "BRIDGE" and "RATSQL".

4.2 COMBINE: Description and Architecture

The RATSQL model augmented with BERT is evaluated on EMA metric, and it showed a very good performance on this metric, however, it is not on the EA metric, simply due to the non-prediction of the values in the SQL query (Values of the conditions in the clauses WHERE and HAVING, the value of the LIMIT clause…). To complete the SQL query, it makes the token "terminal" as values.

So, the idea of COMBINE (Fig. 2) is to combine RATSQL and BRIDGE, to benefit from the advantages of each model, since the first is more performant in EMA (SQL without values), and the second is more accurate in EA (SQL with values). In this sense, we decided to use the two pre-trained models of the two works, knowing that they are trained and validated on the same Sets (Train and Dev) of Spider Dataset.

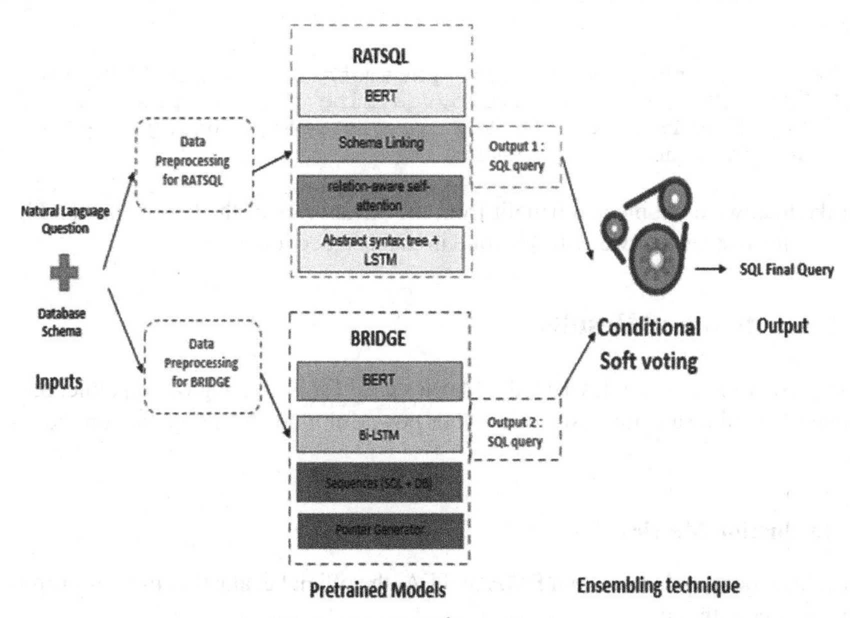

Fig 2. COMBINE's mechanism

So, our principle is as follow: giving the question and the DB schema, we do data preprocessing according to the two components (RATSQL and BRIDGE), then we call their pre-trained models, and based on a "Conditional Soft Voting" (See the procedure pseudo-code below), we return the SQL query generated by one of the two pre-trained model, RATSQL or BRIDGE.

Procedure pseudo-code of the component "Conditional Soft Voting":

```
input:
    sql_ratsql: sql query generated using ratsql model
    sql_bridge: sql query generated using bridge model
output:
    sql_final: select one of the two inputs
init ratsql probabilities list
init bridge probabilities list
determine sql type depending on inputs
case typesql of:
      with_value:
                  set ratsql_prob from probabilities list
                  set bridge_prob from probabilities list
      no_value:
                  set ratsql_prob from probabilities list
                  set bridge_prob from probabilities list
endcase
if ratsql_prob > bridge_prob
      return sql_ratsql
else
      return sql_bridge
return

#### Note that the two parameters "SQL_RATSQL and
SQL_BRIDGE" are the result by calling the two pre-trained
models of RATSQL and BRIDGE respectively, using a given
NL question and a Db schema.
```

Like that, we combine and benefit from the advantages of the two works, and have better performances that we will illustrate in the next section.

5 Evaluation and Results

In the present section, we describe the Metrics used for evaluating our pipeline, results obtained as well as a comparison with other powerful models that operate on the same task.

5.1 Evaluation Metrics

We evaluate our pipeline on both EMA and EA, the official evaluation metrics proposed by the Spider authors:

- **Exact Match Accuracy (EMA):** It checks if the predicted SQL exactly matches the expected one. This is a less important metric because a semantically correct SQL query may differ from the predicted and the expected SQL query.

- **Execution Accuracy (EA):** By executing the predicted SQL query and the expected one on a real DB, and checking if the execution returns are identical. In this evaluation, the models must also predict the correct values of the SQL query and place them in the correct order, not only to predict the sketch as well as the tables/columns.

We note that most works operated on Spider Dataset evaluate their models only on EMA metric since they don't predict values of the SQL query to evaluate on EA metric.

5.2 Results

Table 1 shows how COMBINE performed on the Spider Dev Set EMA, compared to other models located in the first Spider Challenge classes.

As we can observe from the table, our proposed pipeline COMBINE achieves 71.4% in EMA, so it outperforms its both components, BRIDGE v1+BERT by 5.9% and RAT-SQL v3+BERT by 1.7%, achieving top-3 in the challenge using EMA metric, at the time of evaluation (as of Dec 15th, 2020), with small marge compared to top-2 and top-1 works (0.4% and 2% respectively).

Table 1. EMA on the Spider dev Set, compared to the other top-performing approaches as of Dec 15, 2020. * denotes approaches without reference in a publication.

Model	EMA
RAT-SQL + GraPPa [36]	73.4%
RAT-SQL + GAP*	71.8%
COMBINE (ours)	**71.4%**
RAT-SQL v3+BERT	**69.7%**
RYANSQL + BERT [37]	66.6%
BRIDGE v1+BERT	**65.5%**
SmBoP + BART [38]	66.0%
RAT-SQL v2 [39]	62.7%

On the other hand, we evaluated our pipeline also on the EA metric of Spider Dev Set, which is the more difficult metric used by most works. Table 2 shows the result obtained, compared to the other most recent and top approaches.

COMBINE performs very competitively, significantly outperforming its two based components, RATSQL v3+BERT by 25.8%, and BRIDGE v1+BERT by 4.7%, and also most of recently and top proposed works, achieving the SOTA with 70% in EA, on Spider Dev Set, at the time of evaluation (Dec 15th, 2020).

As shown in the two tables, we participated in the two challenges proposed by the Spider's community: EA and EMA metrics. We remark that most of the works participate only in the 2nd challenge which concerns the metric EMA since these works failed to predict values of the SQL query due to the complexity of the task. To predict the value,

Table 2. EA on the Spider dev, compared to the other top-performing approaches as of Dec 15, 2020. * denotes approaches without reference in a publication and - denotes approaches without evaluation on dev Set.

Model	EA
COMBINE (ours)	**70%**
AuxNet + BART *	–
BRIDGE v1+BERT	**65.3%**
RATSQL v3+BERT	**44.2%**
GAZP + BERT [40]	59.2%

the model must be able to both extract DB content, identify and copy values from the user question, and generate numbers (e.g. 2 in "LIMIT 2"). Even if the challenge community encourages to work on both metrics, especially EA, which the SQL query generated, as output, is the most realistic and complete one to be executed on real DBs.

6 Conclusion

We present COMBINE, a pipeline for SQL generation from NL evaluated in the large cross-domain complex Text-to-SQL Spider DataSet. Combining two powerful pre-trained models, COMBINE outperforms them in both EMA and EA on Spider Dev Set, and achieves very good results, better or comparable with recent top approaches operated on the Spider challenge.

As future work, we plane to set up other models for generating other source code such as Python and Java from NL, and integrate our best works to a chatbot or as a plugin in code editors to facilitate programming and assist developers.

References

1. Wang, B., Shin, R., Liu, X., et al.: Rat-SQL: relation-aware schema encoding and linking for text-to-SQL parsers. arXiv preprint arXiv:1911.04942 (2019)
2. Lin, X.V., Socher, R., Xiong, C.: Bridging textual and tabular data for cross-domain text-to-SQL semantic parsing. In: Proceedings of the 2020 Conference on Empirical Methods in Natural Language Processing: Findings, EMNLP 2020, Online Event, 16–20 November 2020, pp. 4870–4888. Association for Computational Linguistics (2020)
3. Zelle, J.M., Mooney, R.J.: Learning to parse database queries using inductive logic programming. In: Proceedings of the Thirteenth National Conference on Artificial Intelligence (AAAI) (1996)
4. Zettlemoyer, L.S., Collins, M.: Learning to map sentences to logical form: structured classification with probabilistic categorial grammars. In: Proceedings of the Twenty-First Conference on Uncertainty in Artificial Intelligence (UAI) (2005)
5. Clarke, J., Goldwasser, D., Chang, M.-W., Roth, D.: Driving semantic parsing from the world's response. In: Proceedings of the Fourteenth Conference on Computational Natural Language Learning (CoNLL) (2010)

6. Liang, P., Jordan, M., Klein, D.: Learning dependency-based compositional semantics. In: Proceedings of the 49th Annual Meeting of the Association for Computational Linguistics: Human Language Technologies (ACL-HLT), Portland, Oregon, USA, pp. 590–599. Association for Computational Linguistics (2011)
7. Yih, W., He, X., Meek, C.: Semantic parsing for single-relation question answering. In: ACL (2014)
8. Matuszek, C., Herbst, E., Zettlemoyer, L., Fox, D.: Learning to parse natural language commands to a robot control system. In: Desai, J., Dudek, G., Khatib, O., Kumar, V. (eds.) Experimental Robotics. Springer Tracts in Advanced Robotics, vol. 88, pp. 403–415. Springer, Heidelberg (2013). https://doi.org/10.1007/978-3-319-00065-7_28
9. Graesser, A.C., Chipman, P., Haynes, B.C., Olney, A.: AutoTutor: an intelligent tutoring system with mixed-initiative dialogue. IEEE Trans. Educ. **48**, 612–618 (2005)
10. Ramakrsihnan, R., Donjerkovic, D., Ranganathan, A., Beyer, K.S., Krishnaprasad, M.: SRQL: sorted relational query language. In: Proceedings of the Tenth International Conference on Scientific and Statistical Database Management (Cat. No. 98TB100243), pp. 84–95. IEEE (1998)
11. Xu, X., Liu, C., Song, D.: SQLNet: generating structured queries from natural language without reinforcement learning. Computing Research Repository. arXiv:1711.04436 (2017)
12. Yu, T., Li, Z., Zhang, Z., Zhang, R., Radev, D.: TypeSQL: knowledge-based type-aware neural text-to-SQL generation. In: Proceedings of the 2018 Conference of the North American Chapter of the Association for Computational Linguistics: Human Language Technologies, pp. 588–594 (2018a)
13. Shi, T., Tatwawadi, K., Chakrabarti, K., Mao, Y., Polozov, O., Chen, W.: IncSQL: training incremental text-to-SQL parsers with non-deterministic oracles. Computing Research Repository. arXiv:1809.05054 (2018)
14. Dong, L., Lapata, M.: Coarse-to-fine decoding for neural semantic parsing. In: Proceedings of the 56th Annual Meeting of the Association for Computational Linguistics, pp. 731–742 (2018)
15. Hwang, W., Yim, J., Park, S., Seo, M.: A comprehensive exploration on WikiSQL with table-aware word contextualization. Computing Research Repository. arXiv:1902.01069 (2019)
16. He, P., Mao, Y., Chakrabarti, K., Chen, W.: X-SQL: reinforce schema representation with context. Computing Research Repository. arXiv:1908.08113 (2019)
17. Yu, T., et al.: Spider: a large-scale human-labeled dataset for complex and cross-domain semantic parsing and text-to-SQL task. In: Proceedings of the 2018 Conference on Empirical Methods in Natural Language Processing, pp. 3911–3921 (2018c)
18. Warren, D.H.D., Pereira, F.C.N.: An efficient easily adaptable system for interpreting natural language queries. Comput. Linguist. **8**(34), 110122 (1982)
19. Androutsopoulos, I., Ritchie, G.D., Thanisch, P.: Natural language interfaces to databases - an introduction. CoRR, cmp-lg/9503016 (1995)
20. Popescu, A.-M., Armanasu, A., Etzioni, O., Ko, D., Yates, A.: Modern natural language interfaces to databases: composing statistical parsing with semantic tractability. In: Proceedings of the 20th International Conference on Computational Linguistics, COLING 04, USA, p. 141es. Association for Computational Linguistics (2004)
21. Simitsis, A., Koutrika, G., Ioannidis, Y.: Précis: from unstructured keywords as queries to structured databases as answers. VLDB J. Int. J. Very Large Data Bases **17**(1), 117–149 (2008)
22. Blunschi, L., Jossen, C., Kossmann, D., Mori, M., Stockinger, K.: Soda: generating SQL for business users. Proc. VLDB Endow. **5**(10), 932–943 (2012)
23. Bast, H., Haussmann, E.: More accurate question answering on freebase. In: Proceedings of the 24th ACM International on Conference on Information and Knowledge Management, pp. 1431–1440. ACM (2015)

24. Zheng, W., Cheng, H., Zou, L., Yu, J.X., Zhao, K.: Natural language question/answering: let users talk with the knowledge graph. In: Proceedings of the 2017 ACM on Conference on Information and Knowledge Management, pp. 217–226. ACM (2017)
25. Song, D., et al.: TR discover: a natural language interface for querying and analyzing interlinked datasets. In: Arenas, M., et al. (eds.) ISWC 2015. LNCS, vol. 9367, pp. 21–37. Springer, Cham (2015). https://doi.org/10.1007/978-3-319-25010-6_2
26. Ferré, S.: Sparklis: an expressive query builder for SPARQL endpoints with guidance in natural language. Semant. Web **8**(3), 405–418 (2017)
27. Affolter, K., Stockinger, K., Bernstein, A.: A comparative survey of recent natural language interfaces for databases. VLDB J. **28**(5), 793–819 (2019). https://doi.org/10.1007/s00778-019-00567-8
28. Dong, L., Lapata, M.: Language to logical form with neural attention. CoRR, abs/1601.01280 (2016)
29. Zhong, V., Xiong, C., Socher, R.: Seq2SQL: generating structured queries from natural language using reinforcement learning. CoRR, abs/1709.00103 (2017)
30. Vaswani, A., et al.: Attention is all you need. CoRR, abs/1706.03762 (2017)
31. Hwang, W., Yim, J., Park, S., Seo, M.: A comprehensive exploration on WikiSQL with table-aware word contextualization. CoRR, abs/1902.01069 (2019)
32. Deriu, J., et al.: A methodology for creating question answering corpora using inverse data annotation (2020)
33. Guo, J., et al.: Towards complex text-to-SQL in cross-domain database with intermediate representation. CoRR, abs/1905.08205 (2019)
34. Yin, P., Neubig, G., Yih, W.T., Riedel, S.: TABERT: pretraining for joint understanding of textual and tabular data. arXiv preprint arXiv:2005.08314 (2020)
35. Herzig, J., Nowak, P.K., Müller, T., Piccinno, F., Eisenschlos, J.M.: TAPAS: weakly supervised table parsing via pre-training. arXiv preprint arXiv:2004.02349 (2020)
36. Yu, T., Wu, C.-S., Lin, X.V., et al.: GraPPa: grammar-augmented pre-training for table semantic parsing. arXiv preprint arXiv:2009.13845 (2020)
37. Choi, D.H., Shin, M.C., Kim, E.G., Shin, D.R.: RYANSQL: recursively applying sketch-based slot fillings for complex text-to-SQL in cross-domain databases. CoRR, abs/2004.03125 (2020)
38. Rubin, O., Berant, J.: SmBoP: semiautoregressive bottom-up semantic parsing. CoRR, abs/2010.12412 (2020)
39. Wang, B., Shin, R., Liu, X., Polozov, O., Richardson, M.: RAT-SQL: relation-aware schema encoding and linking for text-to-SQL parsers. ArXiv, abs/1911.04942 (2019)
40. Zhong, V., Lewis, M., Wang, S.I., et al.: Grounded adaptation for zero-shot executable semantic parsing. arXiv preprint arXiv:2009.07396 (2020)

Recognition of Isolated Gestures for Indian Sign Language Using Transfer Learning

Kinjal Mistree[1(✉)], Devendra Thakor[1], and Brijesh Bhatt[2]

[1] C. G. Patel Institute of Technology, Uka Tarsadia University, Bardoli, India
{kinjal.mistree,devendra.thakor}@utu.ac.in
[2] Dharmsinh Desai University, Nadiad, India
brij.ce@ddu.ac.in

Abstract. Sign language is a visual and complete natural language used by Deaf to communicate with each other and hearing people. The Deaf community finds it very difficult to express their feelings to the hearing people, because of which human interpreters are needed to help them in emergency situations. But always human interpreting service is not available when needed, because in India only around 300 certified sign language interpreters are available. In order to fill this gap, an automated system can be designed that would facilitate recognition of Indian Sign Language (ISL) gestures. This paper describes such an approach that takes ISL image as an input and gives corresponding class as an output. We have shown that use of simple image manipulation techniques and transfer learning give promising result in ISL hand gesture recognition. For 11 categories of ISL words, we have achieved 99.07% accuracy that outperforms accuracy reported by existing approach.

Keywords: Indian Sign Language · Pretrained model · Hand gesture recognition · Image augmentation

1 Introduction

Just as spoken language, sign language is also a natural language, but main difference exists on modality. Sign language can be considered as a collection of gestures, movements, postures, and facial expressions corresponding to letters and words in spoken languages [1]. The Deaf community uses sign language as means of communication with each other and the hearing people. But it becomes difficult to communicate with hearing person if that person doesn't know the sign language. In emergency situations, human interpreters are needed for translation of spoken language to sign language and vice versa. In India, around 300 certified human interpreters are available [2] and they may not be available in emergency situations. In order to meet this shortage, we can design a system that provides an effective alternative in public places, hospitals, railway stations, police stations etc., when human interpreters are unavailable.

Considering grammar of sign language, a lot of variations are present in each country's sign language. Due to this, the techniques used to recognize sign language of other

M. Singh et al. (Eds.): ICACDS 2021, CCIS 1441, pp. 107–115, 2021.
https://doi.org/10.1007/978-3-030-88244-0_11

countries cannot be adopted directly for ISL [3]. Also, very less datasets are available publicly that restricts research motivation in this field. ISL recognition is difficult task as it involves interpretation of visual information of signs.

Pretrained deep learning models have performed remarkably well for object recognition and computer vision tasks [4]. But they require large dataset, so as a solution of that problem image augmentation techniques can be used to increase size of dataset. In this paper, we particularly address an issue of recognizing ISL words in generalized settings, i.e. ISL word recognition by incorporating geometric and photometric variations using data augmentation and pretrained model to get computationally effective result. We have discussed that combination of transfer learning with the concept of horizontal flipping and data augmentation increases classification accuracy.

We have organized this paper as follows: Existing approaches related to ISL isolated sign recognition is described in Sect. 2. Section 3 discusses proposed approach with detailed explanation. The experimental analysis along with results are shown in Sect. 4. Conclusion and future directions are provided in Sect. 5.

2 Previous Work

Sign language translation can be categorised in sign to text translation and text to sign translation. This paper focuses on sign to text translation, which consists of: 1) Sign language recognition and 2) Translating word sequence into English text. For sign language recognition, mainly two approaches exist: 1) Device-based approach and 2) Vision based approach. The problem with the device-based approach is that it obstructs natural movement of body [5]. The signers have to wear gloves and sensors are required to capture position and movement of gestures. So, vision-based approach is better than device-based approach in emergency situations. In this section, we have discussed previous work done using both the approaches.

Performance of various feature extraction methods like Discrete Cosine Transform (DCT), Edge-oriented histogram, Fourier Transform and centroid were compared by Pansare [6] and authors have shown that DCT achieves better result than other methods. In an approach proposed by [7], highest overall accuracy achieved was 90% using k-Nearest Neighbours (kNN) algorithm that was used by authors to classify 30 images from 26 gestures. Rekha et al. [8] presented an ISL recognition system that used Support Vector Machine (SVM) for two handed gestures. The dataset from 26 gestures was collected by authors containing 3 dynamic and 23 static gestures. The static gestures were classified using SVM with classification rate 86.3%. Agrawal et al. [9] used HOG descriptors, shape descriptors and SIFT feature to produce feature vector and multi-class SVM for double handed ISL recognition system. The authors captured 235 images of 36 categories and achieved 93% accuracy.

Adithya et al. [10] presented an algorithm that used Artificial Neural Network (ANN) for 720 fingerspelled signs. They used YCbCr color model for skin color segmentation and achieved 91.11% accuracy. [11] used concepts of genetic algorithm, neural network, particle swarm optimization, evolutionary algorithm to recognize 22 ISL gestures with 10 images for each gesture with 99.96% accuracy. HOG and SIFT feature matrices were used by [7] to classify ISL single and double handed alphabets. Authors achieved 90% accuracy using k-NN classifier. [12] presented work for recognising ISL static signs from large category of images where each class contains approximately 350 images. The authors used image resize and normalization as preprocessing steps and achieved 99.9% accuracy.

[13–17] presented system for ISL recognition in isolated mode that uses Microsoft Kinect as acquisition mode. We have chosen vision-based approach for experiments as it does not hinder natural movement of hands.

The approaches described here use feature extraction techniques to extract useful features of ISL signs. Also, most of the work described here is done with controlled settings of laboratory and specific illumination. No work is reported till date, taking transfer learning on ISL sign recognition into consideration. This motivated us to use the concept of pretrained model in combination with data augmentation for ISL sign recognition.

3 Methodology

This section describes the details of the dataset used by us and systematic explanation of the proposed approach.

3.1 Dataset

We have used repository of training and testing samples created by [18] having 10 static hand gesture categories with each category having approximately 100 images of ISL sign. The images were captured at the Robotics and AI Lab, IIIT-Allahabad, India with static background and under different light illumination conditions. Each static category has approximately 1000 images and were captured at 30 fps, having resolution of 320 * 240 pixels with Sony Handycam. Figure 1 shows few samples from the dataset used by us.

Fig. 1. Samples of still images from IIITA static gesture dataset [18]

3.2 Our Approach

Traditional machine learning techniques work with small amount of data but require same data distribution across the classes. Also, feature extraction techniques are needed to extract useful features from the data. On the other hand, deep learning models require large amount of data to solve problems. Pretrained models are proven efficient to work with small amount of data. Another solution to deal with small amount of data is data augmentation technique which can be used to make dataset inflated that was originally small. To address the variances like different camera settings and environment where signer performs sign, we have combined these two approaches to make generalized settings for ISL recognition. The general framework of proposed approach is shown in Fig. 2.

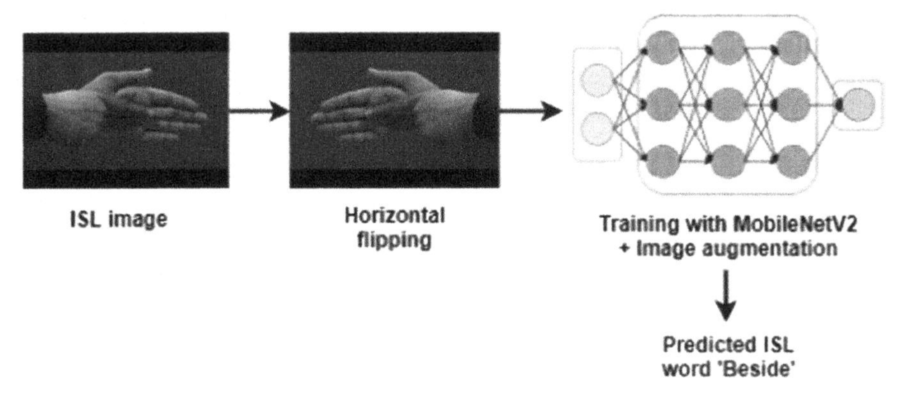

Fig. 2. Block diagram of proposed approach

Each signer is either left-handed or right-handed. To incorporate both left-handed and right-handed signs, we have performed horizontal flipping on each sign. Though horizontal flipping is one of the data augmentation techniques, we have explicitly separated this technique from the other set. The reason behind this is, if we use online data augmentation technique, the modifications in the images will be applied randomly, so not every image will be changed every time.

The resultant images after performing horizontal flipping were given as input to MobileNetV2 model for classification. We have used MobileNetV2 as pretrained model because it is light-weight deep neural network and best suited for mobile and embedded vision applications [19]. We have performed fine-tuning on the MobileNetV2 model by empirically changing the configuration of the top layers of the model in order to get the best recognition accuracy. The output of this model is category of ISL word.

4 Experiments and Results

We have discussed experiments conducted on 1) MNIST sign language dataset [20] and 2) dataset created by [18], in this section.

4.1 Experiment on MNIST Sign Language Dataset

We have performed experiment on MNIST sign language dataset in order to observe the performance of image augmentation on dataset. The American Sign Language (ASL) alphabet dataset of hand gestures has 24 classes of alphabet (excluding alphabets J and

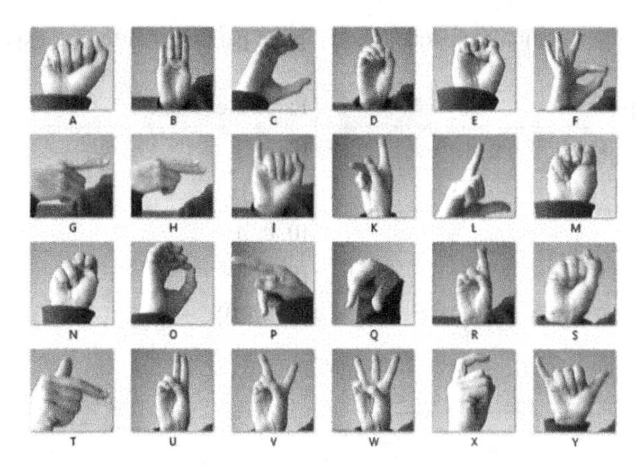

Fig. 3. ASL alphabet in MNIST sign language dataset

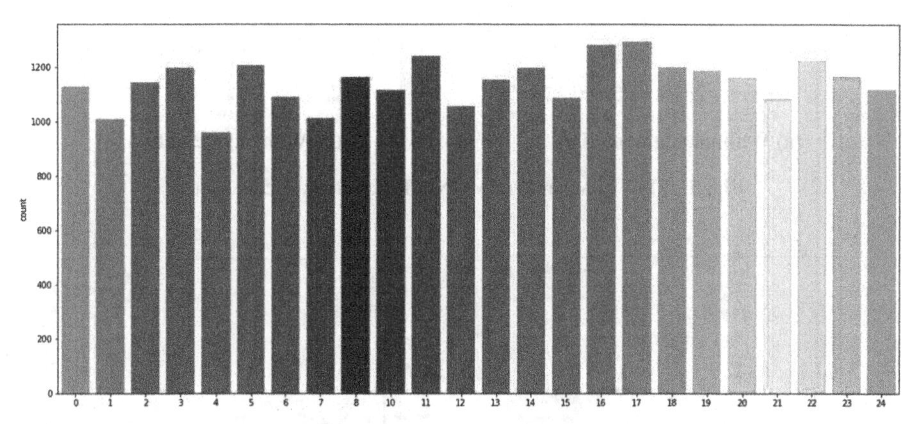

Fig. 4. Count plot of ASL alphabet in MNIST sign language dataset

Z that require motion). Figure 3 shows ASL alphabet and Fig. 4 shows count plot of ASL alphabet from the dataset.

For all the images of the dataset, we have performed photometric and geometric transformations. Table 1 shows types of augmentation techniques with the range of the parameters used for our experiment.

We have trained MobileNetV2 model with 60 epochs and used early stopping on validation loss as termination criteria. When no significant improvement in accuracy is observed after 2 continuous epochs, the training gets stopped. We have empirically changed top layers configuration and kept top 6 layers as trainable as it gave highest accuracy. Figure 5 and 6 shows plots of accuracy and loss without and with using augmentation techniques.

Table 1. Augmentation techniques and range of parameters used for experiments

Augmentation type	Parameter range
Zooming	[0.7, 1]
Rotation	[0, 15]
Vertical shifting	[0, 0.3]
Horizontal shifting	[0, 0.3]
Brightness	[0.2, 1.0]

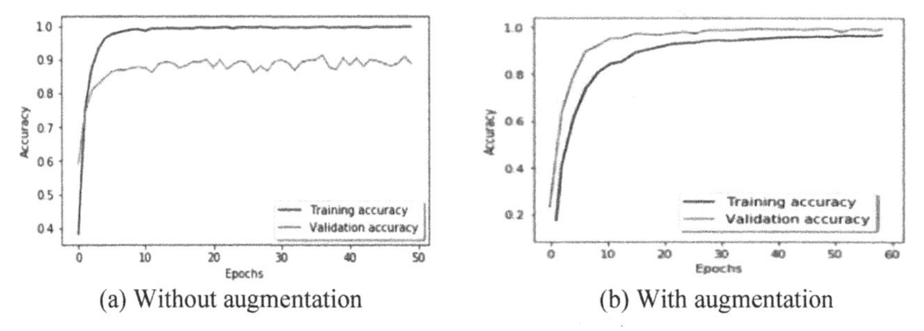

(a) Without augmentation (b) With augmentation

Fig. 5. Plots of accuracy on MNIST sign language dataset

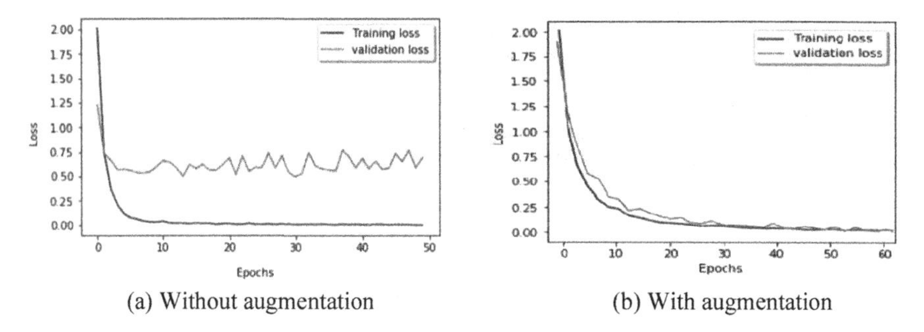

(a) Without augmentation (b) With augmentation

Fig. 6. Plots of loss on MNIST sign language dataset

4.2 Experiment on IIITA Static Sign Language Dataset

We have first performed horizontal flipping on IIITA static ISL dataset by taking 100 images of each ISL word class. Figure 7 shows a sample of original image and image after using horizontal flipping for ISL sign ‘Middle’. Figure 8 shows augmented images after applying image augmentation techniques, as shown in Table 1.

Because of use of horizontal flipping, the total number of images in dataset would get doubled. Among all resultant images, we have used 70% images for training, 15% for validation and 15% for testing. Table 2 shows performance of MobileNetV2 model where we can observe that using simple image augmentation techniques and pretrained model, classification accuracy is improved.

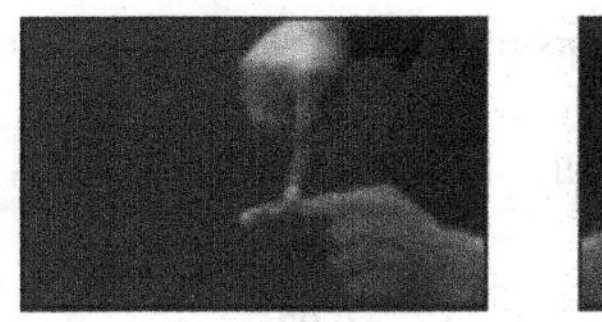
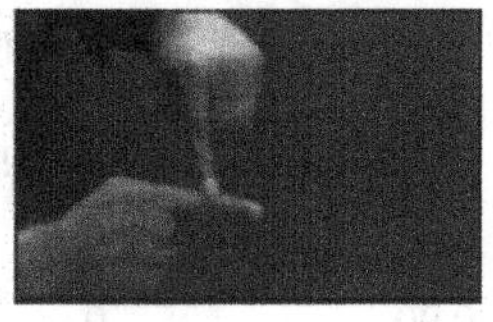

Fig. 7. Augmented image after applying horizontal flipping on ISL gesture 'Middle'

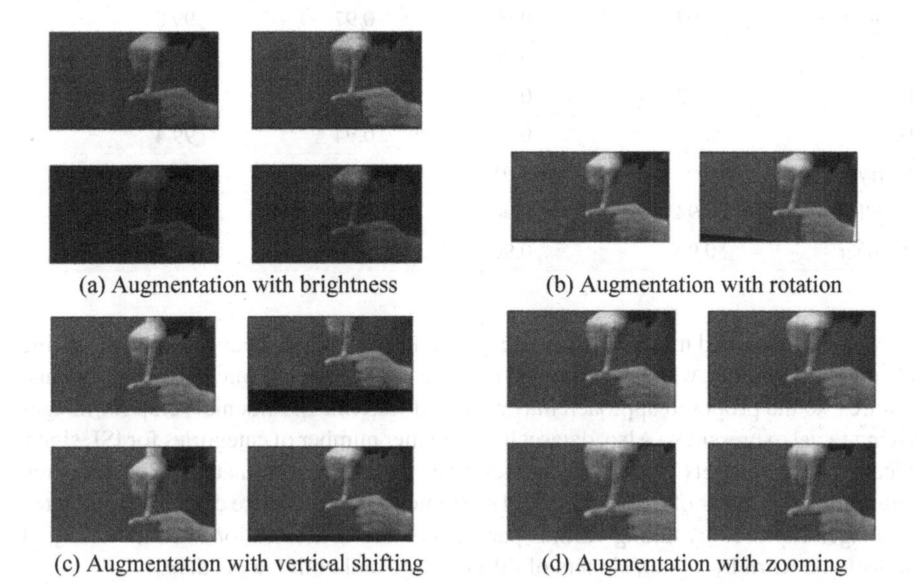

(a) Augmentation with brightness (b) Augmentation with rotation

(c) Augmentation with vertical shifting (d) Augmentation with zooming

Fig. 8. Effect of photometric and geometric transformations on ISL gesture 'Middle'

Table 2. Model performance on ISL hand gesture dataset

Parameters	Without augmentation	With augmentation
Training accuracy	98.91%	97.21%
Training loss	0.0217	0.00904
Validation accuracy	87.6%	98.3%
Validation loss	0.07134	0.00893
Testing accuracy	91.2%	98.102%
Testing loss	0.0497	0.00736

Precision, recall, F1-score and accuracy for each class are calculated to evaluate performance of each ISL word and are shown in Table 3. Our approach was compared with an approach presented by [18]. Average accuracy reported by [18] was 96.33% and our approach gives 99.07% accuracy with generalized settings. It is clear that our approach gives better recognition result with augmented image set of ISL words.

Table 3. Confusion matrix for word classes of IIITA static sign language dataset

Word class	Precision	Recall	F1-score	Accuracy
Aboard	0.92	0.90	0.91	99.3
Anger	0.90	0.96	0.93	99.6
Ascend	0.95	1.00	0.98	100
Beside	0.94	0.99	0.97	99.8
Drink	0.95	0.99	0.96	98.2
Flag	0.92	0.98	0.96	99.8
Hang	0.97	0.91	0.94	99.4
Marry	0.95	0.91	0.93	95.7
Middle	0.94	0.96	0.95	99.8
Prisoner	0.96	0.96	0.96	99.1

We have observed maximum accuracy while keeping top 6 layers as trainable, using MobileNetV2 model, which is presented in Table 2. The dataset contains images of hand gestures, so the proposed approach may require different experimental setup for images having facial expressions. Also, dataset has a smaller number of categories for ISL signs. Because of less variety of gestures model gives good accuracy that may be hampered with a greater number of image classes. The framework shown here can be implemented on images captured by taking various parameters into consideration such as variety of cloths to be worn by signer, signers of different age and both the genders.

5 Conclusion and Future Work

We have empirically shown that combination of simple image manipulation techniques and pretrained models have strong ability to generalize to images via transfer learning. Use of pretrained model with image augmentation techniques works effectively on same data distribution and small dataset. We have proposed a novel approach that works on left-handed and right-handed signs both, for ISL static gesture recognition. The proposed approach gives promising results for each ISL gesture, in comparison with existing approach.

The current work can be extended by implementing ISL video recognition using other pretrained models. In future, we aim to develop an approach that uses mobile platforms to recognize videos of ISL sentences and converts in English text. We are also in the process of creating ISL corpus that can be used to motivate researchers to work in the direction to help Deaf community.

References

1. Durkin, K., Conti-Ramsden G.: Young people with specific language impairment: a review of social and emotional functioning in adolescence. In: Child Language Teaching & Therapy, vol. 26, pp. 105–121 (2010)

2. ISLRTC. http://www.islrtc.nic.in/history-0. Accessed 20 Nov 2020
3. Zeshan, U.: Distinctive features of Indian sign language in comparison to foreign sign languages. In: The Peoples Linguistic Survey of India (2013)
4. Zhao, Z., Zheng, P., Xu, S., Wu, X.: Object detection with deep learning: a review. IEEE Trans. Neural Netw. Learn. Syst. **30**, 3212–3232 (2019)
5. Yu-Bo, Y., David, G., Shan, Z.: Machine learning in intelligent video and automated monitoring. Sci. World J. 1–3 (2015)
6. Pansare, J.R., Dhumal, H., Babar, S., Sonawale, K., Sarode, A.: Real time static hand gesture recognition system in complex background that uses number system of Indian sign language. J. Signal Inf. Process. **03**(03), 364–367 (2013)
7. Gupta, B., Shukla, P., Mittal, A.: K-nearest correlated neighbor classification for Indian sign language gesture recognition using feature fusion. In: International Conference on Computer Communication and Informatics (ICCCI), pp. 1–5 (2016)
8. Jayaprakash, R., Majumder, S.: Shape, texture and local movement hand gesture features for Indian sign language recognition. In: 3rd International Conference on Trendz in Information Sciences & Computing (TISC2011), pp. 30–35 (2011)
9. Agrawal, S.C., Jalal, A.S., Bhatnagar, C.: Recognition of Indian sign language using feature fusion. In: 4th International Conference on Intelligent Human Computer Interaction (IHCI), pp. 1–5 (2012)
10. Adithya, V., Vinod, P., Gopalakrishnan, U.: Artificial neural network-based method for Indian sign language recognition, pp. 1080–1085, IEEE (2013)
11. Hore, S., et al..: Indian sign language recognition using optimized neural networks. In: Balas, V., Jain, L., Zhao, X. (eds.) Advances in Intelligent Systems and Computing, pp. 553–563. Springer, Cham. (2015). https://doi.org/10.1007/978-3-319-38771-0_54
12. Wadhawan, A., Kumar, P.: Deep learning-based sign language recognition system for static signs. Neural Comput. Appl. **32**(12), 7957–7968 (2020). https://doi.org/10.1007/s00521-019-04691-y
13. Mehrotra, K., Godbole, A., Belhe, S.: Indian Sign Language recognition using Kinect sensor. In: Kamel, M., Campilho, A. (eds.) International Conference on Image Analysis and Recognition, pp. 528–535. Springer, Cham. (2015). https://doi.org/10.1007/978-3-319-20801-5_59
14. Kumar, P., Gauba, H., Roy, P.P., Dogra, D.P.: A multimodal framework for sensor-based sign language recognition. Neurocomputing **259**, 21–38 (2017)
15. Kumar, P., Gauba, H., Roy, P.P., Dogra, D.P.: Coupled HMM based multi-sensor data fusion for sign language recognition. Pattern Recogn Lett **86**, 1–8 (2017)
16. Kumar, P., Saini, R., Behera, S.K., Dogra, D.P., Roy, P.P.: Realtime recognition of sign language gestures and air-writing using leap motion. In: Fifteenth IEEE International Conference on Machine Vision Applications (MVA), pp. 157–160 (2017)
17. Kumar, P., Saini, R., Roy, P.P., Dogra, D.P.: A position and rotation invariant framework for sign language recognition (SLR) using Kinect. Multimed. Tools Appl. **77**(7), 8823–8846 (2017). https://doi.org/10.1007/s11042-017-4776-9
18. Nandy, A., Prasad, J., Mondal, S., Chakraborty, P., Nandi, G.: Recognition of isolated Indian sign language gesture in real time. In: Das, V.V., et al. (eds.) Communications in Computer and Information Science, vol. 70. Springer, Heidelberg (2010). https://doi.org/10.1007/978-3-642-12214-9_18
19. Sandler, M., Howard, A., Zhu, M., Zhmoginov, A., Chen, L.C.: MobileNetV2: inverted residuals and linear bottlenecks. In: The IEEE Conference on Computer Vision and Pattern Recognition (CVPR), pp. 4510–4520 (2018)
20. Kaggle. https://www.kaggle.com/datamunge/sign-language-mnist. Accessed 2 Oct 2020

A Study of Five Models Based on Non-clinical Data for the Prediction of Diabetes Onset in Medically Under-Served Populations

Rohit Srivastava[1], Sandeep Kumar[2], Vivudh Fore[3], and Ravi Tomar[1](✉)

[1] School of Computer Science, University of Petroleum and Energy Studies, Dehradun, India
[2] IIMT Group of Colleges, Greater Noida, India
[3] Vivudh Fore, Gurukul Kangri University, Haridwar, Uttarakhand, India

Abstract. Diabetes is an extremely common disease, caused due to increase in the level of sugar. Due to the high cost of Blood tests, it becomes an expensive treatment. The primary objective of this research is to identify a précised data model for the detection of diabetes. The major objectives for the research work can be defined as (i) Identification and examination of the most widely recognized data mining strategies used in present-day Decision Support Systems. (ii) Using data mining methods to find the level of danger of diabetes with the point of improving the nature of care. (iii) Identify an improved structure for diabetic visualization. The process involves KDD implementation for data processing and modeling. Five different data models were compared based on Accuracy and the equivalent ROC curve is plotted to find out the best method for data prediction. The model, which has the best outcome, had an accuracy of 88.56%.

Keywords: Diabetes · Nerve · SMOTE · MLP · SVM · Simple logistics · HbA1c

1 Introduction

This work directed a novel trial to build up the model that can foresee diabetes utilizing just non-clinical boundaries. The major advantage of this work is to plan a prescient model, which does not require a patient to experience blood tests, and diabetes is anticipated utilizing non-clinical boundaries. This analysis is finished keeping in view that not all patients are wealthy to pay a huge amount for blood tests particularly in developing and poor countries.

The work additionally centered on the data misbalancing that is at the current issue of research in AI. In this investigation, the order calculations is done with data rebalancing calculation for expanding the exactness and to reduce the inclination of calculations.

Diabetes influences major portions of the human body like the eyes, kidneys, heart, and nerves [2]. Thus, early prediction of disease can save human lives. Diabetes is classified into 2 types: Type I and Type II Diabetes [3]. In Type I disease, the human

M. Singh et al. (Eds.): ICACDS 2021, CCIS 1441, pp. 116–124, 2021.
https://doi.org/10.1007/978-3-030-88244-0_12

body will not create insulin, and 10% of diabetes are of this type while Type II diabetes is non-insulin-subordinate diabetes caused due to insufficient creation of insulin by the pancreas. Specialists have demonstrated that data mining and AI models, Decision Tree, Support Vector Machine etc. work well in diagnosing illnesses [4]. AI techniques boosts the analysts to recover new realities from huge wellbeing-related informational collections, which improves infection management. As indicated by CDA (Canadian Diabetes Association) somewhere in the range of 2010 and 2020, the normal increment of the infection would be from 2.5 million to 3.7 million [5].

The analysis shows that the research done till now is highly focused on the images and data generated from the different investigation reports and also a percentage of human error was high due to which the correct prediction was not done in the right manner. The research gap was identified as the precision of prediction without using the investigation reports. The models predict the data based on clinical parameters using SMOTE classifier generated from the online dataset. This paper compares the performance of different data mining models using SMOTE [6] and the best model for data prediction is computed.

The proposed work identifies the best model for diabetes prediction. Five different models are compared based on accuracy and ROC value. The best model is identified based on the comparative analysis. The work encompasses non-clinical parameters for identification and prediction of diabetes. The non-clinical parameters makes the usage of the data model largely and to a variety of people.

2 Literature Survey

Most of the examination work done on diabetes have utilized Pima Indian informational collection in their work. Almost every scientist has attempted to assemble a model with an expanded precision for forecasts of diabetes utilizing different data mining techniques [10]. Consolidated PCA (Principal Component Analysis) and Versatile Neuro-fluffy derivation framework (ANFIS) [11] for identification of diabetic ailment, their point was to expand the precision of forecast of diabetes by utilizing their mix. This framework worked in two stages: In stage, I measurement is done utilizing PCA, and in stage II analyze of diabetes is finished utilizing ANFIS [12]. Examination work analyzed different variations of Decision trees, LSSVM, Genetic calculation strategies for planning, and at end of exploration, it was presumed that LSSVM based Gaussian outspread capacity end up being the best among just for a prognostic reason. The model named GA-LSSVM [13] had an exactness of 81.33%.

An overview about the procedures that could be utilized for the planning and information-mining device for the forecast of diabetes was introduced [14], different methods were tried for exactness, and Partial Least Square-Linear Separate Analysis accomplished a most noteworthy exactness of 76.78% [15]. An overview for choosing the best delicate registering device for the forecast of diabetes was done.

They utilized different calculations to accomplish the necessary end. After experimentation, it was ANN with a precision of 89% being the best order device [16]. Their examination work broke down different characterization calculations of data mining to break down the diabetic information; the intention was to pick the best classifier that could foresee diabetes. The dataset utilized here was gathered structure clinical diabetic place. The result of the decision tree calculation with a precision of 67.15% [17] was the top model for anticipation purposes.

The exploration work applied different calculations for the order of diabetes to set up the best procedure for anticipating diabetes; here RBF [18] arrange was demonstrated to be the best classifier for guess having exactness of 80% [10].

The researcher applied different information-digging methods for planning a novel procedure for the forecast of diabetes, the strategies applied were Discriminant investigation, Naïve Bayes, KNN, SVM [15]. The point here was to choose the best classifier for structuring the prescient model. The Discriminant examination procedures had a slight edge on the precision with 76.35 [18] over others for this situation. In their examination work attempted to break down the indications of diabetic ailment to foresee its hazard utilizing data mining grouping calculations.

The model had two stages, in the first stage visit designs identified with diabetes were removed from the database utilizing FP-development [12] calculation and in stage two Decision tree was utilized for the chance expectation of diabetes. The model had a precision of 78% [14]. In their exploration work, they attempted to set up the best expectation model for diabetes forecast among calculated relapse and ANN. The ANN was assessed utilizing its different variations. The exploration was finished up with the ANN model with a precision of 80% being higher than calculated relapse.

3 Methodology

3.1 Dataset Description

After an intensive clinical interview and conversations with area clinical specialists, clinical boundaries were removed from dataset and a model for diabetic forecast utilizing just non-clinical boundaries was generated. The primary point of this analysis is to plan a prognostic model that decreases the prerequisite of blood tests. The boundaries removed are Fasting glucose plasma, Postprandial and HbA1c esteem. The dataset was diminished to only eight qualities: Age, Waist thickness, BMI (Body Mass Thickness), Systolic blood pressure, Diastolic blood pressure, Family history, Gender, and the last one is characterization class. The tool used for modeling is WEKA. The dataset is converted to ARFF (Attribute Relation File Format) which is finally accepted in WEKA (Fig. 1).

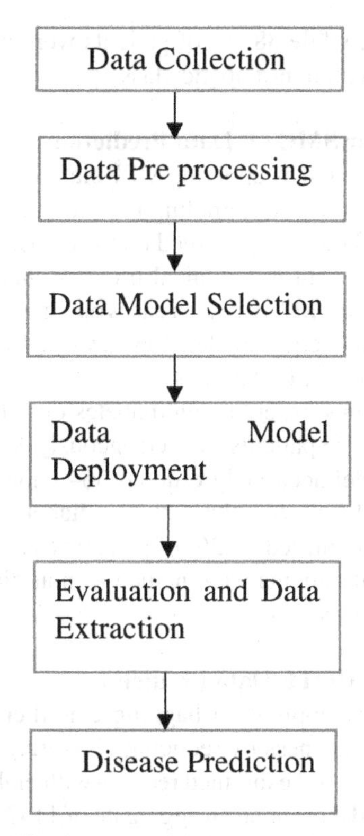

Fig. 1. Flow of diabetes prediction

4 Model Analysis and Result Comparison

(a) Synthetic Minority Over-sampling Technique (SMOTE) – Decision Tree model design for data prediction

Dataset was experimented by applying a decision tree classifier merged with SMOTE for upgrading the grouping precision. J48 calculation was run on the dataset in the wake of applying SMOTE having 1138 inspected records with eleven non-clinical traits as characterized previously. The execution time of the model was 0.04 s. The model arranged an aggregate of 1138 tested examples out of which 872 occasions were arranged effectively framing an exactness of 86.549%, while 181 cases were arranged erroneously giving an error of 14.4509%.

The model had 485 patients experiencing diabetes out of which 353 patients were accurately diagnosed diabetic. 77 patients were inaccurately diagnosed non-diabetic while they were diabetic. The model effectively arranged 535 patients as non-diabetic. Out of 404 however, 73 cases were delegated yes while they are non-diabetic. Discussing the accuracy score, the model effectively identified 84.9% of patients as diabetic (yes class) and they had diabetes, while 88.4% of patients were named as non-diabetic (no class) and they had a place with non-diabetic class.

(b) MLP classifier model for SMOTE Data Prediction
Diabetic dataset was experimented by applying MLP classifier on WEKA, usage of ANN and consolidated it with SMOTE for upgrading the grouping precision by diminishing class irregularity. MLP calculation was pursued on the dataset applying SMOTE having 1142 examined records with eight non-clinical traits as characterized previously. The model arranged an aggregate of 1142 inspected cases out of which 890 cases were characterized accurately shaping an exactness of 88.2158%, while 148 cases were characterized erroneously giving incorrectness of 12.3582%.

The model had 430 patients experiencing diabetes out of which 367 patients were effectively delegated diabetic. 63 patients were erroneously delegated non-diabetic while they were diabetic. The model accurately ordered 564 patients as non-diabetic out of 280 yet 85 cases were named yes while they are non-diabetic. Discussing the exactness score, the model accurately identified 85.2% of patients as diabetic (yes class) and they had diabetes, while 89.2% of patients were named as non-diabetic (no class) and they had a place in non-diabetic class.

(c) Bagging Classifier for SMOTE Data Prediction
Dataset was experimented by applying a bagging classifier joined with SMOTE for upgrading the arrangement exactness by reducing class irregularity. The classifier was pursued on the dataset having 1142 examined records with eight non-clinical qualities as characterized above. The model grouped an aggregate of 1142 inspected occurrences out of which 967 cases were characterized effectively framing an exactness of 88.5642%, while 150 cases were arranged mistakenly giving an error of 12.1618%. This model has a normal effective generally speaking precision rate.

The model had 430 patients experiencing diabetes out of which 360 patients were effectively identified diabetic. 70 patients were erroneously named non-diabetic while they were diabetic. The model accurately grouped 531 patients as non-diabetic out of 304 yet 77 cases were delegated yes while they are non-diabetic. Discussing the exactness score, the model effectively marked 82.4% of patients as diabetic (yes class) and they had diabetes, while 88.4% of patients were identified as nondiabetic (no class) and they had a place with non-diabetic class.

(d) Simple logistic Classifier for SMOTE Data Prediction
Dataset was experimented by applying a Simple Logistic classifier, a weak usage of Logistic relapse, and joined it with upgrading the grouping precision. The classifier was pursued on the dataset applying SMOTE having 1142 tested records with eight non-clinical traits as characterized previously. The model ordered a sum of 1142 inspected cases out of which 857 occurrences were ordered accurately shaping an exactness of 82.345%, while 179 cases were arranged erroneously giving an incorrectness of 18.474%. This model has a normal fruitful by and large precision rate.

The model had 430 patients experiencing diabetes out of which 345 patients were accurately delegated diabetic. 85 patients were inaccurately identified non-diabetic while they were diabetic. The model accurately characterized 512 patients as non-diabetic out of 304 yet 89 cases were delegated yes while they are non-diabetic. Discussing the accuracy score, the model effectively marked 82.0% of patients as diabetic (yes class) and they had diabetes, while 86.0% of patients were marked as non-diabetic (no class) and they had a place with non-diabetic class.

(e) SVM Classifier for SMOTE Data Prediction
We experimented diabetic dataset by applying the Support Vector Machine classifier and joined it with SMOTE for upgrading the arrangement exactness by lessening the class unevenness of our prescient model. The classifier was run on the dataset in the wake of applying SMOTE having 1142 inspected records with eight nonclinical traits as characterized previously. The model characterized a sum of 1142 inspected cases out of which 858 cases were characterized accurately framing an exactness of 83.2144%, while 189 cases were grouped erroneously giving incorrectness of 15.7996%.

This model has a normal effective largely precision rate. The model had 438 patients experiencing diabetes out of which 350 patients were effectively delegated diabetic. 80 patients were inaccurately named nondiabetic while they were diabetic. This outcome gave the model a TP Rate of 0.854. The FP pace of the model is 0.162. The model accurately characterized 524 patients as non-diabetic out of 304 yet 84 cases were delegated yes while they are nondiabetic. Discussing the accuracy score, our model effectively named 78.6% of patients as diabetic (yes class) and they had diabetes, while 86.8% of patients were named as non-diabetic (no class) and they had a place with nondiabetic class (Table 1, Figs. 2, 3).

Table 1. Comparison table for all data models

Data model	Error rate (%)	True positive rate (%)	False positive rate (%)	Accuracy (%)	ROC value (decimal)
Decision tree based model	14.4509	0.865	0.158	86.549	9.3
MLP based data model	12.3582	0.879	0.157	88.215	9.5
Bagging based data model	12.1618	0.883	0.159	88.564	9.6
Simple logistics based data model	18.474	0.842	0.174	82.345	9.2
SVM based data model	15.799	0.854	0.162	83.214	9.1

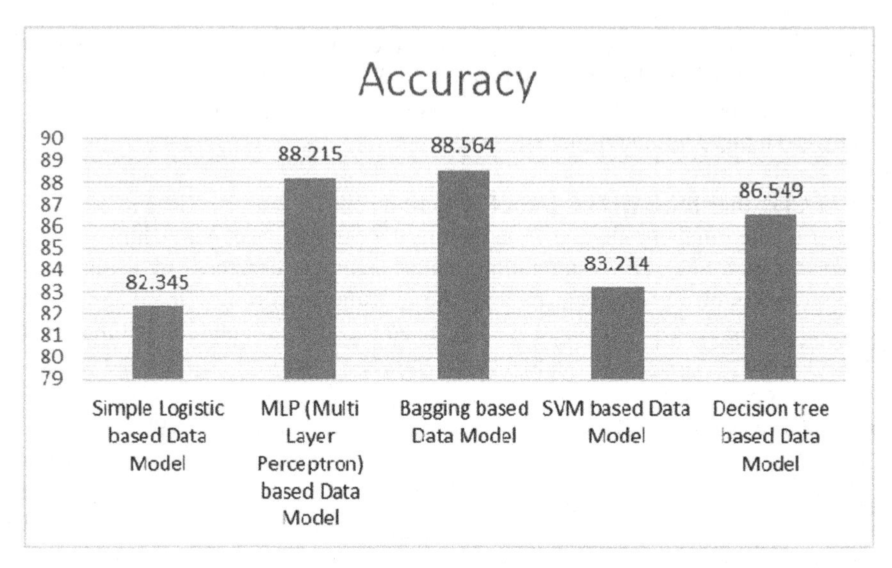

Fig. 2. Accuracy analysis of different models

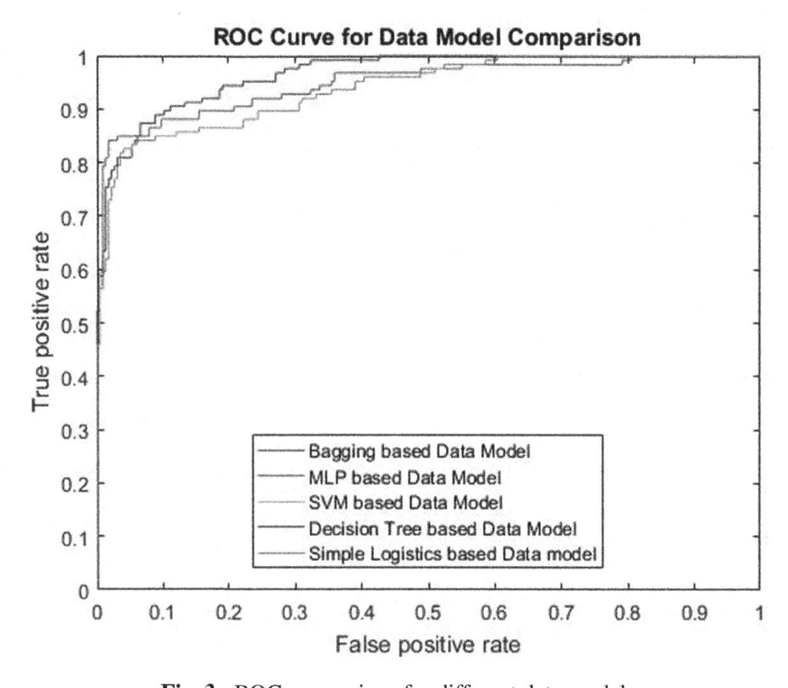

Fig. 3. ROC comparison for different data models

5 Conclusion

In this work, a comparative analysis of five different data model is performed that could predict diabetes on non-clinical boundaries and reduce the requirement for blood tests. Three of the classifiers had practically the same precision: Bagging, MLP and Decision tree however, the bagging classifier has a slight edge over the other two with a precision rate of 88.564% with an error rate of 12.1618%. The model has an amazing score of 9.4 on the ROC scale. The models does not require any blood or lab boundary tests to analyze diabetes. The best-calculated model can be separated from earlier ones for the following reasons:

1. Essential screening features were applied without the prerequisite for the patient to experience blood tests.
2. Data Imbalance was reduced by applying SMOTE rebalancing calculation to upgrade the accuracy of prediction.

Although the accuracy and affectability of best model are on the lower side as lab boundaries are removed, yet the model could predict early anticipation of diabetes without the necessity for experiencing blood tests.

References

1. Vinay, A., Shekhar, V.S., Murthy, K.N.B., Natarajan, S.: Face recognition using gabor wavelet features with PCA and KPCA - a comparative study. Procedia Comput. Sci. **57**, 650–659 (2015)
2. Amer, S., Mahmoud, H., El-Shishtawy, T.: A system for building incorporated medication collaboration metaphysics. Worldw. J. Adv. Comput. Technol. **10**(2), 1–9 (2018). http://www.globalcis.org/ijact/ppl/IJACT3620PPL.pdf
3. Meher, B., Agrawal, S., Panda, R., Abraham, A.: A study on area-based picture combination strategies. Inf. Comb. **48**, 119–132 (2019)
4. Banda, J.M., Callahan, A., Winnenbrug, R.: Plausibility of focusing on drug-drug occasion affiliations found in electronic wellbeing records. Medicat. Saf. **39**(1), 45–57 (2016)
5. Hurrle, S., Hsu, W.H.: The etiology of oxidative pressure in insulin opposition. Biomed. J. **40**(5), 257–62 (2017). https://doi.org/10.1016/j.bj.2017.06.007. PMid: 29179880, PMCid: PMC6138814
6. Juntarawijit, C., Juntarawijit, Y.: Relationship among diabetes and pesticides: a case-control concentrate among Thai ranchers. Ecol. Health Prev. Med. **23**(1), 3 (2018). https://doi.org/10.1186/s12199-018-0692-5. PMid: 29374457, PMCid: PMC5787249
7. Bhatia, K., Arora, S., Tomar, R.: Diagnosis of diabetic retinopathy using machine learning classification algorithm. In: 2016 2nd International Conference on Next Generation Computing Technologies (NGCT), Dehradun, India, pp. 347–351 (2016). https://doi.org/10.1109/NGCT.2016.7877439.
8. Kumar, V., et al.: Video super resolution using convolutional neural network and image fusion techniques. Int. J. Knowl. Based Intell. Eng. Syst. **24**, 279–287 (2020)
9. Yadav, P.: Execution analysis of Gabor 2D PCA feature extraction for gender identification utilizing face, pp. 2–6 (2017)

10. Rezaee, R.: An assessment of grouping calculations for expectation of medication cooperations: identification of the best calculation. Glob. J. Pharm. Investig. **8**(2), 92–99 (2018). https://www.jpionline.org/index.php/ijpi/article/see/255
11. Sneha, N., Gangil, T.: Analysis of diabetes mellitus for early prediction using optimal features selection. J. Big Data **6**(1), 1–19 (2019). https://doi.org/10.1186/s40537-019-0175-6
12. Srivastava, R., Srivastava, P.: A multimodal based approach for face and unique mark based combination for confirmation of human. Int. J. Bus. Anal. **6**(3) (2019). https://doi.org/10.4018/IJBAN.2019070102
13. Srivastava, R.: A score level fusion approach for multimodal biometric fusion. Int. J. Sci. Technol. Res. (2020)
14. Tomar, R., Sastry, H.G., Prateek, M.: A novel protocol for information dissemination in vehicular networks. In: Hsu, C.-H., Kallel, S., Lan, K.-C., Zheng, Z. (eds.) IOV 2019. LNCS, vol. 11894, pp. 1–14. Springer, Cham (2020). https://doi.org/10.1007/978-3-030-38651-1_1
15. Vamathevan, J.: Uses of AI in drug disclosure and improvement. Nat. Rev. Drug Discov. (2019). https://www.nature.com/articles/s41573-019-0024-5
16. Meng, X.-H., Huang, Y.-X., Rao, D.-P., Zhang, Q., Liu, Q.: Correlation of three information digging models for foreseeing diabetes or prediabetes by hazard factors. Kaohsiung J. Med. Sci. **29**(2) (2013)
17. Luo, Y., Zhang, T., Zhang, Y.: Optik a tale combination strategy for PCA and LDP for outward appearance highlight extraction. Opt. Int. J. Light Electron Opt. **127**(2), 718–721 (2016)
18. Zou, Q., Qu, K., Luo, Y., Yin, D., Ju, Y., Tang, H.: Foreseeing diabetes mellitus with machine learning techniques. Outskirts Genet. **9**, 515 (2018). https://doi.org/10.3389/fgene.2018.00515

Representation and Visualization of Students' Progress Data Through Learning Dashboard

Anagha Vaidya and Sarika Sharma(✉)

Symbiosis Institute of Computer Studies and Research,
Symbiosis International (Deemed University), Atur Centre, Model Colony, Pune 411016, India

Abstract. Learning process of students is very important, it motivates the students for acquiring skills of critical thinking and gives a learning experience. Hence, a systematic feedback is required by the students for measuring their learning processes. This measurement is performed through different educational techniques namely learning analytics (LA), Education Data Mining (EDM) and Learning Dashboard (LD). Students' learning measurement can be performed by measuring their log activity, content they read, by tracking assignment solving method etc., however in a traditional learning system, students' learning is measured through the marks they score in the different evaluations. Therefore, these evaluations must be well structured and carefully planned in advanced. Hence the systematic management tool is required which assists the teacher as well as students for their learning progress at any time. In this paper we propose and demonstrate the Learning Dashboard in traditional learning environment which measures the different skills of students through a systematic assessment plan and the results of these assessments are displayed through an analytical report which will be helpful for all stakeholders.

Keywords: Learning dashboard · Learning analytics · Educational data mining · Learning management system · Visualization

1 Introduction

The theoretical frameworks for Learning Dashboard (LD) development started from 2004 and are progressing till date. These dashboards track the students' progress towards the learning objective. It measures the students' progress through how many times he spends his time on learning concept, the approach of online solving the problem, how many time he interact with the peers, list of different action taken by a teacher to facilitate the learning process etc. [8]. LD exhibits the different statistics about learner, learning process with reference to learning context [6, 7]. These measurements are feasible in the online learning environment i.e. is on Learning Management System (LMS).

Though the digitalization is introduced in the education domain still classroom teaching-learning is in practice. In literature, most of the LD designs are available in digital environment only. They are not applicable in direct teaching-learning process. Hence, there is a need of a proper technique for implementing LD in an in-person teaching-learning environment.

M. Singh et al. (Eds.): ICACDS 2021, CCIS 1441, pp. 125–135, 2021.
https://doi.org/10.1007/978-3-030-88244-0_13

All these LD were developed in online teaching learning environment. Both in the traditional teaching learning Environment or in face-to-face teaching learning environment (TLE), the student progression can be tracked through subject assessment. Thus, the assessment is the only measurement parameter which can be used in designing LD in TLE. Therefore, these assessments must be expertly prepared, as they provide the information to students about the knowledge they acquire, the skill they pursue and also identify some invisible obstacles. The assessment result additionally helps the teacher for assessment of their teaching processes, evaluation activities [1]. In TLE hardcopy of the grade-sheet of the result is given to the student at the end of the semester. These subject grades are difficult medium to understand the learning curve of a student. Hence, there should be a tool or technique which display his/her learning curve during the semester. This will guide the students for taking certain action.

Nowadays most of the universities are following the continuous subject evaluation strategy, the results of these evaluations is displayed to the students immediately along with analytical and predictive analysis and their learning curve. This is achieved through visualization and data analytics techniques. The visualization always help the stakeholder to understand and analyze the data easily [2], it encourage learners to engage more in the learning activity, the decision-maker for taking a certain action, and the teachers can interpret their teaching-learning tools and techniques as well as learning material. The tool provides recommended and predictive decision to all.

The assessment is a pillar of the system therefore it is must be well-planned activity from design till the display of results and it is integrated. The literature suggests that learning science plays a vital role in the assessment [9, 4]. It calculates students learning from knowledge acquisition till the application on different level. Hence in the assessment questions should be associated with the learning science.

1.1 Problem Identification

In digital era, technology has great impact on education. With growing number of digital information infused into learning environments which affect on teachings of educators as well as learning of the students. Hence innovative practices need to be introduced in it. Also, the effectiveness of innovative practices needs to be critically analyzed. These analyses can be performed through new emerging fields in education namely learning analytics (LA) and education data mining (EDM). The LA collects the education data through different educational practices, EDM presents meaningful analysis from it. The analysis measurement parameters are the contents of discussion boards, forums, chats, web pages, documents, student's login information [9] in online education forum. These data are collected easily in online education system, substantial data is collected through how the student interacts with the system, their approach towards the problem solving, how they interact with their friends etc. The data is collected through these different interaction and different analytical reports namely predict student performance, learning pattern, abnormal fall in performance, changes in teaching approach, etc. [7, 5] are generated. These reports guide the teacher how to proceed for further interaction. It also guides education decision maker for acting for dropping the course, changing the syllabus etc.

But in direct teaching learning practices, very few processes are available for collecting such type of data. In most of the educational setup education ERP system is set up which collect only the personal data about students and their examination marks. The examination marks are only used for measurement. Therefore, innovations in the assessment process is required. The assessments should be continuous, skill-based, and more structured. Therefore, needs a tool to extract minimal data related to assessments and provide appropriate analysis of the collected data to different education stakeholders.

1.2 Research Gap and Objective

LD development is performed on e-leaning, Learning Management System (LMS), social networking, informal and open learning environment. The stakeholders are teacher and students. The objectives of these tools are identification of risk students, feedback generation and identification of excellent performing students. The measurements criteria of LD are content analysis, login trends, message exchanges, discussion behavior, log activities. The output is presented in visual format with bar graph, scatter plot, pie chart, timeline, and sociogram.

In this research paper researchers attempt to develop conceptual LD model in TLE environment with different measurement parameters. Two major stakeholders namely students and teachers are involved. The students are visualization of the students' progress which help the students for measurements of self-awareness, self-reflection, and assessment with colleagues, working performance in groups, etc. For teacher, it lists the students who are at risk level, feedback of their subject etc., the decision maker of the organization will get the feed feedback of course, introduction of new courses or dropping courses from the syllabus.

The paper is organized as follows: The section two of literature review compiles the relevant literature. The segment three presents the research design. Section four presents the model development process. The section five is about experimental results and discussion. The paper concluded in section six with the limitations and future research area.

2 Literature Review and Theoretical Backgrounds

The theoretical frameworks for LD development started from 2004 and are progressing till date. The first generation of LD was developed in 2007–2010 It displays the student activity in an online course. Its presentations is comprehensive visualizations of how many students attend the courses, how students utilized reading of materials, and how they do the submissions. This is accomplished through data mining technique which also helped teachers to view students on-line exercise works. This output is displayed in the bar chart [5]. In next generation (2010–2012), LDs were developed for a collaborative learning environment and provided recommendations of the learning process, identification of student retention through measurement of learner's performance and presented statistical analysis on learning data. Dashboards in 2012–2015 measure the learning behaviors of the students in the virtual environment, helping the teacher to improve learning methodology, learner status summary. The dashboard introduced the concept of a personalized learning environment [8]. From 2015–2018 some of the LD added the motivation and support as additional characteristics in the development. These dashboards collect the data from the questionnaire and provide reports of planning, monitoring, and regulating activities of each group and behavioral report of motivation, self-regulating strategies [11]. [12] Noticed that majority of the LD design considered all functional aspect, technical features. [7] indicate LDs are developed for on-line learning, Computer Support Collaborative Learning (CSCL), and blended learning environment and provide technical support on learning management systems such as Moodle, Massive online open source courses (MOOC) etc.

These dashboards measure learning process of learners using usual dashboard interaction parameters, a very few dashboards present the theoretical thought of including measurement of the student's motivation and behavior in the learning process. [13] Perform the study on student motivation who are using LD. The paper reported learning outcomes are not improved much, and some of the students noted negative feedback on the motivation. The study indicates that LD designing needs to better understand the design and implementation of the dashboard to engage the students and improve the positive behavior in them [14]. [9] Perform a systematic literature review on the use of LD concerning six parameters of metacognitive, cognitive, behavioral, emotional, self-regulation, and usability. The paper reports the very few LD to suggest a theoretical framework for the implementation of these parameters in their design. Hence there is a gap between the implementation of the educational concept and actual learning. Therefore, the learning science concept must be considered while developing LD.

Learning science is, understanding how knowledge develops and how to support knowledge development. LD must find ways to connect the cognition, metacognition, and pedagogy to help improve learning processes [13]. With a stronger connection to learning sciences, learning analytics can promote effective learning design found that learning science has been guided by three predominant theoretical frameworks: behavioral, cognitive, and contextual. According to [16], the students' measurable outcomes are associated with their learning science concepts. Different parameters (like time spent on the assignment, how student utilized study content, how many times he login for the material view etc.) are used for these measurements. Sophisticated classroom assessment techniques are effective measurement [17]. This continuous classroom assessment is used in the traditional learning teaching environment but not get associated with learning science.

3 Research Study Design

Method

The designing of a tool to determine the extent to which the student leaning is performed along with identification of student at high and low risk etc. The tool reads through excel sheet which is provided by the teacher after the student evaluation. On the collected data (of different evaluations) qualitative method are used for data analysis. A case study approach is followed for addressing the research question.

Participants

The data was collected from management institute. There are four semesters, in first semester fifteen compulsory subjects, from second semester the specialization starts. The institute offer minor and major specialization. The minor specialization is not compulsory, if someone opt it then it is displayed on result sheet. For every subject, the continuous evaluation was performed i.e., the student evaluation was performed throughout the semester with a specific time interval. There are fourteen predefined criteria from the university, these different criteria are considered in internal evaluation. The written exam is conducted for the external evaluation. The subject teacher submits their marks in the excel sheets. This qualitative method is used for data analysis and some machine learning algorithms are used for predictive and prescriptive analysis. Descriptive analytics is performed for learning feedback, self-analysis, and learning ability of the students. Identification of how many students acquire knowledge, how many of them are implementing the knowledge, their strength and weakness in different skills etc. Predictive analytics is performed to find out student at risk level, prediction of number of students will opt the specialization. Prescription analytics is suggest the adding or deleting the course contents, course to be updated or drop from the syllabus. Mapping of courses and their impact in learning resources. Suggest for updating and deleting some criteria in the evaluation processes.

The three stack-holders, student, teacher, and decision makers may use the tools. In the TLE environment how the students have learned a subject can be understood only through the exam. The behavior approach can be evaluated through team work activities in which a teacher has observed the student's interaction within team and problem solving ability and these are scaled on a result sheet with some marks. Hence only these marks help for the experiment designing.

Platform of Tool Development

This tool is developed by using R-shiny. It is an R package that is used for web applications development. The marks maintained in excel sheet with some coding modifications is the input for the tool. This excel sheet provides good performance and scalability. The tool is developed by taking the advantages of statistical and analytical analysis power of R-programming and web development framework features of R-shiny. R-Shiny is a platform as a service (PaaS) for hosting R web applications. It is easy to use, having good response time and delivers real-time output(s) for the given input by using reactive and web development features. It provides customization support for CSS, java scripting, Java Event handling, and database connection in application development which makes the application more dynamic, effective, and efficient.

The construct validity measurement technique is used for validating this tool. For the measurement of the tools, the skills measurement and learning measurements indicators are observed and are linked with different associative parameters of the results. The data of previous four batches are considered for testing. The model evaluated this data and the accuracy of the model validation is up to 60%

4 Procedures

[10] Proposed a model for connection of learning theories with learning analytics. It is an iterative model and identifies the learning environment in which the students are learning, identify the measurement parameters for the measuring the learning outcomes on it select learning theories which guide the data collection.

The model or tool is developed as per the process demonstrated in Fig. 1.

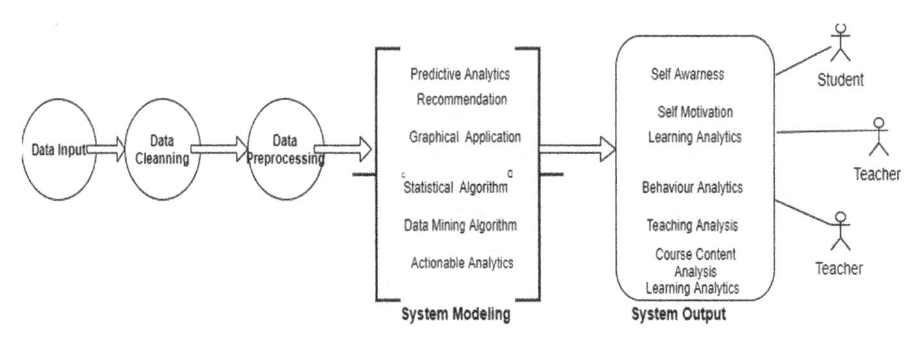

Fig. 1. Model diagram

4.1 Data Input: The Input to the Tool Is

a. Profile of the test: This file includes the test information about the batch, semester, subjects for the semester, and the assessment type.
b. Evaluation type: This file has information of different types of evaluations. The different evaluation types are the class test, assignments, project work, etc. It also includes the rubric in it.
c. Evaluation topic: This file maintains the information of evaluation subject topic per subject wise.
d. Criteria of evolution/skill evolution: Learning can perform a minimum of three ways – mental skill, develop learning attitude, and physical skill. These skills are explain by Benjamin Bloom along with quantitative measurement. This file stores the information of measurement of learning criteria apply for a given test.
e. The teacher submits the teaching plan with Outcome Based Evaluation (OBE) format, sample format is displayed in Fig. 2. The OBE contain the course objective and program objective and evaluation plan. The processes of evaluation are mapped in the Fig. 3.

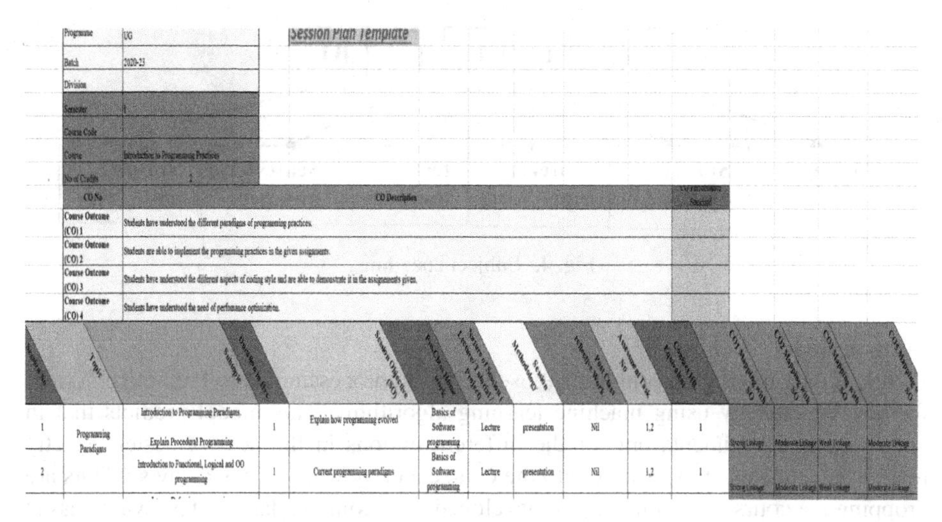

Session Plan Template

Programme	UG
Batch	2020-23
Division	
Semester	I
Course Code	
Course	Introduction to Programming Practices
No of Credits	2

CO No	CO Description
Course Outcome (CO) 1	Students have understood the different paradigms of programming practices.
Course Outcome (CO) 2	Students are able to implement the programming practices in the given assignments.
Course Outcome (CO) 3	Students have understood the different aspects of coding style and are able to demonstrate it in the assignments given.
Course Outcome (CO) 4	Students have understood the need of performance optimization.

1	Programming Paradigms	Introduction to Programming Paradigms. Explain Procedural Programming	1	Explain how programming evolved	Basics of Software programming	Lecture	presentation	Nil	1,2	1	Strong Linkage	Moderate Linkage	Weak Linkage	Moderate Linkage
2		Introduction to Functional, Logical and OO programming	1	Current programming paradigms	Basics of Software programming	Lecture	presentation	Nil	1,2	1	Strong Linkage	Moderate Linkage	Weak Linkage	Moderate Linkage

Fig. 2. OBE for the subject

Assessment Plan

Assessment Task No	Marks Allocated	Description of the task	Assessment Elements / Rubric	Bloom's Level	Weightage	CO1	CO2	CO3	CO4
						Mapping			
Assessment 1	10M	Oral Test	Pragramming Paradigms	III	16.66	Strong Linkage	Strong Linkage	Moderate Linkage	Moderate Linkage
			Coding Conventions, Code Design	VI	16.66	Moderate Linkage	Moderate Linkage	Strong Linkage	No Linkage
Assessment 2	10M	Assignment	Coding Conventions, Code Design	IV	8.33	Strong Linkage	Moderate Linkage	No Linkage	Strong Linkage
			Code Optimizations	IV	8.33	Moderate Linkage	Strong Linkage	No Linkage	No Linkage
			Paradigms	IV	8.33	No Linkage	Moderate Linkage	Strong Linkage	No Linkage
			Case Study	VI	8.33	Moderate Linkage	Moderate Linkage	No Linkage	Strong Linkage

Bloom's Taxonomy	
Level 1	remember
Level 2	Understand
Level 3	Apply
Level 4	Analyse
Level 5	Evlauate
Level 6	Create

Fig. 3. Assessment planner

4.2 Data Cleaning and Preprocessing

The teacher submits the marks in an excel file. Next step clears the blank space from the file, add appropriate data in it. Every subject is evaluated on different criteria such as on class test, assignment etc. Each column needs some encoding technique for the further processing. Example of a subject coding in the sheet is encoded as shown in Fig. 4. This coding criteria give the complete summary information of the exam.

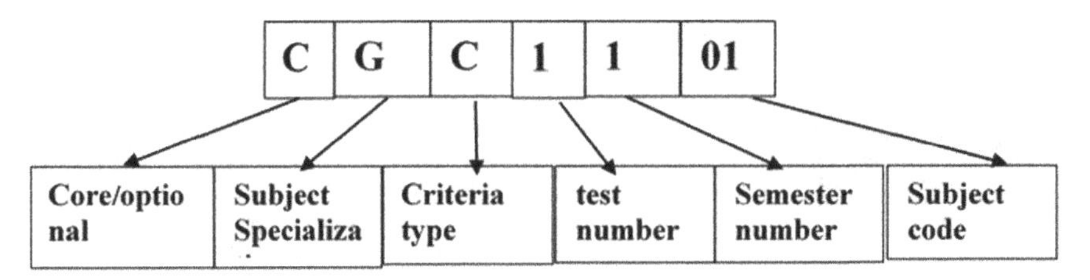

Fig. 4. Subject encoding

System Model

Different data mining algorithms are used after preprocessing data. Predictive system for development by using machine learning algorithms. This model predicts that in what way many students opt for the different options in the next semester, exactly how many students are retaining into the course and by what means many students are dropping the course. The model also developed on recommendary system which assist the students for opting a subject or identify his/her curiosity on a subject. Actionable analytics perform the different statistical analytics and provide report about how much students have understood the subject as per the different measurement criteria. The report suggests certain action for taking extra lectures, assignment etc. By using different data mining techniques, the data is analyzed and provide information for taking certain actions by education decision makers. Sample screen shot is displayed in Fig. 5 and listed the some reports of the system as follow:-

a. Self-refection and self-awareness: This report show the student learning activities and student performance in the academic year. This is a comparison report with themselves. It provides visual information of their learning process which is compare with the whole class.
b. Learning Analytical Report: Different analytical reports like student, subject, teacher, course etc. wise analytical reports are generated which are used to identify the student learning risk.
c. Course Content Analysis: Displays the information about course contents.
d. Behavioral Analysis: Measures the knowledge learning using Blooms taxonomy.
e. Teacher Analysis: The teacher can track the learner progress of the student. This report also provide feedback about their teaching.
f. Analysis Detection in course evaluation: It will show the anomaly into the subject evaluation.
g. Predictive Analysis: Next year's students' dropout rate, probably selection of the subjects etc. will be predicted by the system.
h. Recommendation Report: System will help the students in academic decision making process like selection of new subject, selection of major and minor subject etc.

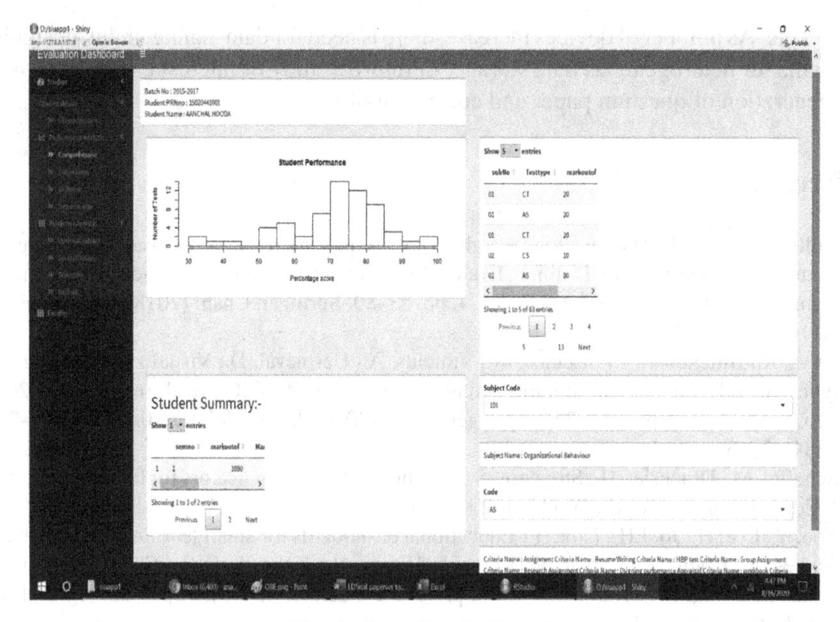

Fig. 5. Learning dashboard.

5 Results and Findings

The tool will be utilized by the three stakeholders namely: student, teacher and institute decision makers. The tool will help the students for evaluating student performance with self, evaluating student performance with the class, evaluating the learning measurement parameter, evaluating skill measurement parameters, behavioral analysis, recommendation of specialization. The Teacher will get the reporting of effectiveness of the evaluation activity, Feedback of the learning. It helps the decision maker for introduction new subject in the syllabus, addition and deletion of the content in the subject, the course continuation in the institute. The data is provided through different visual aids like bar graph risk quadrant, scatterplot, win-lose chart, sociogram etc.

6 Conclusion and Future Work

Different activities are conducted in education environments. These activities are generating a huge data. A meaningful information can be derived from this data. This information can be used by the students for his future career. This paper discussed about how different visualization techniques can be used for the presentation. The student's progression is mapped by using LD. The paper discussed the model of this visualization. The model display student learning process, their learning progression. It helps them for taking future decision regarding subject selection, displaying their strength and weaknesses.

The current version about the Learning Dashboard is developed in a window-based operating system, in future it can be developed to support various other devices like mobiles etc. As number of devices increase, there is need for data management, designing of schema for heterogeneous data storage. In future a study of interest can be to provide auto generation of question paper and correction of it.

References

1. Dolin, J., Black, P., Harlen, W., Tiberghien, A.: Exploring relations between formative and summative assessment. In: Dolin, J., Evans, R. (eds.) Transforming Assessment. Contributions from Science Education Research, vol. 4, pp. 53–80. Springer, Cham (2018). https://doi.org/10.1007/978-3-319-63248-3_3
2. Paiva, R., Bittencourt, I.I., Lemos, W., Vinicius, A., Dermeval, D.: Visualizing learning analytics and educational data mining outputs. In: Penstein Rosé, C., et al. (eds.) AIED 2018. LNCS, vol. 10948, pp. 251–256. Springer, Cham (2018). https://doi.org/10.1007/978-3-319-93846-2_46
3. Scheffel, M., Drachsler, H., Stoyanov, S., Specht, M.: Quality indicators for learning analytics. J. Educ. Technol. Soc. **17**(4), 117–132 (2014)
4. Yoo, Y., Lee, H., Jo, I.H., Park, Y.: Educational dashboards for smart learning: review of case studies. In: Chen, G., Kumar, V., Kinshuk, Huang, R., Kong, S. (eds.) Emerging Issues in Smart Learning. LNET, pp. 145–155. Springer, Heidelberg (2015). https://doi.org/10.1007/978-3-662-44188-6_21
5. Ray, S., Saeed, M.: Applications of educational data mining and learning analytics tools in handling big data in higher education. In: Alani, M., Tawfik, H., Saeed, M., Anya, O. (eds.) Applications of Big Data Analytics, pp. 135–160. Springer, Cham (2018). https://doi.org/10.1007/978-3-319-76472-6_7
6. Brouns, F., et al.: ECO D2.5 learning analytics requirements and metrics report (2015)
7. Schwendimann, B.A., et al.: Perceiving learning at a glance: a systematic literature review of learning dashboard research. IEEE Trans. Learn. Technol. **10**(1), 30–41 (2017)
8. Park, Y., Jo, I.H.: Development of the learning analytics dashboard to support students' learning performance. J. Univ. Comput. Sci. **21**(1), 110–133 (2015)
9. Jivet, I., Scheffel, M., Specht, M., Drachsler, H.: License to evaluate: preparing learning analytics dashboards for educational practice. In: Proceedings of the 8th International Conference on Learning Analytics and Knowledge, pp. 31–40. ACM (2018)
10. Wong, J., et al.: Educational theories and learning analytics: from data to knowledge. In: Ifenthaler, D., Mah, D.-K., Yau, J.-K. (eds.) Utilizing Learning Analytics to Support Study Success, pp. 3–25. Springer, Cham (2019). https://doi.org/10.1007/978-3-319-64792-0_1
11. Mutlu, B., Simic, I., Cicchinelli, A., Sabol, V., Veas, E.: Towards a learning dashboard for community visualization. In: Proceedings of the 1th Workshop on Analytics for Everyday Learning co-located with the 13th European Conference on Technology Enhanced Learning (EC-TEL 2018), Leeds, UK, pp. 1–10. CEUR Workshop Proceedings (2018)
12. Bodily, R., Verbert, K.: Trends and issues in student-facing learning analytics reporting systems research. In: Proceedings of the Seventh International Learning Analytics & Knowledge Conference, pp. 309–318. ACM (2017)
13. Viberg, O., Hatakka, M., Bälter, O., Mavroudi, A.: The current landscape of learning analytics in higher education. Comput. Hum. Behav. **89**, 98–110 (2018)
14. Bennett, L., Folley, S.: Four design principles for learner dashboards that support student agency and empowerment. J. Appl. Res. High. Educ. **1**, 15–26 (2019)
15. Howell, J.A., Roberts, L.D., Mancini, V.O.: Learning analytics messages: impact of grade, sender, comparative information and message style on student affect and academic resilience. Comput. Hum. Behav. **89**, 8–15 (2018)

16. Haendchen Filho, A., Tomazoni, E.K., Paza, R., Perego, R., Raabe, A.: Bloom's taxonomy-based approach for assisting formulation and automatic short answer grading. In: Brazilian Symposium on Computers in Education (Simpósio Brasileiro de Informática na Educação-SBIE), 29, pp. 238–247 (2018)
17. Wilson, L.O.: Anderson and Krathwohl–Bloom's taxonomy revised (2018). https://thesecondprinciple.com/teaching-essentials/beyond-bloom-cognitive-taxonomy-revised. Accessed

Denoising of Computed Tomography Images for Improved Performance of Medical Devices in Biomedical Engineering

Harjinder Kaur, Deepti Gupta, and Mamta Juneja(✉)

Computer Science and Engineering, UIET, Panjab University, Chandigarh, India
mharji94@gmail.com, {deeptigupta,mamtajuneja}@pu.ac.in

Abstract. Image preprocessing is the most significant step for improved segmentation and classification in the Computer Aided Diagnosis (CAD) system. Certain noises such as Gaussian and Poisson destroy the performance of the CAD system. There are various types of medical images such as Computed tomography (CT), Magnetic resonance imaging (MRI), Positron emission tomography (PET) and Ultrasound. CT is the most commonly used imaging modality used in the detection of pancreatic cancer responsible for an increased rate of deaths globally. These images are attenuated during compression, transmission and acquisition which cannot be removed completely. The paper presents the comparative analysis of the filters used for denoising the CT images in order to improve the accuracy of the segmentation and classification for the improved performance of the CAD system. The Peak-signal-to-noise (PSNR), Mean square error (MSE) and Structural similarity index (SSIM) have been calculated for the images denoised using the anisotropic diffusion filter, wavelet filter, bilateral filter, Non-local mean filter (NLM), wiener filter, total variation filter (TV) BM3D filter, median and Gaussian filter. The wavelet filter outperformed the other filters with 28.43, 103.12 and 0.59 values of PSNR, MSE and SSIM respectively. The suggested approach can be embedded in CT scanners to filter out the noises present in images with inbuilt technology and improve the performance of diagnostic devices.

Keywords: Denoising · Pancreatic cancer · Filtering · CT scan

1 Introduction

The formation of tumors that build in the cells of pancreas is known as pancreatic cancer where the cells block their processing enlarging disorderly to form a lump. It produce and circulate in body parts because of which cancerous tumor are known to be destructive. With expansion it starts to influence the significant component of the pancreas that is supplied to further neighbouring blood vessels and organs, afterwards they circulate in distinctive component of the body as metastasis. The probability of the identification of pancreatic cancer at the initial stage is a challenging function but after the investigation process the prevention of tumors to reach at advanced stage is possible.

M. Singh et al. (Eds.): ICACDS 2021, CCIS 1441, pp. 136–146, 2021.
https://doi.org/10.1007/978-3-030-88244-0_14

for all categorize of the cancers, the American cancer society (ACS) analysed, the one-year continuity estimation is around 20%, although the 5-year continuous progression even rest below [1].

Detection is almost an early stage treatment of all types of cancer. The overall effectiveness of the diagnosis determines its performance. The detection process gives us the location, size, and shape of the cancer in the pancreas [2]. Diagnosis of pancreatic cancer can be performed using various imaging modalities which include ultrasonography, endoscopic retrograde pancreatography, CT, and angiography [3]. Investigation of pancreatic cancer comprises three-step phenomenon Detection, Diagnosis and Staging [4]. Amongst all the imaging modalities used for diagnosis, CT is commonly employed with many advantages over other imaging modalities. But, the presence of noise will make it difficult to use these images for a variety of purposes, including medical imaging and satellite imaging. Some techniques are favoured because of their ability to model accurately and work with inexact data [5]. Information in the images due to the existence of noises gives the output in the form of unreliable diagnosis. For correct segmentation and classification of cancerous areas, the denoising of the image is an essential step. There are different categories of noises present in different regions of images such as Rician, Gaussian, Salt and pepper. Thus, denoising helps us to reduce the noise existence in the images for efficient and precise diagnosis [6].

1.1 Common Noises

The noises commonly present in the CT images are Gaussian and Poisson, which are given as follows:

- **Gaussian noise:** Gaussian noise is known to be a statistical noise that consists of a standard distribution of probability density function known as Gaussian distribution that includes only different values for the assessment of the distribution [7]. Mathematically, the Probability density function (PDF) p_g of Gaussian random variable is denoted as:

$$p_g(z) = \frac{1}{\sigma\sqrt{2\pi}} e^{\frac{-(z-\mu)^2}{2\sigma^2}} \tag{1}$$

  Here, the grey level is denoted by z, the mean is represented by the μ and is used to represent the standard deviation. The standard deviation for all sub-windows can be described with the minimum, maximum and reference standard deviation.
- **Poisson Noise:** Poisson noise, also called the quantum noise or the shot noise that is present in the pictures captured by a sensor. X-ray and gamma ray sources transmit the photons per unit time having irregular fluctuations. In the medical term, this noise is the outcome of collected spatial and temporal randomness [8]. The PDF of this noise is the root means square values directly proportional to square root of the image intensity given as

$$p(z) = \frac{e^{-\lambda}\lambda^{x}}{Z!} \tag{2}$$

  For $\lambda > 0$ and $x = 0, 1, 2.....$ Here, λ is mean/variance and Z is number of photons.

2 Filters Used for Denoising

Denoising is very helpful in raising the performance in case of segmentation and classification. Different types of internal or external factors that cause noise in the images can be removed by applying various optimal algorithms so that the better accuracy or precision can be achieved in the CAD system. Filters used for denoising of CT images are as given below.

2.1 Mean Filter

It is responsible for the smoothening of the images after reduction of the variation present in between two adjacent pixels in the image. The mean of the pixels is calculated and substituted instead of each pixel in the original image which falls in the range of the square kernel. The large-sized kernel reduces noise to a greater extent with the addition of the blurring effects [9].

2.2 Block Matching and 3D (BM3D) Filter

Block matching paradigm is relatively hard in computation but uses an aggregation in further steps. It filters out all 2-dimensional image patches in the same way in the 3D block and enlarges the edge contrast. Due to this approach, it is more difficult, moderate and inflexible than normal methods [10].

2.3 Anisotropic Diffusion Filter

Perona-Malik diffusion is the other name for the anisotropic diffusion filter whose main purpose is to eliminate image noise without affecting the important image parts that refer to the edges line including other details. It is based on the non-linear diffusion scheme for placement of the linear diffusion filtering [11].

2.4 Median Filter

The median filter is based on the non-linear filtering algorithm that eliminates the noise from an image. Denoising improves the output of later processing (i.e. edge detection details). The images are obtained after applying a median filter as the window stalk on the input image to recover the centre pixel value with the value obtained after sorting all pixels lying under the mask [12].

2.5 Wiener Filter

Wiener filter is based on mathematical operation. The aim of this filter is to determine a mathematical estimation of unknown signal using the similar form of signal as an input and filtering that known signal to show the evaluation as an output. It minimizes the mean square error between the estimated random movement and desired process [13].

2.6 Non Local Means (NLM) Filter

The NLM filter moderates the pixels of related images by their intensity areas. NLM filter is extremely based on parameters such as R_{search}, R_{sim} and *h*. Where, R_{sim} is the similarity measure used to find similar regions and *h* is the controlling parameter for degree of smoothing. Here, p is the pixel that is being restored for filtering and q shows each pixel in the image [14]. Weights are assigned according to the similarity that is calculated between the neighbourhood assigned as N_p and N_q of the pixels namely *p* and *q* [15].

2.7 Gaussian Filter

Gaussian filter is based on the non-uniformity of the low pass filter that blurs the images in order to remove the noise details with a weighted mask convoluted on the image in a two-way process. First, it implements involution of the initial image among its horizontal direction image and then the horizontally convoluted image with the Gaussian kernel in the vertical direction to achieve the finishing image [16].

2.8 Wavelet Filter

Wavelet filters can be viewed as high pass filters used to eliminate Gaussian noise from the images with the appearance of power computation. To construct the wavelet of initial and noisy signal which consist of collection of impulsive, speckle and Gaussian noise is done with the help of inverse wavelet transform which is further segmented in a similar way throughout its coefficients at the same time of transformation [17].

2.9 Total Variation Filter

Total variation filter given by Rudin, Osher and Fatermi is dependent on partial differential equation algorithm for preserving the edges to remove the noise. It is described in charge of optimization problems for the outcome gained by reducing a local cost function [18].

2.10 Bilateral Filter

The bilateral filter is noise reducing and a non-linear smoothing filter for images. It is used to disintegrate an image into a desired range without generating aureoles using a non-filtering process. It recovers the magnitude of every pixel with the weighted average of magnitude values from nearby pixel values [19].

3 Literature Review

Yuan et al. (2018) suggested a variational degree set to perform a denoising of a rician noised image. Also, the use of multipliers ADMM having fluctuating directions in combination with the suggested approach improved the performance giving better outcomes

[20]. Thereafter, Sharma et al. (2019) applied Sylvester-lyapunov and NLM filtering in place of other filters, as they gave better results than others. The significant details in this case were found to be preserved and the values of PSNR, SSIM were found to be better than other filters [21]. In the same year, Liu et al. (2019) also suggested ADMM filtering for removal of rician noise followed by Maximum a posteriori (MAP) for image estimation. The non-uniformly reachable depth was further filtered and the issues of closed form and Newton's approach was solved with noises. Further, the divergence of intensity eventually estimated the rician noise in improved manner [17]. Further, Oba et al. (2020) suggested an approach to deal with patient's health ensuring clinical safety at the pandemic of Coronavirus disease (COVID-19) for pancreatic evaluation at initial stage of disease. This pandemic has extensively burdened the clinicians influencing the clinical care for other diseases across the world. As a result patients suffering with malignant disorders of pancreas have increased haphazardly, which if attacked by the pandemic may worsen their conditions further and are at high risk of fatality. Thus, an online consultation for patients across the globe having pancreatic disorders has been started with aim to examine their conditions in the time of pandemic employing desired articulations [22]. Recently, Miyasaka et al. (2020) compared the perioperative concerns between Open pancreatic disease (OPD) and Minimally invasive pancreatic disease (MIPD). It is seen that in OPD, mostly harm is associated with belly organs and the resectioning of pancreas may have a significant role in repairing obstinate infection. Moreover, the re-sectioning of multi-organ causes sequelae, such as insufficiency of glucose in heavily weighted persons leading to serious confusions. Now, as the intrusive clinical applications have gained popularity for the malignant growth of pancreas and belly related administration issues. Thus, the preferred remedy for the same is considered to be Minimally invasive pancreatic resection (MIPR) and distal pancreatectomy along with pancreaticoduodenectomy [23].
Further, based on above research studies following gaps and current scenario/study has found:

- Accurate diagnosis of pancreatic cancer needs efficient denoising for removal of noise from CT images.
- Denoising of CT images is considered difficult due to loss of other significant edge details leading in inaccurate diagnosis.
- Very less effective denoising approaches have been used for denoising pancreatic CT images.

4 Materials

4.1 Dataset Description

The cancer imaging archive (TCIA) data collection is accessible for everyone, supported by the National Institutes of Health Clinical Center for the testing purpose. The dataset comprises of 19,000 images of size 10.2 GB. 82 images are used here out of 19000. As, 19000 here specify the total images in the dataset produced by operating on 82 patients having 29 females and 53 males, where each patients scan includes approximately 232 slices. Thus 82 images i.e. per slice of a patient have been used for evaluation [24].

4.2 Experimental Setup

PYTHON 3.7.2 using skimage and scikit libraries has been used for implementing the proposed approach. The experimental study is carried out in an environment with HP PC, Windows 7 Home Basic, 64 - Bit operating system, Intel Core i7 8th generation 8750H processor with maximum bandwidth of 1333 MHz and NB Frequency of 1895.4 MHz and 4.00 GB RAM.

5 Methodology

This section presents the methodology used for denoising CT images using a series of steps shown in Fig. 1 below. Denoising is a significant step for any computer aided diagnosis systems, as it greatly affects the performance of segmentation and classification which are key steps for accurate diagnosis. CT imaging commonly used for pancreatic cancer is found to contain a certain amount of gaussian and poison noise which suppress the significant edge details required for diagnosis. Thus, its removal is of utmost importance and this study aims to suggest a suitable filter for the same. The methodology starts with the introduction of gaussian noise and poison noise in the original input image. These noised images were then denoised using state of the art filters to analyse the best performing. Further, the performance filters were analysed using metrics such as PSNR, MSE, SSIM.

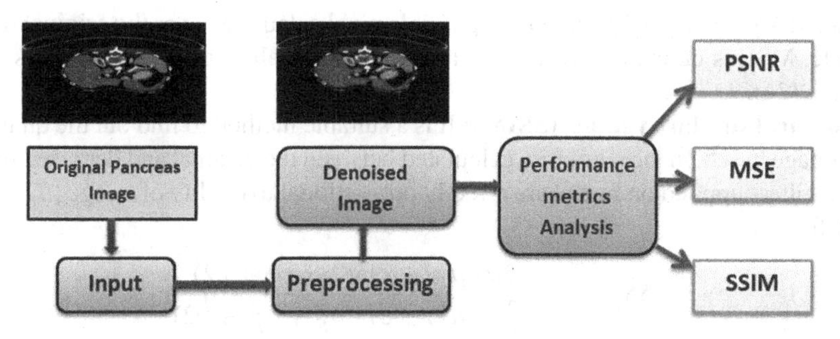

Fig. 1. Flow diagram of methodology

The approach used for the denoising of the CT images is processed in the following steps:

1. The combined gaussian noise and poison noise were added in the input image.
2. Initially, the Gaussian noise level was set to 0.5 and the poisson noise level was set to 0.01.
3. Further the noise in the image was removed using the state-of-the-art filters namely anisotropic filter, wavelet filter, bilateral filter, NLM, wiener filter, total variation filter, BM3D filter, median and Gaussian filter.
4. In order to analyse the execution of the different filters the original and the denoised image were analysed using MSE, SSIM and PSNR.

6 Results and Discussions

This section presents the dataset used followed by performance metrics.

6.1 Denoising Performance Metrics

- **Mean square error (MSE):** It is computed as the average of the squares of the errors, i.e., the average squared difference between the predicted values and the predictor. MSE is given as:

$$MSE = \frac{1}{m \times n} \sum_{i=0}^{m-1} \sum_{j=0}^{n-1} [I(x,y) - I'(x,y)]^2 \tag{3}$$

  In the above equation I(x,y) represents the original image and I'(x,y) refers to the denoised image having intensities x and y with m number of rows and n number of columns. This approach always gives a positive value for the MSE, where the lower value shows improved results [25].
- **Peak signal to noise ratio (PSNR):** There is another performance metric that is usually expressed in a logarithmic way and measured with the help of MSE to identify the correlation between original and denoised image.

$$PSNR = 10log_{10}(\frac{MAX_i^2}{MSE}) \tag{4}$$

  Here, MAXi refers to the maximum possible pixel calculated over the original image. While, MSE is defined in equation 3 above. Higher values of PSNR signifies better results [26].
- **Structural similarity index (SSIM):** It is a suitable method to find out the quality of the image based on the similarity calculated between the original and denoised image. Here, filters applied on image are directly proportional to quality of image [27]. *SSIM* is defined as

$$SSIM = \frac{(2\mu_F \mu_{F'} + c1)(2\sigma_F \sigma_{F'} + c2)}{(\mu_F^2 + \mu_{F'}^2 + c1)(\sigma_F^2 + \sigma_{F'}^2 + c2)} \tag{5}$$

  In the above equations, two different images named as F and F' having contrast values c1 and c2 μ_F and $\mu_{F'}$ representing the mean value of image F and F' whereas σ_F and $\sigma_{F'}$ represents sample variance of images F and F' [28].

6.2 Performance Analysis

This section analyzes the work by different filters for removal of noise from the CT images of pancreas. Figure 2 shows the real image on the outside of the noise and an image perverted by Gaussian noise. Whereas, Fig. 3 represents the images denoised using various traditional approaches.

Further, Table 1 and Fig. 4 presents the comparison of PSNR, MSE and SSIM values for various filters that have been used for the removal of the noise from the pancreas CT images.

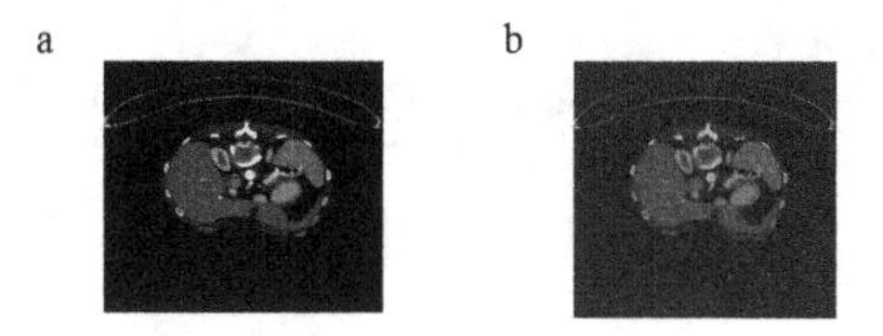

Fig. 2. (a) Original image; (b) Noisy image.

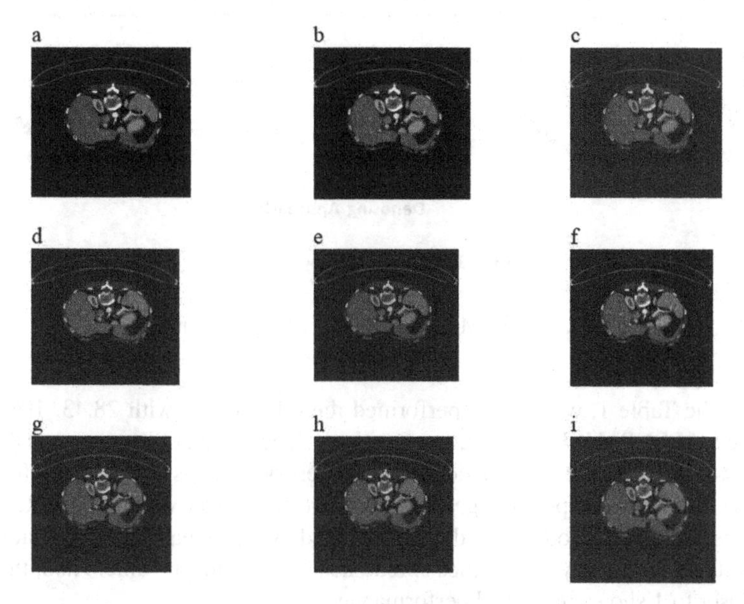

Fig. 3. (a) BM3D filter (b) Anisotropic filter (c) Median filter (d) Wiener filter (e) NLM filter (f) Gaussian filter (g) Wavelet filter (h) TV filter (i) Bilateral filter.

Table 1. Comparison of performance by using different filters

Filters	Performance parameters		
	PSNR	MSE	SSIM
BM3D	26.51	114.31	0.25
Anisotropic	26.53	113.71	0.34
Median	28.23	98.53	0.46
Wiener	27.45	98.82	0.37
NLM	27.41	102.74	0.47
Gaussian	27.18	98.74	0.54
Wavelet	**28.43**	**103.12**	**0.59**
Total variation	26.50	114.51	0.24
Bilateral	28.41	108.04	0.62

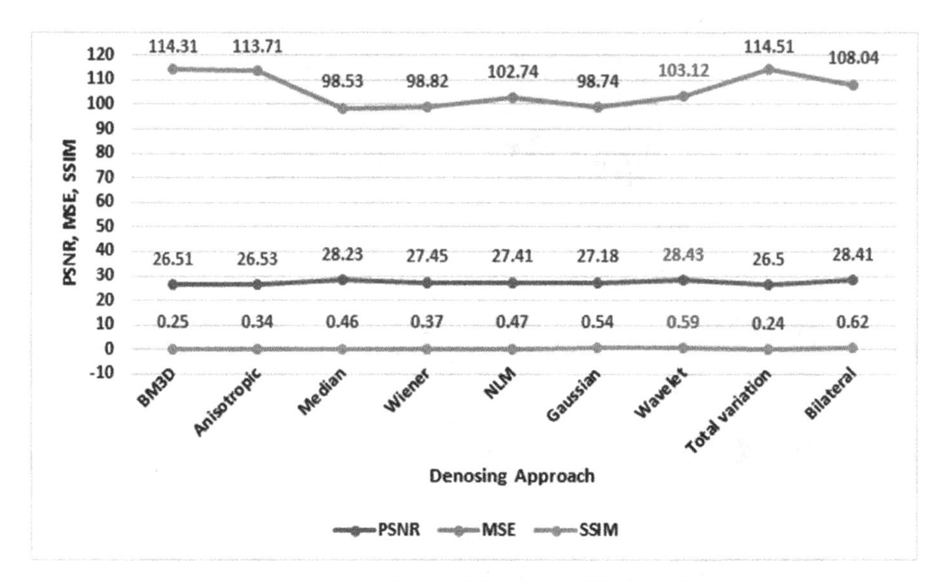

Fig. 4. PSNR, MSE, SSIM plots for different filters

As per the Table 1, wavelet outperformed the other filters with 28.43, 103.12 and 0.59 values of PSNR, MSE and SSIM respectively. The wavelet filter removes the noise in such a way that significant edge details were preserved. Thus in Fig. 4, the filter with large error follows low aspect images and thus gives the lower value of PSNR. Further, larger values of MSE reproduce the data, distributed over its mean and thus minor values were selected as it displays the values spread nearly to its mean value. Also, the SSIM values closer to 1 shows improved performance.

7 Conclusion and Future Scope

The paper represents the comparison of different filters used to denoise the pancreas CT images. Further, the comparative analysis of the traditional filters is performed on the basis of performance metrics such as PSNR, MSE and SSIM. The observations suggested that the wavelet filter achieves 28.43, 103.12 and 0.59 values of PSNR, MSE and SSIM respectively, which outperforms other traditional filters. Further in the future, an effort will be made to improve the values of PSNR, MSE and SSIM for this and other datasets with different other noises using deep learning approaches followed by the segmentation of the required region of interest for increased accuracy of the CAD system. This paper will help doctors for better diagnosis of cancer as well as help industries to equip scanners with inbuilt support of appropriate filters for improved capturing of medical images. Moreover, will aid post processing of medical images captured within unconstrained or noisy environments with optimal filters and thereby improving accuracy, reducing overall processing and diagnosis time.

References

1. Vincent, A., Herman, J., Schulick, R., Hruban, R.H., Goggins, M.: Pancreatic cancer. Lancet **13**(378(9791)), 607–20 (2011)
2. Raimondi, S., Lowenfels, A.B., Morselli-Labate, A.M., Maisonneuve, P., Pezzilli, R.: Pancreatic cancer in chronic pancreatitis; aetiology, incidence, and early detection. Best Pract. Res. Clin. Gastroenterol. **24**(3), 349–58 (2010)
3. Takhar, A.S., Palaniappan, P., Dhingsa, R., Lobo, D.N.: Recent developments in diagnosis of pancreatic cancer. BMJ **329**(7467), 668–673 (2004)
4. Warshaw, A.L., Gu, Z.Y., Wittenberg, J., Waltman, A.C.: Preoperative staging and assessment of resectability of pancreatic cancer. Arch. Surg. **125**(2), 230–233 (1990)
5. Katiyar, A., Katiyar, G.: Denoising of images using neural network: a review. In: Singh, S., Wen, F., Jain, M. (eds.) Advances in System Optimization and Control. LNEE, vol. 509, pp. 223–227. Springer, Singapore (2019). https://doi.org/10.1007/978-981-13-0665-5_20
6. Vrieling, A., et al.: Cigarette smoking, environmental tobacco smoke exposure and pancreatic cancer risk in the European Prospective Investigation into Cancer and Nutrition. Int. J. Cancer **126**(10), 2394–4039 (2020)
7. Breyer, R.J., Mulligan, M.E., Smith, S.E., Line, B.R., Badros, A.Z.: Comparison of imaging with FDG PET/CT with other imaging modalities in myeloma. Skeletal Radiol. **35**(9), 632–640 (2006). https://doi.org/10.1007/s00256-006-0127-z
8. Verma, R., Ali, J.: A comparative study of various types of image noise and efficient noise removal techniques. Int. J. Adv. Res. Comput. Sci. Softw. Eng. **3**(10) (2013)
9. Fan, L., Zhang, F., Fan, H., Zhang, C.: Brief review of image denoising techniques. Vis. Comput. Ind. Biomed. Art **2**(1), 1–12 (2019). https://doi.org/10.1186/s42492-019-0016-7
10. Yahya, A.A., Tan, J., Su, B., Hu, M., Wang, Y., Liu, K., Hadi, A.N.: BM3D image denoising algorithm based on adaptive filtering. Multimedia Tools Appl., 1–37 (2020)
11. Perona, P., Malik, J.: Scale-space and edge detection using anisotropic diffusion. IEEE Trans. Pattern Anal. Mach. Intell. **12**(7), 629–39(1990)
12. Huang, T., Yang, G.J., Tang, G.: A fast two-dimensional median filtering algorithm. IEEE Trans. Acoust. Speech Signal Process. **27**(1), 13–18 (1979)
13. Farhang-Boroujeny, B.: Adaptive Filters: Theory and Applications. Wiley, Hoboken (2013)
14. Manjón, J.V., Carbonell-Caballero, J., Lull, J.J., García-Martí, G., Martí-Bonmatí, L., Robles, M.: MRI denoising using non-local means. Med. Image Anal. **12**(4), 514–523 (2008)
15. Haddad, R.A., Akansu, A.N.: A class of fast Gaussian binomial filters for speech and image processing. IEEE Trans. Signal Process. **39**(3), 723–727 (1991)
16. Diwakar, M., Kumar, P.: Singh AK : CT image denoising using NLM and its method of noise thresholding. Multimedia Tools Appl. **79**(21), 14449–14464 (2020)
17. Primer, A., Burrus, C.S., Gopinath, R.A.: Introduction to wavelets and wavelet transforms. In: Proceedings of International Conference (1998)
18. Rudin, L.I., Osher, S., Fatemi, E.: Nonlinear total variation based noise removal algorithms. Phys. D Nonlinear Phenom. **60**(1–4), 259–68 (1992)
19. Tomasi, C., Manduchi, R.: Bilateral filtering for gray and color images. In null, p. 839 (1998)
20. Haddad, R.A., Akansu, A.N.: A class of fast Gaussian binomial filters for speech and image processing. IEEE Trans. Signal Process. **39**(3), 723–7 (1991)
21. Diwakar, M., Kumar, P., Singh, A.K.: CT image denoising using NLM and its method of noise thresholding. Multimedia Tools Appl. **79**(21), 14449–14464 (2020)
22. Oba, A., et al.: Global survey on pancreatic surgery during the COVID-19 pandemic. Ann. Surg. (2020)
23. Miyasaka ,Y., Ohtsuka, T., Nakamura, M.: Minimally invasive surgery for pancreatic cancer. Surg. Today **51**, 194–203 (2021)

24. Roth, H.R., et al.: Data from pancreas-CT. Cancer Imaging Archive (2016)
25. Botchkarev, A.: Performance metrics (error measures) in machine learning regression, forecasting and prognostics. Properties and typology. arXiv preprint arXiv:1809.03006 (2018)
26. Bhadauria, H.S., Dewal, M.L.: Performance evaluation of curvelet and wavelet based denoising methods on brain computed tomography images. In: International Conference on Emerging Trends in Electrical and Computer Technology (IEEE), pp. 666–670 (2020)
27. Tyagi, V.: Understanding Digital Image Processing. CRC Press, Boca Raton (2018). https://doi.org/10.1201/9781315123905
28. Grover, T.: Denoising of medical images using wavelet transform. Imp. J. Interdiscip. Res. **2**, 541–548 (2016)

Image Dehazing Through Dark Channel Prior and Color Attenuation Prior

Jacob John(✉) and Prabu Sevugan

School of Computer Science and Engineering, Vellore Institute of Technology, Tiruvalam Road, Vellore 632014, Tamil Nadu, India
sprabu@vit.ac.in

Abstract. With an increase in motor transportation in megacities, the pollutants in the air are rising dramatically. This increases the number of unburnt particulates, dust, and smoke being released into the air. Haze is a phenomenon that emerges from such conditions that tends to affect image quality. Two single-channel algorithms have been proposed for dehazing images: The Dark Channel Prior and the Color Attenuation Prior. The Dark Channel Prior is based on the statistic that low-intensity pixels exist in at least one-color channel in outdoor haze-free images. The color attenuation method is based on the parameters obtained via a comprehensive analysis of scene depth of hazy images modeled using a linear model. This paper does a comprehensive analysis of the two techniques to evaluate them objectively. Finally, this paper also overcomes the limitations of the FoHIS, a synthetic fog-generation technique, by integrating a novel approach to finding a depth map automatically.

Keywords: Dark Channel Prior · Color Attenuation Prior · Dehazing · Defog · Haze removal · Image restoration · Computer vision

1 Introduction

With advancements in commercially available image acquisition devices such as autofocusing, high-speed precision control, and higher shutter speeds, external discrepancies become more apparent. Some of these irregularities occurring in captured images occur due to natural factors such as weather, ambient light, and atmospheric disturbances. Fog, hail, rain, and snow are examples of poor weather conditions in the outdoor environment that directly affect or cause pixel intensity changes in images. Fog can reduce the visibility and contrast of a scene. Anwar and Khosla [1] categorize fog into two categories – low fog, which tends to reduce visibility to a range of 300 and 1000 m, and dense fog, with a visibility range of below 100 m. Furthermore, visibility degradation can lead to fatal injuries on the roads. According to an article by Times of India [2], the number of fog-related road accidents has increased by 88%, from 5,886 in 2014 to 11,090 in 2017. Removal of haze can correct the color shift caused by airlight, thereby significantly improving the clarity of an image. However, it is seen as a challenging task as haze removal depends on identifying the scene's "unknown" depth data [3]. This is because the depth map needs to be recovered in cases where very little information is made available [4].

M. Singh et al. (Eds.): ICACDS 2021, CCIS 1441, pp. 147–159, 2021.
https://doi.org/10.1007/978-3-030-88244-0_15

2 Related Works

Two main image dehazing algorithms are Color Attenuation Prior [5] and Dark Channel Prior [6], and they are compared objectively in the further sections. Both of these are single image dehazing methods. Another such category of algorithm is proposed by Fattal in [7]. Fattal's model accounts for the transmission function in addition to the surface shading. An estimation of the scene's albedo is used to calculate the medium's transmission [8]. The transmission function T(x) is calculated using a Gauss-Markov random field (MRF) model [9] and independent component analysis (ICA) model [10]. The underlying assumption in Fattal's model is the uncorrelated relationship between the transmission function and surface shading. Since the input color of the image guides the MRF model, it would fail for gray images [11]. Furthermore, this algorithm would work well with light foggy images as dense haze is typically of high intensity. Tan [12] proposes a single image dehaze algorithm that utilizes the higher contrast characteristic of the haze-free image to maximize the local contrast of a hazy image. However, since the algorithm does not account for color restoration and restores an image having maximum contrast, the enhanced image suffers from color distortion and a halo effect around depth discontinuities.

Tang et al. [13] use the relationship between the haze-relevant features in an image and its true transmission using a Random Forest model. Features are extracted from each patch after the foggy image is divided into several patches. A guided filter [14] is used by Tang et al.'s methods, Color Attenuation Prior and Dark Channel Prior, to smooth and optimize the transmission factor. This algorithm can restore foggy images with homogenous and dense fog because of the learning-based algorithm's intrinsic ability to adapt and create new regression models for each dissimilar weather condition [15]. However, it fails in heavy haze cases with a low signal-to-noise ratio. In addition to this, similar to the above approaches, this approach also suffers from unnecessary boosting of noise. A Bayesian model using a probabilistic method is proposed by Nishino et al. [16]. This model is based on the fact that scene albedo ρ and depth $D(x)$ are two statistically independent components [17]. A factorial Markov random field is used to model the image and accurately estimate the scene radiance. This algorithm eliminates the halo artifacts and produces images with high visibility. However, bayesian defogging might lead to color distortion and edge degradation in images with dense haze. In addition to this, unlike Tan's method, parameters need to be manually set by the user [18]. Multiple other image techniques for defogging images can also be used to restore a non-hazy image. Some of these image defogging approaches use multiple polarization images. This is because path radiance or airlight scattered by the atmosphere is partially polarized [15]. Polarization is also used when the object is assumed to be polarized [19]. Mounted polarizers are also sometimes used by photographers to increase vividness in their images.

Treibitz et al. [20] try to identify whether the use of polarization is justified by the image loss. This approach used one and two polarization images and quantitatively analyzed the change in signal-to-noise ratio (SNR) after image restoration. A polarization filter with different orientations is used to acquire the images of the same scene in various intensities. Schechner et al. [21] have also discussed using two polarization images for image restoration in instant dehazing. This process analyzes the image formation process and takes into account the effect of polarization on atmospheric scattering. The airlight intensity is divided into parallel and perpendicular components and is used to define an airlight model. Two key parameters – the degree of polarization p and the atmospheric light A_∞ along with the transmission T and airlight model $\widehat{V}(x)$ are used to inversely solve the physical model. Furthermore, Schechner et al. also improve the algorithm to account for airlight scattering that occurs on a clear day [22]. This is done by analyzing the polarization-filtered images, but it requires the estimation of airlight parameters. Shwartz et al. [23] have hence proposed a method of blind haze separation that estimates A_∞. This is done by selecting two similar features in an image and using blind estimations accompanied by an ICA model.

3 Proposed Methods

Since hazing in images occurs mainly due to two phenomena, haze can be expressed as the addition of air-light and attenuation [4] as in Eq. (1).

$$\text{Haze} = \text{Attenuation} + \text{Air-light}. \tag{1}$$

Since haze is a depth-dependent noise, the light intensity can further be expressed as the weighted sum of the scene's airlight radiance by the transmission factor, as given in Eq. (2) [23]:

$$I(x) = \mathrm{T}(x)\mathcal{J}(x) + (1 - \mathrm{T}(x))A_\infty. \tag{2}$$

$I(x)$ is the intensity at the coordinates x of an image. This intensity, as perceived by the camera sensor, is of the hazy image. $\mathcal{J}(x)$ denotes the haze-free reflectance or the scene radiance, which is the image this study wishes to recondition. The global atmospheric air light is denoted by A_∞. The transmission factor or $\mathrm{T}(x)$ increases exponentially with depth in a uniform medium. $\mathrm{T}(x)\mathcal{J}(x)$ hence represents attenuation or the global loss of scene radiance due to the distortion of atmospheric light in a dense medium. $\mathrm{T}(x)$ can be described further using Eq. (3).

$$\mathrm{T}(x) = e^{-\mu_s \cdot D(x)}. \tag{3}$$

$\mathrm{T}(x)$ is exponentially dependent on the scattering coefficient of haze, μ_s and the distance from the sensor or the scene depth $D(x)$ [24]. Thus, it can be inferred that scene radiance gets exponentially attenuated with distance [25], in this case, assuming that the scattering coefficient is relatively constant in a homogenous atmospheric condition.

3.1 Color Attenuation Prior

The scattering coefficient is the cross-sectional area σ_s per unit volume ρ_s of a haze particle, or $\mu_s = \sigma_s \rho_s$. Hence, in an ideal case, the range of depth, $D(x)$, can be taken to be $[0, +\infty)$, as the object of interest can appear very distant in the background. This thereby results in Eq. (4), when depth tends to infinity.

$$I(x) = A_\infty, D(x) \to +\infty. \tag{4}$$

Note that in Eq. (4), the intensity, $I(x)$, can appear to approximate to global atmospheric airlight, A_∞ when the depth of the object tends to infinity. T(x) becomes a very small value as $e^{-\mu_s \cdot D(x)} \to 0$ and hence eliminates the attenuation caused due to the medium. By introducing a threshold and ensuring that the pixel should belong to a region of very large depth, Eq. (5) is obtained. This equation and its condition can be used to estimate the depth at which atmospheric light is approximately equivalent to the intensity of a pixel. Thus, this proposes a depth information restoration as suggested.

$$A_\infty = I(x), x \in \{x|\forall u: D(u) \leq D(x)\}. \quad (5)$$

In the Color Attenuation Prior method, supervised learning algorithms were used to extract the depth map of an image. This linear model is trained, and it generates parameters for the scene depth of a hazy image. Additionally, this method tries to estimate T(x) of the concentrated haze to extract scene radiance $\mathcal{J}(x)$ from the hazy image $I(x)$. The transmission factor can be estimated using an atmospheric scattering model. This is because airlight directly impacts hazy areas in images. The scattering of light tends to reduce saturation by reducing brightness. It can be thus inferred, the stronger the influence of airlight, the denser the haze in a section would be. Therefore, low saturation and high brightness regions are often characterized as hazy.

Fig. 1. Hazy image *(left)* and its corresponding graph *(right)* showing the difference between the brightness and saturation of each image with each pixel location

In Fig. 1, as the difference between brightness and saturation increases, the concentration of haze also grows. As the pixel moves further towards the top, there is a visual increase in the thickness of the haze. Furthermore, the accompanying graph also shows the corresponding increase in difference. Thus, this allows the method to state the assumption that the concentration of haze is positively correlated with a scene's depth.

$$D(x) \propto \mathcal{V}(x) - \mathcal{S}(x) \propto C(x). \quad (6)$$

Equation (6) is the statistic defined as Color Attenuation Prior and plays a significant role in the image's dehazing process. This allows the restoration of scene radiance and thus yielding a haze-free image. $C(x)$ is the concentration of the haze at a given pixel in a single-dimensional spatial representation of an image. Furthermore, $\mathcal{S}(x)$ is the saturation and $\mathcal{V}(x)$ is the brightness of the scene. $D(x)$ is the distance between the source and the observer, also known as scene depth. The Color Attenuation Prior method also proposes a linear model used for approximating the concentration of haze from between the saturation and the brightness of the scene. The model is expressed in Eq. (7) as a sum of unknown linear coefficients and a random variable.

$$D(x) = \vartheta_0 + \vartheta_1 \mathcal{V}(x) + \vartheta_2 \mathcal{S}(x) + \varepsilon(x). \tag{7}$$

The variable representing the random error of the model is $\varepsilon(x)$ whereas, the unknown linear coefficients are ϑ_0, ϑ_1 and ϑ_2. A Gaussian density of zero mean and variable σ^2 is used for ε. Thus, $\varepsilon(x) \sim N(0, \sigma^2)$. The training process to learn parameters ϑ_0, ϑ_1 and ϑ_2 is similar to that of Qin et al. [27]. To produce synthetic depth maps and hazy images for training samples, image data was collected using Google Images and Flicker. In this case, 500 images were used to produce 500 training samples for learning. The parameter estimates for the best learning result after 512 epochs are $\vartheta_0 = 0.123$, $\vartheta_1 = 0.960$, $\vartheta_2 = -0.780$, $\mu_s = 1.0$ and $r = 15$. r refers to the scale of the depth map used to overcome the issue of high estimated depth values and to process the raw depth map.

3.2 Dark Channel Prior

The Dark Channel Prior [6] is another statistic based on the realm of haze-free outdoor images. By combining the haze imaging model with this prior, the thickness or opacity of the haze can be then estimated. Thus, this effectively removes the haze and renders a haze-free image. The Dark Channel Prior method uses pixels with very low intensity in at least one of the RBG channels. Such pixels are termed dark pixels. Since dark pixels are directly affected by airlight, they can provide a direct and accurate estimation of T(x) or the haze transmission. Hence the original image can be easily restored as the thickness of the haze is now available. However, it should be noted that dark pixels occur in most local regions apart from the sky-covered areas.

As a reference, the haze imaging Eq. (2) can be geometrically represented as vectors in an RGB color space. Furthermore, Eqs. (2) and (3) can also be used to infer that vector representations of $I(x)$, $\mathcal{J}(x)$ and A_∞ have collinear endpoints and are coplanar. This is further reinforced by Fig. 2 as the endpoints of $I(x)$, $\mathcal{J}(x)$and A_∞ meet along the same line. The transmission ratio T(x) is thus given in Eq. (8).

$$\mathrm{T}(x) = \frac{\|\mathrm{A}_\infty - \mathrm{I(x)}\|}{\|\mathrm{A}_\infty - \mathcal{J}(\mathrm{x})\|} = \frac{A_\infty{}^c - I^c(x)}{A_\infty{}^c - \mathcal{J}^c(x)}. \tag{8}$$

In Eq. (8), the color channel index is $c \in \{r, g, b\}$. $\mathbf{A}_\infty$, $\mathbf{I(X)}$ and $\boldsymbol{\mathcal{J}}\mathbf{(X)}$ are vectors representing their corresponding scalars in the color space. Hence, the ratio of the magnitude of the difference between atmospheric light and intensity and the magnitude of the difference between atmospheric light and scene radiance is T(x). The dark channel can be formally defined with Eq. (9).

$$I^{dark}(x) = \min_{y \in \Omega(x)} \left(\min_{c \in \{r,g,b\}} I^c(y) \right). \tag{9}$$

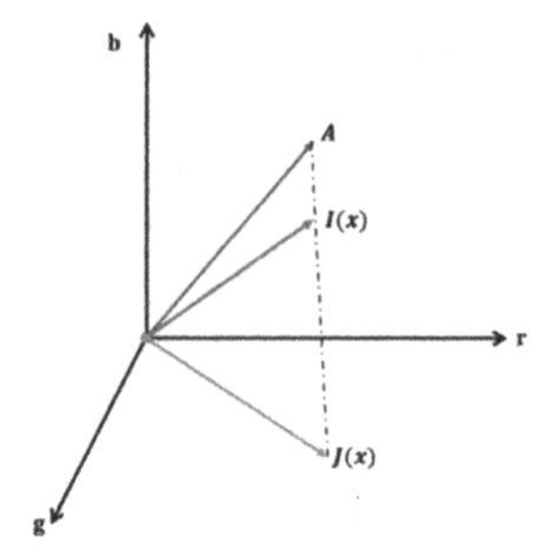

Fig. 2. Geometrical representation of the haze imaging model across the red (r), green (g), and blue (b) channels (Color figure online)

The dark channel for any arbitrary image, **I** is represented as I^{dark}. The local patch, typically of size 15 × 15, is given as $\Omega(x)$. This patch is centered around the pixel of interest. It is used to calculate the minimum intensity pixel in the whole square. Hence, I^c is the color channel of **I,** and the minimum intensity for any of the three colors in the RGB color space, c, is given as input for the next minimum calculation. This is the minimum intensity color at each pixel in the square. This definition can be further extended to define the intensity dark channel in haze-free as low and tends to 0, provided that dark channel regions do not include the sky. Hence, the following observation can be made on a haze-free image, **I** and is the *Dark Channel Prior*:

$$I^{dark} \to 0. \tag{10}$$

Furthermore, the algorithm proposed also considers regions due to other factors such as shadows, colorful objects, and dark objects, which cause low intensity in the dark channels. In order to verify this, a dataset of 100 images was taken from Flickr. This was followed by manually cutting out the sky regions. However, as stated by the algorithm, cropping out the sky regions is not necessary as in an image that is hazy, the atmospheric light is very similar to the color of the sky. This results in a transmission close to 0, which is negligible in further steps. Since the haze-free regions are less bright than the ambient light reflected by haze when the transmission is low, the following steps can be used as a means of atmospheric light estimation. A defined method is used to obtain the dark channel of the image.

- Select the haziest pixels in the image. These are the top 0.1% brightest pixels (also known as the highest intensity values) in the dark channel.
- Among these pixels, the corresponding RGB values in the input image **I** are retained, giving the atmospheric light estimation for the image.
- Finally, to restore the haze-free image, the scene radiance needs to be retrieved. Since the recovered scene radiance is prone to noise, a transmission lower bound, T_0, of typically 0.1, is maintained.

Thus, in regions of very thick haze, a small amount of haze is persevered. Furthermore, the image after haze removal may appear dull because this lower bound isn't as bright as the atmosphere. This scene radiance, $\mathcal{J}(\mathbf{x})$ can be mathematically described in Eq. (11).

$$\mathcal{J}(\mathbf{x}) = \frac{\mathbf{I}(\mathbf{x}) - \mathbf{A}_{\infty}}{\max(\mathrm{T}(x), T_0)} + \mathbf{A}_{\infty}. \tag{11}$$

3.3 Synthetic Fog

Before applying any hazing techniques, the first step is to add synthetic haze patches objectively to each of the 50 images. To do this, a framework named Foggy and Hazy Images Simulation (FoHIS) was used. This was proposed by Zhang et al. in [26]. A method to evaluate the simulation was also introduced. After simulating and adding fog to each of the images, their authenticity was tested using an Authenticity Evaluator for Synthetic foggy/hazy Images (AuthESI). The algorithm is said to perform well. Furthermore, images produced are consistent with subjective judgments. This was done by obtaining a mean opinion score (MOS) from real-life subjects. The method performed better and faster than existing methods, such as Adobe Photoshop and Guo's method [27]. However, it requires a depth map of each image.

OpenCV's StereoBM was used to generate depth maps of images where stereo images of the subject were available. However, for single images, the depth map was obtained using Eigen's method [28] of predicting using multi-scale deep networks. Eigen's method uses two deep network stacks and a scale-invariant error to measure the depth relations of an image. This approach can detect object boundaries well and produces good results. However, depth maps produced lack clarity, making haze simulation poor for some of the images obtained online. The low-resolution images would have to be scaled up to be used as input for FoHIS. Figure 3 shows an example of this process.

AuthESI first constructs natural scene statistic (NSS) features of haze and then fits them to a multivariate Gaussian model (MVG). A 16-feature vector is produced for all patches, which is then fitted using MVG probability density. The final authenticity values (AV) is then computed using the FoHIS image's MVG model and the Bhattacharya distances of the mean and covariance matrices of the natural MVG model. For this experiment's image set, the average AV was 3.9627 for 38 of the images. Outliers were removed due to a very low AV for 12 images. Low-resolution depth map produced by Eigen's method ensures outliers in some cases. The designated images will be used for dehazing.

Fig. 3. Result of FoHIS (*center*) on an image (*left*) and its corresponding depth-map (*right*).

4 Implementation

All computation and analysis were performed in *python 3.6.* Hazing and dehazing were performed on 50 different images collected from Google Images and Flickr. It should be noted that most of the images collected do not have a large sky area in the background. This is because both the algorithms would not perform well in scenarios with large sky areas [15].

4.1 Dark Channel Prior

Extract Dark Channel of the Image. The image is first split into three channels – R, G, B using *cv2.split(image).* The minimum of each of these channels is taken as the dark pixels. The minimum is calculated using *cv2.min(cv2.min(r, g), b).* A 15×15 kernel or a rectangular structuring element is created using the cv2 function, *cv2.getStructuringElement(cv2.MORPH_RECT,(15,s15)).* This structuring element is used for morphological operations. The dark pixels are removed at each 15×15 patch by performing an erosion operation on them. This is done using *cv2.erode(dark_pixels,kernel).*

Airlight Estimation. The image with and without the dark pixels is used for airlight estimation. The haziest pixels are selected based on the brightest 0.1% pixels of the image. The 0.1% number of pixels is calculated by dividing the image size by 1000. This is because $x \times (0.1/100)$ is the same as $x/1000$. Hence we use math library's function and get *int(max(math.floor(image_size/1000),1)).* The max operation is used as the smallest size can only be 1, as we cannot have half a pixel. The brightest 0.1% pixels are extracted from the vector containing all the dark pixels. These pixels are then converted back to their corresponding RGB channels for the atmospheric light.

Transmission Factor Estimation. The transmission factor is calculated based on the Eq. (12) given in the proposed method.

$$T(x) = 1 - \min_{y \in \Omega(x)} \left(\min_{c \epsilon \{r,g,b\}} I^{c(y)} / A^c_\infty \right). \tag{12}$$

Transmission Refinement. A guided filter is first defined using the method proposed by He et al. [14]. OpenCV's official guided filter can also be used. It is defined by *cv.ximgproc.guidedFilter(guide, src, radius, eps[,dst[, dDepth]])*. This guided filter is applied on the grayscale image obtained using the cv2 function, *gray = cv2.cvtColor(image,cv2.COLOR_BGR2GRAY)* and the numpy function, *gray = np.float64(gray)/255*. A radius of 60 and eps or regularization term of 0.0001 is used as arguments.

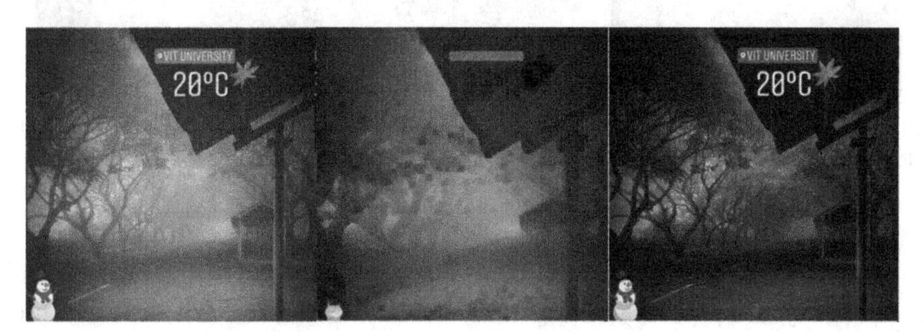

Fig. 4. Input image (*left*), Dark Channel of the hazy image (*center*), and Restored, dehazed image (*right*)

Scene Radiance Recovery. The scene radiance is recovered using Eq. (11). The output of this is the final restored image that can be seen in Fig. 4.

4.2 Color Attenuation Prior

Calculate the Depth Map. The depth map of the image can be calculated using the Eq. (7) and the parameters given in the proposed method. The Hue-Saturation-Value or the HSV format of the image can be obtained using the function *cv2.split(cv2.cvtColor(image, cv2.COLOR_BGR2HSV))*.

Guided Filter on Depth Map. A guided filter is used for defining the depth map extracted from the original hazy image. A radius of 8 and eps or regularization term of 0.2×0.2 is used as arguments. Other recommended radius includes 4, 8, or 18. Eps values of 0.4×0.4 and 0.1×0.1 should also be tested.

Calculate Airlight Using Depth Map. The 0.1% brightest top pixels in the depth map are selected, similar to Dark Channel Prior's method, to estimate the atmospheric light. Equation (5) shows this estimation of airlight. These pixels are then converted back to their corresponding RGB channels. This is the atmospheric light of the image.

Scene Radiance Recovery. The scene radiance is recovered using Eq. (11), presented in the Dark Channel Prior method. The resultant image is shown in Fig. 5.

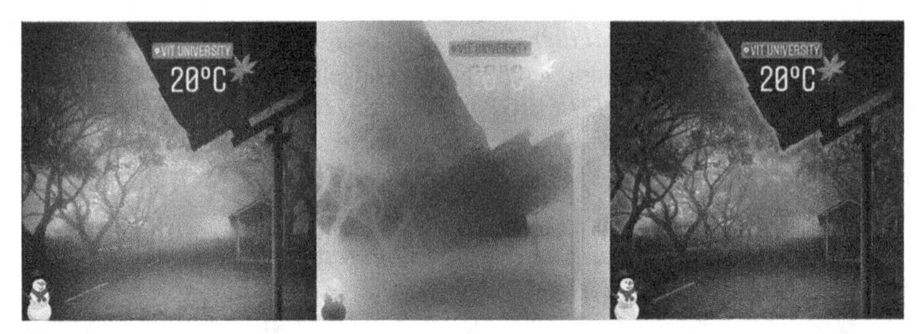

Fig. 5. The original hazy image (*left*), its transmission (*center*), and restored dehazed image after Color Attenuation Prior (*right*)

5 Results and Discussion

To objectively compare the two algorithms, several evaluation methods were used. Table 1 shows the result of comparing the differences between each dehazed image and its corresponding ground truth image. The evaluation metrics used were the peak signal-to-noise ratio (PSNR), mean square error (MSE), weighted peak signal-to-noise ratio (WPSNR), and structural similarity indices (SSIM). Inspired by the image evaluation criteria used by Li et al. in [29], this paper also uses two no-reference image quality assessment models. The first is a blind image integrity notator using DCT statistics (BLIINDS-II), and the second is a spectral entropy-based quality (SSEQ). These are said to represent human quality perception better than other objective measures, including PSNR, WPSNR, and SSIM. The results are obtained by performing a three-pass evaluation on 38 images.

Table 1. Results of all quality assessment tests of the Color Attenuation Prior method and the Dark Channel Prior method.

	Color Attenuation Prior	Dark Channel Prior
MSE	0.007	0.0185
PSNR	38.945	33.183
WPSNR	18.314	14.913
SSIM	0.9177	0.891
BLIINDS-II	48.535	54.313
SSEQ	56.144	59.134

It is evident from Table 1 that Color Attenuation Prior performed better than Dark Channel Prior in 4 evaluation criteria – the MSE is 60%, PSNR is 18%, WPSNR is 23%, and SSIM is 3% higher in Color Attenuation Prior than Dark Channel Prior. However, in the BLIINDS-II and SSEQ evaluation criteria, Dark Channel Prior performed better than the Color Attenuation Prior. Hence it can be inferred that the loss of information in the

Dark Channel Prior is more than Color Attenuation Prior, though it performs marginally better in two evaluation criteria. This is because, under dense and inhomogeneous fog weather, the color restoration of Dark Channel Prior is poor. Furthermore, it should also be noted that both algorithms perform poorly when the images contain a large white area or large sky area.

6 Conclusion

This paper performed a comprehensive analysis between two popular dehazing methods, namely, Color Attenuation Prior and Dark Channel Prior. This paper also suggested integrating a novel method to find depth maps using single images to improve the comparison pipeline. FoHIS requires a depth map for generating synthetic haze for each image. This paper used StereoBM in OpenCV for stereo images, while Eigen's novel method was used to predict a depth map from a single image using multi-scale deep networks. Based on the results obtained, it is apparent that the Color Attenuation Prior method performs better than the Dark Channel Prior method in terms of image dehazing. In most cases, the resultant images in both methods are approximately natural clear images. A key drawback in all polarization dehazing techniques is the dependency on partial polarization due to airlight. Hence, it fails in conditions with dense haze. This is because the impact of these would minimize with a reduction in the degree of polarization. Thus, it would fail, especially in scenarios with moving frames (also termed as videos), which provides a scope for further improvement in dehazing.

References

1. Anwar, M.I., Khosla, A.: Vision enhancement through single image fog removal. Eng. Sci. Technol. Int. J. **20**, 1075–1083 (2017)
2. Kanpoor P.: Over 10,000 lives lost in fog-related road crashes. Times of India (2019). https://timesofindia.indiatimes.com/india/over-10000-lives-lost-in-fog-related-road-crashes/articleshow/67391588.cms
3. Singh, R., Dubey, A.K., Kapoor, R.: A review on image restoring techniques of bad weather images. Int. J. Comput. Appl. **975**, 8887 (2017)
4. Suruti, F.R.F., Balaji, R.: Survey on various dehazing techniques. Int. J. Comput. Sci. Mob. Comput. **5**, 218–223 (2016)
5. Zhu, Q., Mai, J., Shao, L.: A fast single image haze removal algorithm using color attenuation prior. IEEE Trans. Process. **24**, 3522–3533 (2015)
6. He, K., Sun, J., Tang, X.: Single image haze removal using dark channel prior. IEEE Trans. Anal. Mach. Intell. **33**, 2341–2353 (2011)
7. Fattal, R.: Single image dehazing. ACM Trans. Graph. (TOG) **27**, 1–9 (2008)
8. Lee, S., Yun, S., Nam, J.-H., Won, C.S., Jung, S.-W.: A review on dark channel prior based image dehazing algorithms. EURASIP J. Image Video Process. **1**, 1–23 (2016). https://doi.org/10.1186/s13640-016-0104-y
9. Fan, H., Sun, Y., Zhang, X., Zhang, C., Li, X., Wang, Y.: MR image segmentation based on improved variable weight multi-resolution Markov random field in undecimated complex wavelet domain. Chin. Phys. B (2021)

10. Venkatesan, R., Prabu, S.: Feature extraction from hyperspectral image using decision boundary feature extraction technique. In: Bansal, J.C., Das, K.N., Nagar, A.K., Deep, K., Ojha, A.K. (eds.) Soft computing for problem solving. vol 2, pp. 927–940. Springer, Singapore (2020). 10.1007/978-981-15-0184-5_79
11. Tripathi, A.K., Mukhopadhyay, S.: Removal of fog from images: a review. IETE Tech. Rev. **29**, 148–156 (2012)
12. Tan, R. T.: Visibility in bad weather from a single image. In: 2008 IEEE Conference on Computer Vision and Pattern Recognition, Anchorage, Alaska, USA, pp. 1–8. IEEE (2008)
13. Tang, K., Yang, J., Wang, J.: Investigating haze-relevant features in a learning framework for image dehazing. In: 2014 Proceedings of the IEEE Conference on Computer Vision and Pattern Recognition, San Juan, Puerto Rico, USA, pp. 2995–3000. IEEE (2014)
14. He, K., Sun, J., Tang, X.: Guided image filtering. IEEE Trans. Anal. Mach. Intell. **35**, 1397–1409 (2013)
15. Xu, Y., Wen, J., Fei, L., Zhang, Z.: Review of video and image defogging algorithms and related studies on image restoration and enhancement. IEEE Access **4**, 165–188 (2016)
16. Nishino, K., Kratz, L., Lombardi, S.: Bayesian defogging. Int. J. Comput. Vis. **98**, 263–278 (2012)
17. Kratz, L., Nishino, K.: Factorizing scene albedo and depth from a single foggy image. In: IEEE 12th International Conference on Computer Vision, Kyoto, Japan, pp. 1701–1708. IEEE (2009)
18. Bansal, B., Sidhu, J.S., Jyoti, K.: A review of image restoration based image defogging algorithms. Int. J. Graph. Process. **9**, 62–74 (2017)
19. Tyo, J.S., Rowe, M.P., Pugh, E.N., Engheta, N.: Target detection in optically scattering media by polarization-difference imaging. Appl. Opt. **35**, 1855–1870 (1996)
20. Treibitz, T., Schechner, Y.Y.: Polarization: beneficial for visibility enhancement? In: 2009 IEEE Conference on Computer Vision and Pattern Recognition, Miami, Florida, USA, pp. 525–532. IEEE (2009)
21. Schechner, Y.Y., Narasimhan, S.G., Nayar, S.K.: Instant dehazing of images using polarization. Comput. Vis. Recognit. **1**, 325–332 (2001)
22. Schechner, Y.Y., Narasimhan, S.G., Nayar, S.K.: Polarization-based vision through haze. Appl. Opt. **42**, 511–525 (2003)
23. Shwartz, S., Namer, E., Schechner, Y.Y.: Blind haze separation. In: 2006 IEEE Computer Society Conference on Computer Vision and Pattern Recognition, New York City, New York, USA, vol. 2, pp. 1984–1991. IEEE (2006)
24. Ancuti, C.O., Ancuti, C., Timofte, R., De Vleeschouwer, C.: O-HAZE: a dehazing benchmark with real hazy and haze-free outdoor images. In: Proceedings of the IEEE Conference on Computer Vision Pattern Recognition Workshops, Salt Lake City, Utah, USA, pp. 754–762. IEEE (2018)
25. Narasimhan, S.G., Nayar, S.K.: Vision and the atmosphere. Int. J. Comput. Vis. **48**, 233–254 (2002)
26. Zhang, N., Zhang, L., Cheng, Z.: Towards simulating foggy and hazy images and evaluating their authenticity. In: Liu, D., Xie, S., Li, Y., Zhao, D., El-Alfy, E.S. (eds.) ICONIP 2017. LNCS, vol. 10636, pp. 405–415. Springer, Cham (2017). https://doi.org/10.1007/978-3-319-70090-8_42
27. Qin, X., Wang, Z., Bai, Y., Xie, X., Jia, H.: FFA-Net: feature fusion attention network for single image dehazing. In: Proceedings of the AAAI Conference on Artificial Intelligence, New York, New York, USA, vol. 34, pp. 11908–11915. AAAI (2020)

28. Eigen, D., Puhrsch, C., Fergus, R.: Depth map prediction from a single image using a multi-scale deep network. In: Advances in Neural Information Processing Systems, pp. 2366–2374 (2014)
29. Li, B., et al.: Benchmarking single-image dehazing and beyond. IEEE Trans. Process. **28**, 492–505 (2019)

Predicting the Death of Road Accidents in Bangladesh Using Machine Learning Algorithms

Md. Abu Bakkar Siddik(✉), Md. Shohel Arman(✉), Afia Hasan(✉), Mahmuda Rawnak Jahan(✉), Majharul Islam(✉), and Khalid Been Badruzzaman Biplob(✉)

Department of Software Engineering, Daffodil International University, Dhanmondi, Dhaka 1207, Bangladesh
{abu35-1937,arman.swe,afia35-1060,jahan.swe,majharul35-1846, khalid}@diu.edu.bd

Abstract. Road accidents are now a common occurrence in our country. Every year thousands of people die in these accidents and thousands of people are crippled and cursed. Recently the level of road accidents has increased drastically. In this research paper, the authors discuss previous road accident history profoundly and predict death by applying the machine learning algorithm to get appropriate accuracy in Bangladesh. In this study, we had applied four classification models such as Decision Tree, K-Nearest Neighbors (KNN), Naïve Bayes and Logistic Regression to predict the death of road accidents in Bangladesh. The model was constructed, trained, and tested using the data from "Prothom Alo" newspaper, from which we collected 1237 road crash incidents. This research would be helpful for the policymakers and stakeholders related to the road to take the future steps with the highest accuracy of 88% in the Decision tree algorithm.

Keywords: Machine learning · Supervised learning · Classification and road accident

1 Introduction

In recent times, road accidents are considered one of the crucial problems in Bangladesh. Every day when we read the newspapers, we can see a series of accidents and deaths of so many people owing to accidents with buses, trucks, motorcycles and other vehicles. These accidents occur for various reasons. The roads in our country are very narrow. These streets are not straight as there are continuing turns at short distances. Reckless driving is responsible for bringing about accidents. Most of the drivers of our country are careless about the traffic rules and regulations that is why they recklessly drive their vehicles. Most of the drivers are not well concerned about their training and they do not have adequate knowledge of the traffic system. Not only drivers but also the Pedestrians are responsible for street accidents. Sometimes they try to cross the road here and there

M. Singh et al. (Eds.): ICACDS 2021, CCIS 1441, pp. 160–171, 2021.
https://doi.org/10.1007/978-3-030-88244-0_16

that causes the driver to occasionally lose his control to save pedestrians from accidents. Moreover, the condition of all vehicles is not as expected. Thus, many times accidents happen due to driving this faulty vehicle. Defective roads often cause the car to overturn, leading to major accidents.

The number of road accidents has increased highly compared to the last two years (Nischa, annual report 2019). In 2019, there were 1,599 more road accidents than in 2018. In 2018, 3,103 accidents occurred where 4,039 people died and 6,425 injured and in 2017, among the 3,349 road accidents 55,745 people died, and 6,908 were injured (Nischa, annual report 2019). Among 5,516 crashes, at least 7,855 people were killed and 13,330 injured in 2019(Jatri Kalyan Samity, annual report 2020). That means each day almost 21 or more people lost their lives on roads. The number of accidents remained almost the same from the previous year but the death toll increased by 8.07 percent (Jatri Kalyan Samity, annual report 2020). The Samity also claimed 5,514 road accidents occurred in 2018 and 77,221 died and 15,466 were injured in the country. In 2019, at least 5,227 people were killed while 6,953 were injured in 4,702 road accidents across the country (Nirapad Sarak Chai, annual report 2019).

The death toll from road accidents is actually the highest among low-income countries and Bangladesh is one of them (WHO, Global Status Report on Road Safety 2015). The financial cost of road accidents and the damage caused by them in the country is around Rs 40,000 crore a year. Due to these accidents, Bangladesh is losing two to three percent of the gross national product (GDP) per year (Accident Research Institute, BUET 2019). The loss of life and property in the accident and the traffic jam caused by the accident are hampering the economy.

Now, therefore, it is time to do research on road accidents. Researchers are trying to reduce the severity of previous accidents by analyzing the data. Many researchers rely solely on traditional research such as surveys and questionnaires to find appropriate results from such research. Because road accidents are unpredictable so it is challenging to observe them and it is difficult to get 100% accurate data. To overcome this challenge, researchers should follow any advanced method, for which machine learning should be perfect. Nowadays, machine learning is one of the vital fields of Artificial Intelligence from which we can get a comparatively better solution. Firstly, the data about the accidents have to be analyzed very well through machine learning, to determine the severity of an accident. There are a lot of advanced and dynamic methods in machine learning.

In this research paper, the authors analyzed data of road accidents through four supervised learning algorithm techniques such as Logistic Regression, Decision tree, K-Nearest Neighbors (KNN), Naive Bayes Classifier. Authors collected the data from the National Newspaper "Prothom Alo" of Bangladesh and classified this dataset into two classes for death prediction (Yes or No) and took four features to get the appropriate result such as Injured, Accident History, Responsible car and Victim car. The authors of this research paper had done the analysis of last seven years accident data (2013–2019) and applied these techniques to achieve the best solutions. From these techniques, the best accuracy was achieved in the Decision Tree algorithm with an accuracy of 88%, which will later help policymakers to reach better conclusions.

The other part of this research is designed as followed- Sect. 2 described the brief description of previous works that were related to this research paper. Section 3 presents the brief description of methodology of this research paper. Section 4 compares the dataset and results by comparing the four supervised algorithms based on their accuracy. Finally, Sect. 5 presents the future directions of this research and come to end with some important suggestions.

2 Literature Review

In Bangladesh, the authors did not find any death prediction-based research paper in this field. There exist some research papers regarding accident analysis for Bangladesh traffic.

In paper [1], the authors analyzed road accidents more deeply and used four types of machine learning algorithms. Those are Decision Tree, K-Nearest Neighbors (KNN), Naive Bayes and AdaBoost. They have gotten best performance by AdaBoost with 80% accuracy. By using these Fatal, Grievous, Simple injury and Motor Collision features with four catagories. The authors deepy classified brutality of accidents.

Authors proposed a model in the paper [2] that was developed by 8482 road accident incidents of the Lebanese Road Accidents Platform (LRAP) database. Authors used seven independent variables such as fatality occurrence, namely, crash type, injury severity, spatial cluster-ID, and crash time (hour). They used five mostly remarkable machine learning algorithms. And the best performance was gained by SMO. Firstly, the researchers of the paper [3] have analyzed the previously occurred different types of accidents in the accident-prone area. Later they determined the main factors associated with accidents such as weather, pollution, road structure. They predicted the root cause behind the fatality by using machine-learning approaches such as Naive Bayes, Decision Tree, K-Nearest Neighbors, and AdaBoost. Using Machine Learning Algorithm.

The authors of the paper [4], present a survey of various existing work related to accident prediction. These algorithms are Random Forest, Logistic Regression, K-Nearest Neighbors (KNN), and Decision Tree. Analyzed the various accident records, provided an appropriate prediction by applying multiple machine learning algorithms so that it could help from future concern. They used two types of data set: 2.25 million US accident records and 1.6 million UK accident records. For the first dataset, they applied Random Forest, Logistic Regression, and Decision Tree and for others they applied K-Nearest Neighbors (KNN).The prime purpose of this paper [5] is to predict the accident-prone areas by analyzing various factors of accidents. The authors used apriori of data mining technique and K-Means of machine learning concepts to identify the main factors of accidents. The accident-prone areas classified into two categories- High and Low. This research would help to diminish the accidents rate in India and help the Regional Transport Office to impose strict actions such as checking the license of the driver, conducting alcohol checks or always placing traffic police in such areas.

The purpose of the authors of this paper [6] is to implement AI in the various related fields of accidents such as traffic stream forecast, mishap expectation, and traveler place, and so on. The authors used machine-learning techniques such as Logistic Regression, Random Forest, Support Vector Classifier (SVM), Decision Tree, K-Nearest Neighbors (KNN), and K-Means to get better prediction. The proposed system would help to maintain the present traffic system in an appropriate way by determining the accident-prone region or non-accident prone region. This paper [7], present the welfare situation of Bangladesh with regard to accidents rate and fatalities rate, which have occurred every year and in terms of units, registered motor vehicle accidents and fatalities that have occurred over time have been evaluated. The author collected the data from police reports and analyzed it. The author analyzed the yearly registered motor vehicles in terms of accidents. The authors of this paper [8] aimed to predict road crash severity by applying Machine Learning Algorithms. Moreover, they suggested the random undersampling of the majority class (RUMC) technique for carrying imbalance accident data to predict the minority crash severity class appropriately. After high observation of the dataset, the authors proposed for evaluation purpose and for this they applied dynamic four different crash severity predictive models such as Random tree, k nearest neighbor, logistic regression, and random forest to get comparatively better accuracy. This paper [9] focused on applying machine-learning algorithms for classification and predicting injury severity of road accidents in Yemen. They used machine-learning to find out the main factors behind the accidents. Authors collected the data of 156 injured vehicles from two leading hospitals since August- October 2015 from 128 locations. By classifying the dataset into three categories as Severe, Serious, and Minor, they got the best performance in Random Forest Algorithm with 94.84% accuracy where they applied five machine learning classifiers: Decision Tree, Multilayer Perceptron, Support Vector Machine, Naïve Bayes, and Random Forest, based on road user, transport way to the emergency department (ED), accident action, total injured person, road type, and crash type. The researchers of the paper [10] have focused on the traffic system of Bangladesh to identify the acuity of accidents using those algorithms such as Decision Tree, K-Nearest Neighbors (KNN), Naïve Bayes, and Gadabouts. Among these four supervised machine learning algorithms, the best performance was achieved by Gadabouts where 43 thousand data were used and eleven main factors were selected. Also classified the severity of accidents into Fatal, Grievous, Simple Injury, and Motor Collision these four categories.

This research paper [11] was based on the data of west Texas in the United States where a large number of crashes occurred by young and inexperienced drivers. The top three factors for affecting road accident were Road class, speed limit, and first harmful event and others. Authors applied Label Encoder and XGBoost because those will be the best option for considering accuracy. The outcome of this research would be beneficial to promote rural teen driver traffic safety in West Texas in the United States.

In this paper [12], authors used a Polynomial regression algorithm to predict accident severity in India. They collected the data set from a government site of the Ministry of Road Transport and Highway (MORTH). The Author's main emphasis was to predict accident severity appropriately. In paper [13] authors analyzed previously occurred accidents in the locality of India and determined the most accident-prone area. They collected the data of road accidents from many institutions and government websites. To make the predictions of road accidents, they select the factors such as weather, pollution, road structure etc. In this study, we applied the most appropriate machine learning technique for classification. They used a Logistic Regression algorithm to get better accuracy.

The authors of the paper [14] aim to develop Machine learning based models which would identify the injury severity appropriately and which would help to spread awareness and safety among the people. Authors classified the injury severity into five classes.

Moreover, the features included driver behavior, light condition, roadway conditions and weather conditions and others. Authors used a hybrid approach such as K-means, Random Forest algorithm. The dataset has been taken from the IRTAD accident database in India. Authors also used Linear Regression algorithms for validation and comparative. For Fatal Injuries accidents Boyer Moore (BM) algorithm performed 1.08% and 1.9% more than random forest and linear regression. In paper [15], authors proposed a system that used the Image Processing technique, accidents have been detected and information has been collected from the CCTV footage. The information has been processed by Machine learning tools to detect possible accidents. In this paper, the authors aimed to analyze vehicles and their state where the accidents have concluded. There were 800 images in the dataset and the clustering algorithm has performed with an accuracy of 93%. The images have been classified into mild or severe accidents and mail has been sent with location to the nearby hospitals for taking the further steps and the Convolutional Neural Network algorithm has been used for classifying images. As a result, the system will take a quick response to help the people who are involved in accidents.

3 Proposed Methodology

Authors used supervised classification Machine learning algorithms such as Decision Tree, K-Nearest Neighbors (KNN), Naïve Bayes and Logistic Regression. These algorithms had applied based on some features like Injured, Accident History, Victim car and Responsible car (Fig. 1).

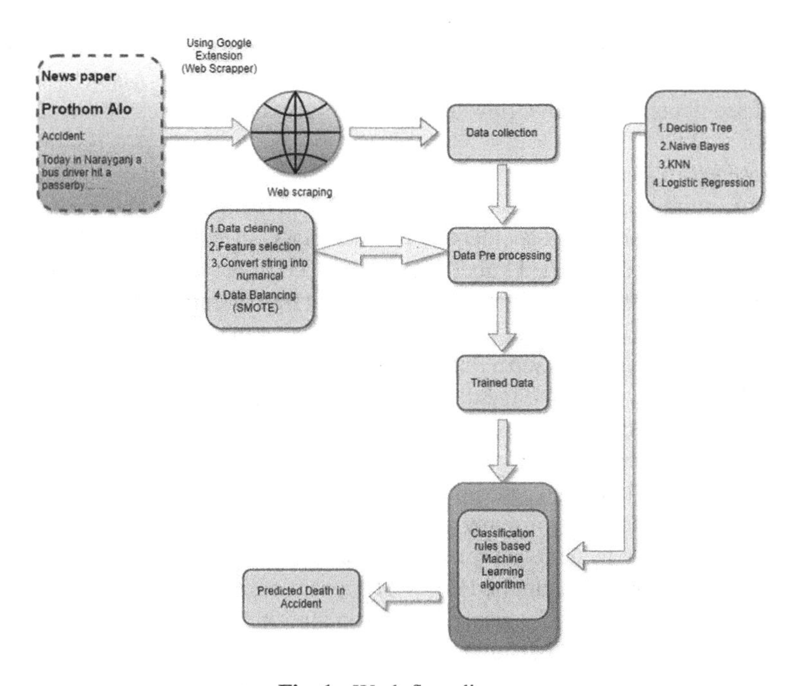

Fig. 1. Work flow diagram

3.1 Data Collection

Appropriate predictions are possible only through accurate datasets. The authors of this research paper had collected the last seven years (2013–2019) road accident data in Bangladesh from a national newspaper in Bangladesh by web scraping using Google extension "Web Scraper" tool. Authors used 70% of data for training the models and remaining 30% for testing. Almost 1237 road accidents data had been used in this research paper. Through which it was possible to get better accuracy from applied models.

Link of the dataset: https://github.com/DipuSiddik/Road-Accident-Data.

3.2 Data Pre-processing

It was challenging to get 100% accurate data of road accidents and besides getting better performance using machine-learning algorithms needs accurate and appropriate data of accidents. This dataset had been collected from "Prothom Alo" News Paper, a leading newspaper in Bangladesh. For that reason, here the dataset was in an unstructured way. The authors prepared this dataset based on some features and converted the dataset from string to numerical. Firstly, we had collected eleven features that were connected to the road accidents and about 13–14% of accident data was inconsistent and these were not informative. Then we selected the top four features using the "Feature Importance" technique that were directly involved in road accidents and which could help to predict death in an accident. When the dataset was ready, notice that our dataset was an imbalanced dataset. Then we used the oversampling technique "SMOTE" for balancing the data. After balancing the data, we used classification machine learning algorithms based on these features such as Injured, Victim Car type, Responsible Car type and Accident History to classify the death (Yes or No) in an accident. By applying the classification algorithm the authors got 88% accuracy in the Decision Tree algorithm.

3.3 Classification Rules Based Machine Learning Algorithms

3.3.1 Decision Tree

Decision Tree algorithm a supervised learning algorithm where it used for solving classification problems. By learning simple decision rules inferred from prior data, a training model has been created using a Decision Tree for predicting the class or value of the target variable. To predict a class label in the Decision tree record we start from the root of the tree.

Firstly, a root node is required for the construction and the best attribute is selected by the acquired approach and the sub nodes are then developed by the decision taken in relation to the status of excellence selected at each node. When each of the nodes have been defined by its achieved condition and then class is taken at the end of the Root node and then it is determined as a leaf. This process of action has sequentially worked until a class has identified at the last stage of the node [1].

3.3.2 K-Nearest Neighbor (KNN)

One of the simplest Supervised Machine Learning algorithms is K-Nearest Neighbor (KNN) and which depends on the similarity of feature. It analyzed distance of all remaining data points from unknown data points and sorted the data points in small to large sizes (Ascending Order) according to the value of distance. The first K-number of points would take from the sorted data point. Out of these K-numbered data points, the class with the most number of points has to identify the unknown data point as that class.

In this research paper, we applied the Euclidean distance formula to measure distance. Euclidean distance is calculated by,

$$d(x, x') = \sqrt{(x_1 - x'_1)^2 + (x_2 - x'_2)^2 + \ldots + (x_n - x'_n)^2} \tag{1}$$

3.3.3 Naive Bayes

Naïve Byes is also a classification technique. Based on several attributes, it predicts the probability of classes. It allocates the new class to the highest probability.

Naïve Bayes works on.

$$\text{posterior probaility, } P(b) = \frac{P(a)*P(a)}{P(b)} \tag{2}$$

Here the probability of "a" being true given that "b" is true.

3.3.4 Logistic Regression

Logistic Regression is a two-class classification and predictive analysis algorithm. When the dependent variables are a binary number 0 and 1, then Logistic Regression can be an appropriate regression analysis to conduct. By applying logistic regression, we can to developed relationship between one dependent binary variable and one or more nominal, ordinal, interval, or ratio-level independent variables. It only works when the data is linearly separate in higher dimensions. As a result, the author has used this algorithm here. Our goal in logistic regression would be to keep the output from the hypothesis within 0 to 1.

If the output is less than 0.5, then the output of our logistic regression will be 0, and if it is equal to or greater than 0.5, then the last output of the logistic regression will be 1. Here the Logistic Regression equations: $g(W^T X)$ = Sigmoid Function and $g(z)$ = Sigmoid/ Logistic Function.

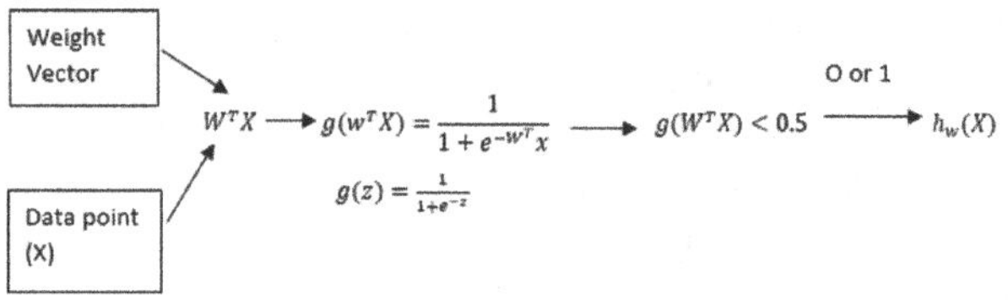

Weight Vector

Data point (X)

$$W^T X \longrightarrow g(w^T X) = \frac{1}{1 + e^{-w^T x}} \longrightarrow g(W^T X) < 0.5 \xrightarrow{\text{0 or 1}} h_w(X) \tag{3}$$

$$g(z) = \frac{1}{1+e^{-z}}$$

4 Dataset Discussion and Result Analysis

4.1 Dataset Discussion

We also did experiments on our dataset's features such as Injured, Responsible car Type, Victim car Type Accident History and Death. We observed from our dataset that injuries in road accidents occurred 906 times out of 1237 times and did not happen in the remaining 331 times (Fig. 2). Moreover, in Fig. 3, we observed that 43% accidents have occurred by passenger cars and the remaining 57% accidents have occurred by freight cars out of 1237 road accidents.

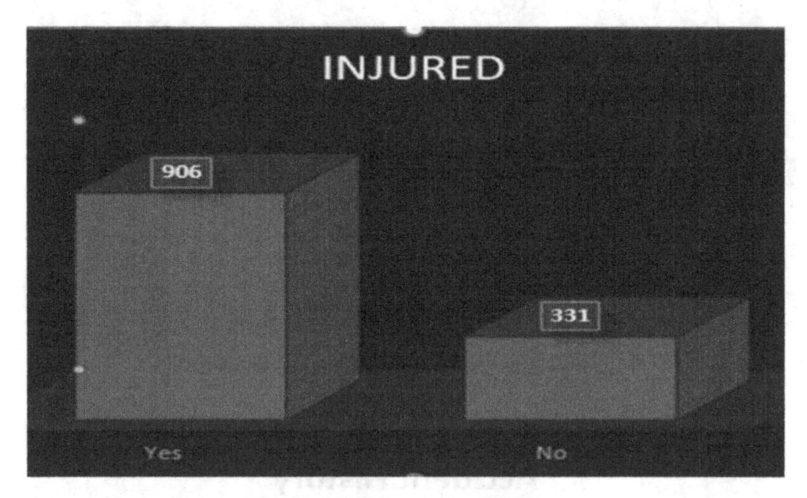

Fig. 2. Injured ratio in accidents.

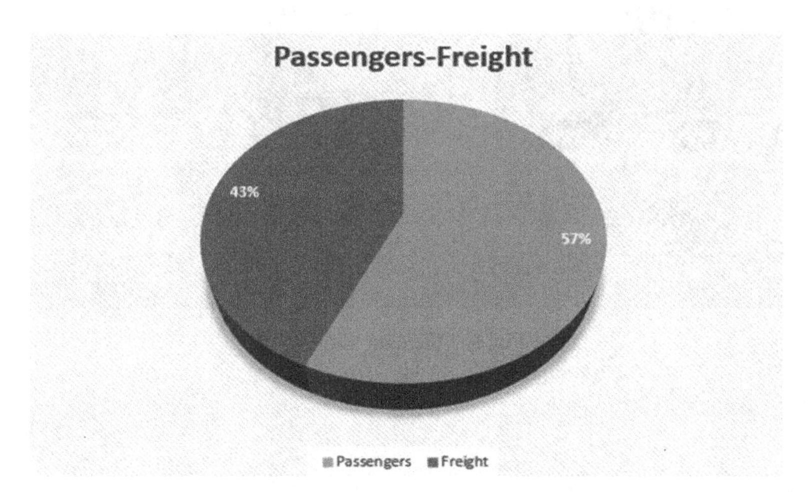

Fig. 3. Responsible car ratio for number of accidents

We also observed that 59% of passenger cars were victims of these accidents and the rest 41% of freight cars (Fig. 4). On the other hand (Fig. 5) 85% of accidents occurred due to collision and the remaining 15% were overturned and others. Which seems to be a big problem. Moreover, we had tried to find out the death ratio from our dataset that among 1237 accidents in our country death happened 1205 times. Therefore, it is alarming for our country (Fig. 6).

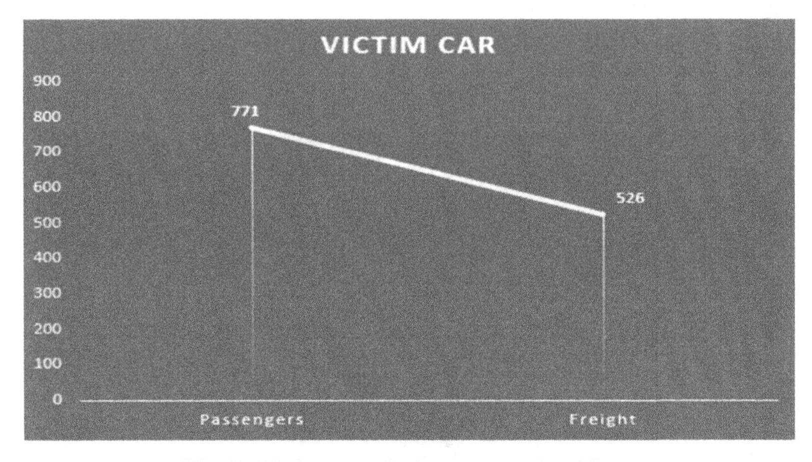

Fig. 4. Victim car ratio for number of accidents

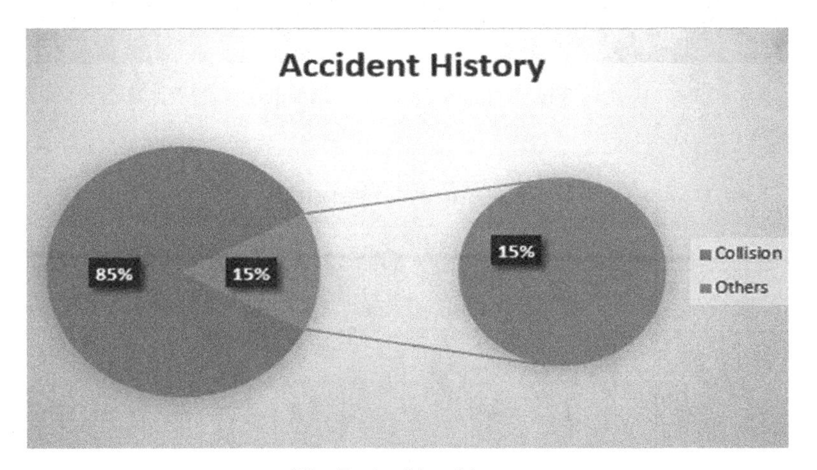

Fig. 5. Accident history

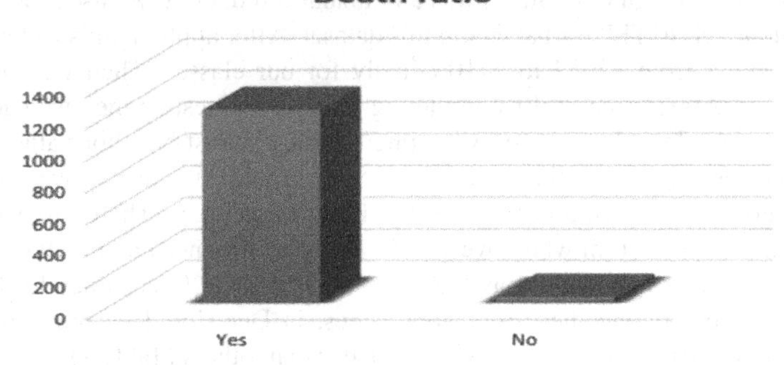

Fig. 6. Deaths ratio in accidents

4.2 Result Analysis

In this research paper, we had classified our dataset into two classes (Yes, No) in death based on four approaches of machine learning algorithms. Among these four machine-learning approaches, we got maximum 88% accuracy in the Decision Tree Algorithm.

Table 1. Evaluation metrics

Algorithm	Accuracy (%)	Precision (%)	Recall (%)	F1 score (%)
Decision Tree	85%	88%	85%	85%
KNN	84%	84%	84%	84%
Naïve Bayes (Bernoulli)	77%	77%	77%	77%
Logistic Regression	83%	84%	82%	82%

After applying K-Fold Cross Validation technique.

Table 2. K-fold cross validation

Algorithm	Result
Decision Tree	88%
K-Nearest Neighbors (KNN)	80%
Naïve Bayes (Bernoulli)	79%
Logistic Regression	82%

We observed that deaths occurred in most of the accidents in our dataset. We classified our dataset into two classes of death (Yes or No) based on four features. We had already mentioned that our dataset contains 1237 road accidents data. Of these 1205 deaths occurred. Though our dataset was an imbalanced dataset, we used the oversampling technique "SMOTE" for balancing this dataset. After applying this technique, our dataset converted from 12037 to 2410 equally for our classes. Then we split it into trains and tests. After successfully completing training and testing then we had applied machine-learning algorithms. First, we applied the K-Nearest Neighbor algorithm and got better performance with the accuracy of 84% (Table 1) and then we applied the Logistic Regression Algorithm and got 83% (Table 1) accuracy. Therefore, we applied the Decision Tree algorithm where we got the best performance among all of the algorithms 85% and finally in Naive Bayes Algorithm, we got 77% accuracy for Bernoulli approach. Besides, among these four approaches, in Decision Tree precision (88%), Recall (85%) and F1 score (85%) was much better than others (Table 1).

After that, we applied K-Fold cross validation technique among our four classification based models and we got the highest accuracy in Decision tree (88%), Logistic Regression (82%), K-Nearest Neighbors (80%), Naïve Bayes (79%) respectively (Table 2). Besides this, authors did observation with the features of the dataset as well as tried to detect their impact to predict the death in a road accident.

5 Conclusions and Recommendations

5.1 Findings and Contributions

In developing countries like Bangladesh, road accidents are a terrible problem. Road accidents are now a major problem around the world. Recently researchers predict that by 2030, road accidents will be the fifth leading cause of death worldwide [4]. Road accidents simultaneously adversely affect social and economic spheres. So now, it is time to plan for our traffic system so that the death rate can reduce. Here, the authors had analyzed the road accident dataset and shown a possibility of death in road accidents with the help of machine learning algorithms. After splitting the dataset into 70% for training and 30% for testing, the best accuracy is achieved by Decision Tree model 85% and got highest precision 88%, Recall: 85%, F1 Score: 85% in Decision Tree classification model based on four features such as Injured, Responsible car, victim car and Accident History. After cross validation the authors got 88% accuracy in the Decision Tree model. In future, it will assist the policymakers and Bangladesh Police for investigation so that the premature death can be identified.

5.2 Limitations

There were some limitations while the author worked on a large number of unstructured dataset and it was challenging to convert the unstructured dataset into a structured way. We know accidents are unpredictable so it was difficult to predict the death by using the previous year's accidents data.

5.3 Recommendations for Future Work

In this paper, deep learning approaches had not been explored with this dataset so it can be used in future with adding some other features to get better acquisition and performance. Moreover, in future it will be very useful and beneficial.

References

1. Labib, Md.F., Rifat, A.S., Hossain, Md.M., Das, A.K., Nawrine, F.: Road accident analysis and prediction of accident severity by using machine learning in Bangladesh. In: 7th International Conference on Smart Computing & Communications (ICSCC) (2019)
2. Ghandour, A.J., Hammoud, H., Al-Hajj, S.: Analyzing factors associated with fatal road crashes: a machine learning approach. National Council for Scientific Research (CNRS) (2020)
3. Dabhade, S., Mahale, S., Chitalkar, A., Gawhad, P., Pagare, V.: Road accident analysis and prediction using machine learning. Int. J. Res. Appl. Sci. Eng. Technol. (IJRASET) **8**(I), 100–103 (2020)
4. Venkat, A., Gokulnath, M., Guru Vijey, K.P., Thomas, I.S., Ranjani, D.: Machine learning based analysis based analysis for road accident prediction. Int. J. Emerg. Technol. Innov. Eng. **6**(02), 31–37 (2020)
5. Sridevi, N., Keerthana, M.V., Pal, M.V., Nikshitha, T.R., Jyothi, P.: Road accident analysis using machine learning. Int. J. Res. Eng. Sci. Manag. **3**(5), 859–861 (2020)
6. Sumanth, P., Anudeep, P.S., Divya, S.: Analysis of machine learning algorithm with road accidents data sets. Int. J. Eng. Manag. Res. **10**(2), 20–25 (2020)
7. Hossen, Md.S.: Analysis of road accidents in Bangladesh. Am. J. Transp. Logist. 1–19 (2019)
8. Fiorentini, N., Losa, M.: Handling imbalanced data in road crash severity prediction by machine learning algorithms, pp. 1–24. Multidisciplinary Digital publishing Institute (MDPI) (2020)
9. Al-Moqri, T., Haijun, X., Namahoro, J.P., Alfalahi, E.N., Alwesabi, I.: Exploiting machine learning algorithms for predicting crash ınjury severity in Yemen: hospital case study. Appl. Comput. Math. 155–164 (2020)
10. Patil, A.S., Bindu, A.N., Nikitha, Y.R.: Smart accident zone detection system. Int. J. Res. Eng. Sci. Manag. **3**(2), 218–221 (2020)
11. Lin, C., Wu, D., Liu, H., Xia, X., Bhattarai, N.: Factor identification and prediction for teen driver crash severity using machine learning: a case study, pp. 1–16. Multidisciplinary Digital publishing Institute (MDPI) (2020)
12. Arora, Y.K., Kumar, S., Tiwari, U.K., Singhal, S., Kumar, V.: Accident severity in India. Int. J. Innov. Technol. Explor. Eng. (IJITEE) **8**(12S3), 55–58 (2019)
13. Lakshmi, T.N., Venugopala Rao, M.S.: Predicting the traffic accident severity using machine learning. Complex. Int. J. (CIJ) **24**(01), 248–253 (2020)
14. Pavan Karthik, G., Sneha, B., Sudalaimuthu, T.: Analysis of road accidents using machine learning. Int. J. Adv. Sci. Technol. **29**(06), 6717–6723 (2020)
15. Nancy, P., Dilli Rao, D., Babuaravind, G., Bhanushree, S.: Highway accident detection and notification using machine learning. Int. J. Comput. Sci. Mob. Comput. (IJCSMC) **9**(3), 168–176 (2020)

Numerical Computation of Finite Quaternion Mellin Transform Using a New Algorithm

Khinal Parmar(✉) and V. R. Lakshmi Gorty

NMIMS, MPSTME, Vile Parle – West, Mumbai 400056, India
{khinal.parmar,vr.lakshmigorty}@nmims.edu

Abstract. In this paper, the authors have defined finite quaternion Mellin transform. The table of finite quaternion Mellin transform for some standard functions is presented. Using hat functions for the approximation of function, we get a stable and efficient algorithm for numerically computing finite quaternion Mellin transform. Efficiency of the proposed method with numerical examples are illustrated. Graphical representation and error analysis is also given.

Keywords: Finite quaternion Mellin transform · Hat function · Numerical computation · Graphical

1 Introduction

Integral transforms provides a powerful operational methods for solving initial value problems and initial-boundary value problems for linear differential and integral equations. Quaternion Mellin transform has a wide range of applications related to the problems in mathematical physics viz. solving boundary value problems and Euler – Cauchy differential equations of quaternion-valued functions. Quaternion operations have extended applications in electrodynamics, instrumentation, potential wedge and general relativity. The analytical evaluations of such transform are rare and complex, so numerical methods are important to study.

In this study, we define finite quaternion Mellin transform. Using hat functions for the approximation of function in the finite quaternion Mellin transform; we get a stable and efficient method for numerically evaluating finite quaternion Mellin transform. Graphical representation and error analysis is also studied.

2 Literature Review

Hjalmar Mellin introduced Mellin transform in 1896. In [13] author studied some properties of fractional order Mellin transform. Authors in [12] introduced a new type of q-Mellin transform called q-finite Mellin transform and studied an inversion formula and q-convolution product. This study was carried forward to a new extension in [9]. Two-dimensional Mellin transform in quantum calculus are studied in [1]. Authors in [3] have discussed the generalized Mellin transform and its properties with examples

M. Singh et al. (Eds.): ICACDS 2021, CCIS 1441, pp. 172–182, 2021.
https://doi.org/10.1007/978-3-030-88244-0_17

and applications. In recent literature [5, 8, 10], some new theorems of Mellin transform are established and applied in various fields.

In 1853, W. R. Hamilton developed quaternions [11]. The necessity of enlarging the operations on three-dimensional vectors to include multiplication and division led Hamilton to introduce the four-dimensional algebra of quaternions. In 1993, Ell [2] introduced the concept of quaternion in integral transform by defining quaternion Fourier transform. In [7], the author introduced the quaternionic Fourier-Mellin transform. In a recent study, the authors in [14] have defined the one-dimensional quaternion Mellin transform.

In [4], the Mellin-transform method for the exact calculation of one-dimensional definite integrals was introduced. It usually requires laborious calculations involving two functions, the generalized hypergeometric function and the Meijer G-function, which may be related to the Mellin-transform method and arise frequently.

The lack of methods defined for numerical computation of finite quaternion Mellin transform in the literature motivated us for the present work.

3 Preliminaries

In quaternions, every element is a linear combination of a real scalar and three imaginary units i, j and k with real coefficients.

Let q be a quaternion defined in

$$\mathbb{H} = \{q = x_0 + ix_1 + jx_2 + kx_3 : x_0, x_1, x_2, x_3 \in \mathbb{R}\} \tag{1}$$

which is the division ring of quaternions, where i, j, k satisfy Hamilton's multiplication rules (see, e.g. [6])

$$ij = -ji = k,\ jk = -kj = i,\ ki = -ik = j,\ i^2 = j^2 = k^2 = ijk = -1. \tag{2}$$

The quaternion conjugate of q is defined by

$$\bar{q} = x_0 - ix_1 - jx_2 - kx_3;\ \ x_0, x_1, x_2, x_3 \in \mathbb{R}. \tag{3}$$

The norm of $q \in \mathbb{H}$ is defined as

$$|q| = \sqrt{q\bar{q}} = \sqrt{x_0^2 + x_1^2 + x_2^2 + x_3^2}. \tag{4}$$

If $f \in L^1(\mathbb{R}; \mathbb{H})$, then the function is expressed as

$$f(x) = f_0(x) + if_1(x) + jf_2(x) + kf_3(x). \tag{5}$$

Definition 1. One-dimensional quaternion Mellin transform (QMT) of $f \in L^1(\mathbb{R}; \mathbb{H})$ (5) exists within the strip $a_1 < Re(s) < a_2$ is defined in [14] as

$$\mathcal{M}_q\{f\}(s) = \tilde{f}(s) = \int_0^\infty f(t)t^{-s-1}dt, \tag{6}$$

where $\tilde{f}(s) = \tilde{f_0}(s) + i\tilde{f_1}(s) + j\tilde{f_2}(s) + k\tilde{f_3}(s)$.

3.1 Hat Functions

Hat functions are defined on the domain [0, 1]. The interval [0,1] is divided into n subintervals $[ih, (i+1)h]$, $i = 0, 1, 2, \cdots, n-1$, of equal lengths h where $h = 1/n$. The family of first $(n+1)$ hat functions is defined as follows [15]:

$$\psi_0(t) = \begin{cases} \frac{h-t}{h}, \; 0 \leq t < h, \\ 0, \; otherwise; \end{cases}$$
$$\psi_i(t) = \begin{cases} \frac{t-(i-1)h}{h}, \; (i-1)n \leq t < ih, \\ \frac{(i+1)h-t}{h}, \; ih \leq t < (i+1)h, \\ 0, \; otherwise; \end{cases} \tag{7}$$
$$\psi_n(t) = \begin{cases} \frac{t-(1-h)}{h}, \; 1-h \leq t < 1, \\ 0, \; otherwise. \end{cases}$$

The function $f \in L^1(\mathbb{R}, \mathbb{H})$ can be approximated as given in [15]

$$f(t) \approx \sum_{i=0}^{i=n} f_i \psi_i(t) = f_0\psi_0(t) + f_1\psi_1(t) + \cdots + f_n\psi_n(t) \tag{8}$$

where $f_i = f(ih)$, for $i = 0, 1, 2 \cdots n$ and $h = 1/n$.

4 Method of Evaluation

Analogous to [12], we define finite quaternion Mellin transform as follows:

Let $a > 0$ and let f be a quaternion-valued function defined on $[0, a]$. The finite quaternion Mellin transform (FQMT) of f is defined as

$$\mathcal{M}_q^a\{f(x)\} = F(s) = \int_0^a a^{-s} f(x) x^{-s-1} dx, \tag{9}$$

within the strip $a_1 < Re(s) < a_2$.

For $a = 1$, we have

$$\mathcal{M}_q^1\{f(x)\} = F(s) = \int_0^1 f(x) x^{-s-1} dx. \tag{10}$$

For this study, we will consider $a = 1$ and the interval will be [0, 1]. We will prepare the table of FQMT of some standard functions for the interval [0, 1] (Table 1).

Now we will write the algorithm for numerical computation of FQMT using hat function. We may approximate $f(x)$ using (8) as:

$$f(x) \approx \sum_{i=0}^{i=n} f(ih)\psi_i(x). \tag{11}$$

Table 1. Table of FQMT.

$f(x)$	$F(s)$ for [0, 1]
x	$1/(1-s)$
e^{-x}	$\Gamma(-s,0)-\Gamma(-s,1)$
xe^{-x}	$\Gamma(1-s,0)-\Gamma(1-s,1)$
$\sin x$	$(\Gamma(-s,0)-\Gamma(-s,1))\sin(-s\pi/2)$
$\cos x$	$(\Gamma(-s,0)-\Gamma(-s,1))\cos(-s\pi/2)$
e^{-x^2}	$\frac{1}{2}(\Gamma(-s/2,0)-\Gamma(-s/2,1))$
$\log x$	$-\frac{1}{s^2}\Gamma(2)$

Using (10) and (11), we get

$$F(s) \approx \int_0^1 \left(\sum_{i=0}^{i=n} f(ih)\psi_i(x)\right) x^{-s-1} dx. \tag{12}$$

Equation (12) can be represented as:

$$F(s) \approx \sum_{i=0}^{i=n} f(ih) \int_0^1 \psi_i(x) x^{-s-1} dx. \tag{13}$$

Now using (7), we get

$$\begin{aligned} F(s) \approx & f(0) \textstyle\int_0^h \left(\frac{h-x}{h}\right) x^{-s-1} dx \\ & + \sum_{i=1}^{n-1} f(ih) \Big[\textstyle\int_{(i-1)h}^{ih} \left(\frac{t-(i-1)h}{h}\right) x^{-s-1} dx \\ & + \textstyle\int_{ih}^{(i+1)h} \left(\frac{(i+1)h-t}{h}\right) x^{-s-1} dx \Big] \\ & + f(nh) \textstyle\int_{1-h}^{1} \left(\frac{t-(1-h)}{h}\right) x^{-s-1} dx. \end{aligned} \tag{14}$$

This algorithm gives us the nth approximate value of finite quaternion Mellin transform of $f(x)$ and is denoted by $F_n(s)$.

The errors between the exact value and numerically approximate value are computed as $E(s) = (Exact\ FQMT\ F(s) - Approximate\ FQMT\ F_n(s))$.

4.1 Program Code

Program code for numerically computing the approximate value of finite quaternion Mellin transform is given below for function as e^{-x} by using Python 3.7.9.

```
import math
import scipy.integrate
import numpy as np
from math import e

def f(x):
  value = math.exp(-x)
    return value
n=100
h=0.01
for s in range(-20,0):
    f1=lambda x:((0.01-x)/0.01)*pow(x,(-s-1))
    i1=scipy.integrate.quad(f1,0,0.01)
    value1=f(0)*i1[0]
    f2=lambda x:((x-((i-1)*0.01))/0.01)*pow(x,(-s-1))
    f3=lambda x:((((i+1)*0.01)-x)/0.01)*pow(x,(-s-1))
    sum=0
    for i in range(1,n-1):
        sum=sum+f(i*h)*(scipy.integrate.quad(f2,(i-
1)*0.01,i*0.01)[0]+scipy.integrate.quad(f3,i*0.01,(i+1)*0
.01)[0])
    value2=sum
    f4=lambda x:((x-0.99)/0.01)*pow(x,(-s-1))
    i2=scipy.integrate.quad(f4,0.99,1)
    value3=f(1)*i2[0]
    M=value1+value2+value3
    print(str(s)+" => " + str(M))
```

5 Numerical Illustrations

Example 1. Consider the quaternion-valued function given as

$$f(x) = e^{-x} + ix + jxe^{-x} + ke^{-x^2} \tag{15}$$

for which we will do the numerical computation of FQMT. The FQMT of $f(x)$ exists in the strip $s \leq -1$. The approximate numerical computation is obtained by using Eq. (14). The approximation is done by considering $n = 100$ as done in [15].

For real component:

The exact finite quaternion Mellin transform is given by

$$F(s) = \Gamma(-s, 0) - \Gamma(-s, 1). \tag{16}$$

Table 2. Calculation of real component.

s	Exact value	Approximate value	Error
−1	0.6322	0.6284	0.0038
−2	0.2643	0.2606	0.0037
−3	0.1607	0.1569	0.0038
−4	0.1140	0.1103	0.0037
−5	0.0878	0.0843	0.0035
−6	0.0713	0.0678	0.0035
−7	0.0599	0.0564	0.0035
−8	0.0517	0.0482	0.0035
−9	0.0454	0.0419	0.0035
−10	0.0404	0.0370	0.0034
−11	0.0365	0.0331	0.0034
−12	0.0332	0.0298	0.0034
−13	0.0305	0.0272	0.0033
−14	0.0281	0.0249	0.0032
−15	0.0262	0.0229	0.0033
−16	0.0244	0.0212	0.0032
−17	0.0229	0.0197	0.0032
−18	0.0216	0.0184	0.0032
−19	0.0204	0.0172	0.0032
−20	0.0193	0.0162	0.0031

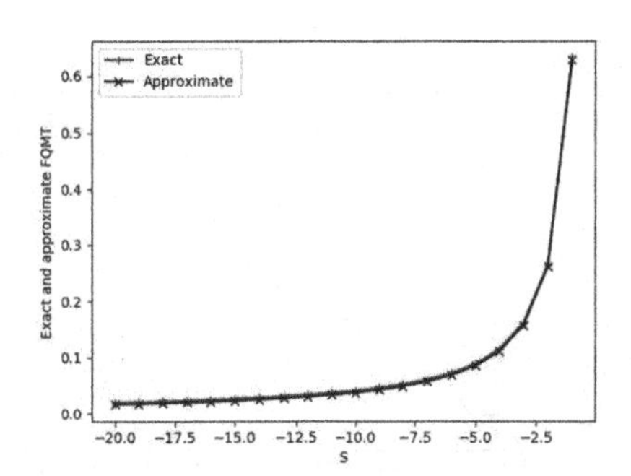

Fig. 1. Graph: real component

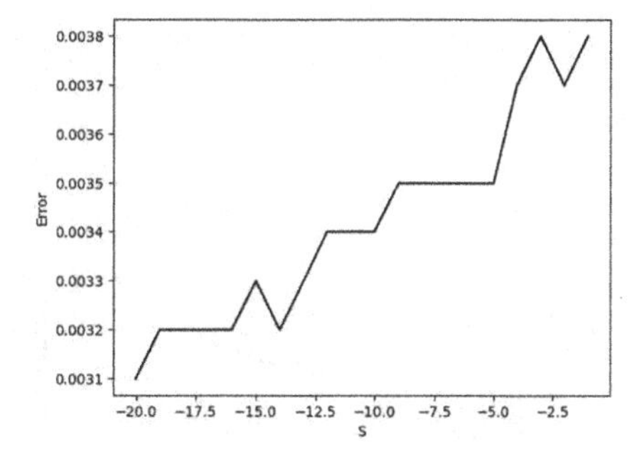

Fig. 2. Error: real component

For i^{th} component:
The exact finite quaternion Mellin transform is given by

$$F(s) = 1/(1 - s). \tag{17}$$

Table 3. Calculation of ith component.

s	Exact value	Approximate value	Error
−1	0.5000	0.4901	0.0099
−2	0.3333	0.3235	0.0098
−3	0.2500	0.2403	0.0097
−4	0.2000	0.1904	0.0096
−5	0.1667	0.1572	0.0095
−6	0.1429	0.1334	0.0095
−7	0.1250	0.1157	0.0093
−8	0.1111	0.1019	0.0092
−9	0.1000	0.0909	0.0091
−10	0.0909	0.0819	0.0090
−11	0.0833	0.0744	0.0089
−12	0.0769	0.0681	0.0088
−13	0.0714	0.0626	0.0088
−14	0.0667	0.0579	0.0088
−15	0.0625	0.0539	0.0086
−16	0.0588	0.0503	0.0085
−17	0.0556	0.0471	0.0085
−18	0.0526	0.0443	0.0083
−19	0.0500	0.0417	0.0083
−20	0.0476	0.0394	0.0082

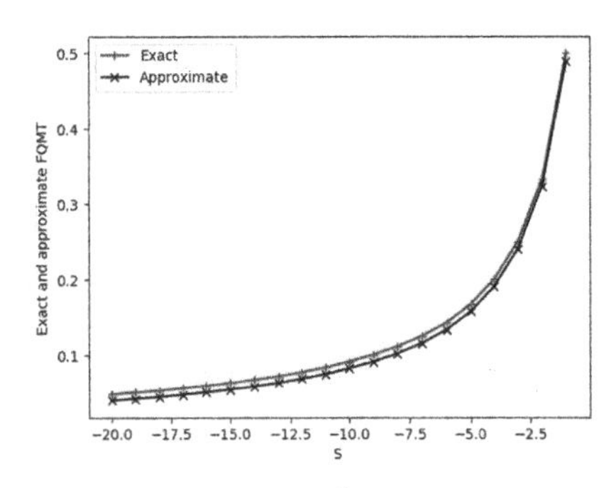

Fig. 3. Graph: i^{th} component

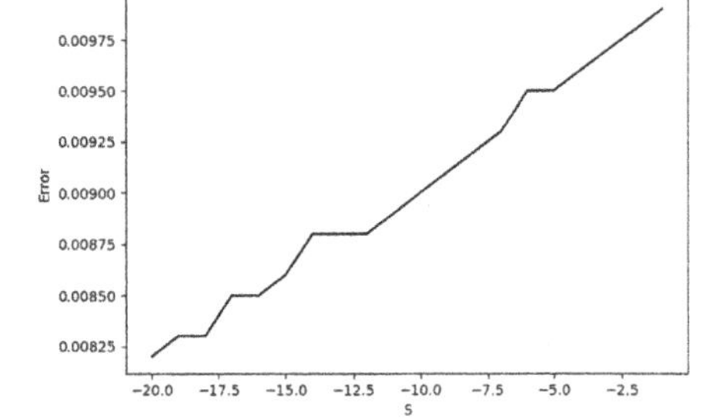

Fig. 4. Error: i^{th} component

For j^{th} component:
The exact finite quaternion Mellin transform is as follows:

$$F(s) = \Gamma(1 - s, 0) - \Gamma(1 - s, 1). \tag{18}$$

Table 4. Calculation of jth component.

s	Exact value	Approximate value	Error
−1	0.2643	0.2606	0.0037
−2	0.1607	0.1569	0.0038
−3	0.1140	0.1103	0.0037
−4	0.0878	0.0843	0.0035
−5	0.0713	0.0678	0.0035
−6	0.0599	0.0564	0.0035
−7	0.0517	0.0482	0.0035
−8	0.0454	0.0419	0.0035
−9	0.0404	0.0370	0.0034
−10	0.0365	0.0331	0.0034
−11	0.0332	0.0298	0.0034
−12	0.0305	0.0272	0.0033
−13	0.0281	0.0249	0.0032
−14	0.0262	0.0229	0.0033
−15	0.0244	0.0212	0.0032
−16	0.0229	0.0197	0.0032
−17	0.0216	0.0184	0.0032
−18	0.0204	0.0172	0.0032
−19	0.0193	0.0162	0.0031
−20	0.0180	0.0153	0.0027

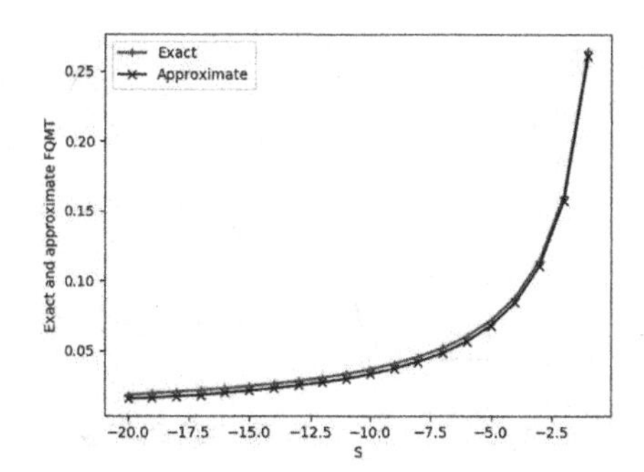

Fig. 5. Graph: j^{th} component

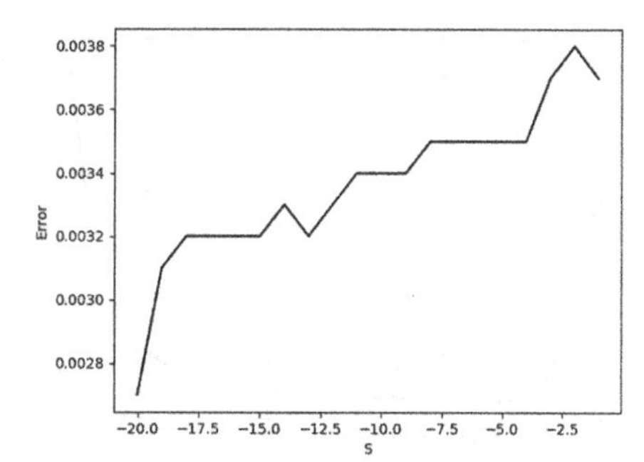

Fig. 6. Error:j^{th} component

For k^{th} component:
The exact finite quaternion Mellin transform is given by

$$F(s) = \frac{1}{2}(\Gamma(-s/2, 0) - \Gamma(-s/2, 1)). \tag{19}$$

Table 5. Calculation of kth component.

s	Exact value	Approximate value	Error
−1	0.7468	0.7431	0.0037
−2	0.3161	0.3123	0.0038
−3	0.1895	0.1858	0.0037
−4	0.1321	0.1285	0.0036
−5	0.1003	0.0967	0.0036
−6	0.0804	0.0767	0.0037
−7	0.0668	0.0632	0.0036
−8	0.0570	0.0535	0.0035
−9	0.0496	0.0462	0.0034
−10	0.0439	0.0405	0.0034
−11	0.0394	0.0360	0.0034
−12	0.0357	0.0323	0.0034
−13	0.0326	0.0292	0.0034
−14	0.0299	0.0267	0.0032
−15	0.0278	0.0245	0.0033
−16	0.0258	0.0226	0.0032
−17	0.0242	0.0210	0.0032
−18	0.0227	0.0195	0.0032
−19	0.0214	0.0182	0.0032
−20	0.0202	0.0171	0.0031

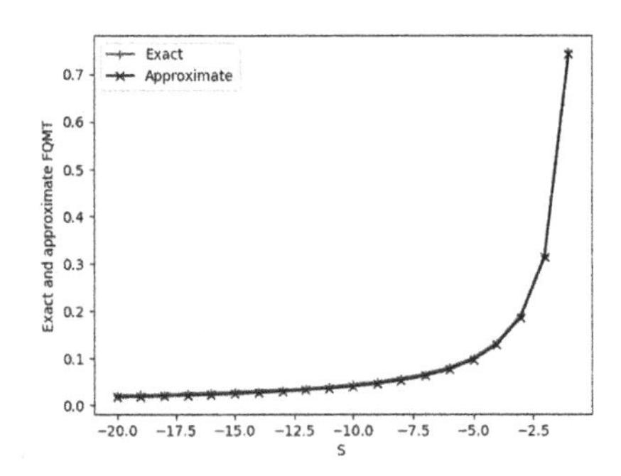

Fig. 7. Graph: k^{th} component

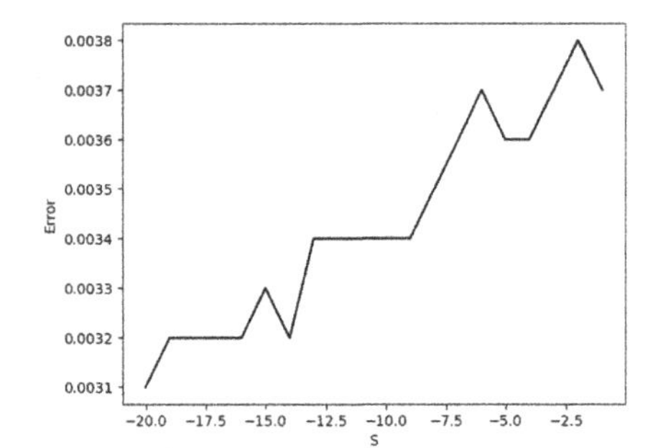

Fig. 8. Error: k^{th} component

The second column of the Table 2, Table 3, Table 4 and Table 5 gives numerical exact values of finite quaternion Mellin transform of all the components of $f(x)$ calculated using the analytical formula (10). The third column of the Table 2, Table 3, Table 4 and Table 5 gives numerical approximate values of finite quaternion Mellin transform of all the components of $f(x)$ calculated using the new algorithm derived in this study.

The comparison between the exact FQMT and approximate FQMT is shown for each component in Fig. 1, Fig. 3, Fig. 5, Fig. 7. The last column of Table 2, Table 3, Table 4 and Table 5 gives the error that we get in approximate values as compared to the exact values. It is evident that for all the values of s the error are appreciably small. The error between the exact and the approximate value is shown in Fig. 2, Fig. 4, Fig. 6, Fig. 8. respectively. Thus the algorithm derived in this study is efficient in numerically computing finite quaternion Mellin transform.

6 Conclusion

Our concern in this article has been to establish an algorithm to numerically compute finite quaternion Mellin transform efficiently. For this aim, first we have developed a finite quaternion Mellin transform. Then a new algorithm is obtained to numerically compute finite quaternion Mellin transform using hat function. The program code for this algorithm is given. The efficiency of the proposed method is illustrated by a numerical example where we have calculated the values for finite quaternion Mellin transform of a function by using the exact analytical formula and numerical approximate algorithm. Graphical representation and error analysis is also studied.

Acknowledgments. The authors would like to appreciate the efforts of editors and reviewers. This study received no specific grant from the government, commercial, or non-profit funding organizations.

References

1. Brahim, K., Riahi, L.: Two dimensional Mellin transform in quantum calculus. Acta Math. Sci. **38**, 546–560 (2018)
2. Ell, T.A.: Quaternion-Fourier transforms for analysis of two-dimensional linear time-invariant partial differential systems. In: Proc. of the 32nd Conf. on Decision and Control, pp. 1830–1841. IEEE (1993)
3. Erdoğan, E., Kocabaş, S., Dernek, N.: Some results on the generalized Mellin transforms and applications. Konuralp J. Math. **7**, 175–181 (2019)
4. Fikioris, G.: Integral evaluation using the Mellin transform and generalized hypergeometric functions: tutorial and applications to antenna problems. IEEE Trans. Antennas Propag. **54**, 3895–3907 (2006)
5. Fortin, J., Giasson, N., Marleau, L., Pelletier-Dumont, J.: Mellin transform approach to rephasing invariants. Phys. Rev. D **102**, 036001 (2020)
6. Hitzer, E.: Quaternion Fourier transform on quaternion fields and generalizations. Adv. Appl. Clifford Algebras **17**, 497–517 (2007)
7. Hitzer, E.: Quaternionic Fourier-Mellin transform. arXiv preprint arXiv:1306.1669 (2013)
8. Ilie, M., Biazar, J., Ayati, Z.: Mellin transform and conformable fractional operator: applications. SeMA J. **76**(2), 203–215 (2018). https://doi.org/10.1007/s40324-018-0171-3
9. Jain, P., Basu, C., Panwar, V.: Finite Mellin transform for (p, q) and symmetric calculus. J. Pseudo-Differential Oper. Appl. **11**, 1595–1620 (2020)
10. Laurita, C.: A new stable numerical method for Mellin integral equations in weighted spaces with uniform norm. Calcolo **57**(3), 1–26 (2020). https://doi.org/10.1007/s10092-020-00374-6

11. Lewis, A.C.: Chapter 35 – William Rown Hamilton, Lectures on quaternions (1853), Landmark Writings in Western Mathematics. Elsevier Science (2005)
12. Nefzi, B., Brahim, K., Fitouhi, A.: On the finite Mellin transform in quantum calculus and application. Acta Math. Sci. **38**, 1393–1410 (2018)
13. Omran, M., Kiliçman, A.: On fractional order Mellin transform and some of its properties. Tbil. Math. J. **10**, 315–324 (2017)
14. Parmar, K., Lakshmi Gorty, V.R.: One-dimensional quaternion Mellin transforms and its applications. Proc. Jangjeon Math. Soc. **24**, 99–112 (2021)
15. Tripathi, M., Singh, B.P., Singh, O.P.: Stable numerical evaluation of finite Hankel transforms and their application. Int. J. Anal. **2014**, 1–11 (2014). https://doi.org/10.1155/2014/670562

Predictive Modeling of Tandem Silicon Solar Cell for Calculating Efficiency

S. V. Katkar[1](✉), K. G. Kharade[1](✉), N. S. Patil[2], V. R. Sonawane[2](✉), S. K. Kharade[1](✉), and R. K. Kamat[1](✉)

[1] Shivaji University, Kolhapur, Maharashtra, India
rkk_eln@unishivaji.ac.in
[2] MVP's KBT College of Engineering, Nashik, Maharashtra, India

Abstract. With the increase in the population of the world, there is an increase in energy requirements. We are making maximum use of a conventional source of energy, i.e., Sun. But even after, there is a considerable loss of power due to the limited capability of energy storage devices and the little energy conversion devices. In this paper, we have focused on energy conversion devices, namely solar cells. We have considered the most popular solar cell called Tandem Silicon Solar cell out of various solar cells. Tandem Silicon is a second-generation solar cell and it has a lower manufacturing cost. We have developed a predictive model using Artificial Neural Network by integrating the Matrix Laboratory (MATLAB) tool. Each model consists of several predictors or variables. Therefore, by collecting the data for relevant variables, a statistical model can be developed. By selecting the right metrics, the output of a predictive model is measured and compared. In this investigation, we have used an artificial neural network for modeling the properties of Tandem Silicon solar cells. For training the neural network, we have used three input parameters and one output parameter. Short circuit current, open-circuit voltage, Fill Factor is considered input of the neural network, and efficiency is regarded as the neural network's output. ANN is the optimal technology for modeling the solar cell.

Keywords: ANN · MATLAB · Modeling · Simulation · Silicon solar cell

1 Introduction

People worldwide use large amounts of energy every day for different purposes like heating, cooling, transportation, and cooking. When this energy is in the nonrenewable form, then this causes air and water pollution. So we can avoid this pollution by using a renewable source of energy [2]. Fossil fuels are responsible for the pollutions, and they will be exhausted over the next few years. To find an essential solution to accomplish the energy requirement is a must. The ever-increasing energy requirement of the world can be fulfilled by using solar energy [4]. It is cost-effective as well as a copious source of energy. A silicon-based solar cell is most demanding because it gives high efficiency, which is better than other solar cells. In this study, we concentrate on improving the

M. Singh et al. (Eds.): ICACDS 2021, CCIS 1441, pp. 183–194, 2021.
https://doi.org/10.1007/978-3-030-88244-0_18

efficiency of a solar cell [6]. Finding the highest efficiency solar cells using experiments in the lab takes lots of effort, workforce, money, and time-consuming processes. We have repeated this process until we get an accurate result. When we simulate this solar cell by changing values and finding out correct deals and using these values in the experiment reduces time, saves money, and gives a perfect result in a short period. A Tandem solar cell is a layered solar cell or stack of different solar cells. These are individuals or connected in series and are also known as multi-junction solar cells. Other types of tandem solar cells are grouped according to the material they used. Tandem solar cells are grouped into three types.

- Organic solar cells
- Inorganic solar cells
- Hybrid solar cells

In organic solar cells contains polymer materials that have a power conversion efficiency of less than 10%. Inorganic solar cells are used into space applications like a satellite. It gives high efficiency, but it is costly. Hybrid solar cells are the third type of tandem solar cells. This type of solar cell provides high efficiency at a low cost. Perovskite solar cells are included in this type of solar cell, which is used for industrial purposes.

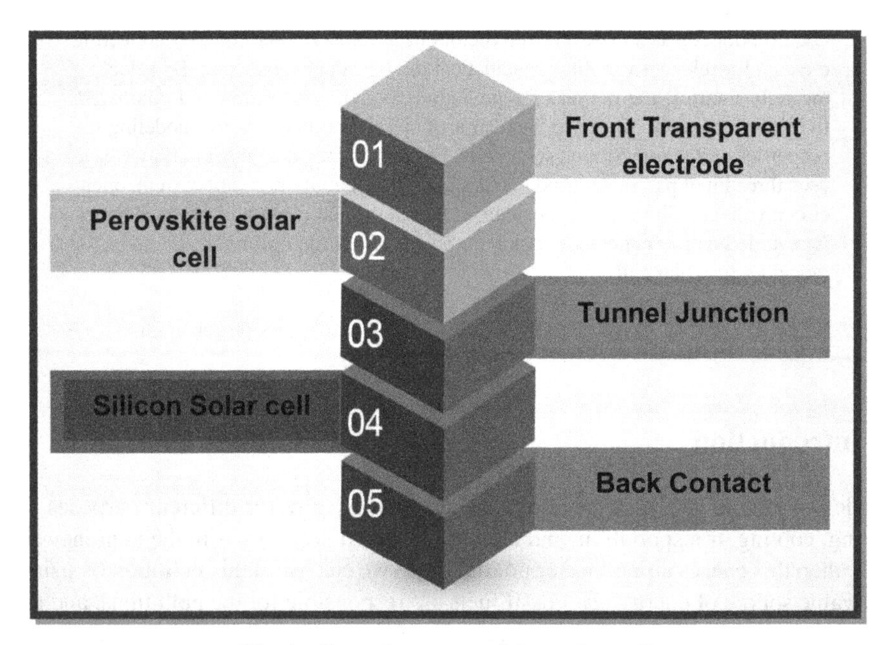

Fig. 1. General structure of the tandem cell

Above Fig. 1 shows the structure of the tandem cell. It looks like a stack. The stack's element contains the Front transparent electrode, Perovskite solar cell, Tunnel junction, Silicon solar cell, and the last part is back contact. Our research focuses on

Tandem silicon solar cells. This solar cell provides the highest efficiency at a low cost. In this research, we have used an artificial neural network. Here we use MATLAB for simulation. There are different learning approaches in artificial intelligence like supervised learning, unsupervised learning, and reinforcement learning. We prefer a supervised learning approach for modeling and simulation of Tandem silicon solar cells [7]. Supervised learning gives you a powerful method for using machine language to identify and process data. You can use labeled data, and a data set categorized to infer a learning algorithm by supervised learning. By integrating machine learning algorithms, the data set is used to predict other unlabeled data. The goal is to make sense of data for particular measurements in supervised learning. The supervised learning process in neural network algorithms is enhanced by continuously monitoring the model's resulting performance and fine-tuning the system to get closer to its target accuracy. Two factors that depend on the degree of accuracy obtainable are the available data, and the algorithm used [10].

We used different algorithms like Scaled Conjugate Gradient and Levenberg-Marquardt algorithm to train the MATLAB data using an artificial neural network [32]. But the result generated by the Levenberg-Marquardt algorithm is more optimized than the Scaled Conjugate Gradient algorithm. So we train our data by using the Levenberg-Marquardt algorithm. The Levenberg-Marquardt calculation joins two minimization techniques: the gradient descent strategy and the Gauss-Newton strategy. The squared mistakes' total is decreased by refreshing the parameters in the steepest-plummet bearing [23]. In the Gauss-Newton strategy, the whole of the squared errors is reduced by accepting the least-squares work as locally quadratic and finding the base of the quadratic. The Levenberg-Marquardt system acts progressively like an inclination plummet technique when the parameters are long from their ideal worth. They work increasingly like the Gauss-Newton technique when the parameters are near their excellent value. This report depicts these techniques and shows programming to tackle nonlinear least-squares bend fitting issues [14].

2 Literature Review

This combinatorial computational science approach empowers hypothetical high-throughput screening for materials plan. Koyama et al. has explored the combinatorial computational chemistry to design various chemical compounds, especially the Lithium-Ion battery [5]. Suzuki et al. have earlier adopted a similar approach to creating Lithium-ion batteries [20]. Kubo et al. have reported the use of computational chemistry for High-Throughput Screening of Metal Sulfide Catalysts for the CO Hydrogenation Process [9]. Martin A. Green [10] said that the latest explosion insists on thin-film solar cell modules. Because of the deficiency of silicon supply thin-film, solar cells have great demand in the recent market. Using thin-film modules instead of silicon reduces up to 50% of development cost [11]. Kishorkumar V. Khot [7] investigates that using the cost-effective chemical bath deposition method of bismuth sulfoselenide thin films has been synthesized. (Bi2(S1 − xSex)3) thin films are nanocrystalline. By measuring a solar simulator's performance, thin-film 0.3845% efficiency has been found [8]. According to Mohan Singh and Harminder Kaur (2016) suggested, the artificial neural network approach plays a vital role in prediction. ANN is inspired by human beings' biological neurons, which work similarly as the human brain works. For a material scientist, it helps solveproblems. ANN provides different techniques that give optimum results for the issues. This research mainly focuses on the methods that help create ANN for material science [12]. S. Khodakarimi et al. reported that polymer solar cells had used bulk heterojunction to study the charge convey mechanism. This monte Carlo simulation approach was used. The simulated parameters used are the diffusion coefficient and a short circuit current [6]. According to H. Parmar [12], the source of converting energy into a photovoltaic electricity system is most beneficial. Solar energy is the most sustainable source of energy. It produces energy, i.e., electric power, in requisites of current and voltage. The photovoltaic system helps create electrical energy, which is environmentally friendly. The researcher has been performed the modeling of a photovoltaic system using an artificial neural network. ANN is used to train the network, which helps reduce the genuine biological efforts for developing the result. Artificial intelligence was used to solve real-world problems with minimum efforts, for this artificial neural network was used to reduce human work. This neural network helps predict cases for different climatic conditions [13]. Jakub Mozaryn [1] reports the neural network approach for predicting energy storage parameters. Here for forecast, they consider the parameters which are used for recharging the energy storage cycle. They also present the experimental results for the proposed methodology [1, 3]. K.G. Kharade et al. [13] shows the use of MATLAB for simulation purpose. He simulated CZTS solar cell by implementing artificial neural network. By changing the number of hidden neuron they train the neural network and observed the results. At hidden neuron number five they find approximate results. For training the neural network they use three input yields and one output yield. After training he found that predicated results and experimental results have hairline difference. So he conclude that the this tool is most useful for predicating purposes because it reduces cost of material wastage which is used for trial and error basis. Also it reduces the man power and time [4]. S.V. Katkar et al. [16] represents the use of soft computing approach for modeling as well as simulation of solar cell. They use solar material i.e. cadmium sulpho selenide. They use its properties for simulation purpose like band gap, thickness of material, crystalline size, microstrain density [17]. S.V. Katkar et al. [16] developed java framework for finding unknown values of solar cell. Here they prefer java for creating application because java has an advantage like you write your code ones and run it everywhere i.e. it is platform indepedant. It is secure and has an automatic garbage collector so they use this language for their research. This application is useful for material scientist for finding the missing values. This application reduces the efforts taken by material scientist during their experiments and also causes to save time and money [18]. T.D. Dongale et al. [20] reports the modeling the thermister by implementing the neural network in MATLAB. For modeling nickel magnise oxide material was considered for thermister. Here nickel content, room temperature and oxalic acid properties were considered as feed in parameters, nickel and magnese acetate were considered as feed out parameters for neural network. Here they observed that at minimum number of hidden neuron they found optimized results.

3 Research Methodology

During this research, data were collected from exploratory databases availed by the various material scientist. Collected data was refined based on required parameters. To train the neural network supervised learning approach was used. The reason for the selection of the supervised learning approach was the parameter list decided during the research. The data set was trained with different supervised learning algorithms, and the result was compared with exploratory data [15]. Levenberg–Marquardt (LM) algorithm was finalized for further data set training based on the product. The dataset was trained, and results were generated and considered as a predictive result. This result was validated again with the actual product, and a conclusion was made.

4 Artificial Neural Network Modeling of Tandem Silicon Solar Cell

Here we have implemented an artificial neural network of modeling the properties of the solar cell. This network is made up of hidden layers and output layers. The hyperbolic tangent sigmoid transfer function and the linear transfer function are used for hidden layers and output layers. For training the given architecture to the feed-forward algorithm, Levenberg Marquardt is used. The network consists of a hidden layer and a layer of outputs [17]. The hyperbolic tangent sigmoid transfer function is used for the hidden layer, while the output layer is used with the linear transfer function. The feed-forward algorithm Levenberg Marquardt is used for the preparation of the present architecture (Table 1).

Table 1. Tabular representation of input and output parameters [21]

Name of the Solar Cell	Input Parameter			Output Parameter
	Short Circuit Current	Open circuit Voltage	Fill Factor	Efficiency
Silicon Solar cell	7.7567	1.3944	0.6924	7.4902
	7.9065	1.3946	0.6924	7.6357
	8.0540	1.3954	0.6925	7.7831
	8.0445	1.3953	0.6925	7.7735
	8.0540	1.3954	0.6925	7.7831
	4.9121	1.3887	0.6920	4.7211
	3.4676	1.3820	0.6916	3.3145
	2.6244	1.3757	0.6912	2.4957
	8.0540	1.3954	0.6925	7.7831
	9.2982	1.3986	0.6927	7.7831

5 Performance of System at Diverse Numbers of Hidden Neurons

The following observations are found at different numbers of hidden neurons by varying Training percentage, validating percentage, and testing percentage. The following things were observed that at different number of neurons for calculation of the training dataset by considering the correlation coefficient is '1' at the time of the validation dataset is 1, and the testing dataset is 1 [18, 27] (Table 2).

- The all-inclusive correlation coefficient is 0.9925 if at '5' hidden neurons
- The all-inclusive correlation coefficient is 0.97165 at '10' hidden neurons
- The all-inclusive correlation coefficient is 0.99766 at '15' hidden neurons
- The all-inclusive correlation coefficient is 0.98655 at '20' hidden neurons

Table 2. Implementation of the system at various number of hidden neuron

No. of Hidden Neurons	Training Percentage (%)	Dataset	MSE	Validation Percentage (%)	Testing Percentage (%)	Correlation Coefficient	Average correlation coefficient
5	65%	Training	$3.70661e^{-8}$	20%	15%	$9.99999e^{-1}$	0.99215
		Validation	1.77516e-1			$9.99667\ e^{-1}$	
		Testing	1.67619e-1			$1.0000\ e^{-0}$	
10	75%	Training	8.42147e-4	10%	5%	9.99900e-1	0.99125
		Validation	2.00315e-0			$1.0000\ e^{-0}$	
		Testing	1.85202e-0			$1.0000\ e^{-0}$	
15	70%	Training	1.80338e-26	20%	10%	$9.99999e^{-1}$	0.99437
		Validation	4.53615e-9			$1.0000\ e^{-0}$	
		Testing	7.70935e-1			$1.0000\ e^{-0}$	
20	85%	Training	1.90935e-5	10%	5%	$9.99999\ e^{-1}$	0.99578
		Validation	7.70420e-2			$1.0000\ e^{-0}$	
		Testing	3.97775e-1			$9.99999\ e^{-1}$	

The above table shows the number of hidden neurons with its training percentage, validation percentage, testing percentage, Mean square error, and correction coefficient [28]. It also shows the average correlation coefficient of each number of hidden neurons. The value of mean square error and mu () of the network is very negligible, which shows ANN's effectiveness in modeling Bi2(S1-xSex)3 thin-film solar cells [19]. The following figure shows the fitting function view of the neural network. Here we have found accurate results at 20 number of the hidden neuron. diagram shows the three input parameters and one output parameter. The middle part shows the hidden layer and output layer. Here weight are given to the activation function i.e. sigmode activation function and then it generated the results woth respect to the number of hidden neurons. The value of hidden neuron '20' gives the most significant result of Tandem silicon solar cells [26] (Fig. 2).

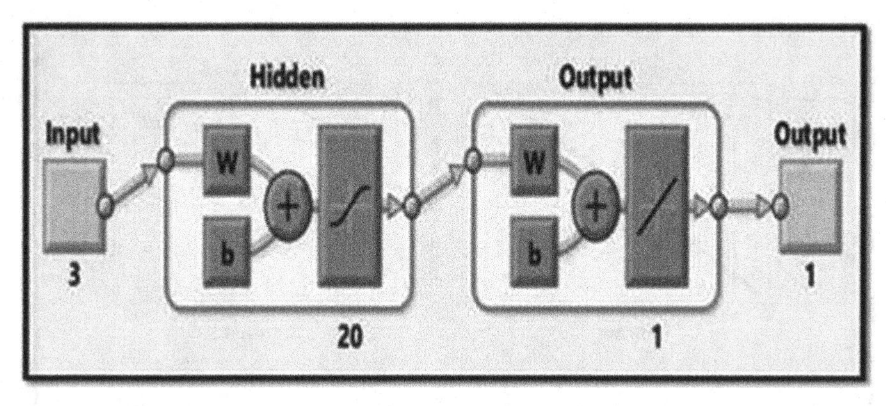

Fig. 2. Function fitting view of a neural network for 20 hidden neurons

Table 3. Predicted efficiency of silicon solar cell at hidden neuron value '20'.

Exploratory efficiency	Predicated efficiency	Difference between exploratory output and predicated output
10.5638	10.56200	0.0018
10.5638	10.55800	0.0058
10.5559	10.55110	0.0048
10.5557	10.55556	0.00014
10.5556	10.55489	0.00071
9.8855	9.885500	0.0000
9.0090	9.00400	0.0050
7.7831	7.77450	0.0050
7.7831	7.76280	0.0086

The Table 3 showcases the hairline difference between exploratory output and predicated output, and it confirms the said model works best for this material [29].

The following graph shows that at "20" number of hidden neurons, Tandem silicon solar cell [22].

Figure 3 represents the correlation coefficient of the neural network. Here for training dataset correlation coefficient is one, for validation 1, testing 1 and overall correlation coefficient is 0.99578 which is closer to one. which shows the optimization of the results [25] (Fig. 4).

Figure 5 shows the information about the error found at hidden neuron 20 blue color represents the training dataset, green color indicates the validation dataset, red color indicates the testing dataset and yellow color indicates the zero error phase [24].

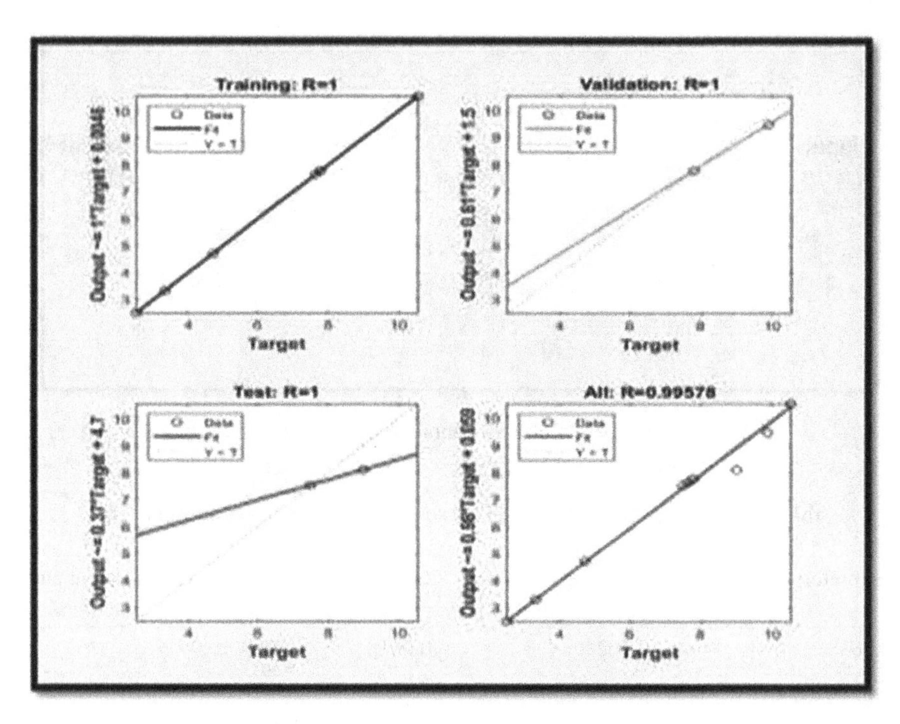

Fig. 3. Correlation coefficient as an outcome for the given network

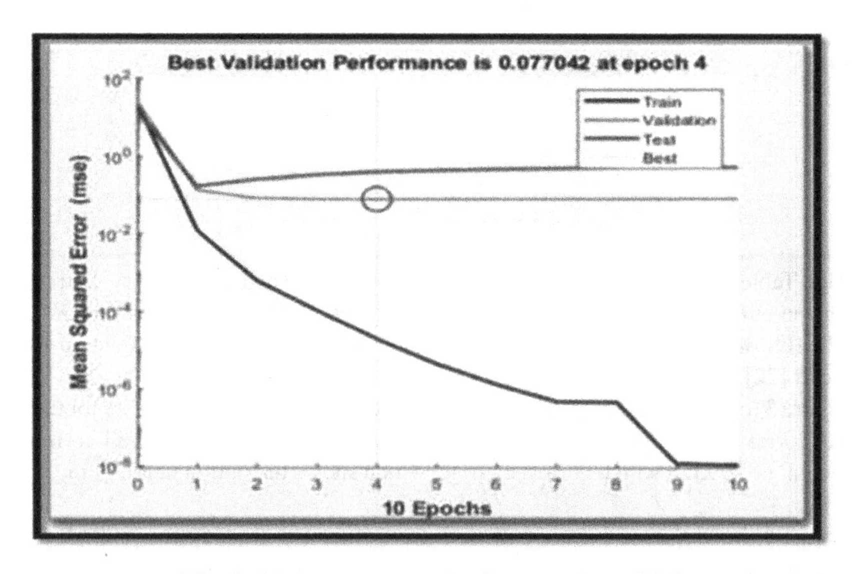

Fig. 4. Means square error for the network provided

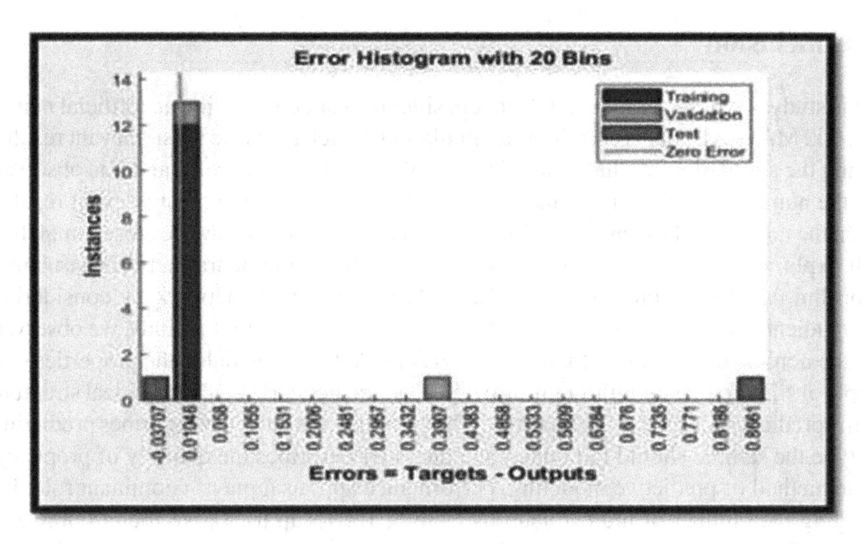

Fig. 5. Error histogram for the tandem silicon solar cell at hidden neuron "20".

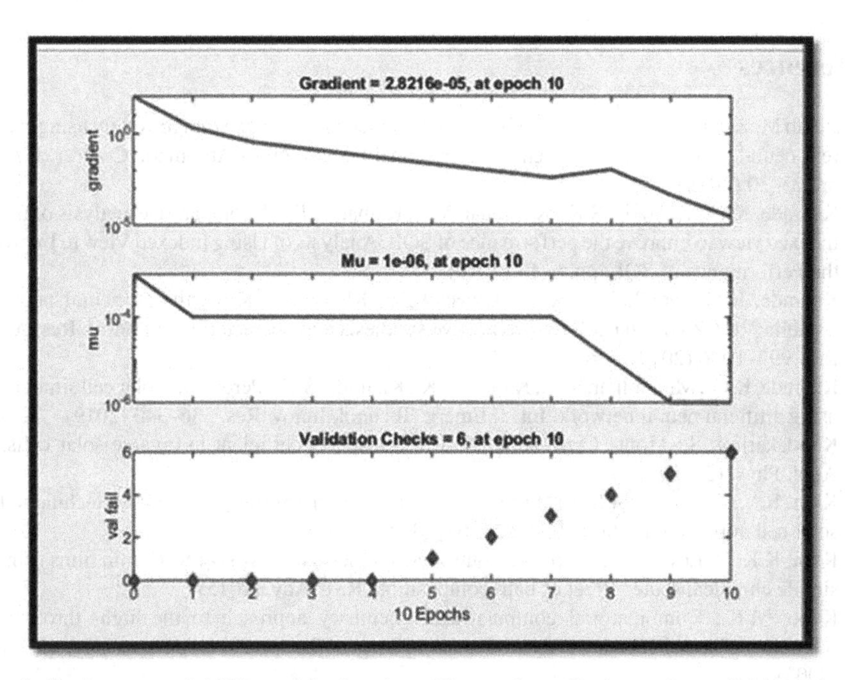

Fig. 6. Performance of ANN model of tandem silicon solar cell when values of hidden Neuron "20" for gradient, mu (), and validation checks parameters of the network

Figure 6 represents the performance of tandem silicon solar cell. Here value of gradient is 2.821e−05 at epoch 10. The momentum update i.e. mu value is 1e−06 at epoch 10. Validation is 6 at epoch 10. The mu value is negligible which shows the effectiveness of the neural network.

6 Conclusion

In this study, we have simulated the tandem silicon solar cell by using an artificial neural network. MATLAB tool is used for this simulation, which gives the most relevant results. During the simulation, we have varied the number of hidden neurons, and it is observed that the number of hidden neurons should be increased to get the most relevant results. When the number of hidden neurons was '20', the obtained results were closely matched with exploratory outcomes. The following figure shows the neural network structure, algorithm used for training the network, and progress of the network by considering mu, gradient, validation checks, and performance parameters. In this study, we observed that the nonlinear features of Tandem silicon solar cells. We modeled the properties of a couple of silicon solar cells through an artificial neural network. ANN is an ideal structure for the prediction of solar cell properties. The classification category describes predicting the type the sample should fall under, and the latter describes the quantity of prophecy. In the method of predictive modeling, performance appraisal plays a dominant role. By changing the number of hidden neurons from 5, 10, 15 up to 30, we found optimized results at 20 hidden neurons. Here we found the mean square error and mu of the network were minimal, confirming the neural network's effectiveness.

References

1. Jakub Mozaryn, A.C.: Prediction of selected parameters of energy storage system using recurrent neural network. In: Conference International Federation of Automatic Control (IFAC), pp. 23–30 (2018)
2. Kharade, K.G., Kharade, S.K., Kumbhar, V.S., Kamat, R.K.: A comparative analysis of using indexed view to improve the performance of SQL. Analysis of Using Indexed View to Improve the Performance of SQL, pp. 6–13 (2020)
3. Kharade, K.G., Mudholkar, R.R., Kamat, R.K., Kharade, S.K.: Artificial neural network modeling of CZTS solar cell for predicative synthesis and characterization. Int. J. Res. Anal. Rev. 997–1006 (2019)
4. Kharade, K.G., Mudholkar, R.R., Kamat, R.K., Kharade, S.K.: Perovskite solar cell simulation using artificial neural network. Int. J. Emerg. Technol. Innov. Res. 336–340 (2019)
5. Khodakarimi, S.: Monte Carlo simulation of transport coefficient in organic solar cells. J. Appl. Phys. (2016)
6. Khot, K.V.: Synthesis of SnS2 thin film via non vacuum arrested precipitation technique for solar cell application. Mater. Lett. 23–26 (2016)
7. Khot, K.K.V.: Low temperature and controlled synthesis of Bi2(S1-x Sex)3thin films using a simple chemical route: effect of bath composition. RSC Adv. (2015)
8. Kubo, M.K.: Combinatorial computational chemistry approach to the high- throughput screening of metal sulfide catalysts for CO hydrogenation process. Energy Fuels, 857–861 (2003)
9. Mahfoud, A.M.: Effect of temperature on the GaInP/GaAs tandem solar cell performances. Int. J. Renew. Energy Res. 629–634 (2015)
10. Green, M.: Thin-film solar cells: review of materials, technologies and commercial status. J. Mater. Sci. Mater. Electron. **18**(S1), 15–19 (2007). https://doi.org/10.1007/s10854-007-9177-9
11. Md.W.Shah, R.L.: Design and simulation of solar PV model using Matlab/Simulink. Int. J. Sci. Eng. Res. (2016)

12. Parmar, H.: Artificial neural network based modelling of photovoltaic system. Int. J. Latest Trends Eng. Technol. 50–59 (2015)
13. Kharade, S.K., Kharade, K.G., Katkar, S.V., Kamat, R.K.: Simulation of dye synthesized solar cell using artificial neural network. Emerg. Trends Eng. Res. Technol. **1**, 73–86 (2020)
14. Kharade, S.K., Kamat, R.K., Kharade, K.G.: Artificial neural network modeling of MoS2 supercapacitor for predicative synthesis. Int. J. Innov. Technol. Explor. Eng. 554–560 (2019)
15. Katkar, S.V., Kharade, K.G., Kharade, S.K., Kamat, R.K.: An intelligent way of modeling and simulation of WO3 for supercapacitor. Recent Stud. Math. Comput. Sci. 109–117 (2020)
16. Katkar, S.V., Kamat, R.K., Kharade, K.G., Kharade, S.K., Kamath, R.S.: Simulation of Cd(SSe) solar cell using artificial neural network. Int. J. Adv. Sci. Technol. 2583–2591 (2019)
17. Katkar, S.V., Dongale, T.D., Kamat, R.K.: Calculation of electrical parameters of solar cell using java based framework. J. Sci. Technol. (2017)
18. ShanZhu: Artificial neural network enabled capacitance prediction for carbon-based supercapacitors. J. Mater. Res. 294–297(2018)
19. Suzuki, K.K.: Combinatorial computational chemistry approach to the design of cathode materials for a lithium secondary battery. Appl. Surf. Sci. 629–634 (2002)
20. Dongale, T.D., Kharade, K.G., Mullani, N.B., Naik, G.M., Kamat, R.K.: Artificial neural network modeling of NixMnxOx based thermistor for predicative synthesis and characterization. J. Nano Electron. Phys. 1–4 (2017)
21. Dongale, T.D., Katkar, S.V., Khot, K.V., More, K.V., Delekar, S.D., Bhosale, P.N.: Simulation of randomly textured tandem silicon solar cells using quadratic complex rational function approach along with artificial neural network. J. Nanoeng. Nanomanuf. 103–108 (2016)
22. Waseem Raza, F.A.: Recent advancements in supercapacitor technology. J. Nano Energy 441–473 (2018)
23. Omotayo, T., Bankole, A., Olanipekun, A.: An artificial neural network approach to predicting most applicable post-contract cost controlling techniques in construction projects. Appl. Sci. **10**(15), 5171 (2020). https://doi.org/10.3390/app10155171
24. Elçiçek, H., Akdoğan, E., Karagöz, S.: The use of artificial neural network for prediction of dissolution kinetics. Sci. World J. **2014**, 1–9 (2014). https://doi.org/10.1155/2014/194874
25. Sarbayev, H., Yang, M., Wang, H.: Risk assessment of process systems by mapping fault tree into artificial neural network. J. Loss Prev. Process Ind. **60**, 203–212 (2019). https://doi.org/10.1016/j.jlp.2019.05.006
26. Ok, S.C., Sinha, S.K.: Construction equipment productivity estimation using artificial neural network model. Constr. Manag. Econ. **24**(10), 1029–1044 (2006). https://doi.org/10.1080/01446190600851033
27. Kocabas, F., Korkmaz, M., Sorgucu, U., Donmez, S.: Modeling of heating and cooling performance of counter flow type vortex tube by using artificial neural network. Int. J. Refrig. **33**(5), 963–972 (2010). https://doi.org/10.1016/j.ijrefrig.2010.02.006
28. Daryasafar, A., Ahadi, A., Kharrat, R.: Modeling of steam distillation mechanism during steam injection process using artificial intelligence. Sci. World J. **2014**, 1–8 (2014). https://doi.org/10.1155/2014/246589
29. Dogan, E., Sengorur, B., Koklu, R.: Modeling biological oxygen demand of the Melen River in Turkey using an artificial neural network technique. J. Environ. Manage. **90**(2), 1229–1235 (2009)
30. Eynard, J., Grieu, S., Polit, M.: Wavelet-based multi-resolution analysis and artificial neural networks for forecasting temperature and thermal power consumption. Eng. Appl. Artif. Intell. **24**(3), 501–516 (2011)
31. Bezerra, E.M., Bento, M.S., Rocco, J.A.F.F., Iha, K., Lourenço, V.L., Pardini, L.C.: Artificial neural network (ANN) prediction of kinetic parameters of (CRFC) composites. Comput. Mater. Sci. **44**(2), 656–663 (2008)

32. Fernández, E.F., Almonacid, F., Sarmah, N., Rodrigo, P., Mallick, T.K., Pérez-Higueras, P.: A model based on artificial neuronal network for the prediction of the maximum power of a low concentration photovoltaic module for building integration. Sol. Energy **100**, 148–158 (2014)
33. Rao, G.N., Kumari, K.A., Shankar, D.R., Kharade, K.G.: A comparative study of augmented reality-based head-worn display devices. Mater. Today Proc. (2021)

Text Summarization of an Article Extracted from Wikipedia Using NLTK Library

K. G. Kharade[1(✉)], S. V. Katkar[1(✉)], N. S. Patil[2], V. R. Sonawane[2(✉)], S. K. Kharade[1(✉)], T. S. Pawar[2(✉)], and R. K. Kamat[1]

[1] Shivaji University, Kolhapur, Maharashtra, India
rkk_eln@unishivaji.ac.in
[2] MVP's KBT College of Engineering, Nashik, Maharashtra, India
pawar.tejaswini@kbtcoe.org

Abstract. Internet is an excellent source of information, where you can get information on all the topics. But due to the large availability of content, it becomes a challenging job to extract exact information. A text summarization system's main objective is defining and presentingthe most relevant information from the given text to the end-users. Nowadays, the data is available in a considerable quantity. It becomes difficult for the user to deal with exact information. It's not possible to read all the information and make a conclusion for specific data. With text summarization, a significantcontent of data is converted into a short set of information. If we talk specifically about Wikipedia's information, almost all the areas areopen on this website. If we search for a specific keyword, it will provide huge details. Text summarization will convey the same extensive information by converting it into small pieces without losing its message. We have taken data from Wikipedia and applied summarization techniques to reduce the content without changing its meaning for demonstrating the concept during this research. Abstractive methods generate an internal semantic representation of the original content. For several years, text summarization has been an active area of study. By summarizing parts of the source document, abstraction can transform the removed content to condense at more strongly than extraction.

Keywords: Abstractive summary · Extractive summary · NLTK · Python · Text summarization · Summarization methods

1 Introduction

In recent years, the necessity for summarization can be seen in many contexts such as newspaper articles, business documents, search engine results, medical summaries, and online portals to find the most relevant content and track down treatment options. Even as the internet's information continues to grow, sub-branching has expanded natural language processing limits. Clear information is valuable in search [15] (Fig. 1).

Using the machine learning algorithm, you summarize critical sentences from the document and incorporate all relevant information. For one of the main findings, extractive and synthesize [7]. The text summation function is vital to further business analysis,

M. Singh et al. (Eds.): ICACDS 2021, CCIS 1441, pp. 195–207, 2021.
https://doi.org/10.1007/978-3-030-88244-0_19

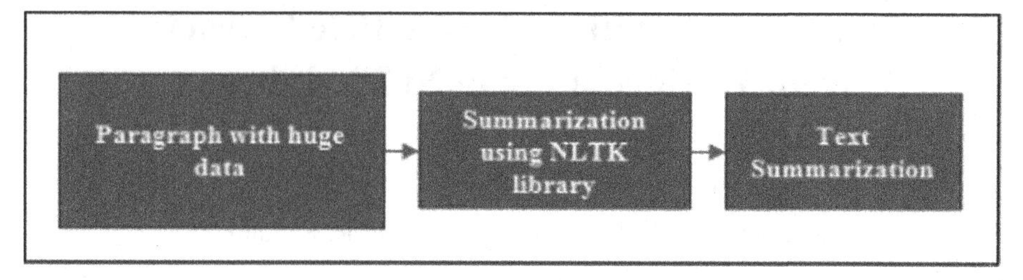

Fig. 1. Text summarization mechanism

marketing, university, research, and teacher functions. It has been noted that because the executive requires a simplified summarization, details in the time given, it is possible to process a limited amount of data [1].

Text summarization creates a brief and fluent description while maintaining the content and overall meaning of critical details. Various automated text summarization approaches have been developed and widely applied in different domains [4]. The method of extracting salient information from an authentic text document is Text Summarization. In this process, the data collected is reported and provided to the user as succinct. To grasp and explain the content of the text is very important for humans. Text Summarization is essential because of its numerous types of applications, such as book summaries, digests, stock markets, news, highlights, scientific abstracts, newspaper articles, magazines, etc. [5]. There are several resources available for text summarization. However, such devices primarily targeted news or essential documents and did not consider scientific articles' characteristics, i.e., their length and complexity [3, 10]. Many comprehensive facts, information server figures, and data and "information overload" are becoming a concern for individuals [6] (Fig. 2).

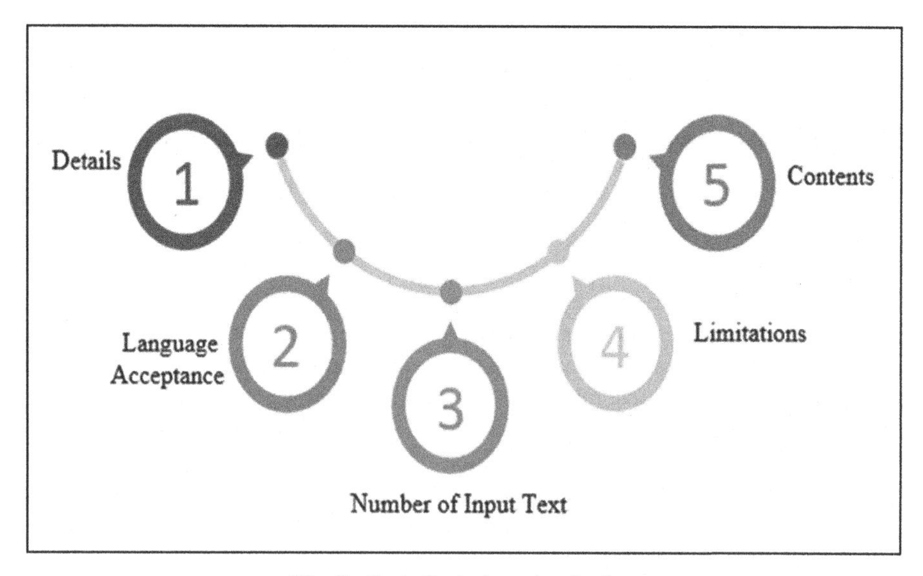

Fig. 2. Basis for text summarization

2 Literature Review

Ujjwal Rani [13] identified that, in bulky data dimensions, rapid production is essential for proficient collection, storage, and supervision. Sometimes, it's difficult to locate correct data because of the enormous quantity of information that you have to put away or in order to find it. In this instance, the expansion of data in different formats, regardless of the usefulness of mining information, is a direct consequence of technology such as big data [8]. Aysa Siddika Asa et al. (2017) reveals that tremendous volumes of data in databases and other data repositories result from automated data processing tools and mature database technology. With the growing amount of online content, locating important information for users is becoming increasingly difficult. The collection of information typically returns a large number of documents in the order of approximate relevance. Users do not read each record to find onethat is useful [2]. Neelima G et al. [10] reveals that we entered the information technology age with the internet's development. Knowledge and information are generated at a tremendous rate every day [9]. N. Moratanch [4] stated that there are several text summarization methods. One of which is getting essential data from authentic text. In-text summarization, the extracted information is presented as concisely as possible to the user as a summary. To do your best, you must understand the text's content and give it a due description. Abstractive and extractive summarization [3]. Goncalves [2] focused on the emergence of big data has caused an explosion in textual data. A considerable amount of valuable information and knowledge can be found in this volume. This increasingly available document supply of documents has called for deep and sophisticated NLP research [17]. It's no secret that automatic text summarization the process of producing concise and fluent summaries without human assistance [7]. Ruchika Aggarwal et al. [5] identified that even though this being an old challenge, increasingly modern work concentrates on biomedical patterns, instructional areas, item analyses, and web publications. It is because of the volume of data on the internet. There is a dire need for automated summarization in NLP research. It consists of summarizing one or more texts [14]. An extractive summary's primary motivation is to choose different words or paragraphs from the original text [5]. Mehdi Allayari et al. discussed the goal of automatic text summarization is to have a succinct and flowing description. More and more methods for automatic text summarization have emerged in recent years. The primary function of search engines is to provide a preview of documents. A typical example of a knowledge condensed headline is news websites that quickly summarize news to discover the information [11]. Internet users receive all of the document's content from the summarizer within a short time period of time and have the option of reading the full article. Some methods such as clustering, optimization, information, and item-based all result in extractive summaries. These methods don't recognise the duplicates that occur in the summaries [12]. Yash Asawa et al. [6] extraction is a process that detects the subset of the data, providing a condensed representation of the main ideas. In the internet era, users want to read as much information as possible, as webpages with text can be highly redundant [13]. Internet users receive all of the document's content from the summarizer within a short time period of time and have the option of reading the full article. There are several kinds of extraction-based, such as text and item, information, and generate-clustering, and word cluster-based approaches [6].

3 Need of Text Summarization

Now that information in digital spaces is largely non-structured, requiring tools that allow people to extract key points easily are more crucial than ever. Text segmentation is the challenge of condensing a long document into a concise, accurate, fluent summary. We enjoy tremendous quantities of data at our fingertips. But the majority of this information is unnecessary, nonessential, and will only serve to confuse. When summarising a text, extract the important ideas from it. With the exception of having the summarization turned on, however, however, that is not the case [16]. The software can now reduce a text of 5,000 words to an engaging sound-bite in less than five seconds. This is true regardless of the language. It is also possible that you must condense a text which you can't understand into a few simple words [18]. Text extraction is entirely language-independent. Since most people's ability is limited to speaking only one language, it exceeds that of human beings. The summarization softwares cover this gap and make the task simple for those who don't need to have to work hard. Both text to text and text to English conversion is in a single program.

4 Methods of Text Summarization

i) **Extractive Text Summarization:** Picking out the most critical points and passages from the writings. The piece is then fed into a data compactor that processes all the essential lines to get a single output. Therefore, every word and line of the document is part of the summary itself. Extractive summarization extracts the basic sentences from the source. It can be achieved in 3 phases;

   a) **Preprocessing step:** In this step, common words like "a", "an", "the" will be eliminated.
   b) **Steaming:** Using the root of each word is acquired through word-stretching
   c) **Part of speech tagging:** Classifying and isolating words from the process of identifying and then classifying them based on the word category they have (nouns, verbs, adverbs, adjectives) [19]

ii) **Abstractive Text Summarization:** Summarization is based on deep learning. Thus, so it creates new phrasings and concepts, just like we have in the document. Since it is harder to obtain information than what is extracted, it is much more difficult to analyze it. An abstractive summary begins with abstract statements or verbal summaries of information not found in the document. The term "abstractive summarization" refers to writers as non-to-selective [20]. Abstractive summarization techniques involve understanding the source text and packing it into fewer sentences. It uses language methods to examine and summarize (Aysa Siddika Asa) the text, then identifies the new concepts and terms to provide the most exact representation.

5 Benefits of Text Summarization

Text summarization produces a dense summary of the summary highlighting the crucial points while maintaining the text's bulk. The most highlighted benefits of text summarization are;

i) To reduce reading time
ii) It'll take a long time to read the entire article, which takes effort, but it's critical to understand what is and isn't being said. Generally, a 500-word essay can take 15 min to read. split-second summary text summarization. By decoupling the theory from data, the user can extract only a fraction of the information but still gets the critical information and actual results [21].
iii) To make the determination procedure simpler during the investigation of reports
iv) Programmed or summary creation frameworks empower business content archive development
v) More significant than the ability of most humans, many summarizations software's manual editing is unnecessary for linguistic summarizers; since they use linguistic models, they can handle texts in most languages [22]
vi) To increase the adequacy of ordering.
vii) Text summarization calculations are more accurate than the calculations made by a human being [25]
viii) That some programs have a feature that others don't can rank-order a word that lists it [23].
ix) Personalized information can be provided in case of a question-answer system.

6 Applications of Text Summarization

Text summarization tools generate the abstracts, which might contain relevant keywords or provide machine-generated keywords. Depending on the kind of summary, there are several kinds of summaries.

i) **Text Compactor:** There is no charge for purchasing the Text Compensator tool. it is accessible to anyone for use. The data and information is packaged in an appropriate way that increases the usefulness. Everyone, including the weak readers, can benefit from these general numbers. Since there is little distinction between them, any diligent student, an amateur or professional, or even a teacher can utilize its facilities [24].
ii) **Anchor Text:** Anchor text is a readable text that has a hyperlink. When they click on the hyperlinked text, they can still reach it. It fits in with the whole page's record, so it is accepted as part of the complete data set, plus it can be seen by everyone, which also makes it unique.
iii) **Video scripting:** Videos are increasingly considered to be a marketing tool of the highest importance. Professional networks like LinkedIn and other video-oriented platforms are now beginning to host videos instead of our friends and families. More or less scripting is required on the type of video. Summarization can serve to assemble multiple sources of research for script development [26].
iv) **The Text Tool:** If you create an image in a paint application, you should use the standard text tool. It takes up the entire canvas, compresses the output. You can also use paragraph spaces to organize summarized text to conform to the typewriter style. The font and color, and weight are chosen according to the type of input.

v) **Open Text Summarizer:** Many people use the Open Summarizer because it can do the same paraphrasing job as a closed source implementation. If sentences are drawn from a larger body of text, the computer is the program that determines the importance of sentences. Those that have greater significance are selected and those that have less importance neglected. It works for all distributions of Fedora, too. More than twenty-five languages are provided by Open Summarizer. There are several different options for configuring these languages, so experiment to see which ones work best for you. There is a consensus that various academic publications have dubbed it creative and advanced. The distribution is split between two parts: the command line and the library [27].

vi) **Media Monitoring:** Information overload is a real problem. summarization represents the opportunity to present continuous data in discrete units

vii) **Search Marketing and SEO:** It is imperative to have a well-rounded understanding of what your competitors are talking about in their content regarding search engine optimization. Since Google has improved its algorithm and has shifted towards search results' topical authority, this has become even more relevant. Creative thought: A powerful technique for looking at shared themes in numerous search results is multi-documentation [28].

viii) **Social Media Marketing:** White papers can also be condensed to make them more useful and shareable on social media sites like Twitter and Facebook. It would open the door for other content reuse by the businesses.

ix) **Financial Research:** The most substantial expense in financial investment firms is acquiring and processing information, including automated stock trading. If you have to do economic analyses every day, you will hit a wall [29].

x) **Internal Document Workflow:** Because large companies have large databases, they produce large amounts of under-utilized knowledge. They should learn how to use tools that help them be more efficient and therefore reuse more of their ability. Better comprehension will enable analysts to get a bird's-eye view of the firm's past activities and allow them to stitch them together into a coherent presentation

xi) **Newsletters:** Most newsletters tend to have a teaser on the front end of introducing new items to the reader and supplying the most exciting and essential information on the rest of the newsletter's topics on the back. It would also make it possible for organizations to include more summaries in newsletters helpful in the mobile [30].

xii) **Question Answering and Bots:** The workplace and the home are finding personal assistants as necessary as functional uses, thus being covered in both. Mostly, assistants are limited to performing a small number of functions. We could do a lovely job of summarization if we needed to. A summarist can build a complete response in a compilation form by going with the most recent documents.

xiii) **Medical Cases:** Because of telehealth's proliferation, there is an increasing need to deal with cases where everything is digital. Telemedicine networks can only ensure a more accessible and open healthcare system. Telescale operations can be critical in the medical case resolution chain when summarizing and routing these medical cases to the appropriate practitioner.

xiv) **E-learning and Class Assignments:** To add more interest to their lectures, many teachers use case studies and current events. By distilling information into more digestible summaries, teachers can speed up their update process of discovering new material [31].

7 Research Methodology

To perform the text summarization, we have selected the data available on Wikipedia by searching for 'INDIA.' The data might get changed on Wikipedia from time to time. But have proposed a method that will work on a real-time basis. It will simply summarize the information at the moment you apply this technique. Here we have not focused on the information available on Wikipedia for a specific word. We have focused on the process of summarization. The primary intention of the overview to minimize user data without changing its meaning. With this method, a user can go through ample information set with a minimum time. The present study returns 12389 words of data. 12389 words have been treated as extensive data, and it takes more time to read the data. We have applied the text summarization technique by using python to reduce the number of words.

8 Design and Development of System

Step 1: Install nltk Packages by using the following command

```
!pip install nltk
```

Step 2: Download stopwords by using the following command

```
import nltk
nltk.download("stopwords")
```

Step 3: Import packages, namely nltk, stopwords, PorterStemmer, WordTokenize, sent Tokenize, BeautifulSoap, urllib for further execution

```
import nltk
nltk.download("stopwords")
from nltk.corpus import stopwords
from nltk.stem import PorterStemmer
from nltk.tokenize import word_tokenize, sent_tokenize
import bs4 as BeautifulSoup
import urllib.request
```

Step 4: To retrieve the information available on Wikipedia for the word "India", text_read in a variable which stores the original text provided by the search result, and text_parsed stores the word in markdown format by using specified libraries

```
original_data = urllib.request.urlopen('https://en.wikipedia.org/wiki/India')
text_read = original_data.read()
text_parsed = BeautifulSoup.BeautifulSoup(text_read,'html.parser')
```

Step 5: Retrieve all the paragraphs by using <p> tag and store it using the text_content variable

```
paragraphs = parsedData.find_all('p')
print(paragraphs)
```

Step 6: Parse the content received from url and store in a variable

Step 7: During these processes, we have scrapped the data and have some simple text cleaning mechanisms before proceeding with the text summarization technique. Process the data to remove stopwords, reduce them to the its rootform, and create a dictionary for the word frequency table.

The frequency table consists of frequency for sample words is mentioned in the following table

Table 1. The frequency table for the paragraph selected

Words	Frequency count	Frequency
India	304	High
The	968	High
Indian	116	High
World	45	Medium
For	58	Medium
Empire	22	Medium
Architecture	14	Medium
Created	9	Low
Followed	7	Low

Step 8: To find the weighted frequencies of the sentences

```
def _calculate_sentence_scores(sentences, wordFrequency) -> dict:

    sentenceWeight = dict()

    for sentence in sentences:
        sentence_wordcount = (len(word_tokenize(sentence)))
        sentence_wordcount_without_stop_words = 0
        for word_weight in wordFrequency:
            if word_weight in sentence.lower():
                sentence_wordcount_without_stop_words += 1
                if sentence[:7] in sentenceWeight:
                    sentenceWeight[sentence[:7]] += wordFrequency[word_weight]
                else:
                    sentenceWeight[sentence[:7]] = wordFrequency[word_weight]

        sentenceWeight[sentence[:7]] = sentenceWeight[sentence[:7]] / sentence_wordcount_without_stop_words

        return sentenceWeight
```

To judge the essay, we'll be looking at the frequency of each word in the text. To this result, we'll add the frequencies of each significant word. To ensure that long sentences don't get disproportionately high scores, we divided each sentence's score by the number of words.

Step 9: To calculate the threshold of the sentences.

```
def _calculate_averageScore(sentenceWeight) -> int:

    sum_values = 0
    for entry in sentenceWeight:
        sum_values += sentenceWeight[entry]

    averageScore = (sum_values / len(sentenceWeight))

    return averageScore
```

If you want the average grade to be created from the sentences, generate some sentences and calculate their average grade. In this way, we can avoid the selection of sentences that have an average score under this threshold

Step 10: To Get the summary

```
def _get_articleSummary (sentences, sentenceWeight, threshold):
    sentenceCounter = 0
    articleSummary = ''

    for sentence in sentences:
        if sentence[:7] in sentenceWeight and sentenceWeight[sentence[:7]] >= (threshold):
            articleSummary += " " + sentence
            sentenceCounter += 1

    return articleSummary
```

Step 11: Process the summary

```
import nltk
def _run_articleSummary (article):

    wordFrequency= _create_dictionary_table(article)
    sentences = sent_tokenize(article)
    sentence_scores = _calculate_sentence_scores(sentences, wordFrequency)
    threshold = _calculate_averageScore(sentence_scores)
    articleSummary = _get_articleSummary (sentences, sentence_scores, 1.5 * threshold)

    return articleSummary
```

Step 12: Download punkt for the outcome

PointSentenceTokenizer uses an unsupervised algorithm to construct a model for abbreviation terms, collections. Terms that start sentences to split a text into a list of corrections.

Step 13: get the final summary as an outcome of the summarization

```
if __name__ == '__main__':
    results = _run_articleSummary (articalContents)
    print(results)
```

After this step you will receive the result as shown below.

Words	Frequency
India	12
The	25
Indian	6
World	1
For	1

9 Result Analyses

We have searched the word INDIA on Wikipedia, which returns the 12389. If you search the same keyword simultaneously, getting a different number of words produced by the search. It does n't matter how many stories have been returned at a specific moment. Our system will work dynamically and It will summarize the result. Every time it will give you summarized data of updated information on Wikipedia. To perform this task, we have integrated python and the Natural Language Toolkit (NLTK) library. Inpython, withtheintegrationofthe NLTK library, text summarization has been achieved. As shown in Table 1, most occurred words have been calculated, minimizing the overall data. After text summarization, the same message has been transferred to the end-user (Fig. 3).

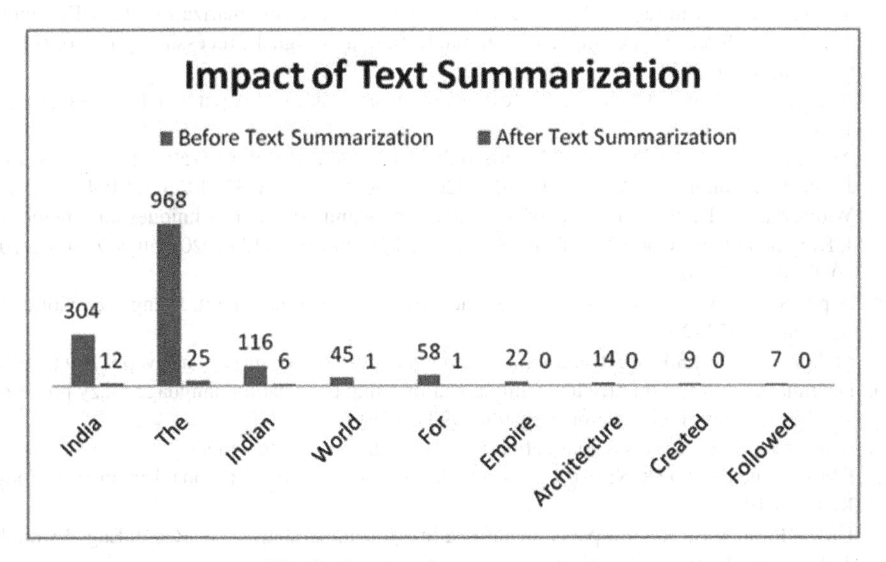

Fig. 3. Impact of text summarization

10 Conclusion

Since a lot of information is available on the webserver, it is impossible to read each available document to determine the record and know whether it is a required document. Hence, a description of these documents will help the reader decide whetherthey are available for misappropriating. Some diverse models can find the essential material from the database to produce an extended text description. In this research paper, we haveintegrated the NLTK library for text summarization. The outcome of this processis the summary of effective content without changing its meaning. Ultimately users can extract enormous information in the form of an outline. Extended range is essential to business analysts, marketing executives, governments, students, researchers, and teachers. The executive is seen as having to be summarized to allow a maximum amount of information to be processed within a limited time frame. This paper delves into the details of both the extractive and the abstractive approaches and their overall performance. The methods and techniques used, with their strengths and shortcomings.

References

1. Gaikwad, D., Mehender, C.: A review paper on text summarization. Int. J. Adv. Res. Comput. Commun. Eng. 154–160 (2016)
2. Goncalves, L.: Automatic Text Summarization with Machine Learning — An overview: 12 April 2020 (22 March 2021)
3. Allahyari, M., et al.: Text summarization techniques: a brief survey. Int. J. Adv. Comput. Sci. Appl. **8**(10), 397–405 (2017). https://doi.org/10.14569/IJACSA.2017.081052
4. Moratanch, N., Chitrakala, S.: A survey on extractive text summarization. In: EEE International Conference on Computer, Communication, and Signal Processing, pp. 1–6. IEEE, Chennai, India (2017)
5. Aggarwal, R., Gupta, L.: Automatic text summarization. Int. J. Comput. Sci. Mobile Comput. 158–167 (2017)
6. Asawa, Y., Balaji, V., Dey, I.I.: Modern multi-document text summarization techniques. Int. J. Recent Technol. Eng. **9**(1), 654–670 (2020). https://doi.org/10.35940/ijrte.A1945.059120
7. Widyassari, A.P., et al.: Review of automatic text summarization techniques and methods. J. King Saud Univ. Comput. Inform. Sci., p. S1319157820303712 (2020). https://doi.org/10.1016/j.jksuci.2020.05.006
8. Arpita Sahoo, A.K.: Review paper on extractive text summarization. Int. J. Eng. Res. Comput. Sci. Eng. 46–52 (2018)
9. Mehdi Allahyari, S.P.: Text summarization techniques: a brief survey. arXiv, pp. 1–9 (2017)
10. Neelima, G.V.M.: Extractive text summarization using deep natural language fuzzy processing. Int. J. Innov. Tech. Explor. Eng. 990–993 (2019)
11. Shai Erera, M.S.-S.: A Summarization System for Scientific Document
12. EMNLP and the 9th IJCNLP, pp. 211–216. Association for Computational Linguistics, Hong Kong (2019)
13. Ujjwal Rani, K.B.: Review paper on automatic text summarization. Int. Res. J. Eng. Technol. 3349–3354 (2019)
14. Lloret, E., Romá-Ferri, M., Palomar, M.: COMPENDIUM: A Text Summarization System for Generating Abstracts of Research Papers. In: Muñoz, R., Montoyo, A., Métais, E. (eds.) Natural Language Processing and Information Systems, pp. 3–14. Springer Berlin Heidelberg, Berlin, Heidelberg (2011). https://doi.org/10.1007/978-3-642-22327-3_2

15. Nageswara Rao, G., ArunaKumari, K., Ravi Shankar, D., Kharade, K.G.: A comparative study of augmented reality-based head-worn display devices. Mater. Today Proc., S2214785320401178 (2021). https://doi.org/10.1016/j.matpr.2020.12.400
16. Wells, M.: 3 Advantages of Automatic Text Summarization. https://ezinearticles.com/?3-Advantages-of-Automatic-Text-Summarization&id=3465270 (22 Dec 2009)
17. Kharade, K.G., Kharade, S.K., Kumbhar, V.S.: A comparative study of traditional server and azure server. J. Adv. Sci. Technol. **13**(1), 329–331 (2017)
18. Ullah, S., Islam, A.B.M.A.A.: A framework for extractive text summarization using semantic graph based approach. In: Proceedings of the 6th International Conference on Networking, Systems and Security – NsysS 2019, pp. 48–56. https://doi.org/10.1145/3362966.3362971. (2019)
19. Kharade, K.G., Kamat, R.K., Kharade, S.K.: Online library package to boost the functionality and usability of the existing libraries. Int. J. Future Revol. Comput. Sci. Commun. Eng. **5**(8), 5–7 (2019)
20. Kharade, S.K., Kamat, R.K., Kharade, K.G.: Simulation of dye synthesized solar cell using artificial neural network. Int. J. Eng. Adv. Technol. 1316–1322 (2019)
21. Dongale, T.D., Kharade, K.G., Mullani, N.B., Naik, G.M., Kamat, R.K. Artificial neural network modeling of Ni sub (x) Mn sub (x) O sub (x) based thermistor for predicative synthesis and characterization (2017)
22. Floydhub: A Gentle Introduction to Text Summarization in Machine Learning. https://blog.floydhub.com/gentle-introduction-to-text-summarization-in-machine-learning/ (15 April 2019)
23. Stopher, D.: Benefits Of Automatic Text Summarization | North East Connected. https://neconnected.co.uk/benefits-of-automatic-text-summarization/ (27 Oct 2020)
24. Elrefaiy, A., Abas, A.R., Elhenawy, I.: Review of recent techniques for extractive text summarization. J. Theor. Appl. Inf. Technol. 7739–7759 (2018)
25. Understanding Automatic Text Summarization-2: Abstractive Methods | by Abhijit Roy | Towards Data Science. https://towardsdatascience.com/understanding-automatic-text-summarization-2-abstractive-methods-7099fa8656fe (n.d.). Retrieved 3 April 2021
26. Verma, N., Tiwari, A.: A survey of automatic text summarization. Int. J. Eng. Res. Technol. **3**(6) (2014)
27. Kharade, S.K., Kamat, R.K., Kharade, K.G.: Artificial Neural Network Modeling of MoS2 Supercapacitor for Predicative Synthesis
28. Kharade, S.K., Kharade, K.G., Kamat, R.K., Kumbhar, V.S.: Setting Barrier to Removable Drive through Password Protection for Data Security
29. Kharade, K.G., Kamat, R.K., Kharade, S.K.: Potential of India to Boom With the Ease of E-Commerce
30. Kharade, S.K., Kharade, K.G., Kamat, R.K.: Opportunities and Challenges of ICT in Rural Development
31. Katkar, S.V., Kamat, R.K., Kharade, K.G., Kharade, S.K.: Inclusion of .NET Framework for Calculating Electrical Parameters of Solar Cell

Grapheme to Phoneme Mapping for Tamil Language

M. Geerthana Anusha(✉), D. Govind, and Vijay Krishna Menon

Center for Computational Engineering and Networking, Amrita School of Engineering, Amrita Vishwa Vidyapeetham, Coimbatore, India
cb.en.p2cen19003@cb.students.amrita.edu,
{d_govind,m_vijaykrishna}@cb.amrita.edu

Abstract. In this paper, a Deep learning-based approach is analyzed to improve the accuracy of the Language Modelling of a system that is a part of ASR. The main intent is to build an End to End Neural Network Language Model. Generally, ASR works in three different modes. The modes are Acoustic Modeling, Lexicons, and Language Modeling. In ASR, G2P models are vital in the case of analyzing the Out Of Vocabulary words. Mostly these G2P models were fabricated using the traditional approaches such as Bi-gram, n-gram language models. The traditional approaches are deceptive in aiding the meanings, long term dependencies of the words and are also impotent in rendering the phonemes of new words. Thus Seq2Seq G2P models exceed the state of the art of the traditional approach. As we all know, the ASR (Speech to Text) system is widely used in AI applications such as voice dialing, system control, navigation, etc. The method which is applied here is the Statistical based one. A method other than the Rule-based approach is trialed out. A suitable neural network of LSTM and BI-LSTM is proposed which enhances the accuracy of the system.

Keywords: Language modelling · ASR · Deep learning-based · Rule-based · Neural network · Seq2Seq

1 Introduction

In order to promote the accuracy of the ASR system, g2p converters are exploited. The ASR system is well known for its audio form of wave transformation to respective text format. The Statistical method employed in ASR includes Acoustic Modelling, Lexicons (Search Space), and Language Modelling. Acoustic Modelling includes HMM-GMM and HMM-DNN. Language Modelling is of distinct fashion such as n-gram, Bi-gram, Tri-gram, Deep-Learning Language models [1]. Finally, Decoding is accomplished in ASR. The drawback of the n-gram language model is that it is unable to capture word order, long-term constraints, lacking in predicting the upcoming words/letters [2,4,6,10]. To strengthen the language models, the Neural network model is manifested. In this work, the Deep-Learning Language Model approach is established.

M. Singh et al. (Eds.): ICACDS 2021, CCIS 1441, pp. 208–217, 2021.
https://doi.org/10.1007/978-3-030-88244-0_20

Phonemes play a major role in the SR system. Lexicons contain the pronunciation of the words. For the essential transformation, lexicons of specific languages have to be learned. This learning part comes in the Language modeling part of ASR. Even Though the lexicons are generated using the rule-based approach for the Indian Languages, it's impossible to capture all the words of the language that are being used in our day-to-day life. Another method which is a deep neural network-based one is introduced and also tried to overcome this insufficiency. Thus the G2P model is evolved.

2 Related Works

The recurrent Neural Network Language model is depicted. The RNN language model narrates that perplexity shrinks to 50% in contrast with the backoff language model [6]. The current language models are explored. The Language model has to toil a large corpus, vocabulary also the structure of language. Here RNN language models such as CNN or LSTM are probed. The state of art Perplexity is diminished to 30 and 23.7 as opposed to the traditional Language Model [4]. In the n-gram Language model, it is arduous to reap the semantics. But neural network-based models fulfill the demand. The precedence of Neural Network Language Modelling is that one and the same model can be employed in forecasting abundant signals not alone languages [7].

Initially the SR model [12] is created for the Punjabi language using the Kaldi toolkit. The language model used here is the N-gram method. To assess and obtain the accuracy, both monophone and triphone models were generated. WER (Word Error Rate) is the metric used to evaluate the system [3]. ASR system acquires data as acoustic and language data. Features like MFCC and PLP were extracted. Ubuntu is used as a medium to conduct the experiments. GMM-HMM model was employed with a context-dependent triphone system. ASR performance is measured. MFCC features accomplished over the PLP features. The Triphone Model gives the maximum accuracy [3].

For Hindi g2p conversion, different methods were established. G2P conversion requires rules. To improve the accuracy, several methods such as Rule-based [14], and Decision tree-based methods were compared [13]. In a Rule-based g2p system, for each grapheme/word, the system generates its corresponding phones. Since the Rule-based system has its own limitations as at different colloquial language, the rules will be useless. The new method as a decision tree-based g2p system is developed and compared with a Rule-based system. Based on the context, decision trees were designed to obtain the variations in phonemes. In this experiment, the decision tree-based g2p system shows the maximum accuracy than the rule-based g2p system [8].

Almost the Indian Languages are bounded with phonetics which consists of nearly 35 to 38 consonants and 15 to 18 vowels. The traditional languages vary in their slang. Thus using phonetics an effort is carried out to unite the languages. These phonetic variations are represented by a common label set. To attain the aim, the unified parser comes into the role. The main purpose of the parser is to develop the phones by converting the Universal coded character set to the common label set [1].

The unified parser builds on rules. The limitations of the parser are that in case of finding syllables [9] and appealing certain rules to the languages. Language-specific rules were developed to address the special cases. Rule-based unified parser procured 80–95% of accuracy [1].

To assist the strangers to speak English, g2p predictions were held. There are two stages: 1) Initially plot words with their phonetics and align them as a one-hot vector 2) G2P model was designed [16].

The illustrated model explains seq2seq mapping [15]. In this sequence to sequence mapping [5], for each character, the mapping is learned. The Neural Machine Translation resembles seq2seq models [11]. English words were translated to their phonetics look like NMT. A recurrent Neural Network is chosen to execute the observation [16].

3 Graphemes and Phonemes

Tamil has its own phonetics. There are 12 vowels, 18 consonants, and a special character in Tamil. These vowels and consonants together lead to a flourish of 216 letters. As a whole, there are 247 characters existing in Tamil. Table 1 shows Tamil graphemes and their corresponding phonemes. The phonemes (pronunciation) of these Tamil graphemes are bloomed out of the Rule-based Unified Parser hoarded from Speech and Music Technology Lab IIT Madras supported by Ubuntu.

Table 1. G2P: vowel and consonants

Grapheme	Phoneme	Grapheme	Phoneme
	a		k
	aa		ng
	i		c
	ii		nj
	u		tx
	uu		nx
	e		t
	ee		n
	ai		p
	o		m
	oo		y
	ou		r
	hq		l
	j		w
	s		zh
	sx		lx
	rx		nd

4 Deep Learning-Based g2p System

To develop a statistical-based g2p conversion system. The statistics of the context is learned in which various phonemes occur which is learned by a Deep Neural Network model. Initially, the lexicons are shaped from the language-specific unified parser using ubuntu. The lexicons hold character/word (graphemes) and their corresponding pronunciations/sound (phonemes). It contains 35,185-word pairs such that pronunciations of unique Tamil words are generated and stored as lexicons (Fig. 1).

```
அ       a
அக      a k a
அகதிகள் a k a t i k a lx
அகதியானேன்      a k a t i y aa nd ee nd
அகத்தியனாரின்   a k a t t i y a nd aa r i nd
அகப்படவில்லை    a k a p p a tx a w i l l ai
அகப்படாமல்      a k a p p a tx aa m a l
அகப்பட்டதும்    a k a p p a tx tx a t u m
அகப்பட்டுக்     a k a p p a tx tx u k
அகமதாபாத்       a k a m a t aa p aa t
அகமத்   a k a m a t
அகமெம்னானின்    a k a m e m nd aa nd i nd
அகமெம்னானைக்    a k a m e m nd aa nd ai k
அகம்பாவம்       a k a m p aa w a m
அகலிகையை        a k a l i k ai y ai
அகழியினால்      a k a zh i y i nd aa l
அகழியின்a k a zh i y i nd
அகற்ற   a k a rx rx a
அகற்றப்பட       a k a rx rx a p p a tx a
அகற்றாமல்       a k a rx rx aa m a l
அகற்றி  a k a rx rx i
அகற்றுவதாக      a k a rx rx u w a t aa k a
```

Fig. 1. Preview of lexicon

4.1 Methodology

The lexicons seem to be similar to the sequence to sequence mapping in NLP. Machine Translation is an application of NLP. In the neural machine translation, one language is being transformed into another language. In this work, grapheme is mapped to its corresponding phonemes which resemble sequence to sequence mapping of NMT.

4.2 Dataset

Dataset is collected from Speech and Music Technology Lab IIT Madras. Lexicon is created using Unified parser rules. The lexicon has grapheme and its corresponding phoneme. The tab space separates words and pronunciations in every line. Then the data is utilized for training, validation, and testing.

4.3 Data Pre-processing

The dataset needs to be pre-processed such that seq2seq mappings work on it. The word pairs are stored in a list. One hot representation is enumerated.

4.4 RNN

Recurrent Neural Network is employed to recover all the characters/words at each time step. Thus RNN is the appropriate deep learning approach to seq2seq mapping here. LSTM and BiLSTM are used (Fig. 2).

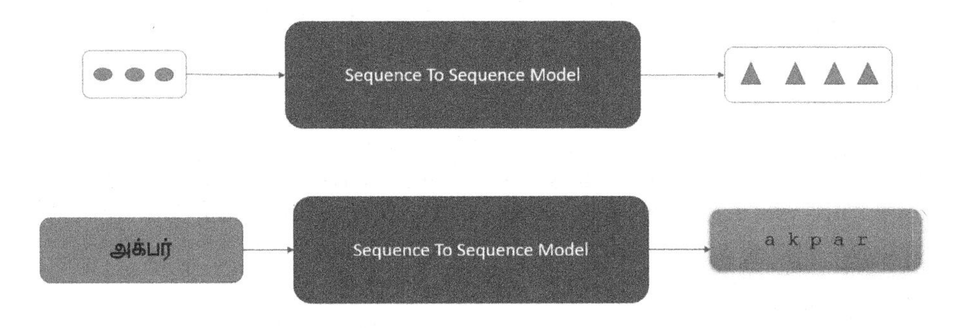

Fig. 2. Seq2Seq model

The words are of variable length. The characters with their one-hot encoded representation are taken within the RNN. The network comprises Encoder and Decoder in its architecture. The encoder has its input layer, LSTM/BiLSTM layers. Initially, the words are transformed to vectors. These vectors are taken as input at the character level. At last, the entire context from the hidden state is taken as input in the Decoder. The decoder has its input layer, LSTM layers. The model is contrived.

4.5 Training

The training set contains 31500 words and the validation split is 0.1. The model is trained with totally different latent dimensions such as 256,512,768, and 1024. The loss function used is categorical crossentropy. The grapheme (a written symbol) is taken as encoder input and also the corresponding phoneme (sound) is obtained at the decoder output. Similarly, the training process is administered (Fig. 3).

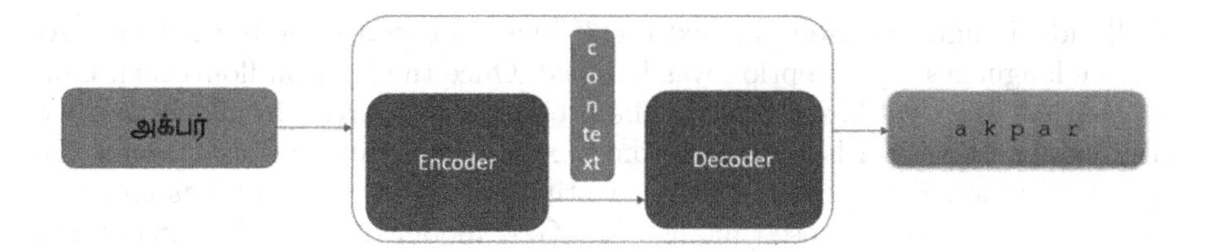

Fig. 3. Working principle of LSTM

4.6 Inference

The inference is utterly different from training. In Training the target knowledge (data) is already out there for the encoder to predict the successive character whereas in Inference the decoder needs to predict its upcoming character in a sequence. To spot the maximum probability (most likely) of the characters argmax function is employed. The prediction depends on the decoding part (Fig. 4).

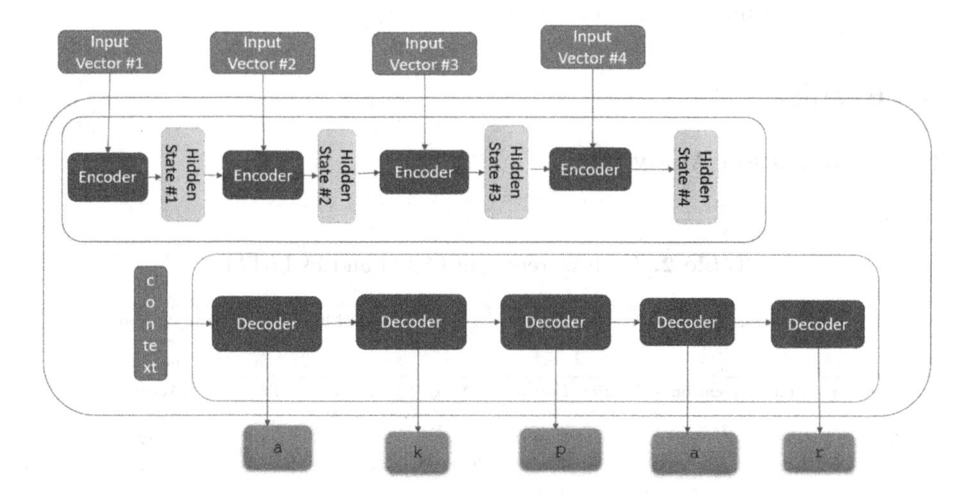

Fig. 4. At each time step

5 Experiments

The dataset contains Tamil sentences spoken by disparate speakers. These Tamil sentences are extracted from the text file. The unique words are carved out of these Tamil sentences using Ubuntu. The pronunciations were invented for these unique words which constitute the Dictionary (Lexicons). Lexicons are generated using the unified parser rules that are collected from the IITMadras. The unified parser is embodied with notable rules. The unified parser comprises a Label set of all Indic Languages. Here the text (UTF-8) is being converted to the labels. As per the languages, an exception was handled. Once the Lexicon flourished, Language Modelling will have thrived. The proposed work arises from the Seq2Seq G2P model. 31500 words are taken from the corpus and are used for training the model. 3500 words were considered as testing data. Python is used as an interface to develop the proposed model. The G2P model blooms with LSTM and BI-LSTM. RNN architecture comprises the Encoder and Decoder parts. After standardizing all the Hyperparameters for instance Latent Dimension, Learning Rate, and Decay. The latent dimension is the imitation of squeezed data varies with 256, 512, 768, and 1024. The Learning rate proclaims the model's prudence. The prediction worked well and the accuracy was obtained. The metric used to evaluate the performance of the model is Word Error Rate. WER is formulated as below

WER = (S + D + I)/N
WER ⟶ Word Error Rate
S ⟶ Number of Substitutions
D ⟶ Number of Deletions
I ⟶ Number of Insertions
N ⟶ Total number of words in the reference

6 Results

6.1 g2p System Analysis

Table 2. Analysis report of LSTM and Bi-LSTM

	Hyperparameter tuning				Word level (in percent)		Phoneme level (in percent)	
G2P system	Latent dimension	Validation split	Learning rate	Decay	Accuracy	WER	Accuracy	WER
LSTM	256	0.05	0.01	0.001	66.93	33.06	88.32	15.39
	512	0.05	0.01	0.001	86.85	13.15	95.72	5.83
	768	0.05	0.01	0.001	5.84	94.16	71.74	44.5
	1024	0.05	0.01	0.001	92.96	7.04	97.81	2.95
Bi-LSTM	256	0.2	0.01	0.01	90.85	9.15	97.66	2.98
	512	0.2	0.01	0.01	41.48	58.52	82.92	26.45
	768	0.2	0.01	0.01	10.94	89.06	17.2	82.8
	1024	0.2	0.01	0.01	12.9	87.1	20.2	85.88

The system has experimented with the standardized latent dimensions such as 256, 512, 768, 1024. In this case, both LSTM and BI-LSTM succeed easily with the varied hyperparameters. Once the tuning is effectively done, the system acquired knowledge. Here the experiment is exposed with different validation split like 0.05, 0.25, 0.5, 0.75, 0.1, 0.2, 0.25. The LSTM manifested valuable accuracy at its 0.05 validation split and in the case of BI-LSTM, it's at 0.2 splits (Table 2).

Then the several learning rates are also consummated with 0.1, 0.01, 0.001, 0.2. Now LSTM and BI-LSTM are worthwhile at their 0.01 rate. There are dropouts which are nearly 5–10%. After executing all the tunings, the accuracy and WER are calculated at its word level as well as phoneme level. LSTM rendered beneficial results with a latent dimension of 1024 whereas Bi-LSTM furtherance at its 256. Table 3 illustrates it.

6.2 Performance of LSTM and BiLSTM

LSTM is a sort of RNN (Recurrent Neural Network) and it's adept in seizing the long-term dependency. LSTM recollects the information to figure out the upcoming terms (words/characters). In the proposed Seq2Seq model, predicting the next character is indispensable in a sequence. Thus LSTM evinces that despite input and output are of variable size, the LSTM models foresee the imminent sequence. LSTM comprises Encoder and Decoder parts. The encoder takes the input and converts it into a constant length vector using one-hot encoding and the Decoder takes up these constant length vectors and confers the prophecies sequence. To procure the maximum probability argmax function is employed. Hence, the requisite LSTM model is evolved.

Bi-LSTM is a supplement of LSTM. Bi-LSTM as the name Bidirectional LSTM uses 2 LSTM. Two LSTM are used in learning the input and the amended replicate of the input characters in orchestration. One LSTM grabs input in one direction while the other LSTM clutches the input in another direction. Bi-LSTM Encoder and Decoder toils as of LSTM. Thus the desideratum Bi-LSTM model is designed.

From the result, it is perceptible that LSTM transcends Bi-LSTM. In the case of the Tamil Language, the preceding characters alter the sound of the succeeding characters [14]. LSTM gazes at the forthcoming characters whereas Bi-LSTM glances at both fore and rear of the characters. Hence LSTM attains a minimum Word Error Rate. WER is calculated using HTK in Ubuntu.

Table 3. LSTM vs BiLSTM

G2P system	Word level accuracy	Phoneme level accuracy
LSTM	92.96	97.81
BiLSTM	90.85	97.66

6.3 Graphical Illustration

The Performance of LSTM and Bi-LSTM is described. The experiment is conducted. LSTM and Bi-LSTM are scrutinized at perceptible Latent dimension which is the number of hidden state units. The dimensions of 256, 512, 768, and

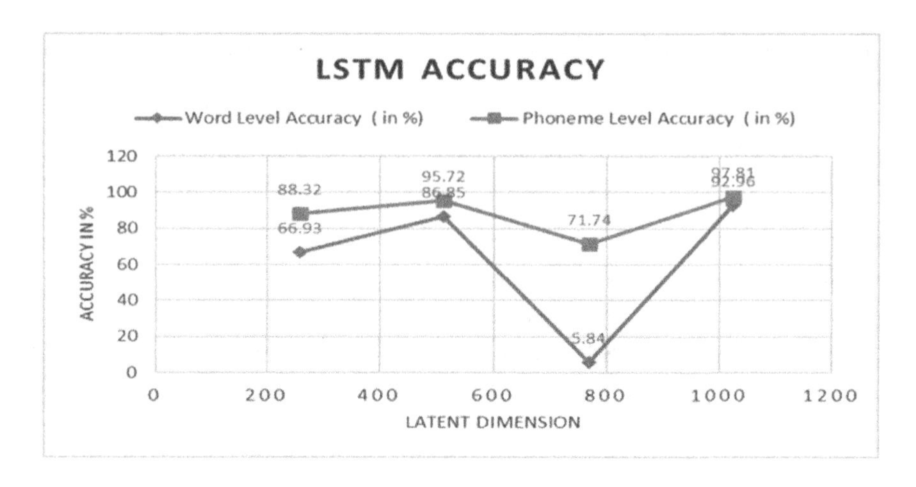

Fig. 5. Learning outcome of LSTM

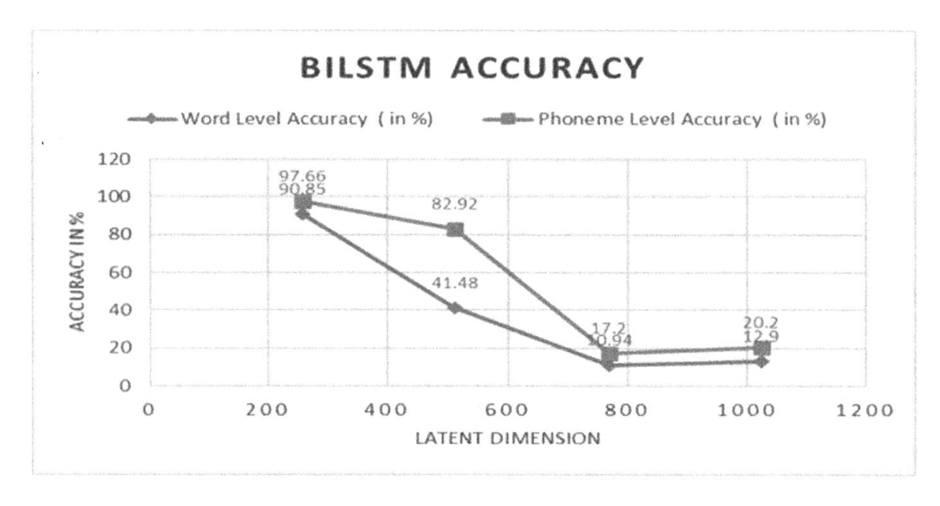

Fig. 6. Learning outcome of Bi-LSTM

1024 are inspected. In LSTM the 1024 units accomplished the least Word Error Rate. In Bi-LSTM the 256 units attained the tiniest WER. LSTM and Bi-LSTM are analyzed. LSTM surpassed Bi-LSTM (Figs. 5and 6).

7 Conclusion

The deep learning-based g2p system has been designed. Both LSTM and BiLSTM models are developed and compared to accumulate the foremost effective Language modeling for the ASR system. Thus the graphical representation clearly explains the accuracy of the deep learning approach. In future work, the fruitful ASR system can be built using this Language Model.

Acknowledgment. I am thankful to Dr. D. Govind and Mr. Vijay Krishna Menon for the worthwhile discussions. Also, I would like to show my gratitude to my department who gave me an opportunity to explore new research ideas. My humble pranam to Mata Amritanandamayi for her support.

References

1. Baby, A., Murthy, H.A., et al.: A unified parser for developing Indian language text to speech synthesizers in tsd, In: Sojka, P., et al. (eds.) TSD 2016, LNAI 9924, pp. 514–521 (2016)
2. Bengio, Y., Ducharme, R., Vincent, P., Janvin, C.: A neural probabilistic language model. J. Mach. Learn. Res. **3**, 1137–1155 (2003)
3. Guglani, J., Mishra, A.N.: Continuous punjabi speech recognition model based on Kaldi asr toolkit. Int. J. Speech Technol. **21**(2), 211–216 (2018)
4. Jozefowicz, R., Vinyals, O., Schuster, M., Shazeer, N., Wu, Y.: Exploring the limits of language modeling. arXiv preprint arXiv:1602.02410 (2016)
5. Kumar, S.S., Kumar, M.A., Soman, K.: Sentiment analysis of tweets in malayalam using long short-term memory units and convolutional neural nets. In: International Conference on Mining Intelligence and Knowledge Exploration, pp. 320–334. Springer (2017)
6. Mikolov, T., Karafiát, M., Burget, L., Černockỳ, J., Khudanpur, S.: Recurrent neural network based language model. In: Eleventh Annual conference of the International Speech Communication Association (2010)
7. Mikolov, T., Kombrink, S., Burget, L., Černockỳ, J., Khudanpur, S.: Extensions of recurrent neural network language model. In: 2011 IEEE International Conference on Acoustics, Speech and Signal Processing (ICASSP), pp. 5528–5531. IEEE (2011)
8. Nallasamy, U., Kumar, C., Srinivasan, R., Swaminathan, R.: Decision tree learning for automatic grapheme to phoneme conversion for Tamil. In: Proceedings of SPECOM (2004)
9. Rabiner, L.: Fundamentals of Speech Recognition, Prentice-Hall, Upper Saddle River (1993)
10. Schwenk, H.: Continuous space language models. Comput. Speech Lang. **21**(3), 492–518 (2007)
11. Selvin, S., Vinayakumar, R., Gopalakrishnan, E., Menon, V.K., Soman, K.: Stock price prediction using lISTM, RNN and CNN-sliding window model. In: 2017 International Conference on Advances in Computing, Communications and Informatics (ICACCI), pp. 1643–1647. IEEE (2017)
12. Shivakumar, K., Aravind, K., Anoop, T., Gupta, D.: Kannada speech to text conversion using CMU sphinx. In: 2016 International Conference on Inventive Computation Technologies (ICICT). vol. 3, pp. 1–6. IEEE (2016)
13. Vidyapeetham, A.V., Ettimadai, C.: Grapheme to phone conversion for Hindi
14. Vidyapeetham, A.: Automatic grapheme to phone converter for Tamil using rules
15. Viswanathan, S., Anand Kumar, M., Soman, K.P.: A sequence-based machine comprehension modeling using LSTM and GRU. In: Sridhar, V., Padma, M.C., Rao, K.A.R. (eds.) Emerging Research in Electronics, Computer Science and Technology. LNEE, vol. 545, pp. 47–55. Springer, Singapore (2019). https://doi.org/10.1007/978-981-13-5802-9_5
16. Yolchuyeva, S., Németh, G., Gyires-Tóth, B.: Grapheme-to-phoneme conversion with convolutional neural networks. Appl. Sci. **9**(6), 1143 (2019)

Comparative Study of Physiological Signals from Empatica E4 Wristband for Stress Classification

Varun Chandra(✉), Ankit Priyarup, and Divyashikha Sethia

Delhi Technological University (DTU), New Delhi, India
{varunchandra_2k17se124,ankitpriyarup_2k17se18,divyashikha}@dtu.ac.in

Abstract. The injurious effects of mental stress on the human body and mind are well known. Many researchers have focused on developing stress monitoring systems using physiological signals obtained from the body to alleviate stress. This study aims to provide a comparative analysis of four physiological signals – Electrodermal Activity (EDA), Heart Rate (HR), Skin Temperature (SKT), and Blood Volume Pulse (BVP), recorded using the Empatica E4 Wristband, in building stress classification models. We collect a dataset on 21 participants comprising their physiological signals while they perform a mental arithmetic task, which acts as a stress inducer. We compare the classification accuracy of machine learning classifiers trained on feature sets built using the four signals taken one at a time, two at a time, three at a time and all together. We achieve the highest accuracy of 99.92% using all the four signals. When we consider three signals at a time, EDA, HR, and BVP feature set achieves the best accuracy of 99.88%, and taking two signals at a time, EDA and HR feature set obtains the best accuracy of 99.09%. This paper can act as a guide for a manufacturer to select an optimal set of physiological signals for building efficient stress monitoring systems.

Keywords: Electrodermal Activity (EDA) · Heart Rate (HR) · Skin Temperature (SKT) · Blood Volume Pulse (BVP) · Empatica E4 Wristband

1 Introduction

It is a well-known fact that one of the most important determiners of health is the presence of mental stress. Stress management is of utmost importance in today's hectic work-driven lifestyle. Stress is known to be a leading cause of many psychological and psychosomatic illnesses. Many medical professionals associate stress with chronic illnesses like diabetes and stroke. However, stress management can happen only with the detection of mental stress. Many researchers have worked on stress detection using physiological signals like Electroencephalogram (EEG), Electrocardiogram (ECG), Heart Rate (HR), Heart Rate Variability (HRV) [1] and have achieved promising results.

M. Singh et al. (Eds.): ICACDS 2021, CCIS 1441, pp. 218–229, 2021.
https://doi.org/10.1007/978-3-030-88244-0_21

Among the physiological signals, Electroencephalogram (EEG) and Electrocardiogram (ECG) has been the most widely used features in stress detection and classification tasks achieving very high accuracies [2–4]. However, recording and working with these signals requires a controlled and sophisticated environment, which can be difficult to create. An alternative can be a commercially available smartwatch or wristband based device that can be worn all day without fatigue. Several wristband-based devices are popular for research purposes, for example, Empatica E4, Garmin Vivosmart, Zephyr BioHarness 3.0, iMotions Shimmer 3+, and Biopac Mobita [5]. These devices can measure features like Galvanic Skin Response (GSR), Heart Rate (HR), Skin Temperature (SKT) with high precision. Numerous studies have used these devices to build efficient stress assessment models. It is imperative to build stress monitoring systems that are highly portable, proficient, and comfortable to wear for daily usage.

This paper provides an in-depth analysis of developing stress classification models using one of such smart wearable devices, the Empatica E4 Wristband. The E4 Wristband is a medically certified sensor device that measures four physiological signals viz., Electrodermal Activity (EDA), Heart Rate (HR), Blood Volume Pulse (BVP), and Skin Temperature (SKT). It also supports an accelerometer to capture motion-based activity. In this study, we perform an extensive exploration of these four physiological signals provided by the wristband to build stress classification models. For this purpose, we collect a dataset of 21 students from a university while they perform a mental arithmetic task wearing the E4 Wristband. The mental arithmetic task serves as a stress-inducing agent. It induces three levels of stress response in a participant – no stress (rest state), mild stress, and high stress. This research aims to provide a comparative analysis of the four physiological signals (EDA, HR, SKT, and BVP) gathered from the E4 Wristband to develop stress classification models. After performing the data processing and feature extraction, we form 15 feature sets corresponding to the 15 ways of selecting the four signals, each set containing features from a distinct combination of signals. We train four machine learning classifiers on these feature sets and present our results.

The rest of the paper is organized in the following manner. Section 2 gives an overview of the literature survey related to this work. Section 3 presents the methodology describing the data processing steps followed by the study. Section 4 explains the techniques used in experimentation. Section 5 presents the results of this study, followed by a discussion in Sect. 6. The last section presents the conclusion and the scope of future work.

2 Related Work

Sandulescu et al. [6] presented a machine learning approach to continuously monitor a person's state and classify it as 'stressful' or 'non-stressful'. They used the BioNomadix module from Biopac, a medical wristband that can record electrodermal activity (EDA) and the blood volume pulse (BVP) signals. The Trier Social Stress Test (TSST) [7] was used to evoke stress in the participants.

They collected data on 5 participants and trained an SVM model on each participant's dataset to detect stress. They achieved accuracies in the range of 73–83% in detecting stress.

Schmidt et al. [8] compiled a publicly available dataset known as Wearable Stress and Affect Detection (WESAD) dataset. The data collection process used the RespiBAN Professional, a chest-worn device, and the Empatica E4 wristband to record physiological signal data of 17 subjects. Using the RespiBAN, four signals were recorded – Electrocardiogram (ECG), Electromyogram (EMG), Electrodermal Activity (EDA), and Temperature (TEMP). The Empatica E4 recorded the BVP, EDA, TEMP, and acceleration (ACC). The study aimed to build a multimodal high-quality data for stress detection tasks. The dataset recorded the signals of the participants in three affective states viz., neutral, stress and amusement. The authors also built classification models for both the binary case (stress vs. non-stress) in which up to 93% accuracy was achieved, and the three-class case (stress vs. amusement vs. non-stress) in which up to 80% accuracy was achieved.

Since most stress studies are conducted in constrained settings, some studies also focus on unrestricted real-life environments. Gjoreski et al. [9] proposed a technique for continuous detection of stress using data collected from both laboratory environment and real-life environment using the E4 Empatica Wristband. The laboratory data was labeled by giving a mental arithmetic task to the participants, which elicited a stress response. For the real-life environment, a mobile application was used, which registered log of stressful events at random periods of the day. A context-based stress detector using both datasets was developed, which achieved the highest accuracy of 92% in detecting stress.

Recently, Can et al. [10] developed a stress detection system that used physiological and context-based data for stress detection. They created a dataset of 16 participants, recording 1440 h of EDA and HRV signals using E4 Empatica Wristband and noting their perceived stress levels. They also recorded contextual information such as weather, physical activity level, and activity type. They found an increase in the perceived stress level classification accuracies when they incorporated the weather and activity-type data to train their models, suggesting the importance of contextual data in stress classification. The maximum accuracy of 81% was achieved on 2-class stress level classification using the HRV data and the weather data.

Mozos et al. [11], along with recording physiological data, they also recorded social activity of the participants which included their voice and movement data for the task of stress detection. To measure the physiological data they used the Biopac Wearable Sensor to record raw EDA and PPG signals. They also used a sociometric sensor, a device worn around the neck, to measure the social response in the form of speech, body movement and proximity with other sensors. Their experimental setup was based on the TSST stress test [7] conducted on 18 volunteering students. For each participant different sets of features were selected using the Correlation Based Feature Selection (CFS) method and each feature's discrimination ability was also analysed. They trained Ada Boost classifier for each participant separately and obtained classification accuracies in the range of 81% to 99%. The main contribution of this paper was to analyse the capability of each sensor modality to detect stress which would help in the selection of sensors in the future.

In this paper, we use the Empatica E4 Wristband to collect physiological data for the task of stress classification. Since this study was conducted during the COVID-19 pandemic, most of the public and private institutions in India were under lockdown, due to which, we had to restrict our study to a cohort of 21 participants only, considering the health and safety of the participants. However, this does not degrade the quality of the research as many successful stress studies [12–14] have been conducted on lesser number of participants as well. To engender stress in participants, we use a mental arithmetic task based on the Montreal Imaging Stress Task [15]. Many studies [2,3,12,16] have used the MIST to study stress and its effects, thus highlighting its capability to induce stress.

3 Methodology

3.1 Data Collection

Participants: This study collects data on 21 engineering students of Delhi Technological University, India. All the participants are between 17 to 25 years of age. Before conducting the procedure, each participant is asked for his/her consent. It is ensured that all the participants are healthy and devoid of any physical or psychological illnesses before conducting the test. The study's data acquisition procedure is approved by the Institute of Human Behaviour and Allied Sciences, New Delhi.

Equipment: This paper uses the E4 Empatica Wristband to record four physiological signals from a participant viz., Electrodermal Activity (EDA), Heart Rate (HR), Skin Temperature (SKT), and Blood Volume Pulse (BVP). The E4 Wristband is a real-time data acquisition device. It is a medical-grade device for recording and unobtrusive monitoring of physiological data. The E4 comprises four sensors:

1. **Photoplethysmography (PPG):** Photoplethysmography is a non-invasive, optical technique used to measure the volumetric changes in arterial blood resulting from the heart cycles. This sensor measures blood volume pulse (BVP) from which other cardiovascular features like heart rate (HR) may be derived. The sampling frequency of this sensor to record blood volume pulse 64 Hz, and for heart rate, the sampling frequency 1 Hz. The sensor also uses an in-built motion artefact removal algorithm to remove unwanted signals from the data.
2. **Electrodermal Activity (EDA):** Electrodermal Activity sensor measures the variations in skin conductance. For this purpose, a small amount of alternating current is passed through the skin between the two silver-plated electrodes placed at the strap of the wristband. The skin conductance is recorded at a sampling frequency 4 Hz.

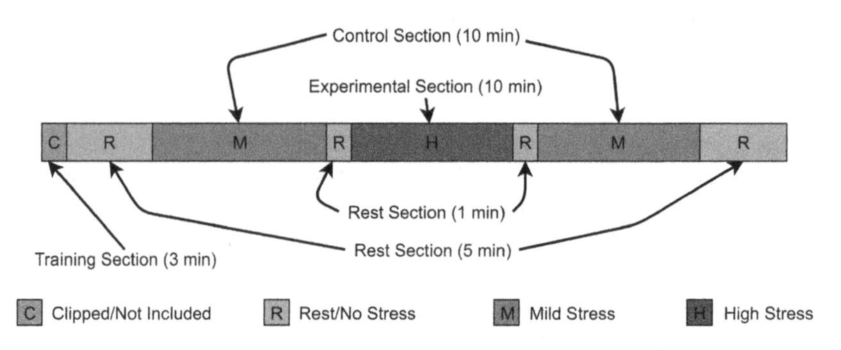

Fig. 1. Overview of the test.

3. **Infrared Thermopile:** A thermopile sensor is a device that can measure an object's temperature by detecting the infrared energy emitted from the object. The infrared thermopile featured by E4 records the skin temperature of the wrist region with a sampling frequency 4 Hz.
4. **3-Axis Accelerometer:** The device also features an accelerometer for recording motion-based activity with high sensitivity across X, Y, and Z axes at a sampling frequency 32 Hz.

Procedure: For the purpose of inducing a stress response in the participants, this study uses a mental arithmetic task based on the standard Montreal Imaging Stress Task (MIST) [15]. The MIST is a computer-based test containing arithmetic questions of varying difficulty levels. There are four sections in the test, i.e., training, rest, control, and experimental. The training section is kept to make the participant familiar with the controls of the test. Each question appearing in the test has a single-digit answer ranging from 0 to 9. The participant answers the questions using a rotatory dial which selects digits from 0 to 9 displayed on the screen. The dial can be rotated using the left and right arrow keys of the keyboard. The rest section does not contain any questions. In this section, the participant is allowed to relax. This section records the baseline physiological signals of a participant. The control section is timed and comprises several arithmetic questions of varying difficulty levels. The questions are adaptive, which means that the difficulty of the questions increases as the participant answers more questions correctly. The participant is instructed to correctly answer as many questions as possible, which introduces stress in the participant. The experimental section contains the adaptive arithmetic questions with the same difficulty levels as the control section. In this section, each question is timed, and the participant has to answer each question in that duration. If the participant fails to answer in time or answers incorrectly, the duration to answer the next question increases. If the participant correctly answers three questions, the duration of the next question decreases. This section also evokes a social threat in the participant. The test shows the participant's current performance score compared with an artificial average performance score of all the

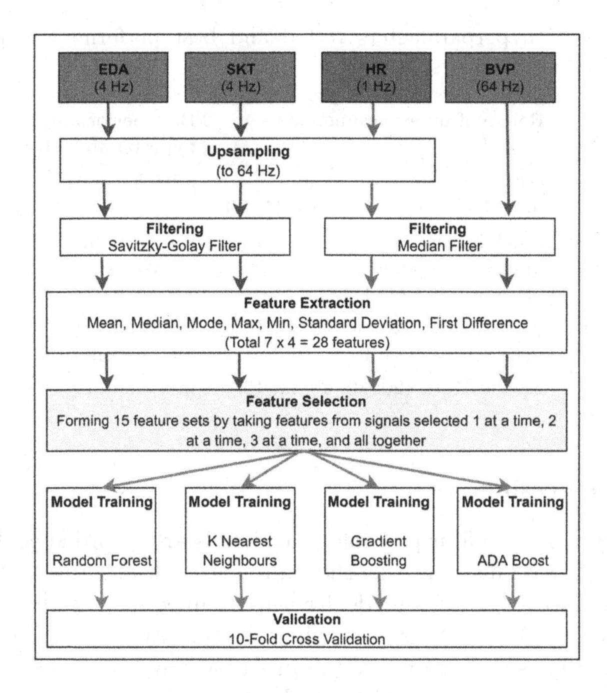

Fig. 2. Flowchart of steps used in data processing and experimentation.

participants on the top of the screen. The artificial global average shown is deliberately kept higher than the participant's capability to score. The participant is instructed to score at least more than the global average to qualify. This helps in eliciting a greater amount of stress in the participant (Fig. 2).

In this paper, we keep two rest sections, two control sections and one experimental section, with a training section in the beginning. In addition, between control and experimental sections, there are small rest sections of 1 min. The whole test takes 45 min to complete.

The physiological signals recorded during the rest sections are labeled as the rest state of the participant or a state with the least amount of stress, the control section is labeled as a state of moderate stress, and the experimental section is labeled as a state of high stress of the participant since this section induces the highest amount of stress. We clip the training section of the test during our analysis, because this section is unlabelled and does not give any valuable information about the psychological state of the participant. Figure 1 shows an overview of the mental arithmetic test.

Before the test, we ask the participant for his/her consent. During the test, the participant is instructed to minimize his/her movements to reduce the artefacts due to motion. After the completing the test, we advice the participant to use techniques like meditation, yoga, or music to return to their normal state.

Table 1. Range of hyperparameters tested and best performing hyperparameters obtained for different classifiers.

Classifier	Range of hyperparameters	Best performing hyperparameters
Random Forest	criterion: ['entropy', 'gini'], max_depth: ['None', 5, 10, 15, 20]	criterion = 'gini', max_depth = 'None'
K-Nearest Neighbours	n_neighbours: [1, 3, 5, 7, 9, 11], distance: ['euclidean', 'manhatten', 'minkowski']	n_neighbours = 1, distance = 'minkowski'
Gradient Boosting	n_estimators: [20, 30, 40, ..., 150], max_depth: [2, 4, 6, ..., 16]	n_estimators = 100, max_depth = 5
ADA Boost	n_estimators: [20, 30, 40, ..., 150]	n_estimators = 100

3.2 Data Processing

Upsampling: Since the four physiological signals are recorded at different sampling frequencies, we must equalize the sample size of all signals before processing. There are two techniques to deal with the unequal sampling frequency of signals viz., upsampling and downsampling. In this study, we have used upsampling, which increases the number of samples of a signal. Among all the features, the blood volume pulse is recorded at the highest sampling frequency 64 Hz. Hence, all the other features i.e., electrodermal activity, heart rate, and skin temperature, are upsampled 64 Hz.

Filtering Noise: Before feature extraction, the foremost step is to reduce noise from the signals. Noise reduction is a widely used pre-processing step when dealing with signals, as it improves the results of later processing steps by removing unwanted components from the signal. For SKT and EDA signals, we use the Savitzky-Golay filter [17], which tends to smoothen the signals without altering their tendency. The HR and BVP signals are filtered using a median filter. A median filter goes through the signal entry by entry and replaces each value by the median of neighboring entries. This is achieved by defining a window size of entries. This paper uses a window size of 4 entries.

Feature Extraction: From each physiological signal, we calculate 7 statistical features viz., *mean*, *median*, *mode*, *standard deviation*, *minimum*, *maximum*, and *first derivative*. For calculating these features we use a window size of 4 for each signal.

4 Experimentation

For performing an extensive comparative analysis of the physiological signals, we build 15 sets of features obtained by taking different combinations of the four signals. The 15 sets are built by taking the four signals one at a time, two

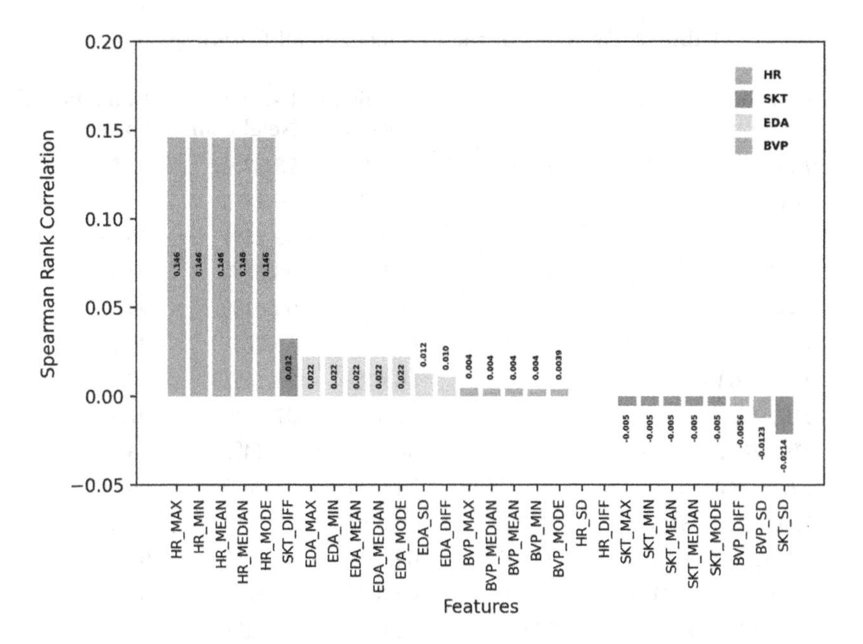

Fig. 3. Spearman rank correlation (ρ) of all the features with respect to the state variable.

at a time, three at a time, and all together. Each set contains 7 statistical features from each signal of that set. We select four machine learning classifiers, i.e., Gradient Boosting, ADA Boost, K-Nearest Neighbours, and Random Forest for building the classification models. These classifiers have been widely used in stress classification tasks as these are computationally inexpensive, easily implementable and suitable for real-time stress monitoring systems. We implement these classifiers using the scikit-learn module of Python, and train them on each set of features.

Before training the models, we also perform hyperparameter optimization of each classifier to search for the best performing hyperparameters. Table 1 shows the range of hyperparameters tested and the best performing hyperparameters for the four classifiers. We use the 10-fold cross validation technique to prevent overfitting of the classifiers and provide an unbiased evaluation of the models.

5 Results

We analyze the correlation of all the features with the response variable which in this case is the state variable denoting the three stress levels. We calculate the *Spearman Rank Correlation* (ρ) value of each feature with the state variable. The Spearman rank correlation is a non-parametric measure of statistical dependence between two variables. It describes how well a monotonic function can depict the relationship between two variables. Figure 3 shows the value of the Spearman correlation for all the features with respect to the state variable.

Table 2. Accuracy of the classifiers on all feature sets.

Feature Set	ADA Boost	Gradient Boosting	K-Nearest Neighbour	Random Forest
EDA	50.82%	57.14%	73.87%	77.54%
HR	50.52%	53.97%	76.18%	76.75%
SKT	48.86%	54.17%	52.25%	60.78%
BVP	38.57%	38.97%	93.58%	93.75%
EDA, HR	53.07%	62.83%	99.01%	99.09%
EDA, SKT	51.63%	62.59%	89.75%	91.55%
EDA, BVP	51.56%	56.91%	94.15%	97.45%
HR, SKT	55.41%	62.42%	97.41%	97.49%
HR, BVP	50.53%	54.08%	95.44%	98.34%
SKT, BVP	48.41%	54.05%	94.06%	96.26%
EDA, HR, SKT	55.72%	66.39%	99.30%	99.41%
EDA, HR, BVP	53.07%	62.68%	96.73%	99.88%
EDA, SKT, BVP	51.59%	62.37%	94.58%	99.08%
HR, SKT, BVP	54.97%	62.00%	96.51%	99.75%
EDA, HR, SKT, BVP	55.72%	66.75%	97.19%	99.92%

We observe that the highest ρ value of 0.162 is obtained by five HR features, indicating a strong positive relationship of the heart rate signal with the state variable. It denotes that the participant's stress level increases with a corresponding increase in the heart rate. The SKT and the BVP features show varied ρ values. Features of both the signals show negative and positive correlations and hence fail to provide any particular trend with the state variable. However, all the EDA features show a consistent positive correlation with the state variable, following the same trend as HR.

This paper trains four machine learning classifiers viz., ADA Boost, Gradient Boosting, K-Nearest Neighbours, and Random Forest on the 15 sets of features. To evaluate and compare the four classifiers on the data, we use the 10-fold cross validation, which employs accuracy as the principal performance metric. Table 2 presents the validation results of the four classifiers on all the feature sets. The results are obtained after extensive hyperparameter optimization.

When we consider signals taken one at a time or feature sets of individual signals, we observe that BVP features give the highest accuracy of 93.75%, followed by feature sets of EDA, HR, and SKT signals. This result contrasts with the inference obtained from considering the Spearman correlation values of the BVP features. Although, Spearman correlation of HR features is the highest, BVP features achieve the highest accuracy in stress classification.

Feature sets built when two signals are taken at a time obtain the best accuracy of 99.09% using EDA and HR features. Among the feature sets containing features from three signals, the set corresponding to EDA, HR, and BVP signals obtains the best accuracy of 99.88%. Finally, when we consider the set containing all the features of the four signals taken together, we get an accuracy of 99.92% which is the highest achieved accuracy of this study.

The evaluation of the four classifiers reveals that Random Forest classifier gives the best classification accuracy in all the feature sets, followed by KNN classifier performing just slightly worse than Random Forest. Gradient Boosting and ADA Boost obtain the worst accuracies, with Gradient Boosting performing just slightly better than ADA Boost.

6 Discussion

The results show that as the number of physiological signals increases, our classification accuracy obtained using the Random Forest classifier also increases. When we consider features from individual signals, the best accuracy achieved is 93.75%. We see a significant improvement in the results for feature sets containing features from more than or equal to two signals as accuracy becomes more than 99%. We infer from the analysis that it is best to use at least two physiological signals for the stress classification task. In the case of the feature set of two signals, our results show that features of EDA and HR signals perform the best. Adding BVP signal to this pair produces a more significant increase in the accuracy than if SKT is added. Also, eventhough, the best classification results of our study are obtained by considering all the four signals taken together, there is only a slight reduction in accuracy when considering models built by taking two signals at a time (EDA, HR), or three at a time (EDA, HR, BVP).

This study can serve as a guide for a developer to select the optimum set of physiological signals to build stress monitoring systems. The development of stress monitoring systems requires the incorporation of expensive biological sensors into the devices. Our analysis shows that using only three signals (EDA, HR, and BVP) obtained from two sensors – photoplethysmography and electrodermal activity sensor, we can achieve classification results that are very close to the results achieved using all the four signals obtained using three sensors (an additional infrared thermopile sensor for SKT). A developer can use this analysis to reduce the cost of manufacturing a stress monitoring system by selecting a minimum number of sensors without compromising the proficiency of the system.

7 Conclusion and Future Work

This research presents a detailed comparative study of four physiological signals, namely, Electrodermal Activity (EDA), Heart Rate (HR), Skin Temperature (SKT), and Blood Volume Pulse (BVP), recorded using the Empatica E4 Wristband, in the task of stress classification. The best stress classification model proposed by this research, is built by using the features of all the four signals taken together. However, the study suggests that, even if two signals are chosen - EDA and HR, we can build equally competent and proficient stress classification models. With only EDA and HR features we achieve more than 99%

accuracy, which suggests the importance of these two signals for building stress classification models. This study also serves as a helpful guide for the selection of physiological signals when building stress classification systems.

Although the work presents a comparative study of the four widely used signals for stress detection, an extended analysis is required which covers other physiological signals like Electroencephalogram (EEG) and Electrocardiogram (ECG) also, so that a complete comparison of signals can be drawn. Since the study is performed in a controlled environment using just a mental arithmetic task to induce stress, we need to expand our data collection process to a more real-life environment so that it becomes relevant for stress classification systems that are more suited for daily-life tasks.

References

1. Giannakakis, G., Grigoriadis, D., Giannakaki, K., Simantiraki, O., Roniotis, A., Tsiknakis, M.: Review on psychological stress detection using biosignals. IEEE Trans. Affect. Comput. (2019)
2. Al-shargie, F.M., Tang, T.B., Badruddin, N., Kiguchi, M.: Mental stress quantification using EEG signals. In: Ibrahim, F., Usman, J., Mohktar, M.S., Ahmad, M.Y. (eds.) International Conference for Innovation in Biomedical Engineering and Life Sciences. IP, vol. 56, pp. 15–19. Springer, Singapore (2016). https://doi.org/10.1007/978-981-10-0266-3_4
3. Priya, T.H., Mahalakshmi, P., Naidu, V.P.S., Srinivas, M.: Stress detection from EEG using power ratio. In: International Conference on Emerging Trends in Information Technology and Engineering (ic-ETITE), pp. 1–6. IEEE (2020)
4. Pourmohammadi, S., Maleki, A.: Stress detection using ECG and EMG signals: a comprehensive study. Comput. Methods Programs Biomed. **193**, 105482 (2020)
5. Nath, R.K., Thapliyal, H., Caban-Holt, A., Mohanty, S.P.: Machine learning based solutions for real-time stress monitoring. IEEE Consum. Electron. Mag. **9**(5), 34–41 (2020)
6. Sandulescu, V., Andrews, S., Ellis, D., Bellotto, N., Mozos, O.M.: Stress detection using wearable physiological sensors. In: Ferrández Vicente, J.M., Álvarez-Sánchez, J.R., de la Paz López, F., Toledo-Moreo, F.J., Adeli, H. (eds.) IWINAC 2015. LNCS, vol. 9107, pp. 526–532. Springer, Cham (2015). https://doi.org/10.1007/978-3-319-18914-7_55
7. Birkett, M.A.: The Trier Social Stress Test protocol for inducing psychological stress. J. Vis. Exp. JoVE (56) (2011)
8. Schmidt, P., Reiss, A., Duerichen, R., Marberger, C., Van Laerhoven, K.: Introducing wesad, a multimodal dataset for wearable stress and affect detection. In: Proceedings of the 20th ACM International Conference on Multimodal Interaction, pp. 400–408 (2018)
9. Gjoreski, M., Gjoreski, H., Luštrek, M., Gams, M.: Continuous stress detection using a wrist device: in laboratory and real life. In: Proceedings of the 2016 ACM International Joint Conference on Pervasive and Ubiquitous Computing: Adjunct, pp. 1185–1193 (2016)
10. Can, Y.S., et al.: Real-life stress level monitoring using smart bands in the light of contextual information. IEEE Sens. J. **20**(15), 8721–8730 (2020)
11. Mozos, O.M., et al.: Stress detection using wearable physiological and sociometric sensors. Int. J. Neural Syst. **27**(02), 1650041 (2017)

12. Minguillon, J., Perez, E., Lopez-Gordo, M.A., Pelayo, F., Sanchez-Carrion, M.J.: Portable system for real-time detection of stress level. Sensors **18**(8), 2504 (2018)
13. Huysmans, D., et al.: Unsupervised learning for mental stress detection-exploration of self-organizing maps. In: 2018 Proceedings of Biosignals, vol. 4, pp. 26–35 (2018)
14. Vila, G., Godin, C., Charbonnier, S., Labyt, E., Sakri, O., Campagne, A.: Pressure-specific feature selection for acute stress detection from physiological recordings. In: 2018 IEEE International Conference on Systems, Man, and Cybernetics (SMC), pp. 2341–2346. IEEE (2018)
15. Dedovic, K., Renwick, R., Mahani, N.K., Engert, V., Lupien, S.J., Pruessner, J.C.: The montreal imaging stress task: using functional imaging to investigate the effects of perceiving and processing psychosocial stress in the human brain. J. Psychiatry Neurosci. **30**(5), 319 (2005)
16. Jun, G., Smitha, K.G.: EEG based stress level identification. In: 2016 IEEE International Conference on Systems, Man, and Cybernetics (SMC), pp. 003270–003274. IEEE (2016)
17. Savitzky, A., Golay, M.J.E.: Smoothing and differentiation of data by simplified least squares procedures. Anal. Chem. **36**(8), 1627–1639 (1964)

An E-Commerce Prototype for Predicting the Product Return Phenomenon Using Optimization and Regression Techniques

Vidya Rajasekaran(✉) and R. Priyadarshini(✉)

B.S. Abdur Rahman Crescent Institute of Science and Technology, Chennai, Tamilnadu, India

Abstract. E-Commerce product returns are considered as a major disease and is also a very challenging issue that greatly impacts the revenue of the E-Commerce firm. Most of the E-Commerce firms consider 10% of the return rates to be normal, but in cases when the product return rate exceeds 10%, the investigation to such cases are further needed. The rising of the return rate is considered as a big threat to the e-commerce industry, which needed to be slowed down. Predictions can be carried out to overcome the return issues in advance and measures can be taken to decrease the return rate. In this paper, the work focuses on developing a prototype model for predicting the return rate of any particular product in advance. In the existing system the return volume management is predicted based on the dependent variable of the manufacturer's production process and their resources alone. Our work focuses on finding the return rate by including a few more parameters which in turn enhances the prediction accuracy. The proposed work is tested on different Machine Learning algorithms for optimizing the results. The results can be applied to overcome the major return and loss percentage by the e-commerce industry to enhance their future revenue.

Keywords: Data analytics · Return rate · Prediction prototype · Metaheuristic · Linear regression · Gradient boosting · Random forest · Machine learning

1 Introduction

In recent years, the e-commerce industry is growing rapidly. Most of the online shopping takes place through the e-commerce system. The major challenge faced by the e-commerce firms is the product return. It affects the revenue of the e-commerce business [6, 11]. The e-commerce company should predict the return rate of the product in order to avoid the financial and the operational costs [14]. The vendor should also predict the return rate in order to avoid the unnecessary losses. So a prediction system becomes mandatory for stopping any kind of losses. The shopping habits for each and every individual customer differs. Each customer has different behavior, places different orders; spend money on different kinds of products. There are also issues where the same product is liked by a particular customer and disliked by another. So this is a biggest challenge for both the vendors and the e-commerce firms. Nowadays social media is being used by the customers where anyone can comment on any product easily which

M. Singh et al. (Eds.): ICACDS 2021, CCIS 1441, pp. 230–240, 2021.
https://doi.org/10.1007/978-3-030-88244-0_22

cuts down revenue of the product easily. Also providing a return option for the product is one of the successes for e-commerce firms, where the customer enjoys a fully satisfied shopping experience. Accepting returns is must for the e-commerce company, but high number of returns for the same product is a serious issue which should be taken into consideration by the e-commerce firm and the vendor [9]. The main reason behind the return is the quality of product or unsatisfied customer. Each product returned affects the revenue and brings an unhappy customer. To improve the profit and sales, the e-commerce system collects and stores the data of customers purchase behavior. This kind of data plays a huge role in decision making, adding a great value to the e-commerce business by improving the sales. The products with high returns can be removed from the catalog by the vendors and the e-commerce firm can suggest idea for improving the sales of the product by providing better quality and ways to satisfy the customer [15]. The return rate of a product is calculated as the percentage of product returns accepted to total sales. [7, 8, 13] understanding the whole supply chain mechanism and their process is mandatory for working on the return issues. A detailed study should be made so that the losses can be understood and the necessary steps can be taken. Sometimes [10, 12] the e-commerce firm also considers the return option to be necessary or not. If the operational cost of return rate exceeds the product cost the firm makes a better decision on declining the return and pay back the product cost to the customer, so that they at least save the unnecessary shipping cost and earn the customer satisfaction.

The primary motive of this paper is to develop an effective model for predicting the return of products. For this research we use a hosted dataset. We developed the model using time, product id and feedback rating. We use a metaheuristic approach to optimize our results based on Linear Regression machine learning method. We summarize our results based on the feedback rating of the products.

2 Literature Review and Real-Time Surveys on Returns in e-Commerce

Similar research work is proposed in [1], where the returns were considered based on the sales, time, product, retailer, production process and resources, multiproduct effect and historical returns. The feedback was not considered, which is an important factor to predict the future returns, since the products with positive feedback will make less number of returns and negative feedback will increase the number of returns. Nearly, 30% of the products ordered online are returned, when compared to the Brick-and-Mortar stores with 8.89% of return [2]. The consumers expect an easy-return policy so that 92% of consumers will get the tendency to buy something again. Free return shipping is expected by 79% of consumers. Nearly 49% of retailers deal with free return shipping where 67% of consumers check the returns policy before placing an order. 62% of consumers are willing to shop online if there are chances to return an item in-store. 58% of consumers want a return policy without asking any reason for return and 47% want an easy-to-print return label. The chances of online purchase offering free return shipping will be 27% for any product that costs more than $1000 and reduces to 10% if free return shipping is not offered. The top three reasons for product return:

1) 23% of return occurs because of receiving wrong item,

2) 22% of the return takes place because of the actual product received looks different from the product ordered,
3) 20% of the return takes place because of receiving the product in damaged condition.

In Barclaycard research [3], focus was made on the serial returners. Consumers make orders on several numbers of items, where they only keep the needed items and return over the rest of the products. Nearly 30% of consumers purposely over-purchase and later return unwanted items. Also 19% of consumers confessed to ordering different varieties of the same item and they can select the items when delivered. A survey conducted by UPS in 2019 states that e-commerce consumers investigate their buying and settle to shops with translucent strategies. It reported that 36% of online shoppers had made a return in the previous three months. 73% of the consumers have stated that their experience in return policies will definitely disturb their idea on buying repeatedly from the vendor or the e-commerce firm. [4, 5] by knowing the returns in advance, unnecessary processing of the product to be shipped and delivered out can be minimized. Also the products which make a very high return rate need to be taken into serious consideration and necessary steps need to be carried out. Not all the returns expect refunds, some return orders also demand for exchanges.

Analyzing the literature and the current issues, the need for this kind of prediction systems are termed to be mandatory in running successful e-commerce businesses.

3 Empirical Research

In this section, the empirical research comprises the below three Sect. 3.1 describes the operational process, Sect. 3.2 describes the data and Sect. 3.3 explains mechanisms for optimizing the results.

3.1 Quantitative Operational Process

In this work we set up an experiment using the quantifiable data to predict the future returns of a product or service. Several vendors are connected to the e-commerce firm in order to sell their products online. Different consumers from various locations place orders to the e-commerce firm. When a consumer places an order for a particular product, the order request is sent to the particular vendor and the shipment is processed. This is termed as successful order sales.

Returns. Returns are considered as a major service provided by the e-commerce companies to retain their customers. The products are returned back from the customers for several reasons. This becomes a hectic job for the vendors and the e-commerce industries to proceed the return process again, since its affects the revenue for both of them. If the product returns are too high, actions need to be taken for reducing return percentage. Vendors and the e-commerce firms need to concentrate more on the return volume to enhance their revenue. The flow of HMRA approach for return volume phenomenon is shown in Fig. 1.

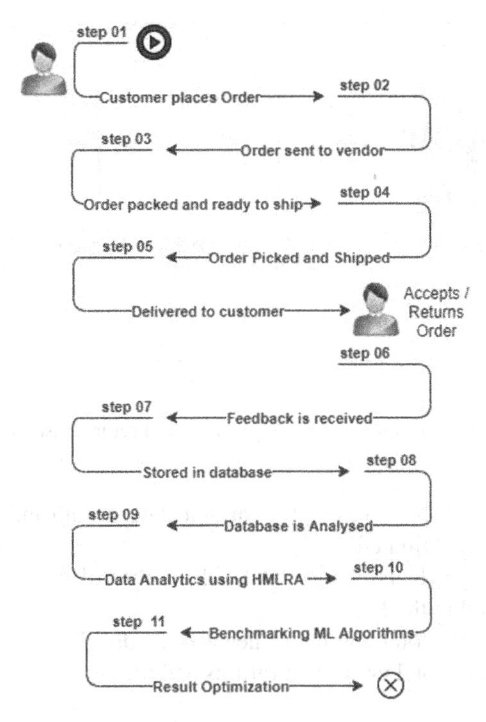

Fig. 1. Flow of HMRA approach for return volume prediction

3.2 Understanding the Data

In this research, we use a dataset with several attributes containing information about the product sales detail. Each and every product is assigned with a unique product_id, customer_id, manufacturer name, return information. For every product sold through the e-commerce site feedbacks are received from the customer and they are stored in the dataset. We use these values altogether and perform computations for predicting the future return rate of any particular product.

3.3 Variables Used in Optimizing Results

We focus on the time, product and feedback variables in particular for predicting the return rate of the products and store the result in $\Delta_{tpf.}$ The subscripts t, p, f denote the time period, Product_id and feedback rating. We use the descriptive data to make the prediction analytics. Diagnostic analytics technique is used to analyze the cause for increased return rate and the prescriptive analytics approach is used for recommending the measures to be carried out for minimizing the return rates.

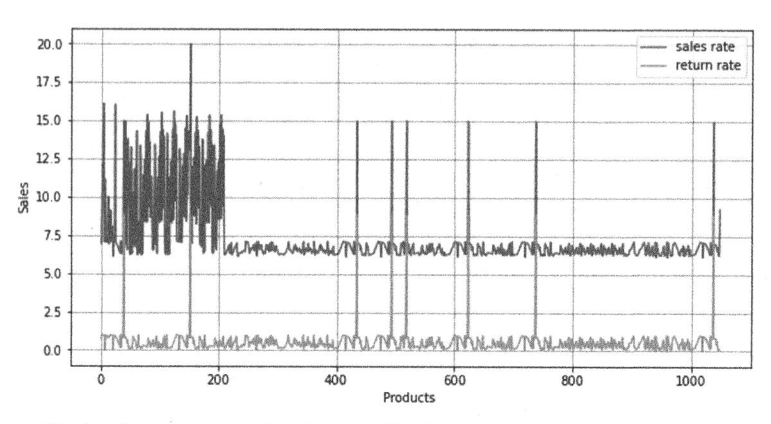

Fig. 2. Graph representing the overall sales and return based on products

Time (t). The time is denoted using the variable date, month and year of the purchase is made and the return is initiated.

Product (p). The products denote the total number of products sold in different categories as illustrated in the Fig. 2.

Feedback (f). It is the total number of negative feedback received which is directly related to the return. The total number of returns and the negative feedback received are illustrated below in Fig. 3.

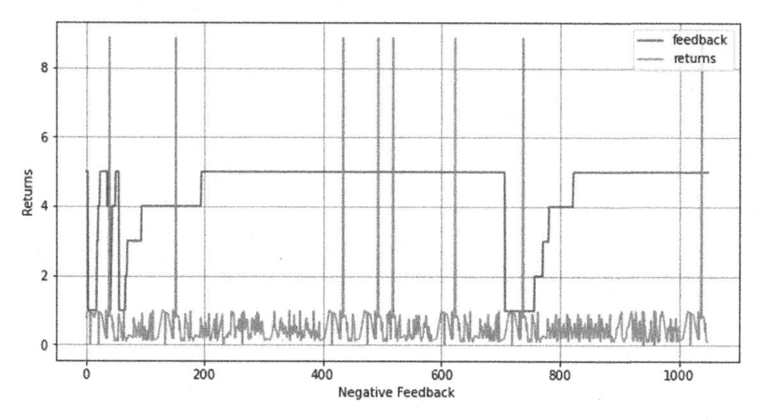

Fig. 3. Graph representing the overall returns based on the negative feedback

4 Hybrid Metaheuristic Based Regression Approach (HMRA)

1. Start
2. Declare input variables as σ_{p_id}, τ_{p_id}, υ_{p_id}, φ_{p_id}
3. Find Count of (A_{manuf_return}) using (σ_{p_id} / τ_{p_id}) * 100
4. Find Count of ($B_{prod\_feedback}$) using (υ_{p_id} / φ_{p_id}) * 100
5. Compute [(A_{manuf_return} + $B_{prod\_feedback}$)/2]/100
6. Store and print the result in Δ_{tpf}
7. Prediction of results

   if Δ_{tpf}= 0 then

   Print ("Type I : Chances of return are low")

   elseif Δ_{tpf}> 0 and <= 0.25 then

   Print ("Type II : Chances of return are moderate ")

   elseif Δ_{tpf}> 0.25 and <= 0.5 then

   Print ("Type III : Chances of return are high")

   elseif Δ_{tpf}> 0.5 and <= 0.75 then

   Print ("Type IV : Chances of return are very high ")

   elseif Δ_{tpf}> 0.75 and <= 1 then

   Print ("Type IV : Chances of return are extreme ")
8. End

5 Experimental Illustration

The experimentation is carried out to predict how much likely a particular product will be returned by the customer using a sample dataset. In this work the return score of any product Δ_{tpf} over a particular time period is calculated by summing the resultant values of the below outcomes of A_{manuf_return} and $B_{prod\_feedback}$ as shown in Eq. (1). Here,

1. Return value of a particular product for the manufacturer (A_{manuf_return})
2. Feedback score of the particular product ($B_{prod\_feedback}$)

$$\Delta_{tpf} = A_{\text{manuf\_return}} + B_{prod\_feedback} \tag{1}$$

5.1 Manufacturer Return Percentage (A_{manuf_return})

Several products are shipped from different manufacturers at different periods of time based upon the orders placed by consumers. While some products are returned back demanding for exchange or refund. The products returned back to the manufacturer are calculated in percentage as, Manufacturers return percentage (A_{manuf_return}). For each product with respective id, it is obtained from the values of the total number of times a particular product is returned back to the manufacturer (σ_{p_id}) by the total number of times a particular product is shipped from the manufacturer (τ_{p_id}) as stated in Eq. (2).

$$A_{manuf\_return} = \frac{\sigma_{p\_id}}{\tau_{p\_id}} \times 100 \tag{2}$$

Products Shipped from Manufacturer ($\sigma_{\mathbf{p\_id}}$). The total number of products dispatched from any particular manufacturer are extracted from the database and represented in three columns as product_id, the manufacturer name and the last column represents the total count of the products shipped from the manufacturer. The product from the manufacturer K-Y is considered to illustrate our experimentation. The K-Y manufacturer has shipped products that are represented with the product_id:AV16khLE-jtxr-f38VFn. The total number of dispatched count is found to be 27 for the particular product which is marked as $\sigma_{\text{p\_id}}$.

Products Returned Back to the Manufacturer ($\tau_{\mathbf{p\_id}}$). The outcome for the total number of products returned back from the consumers for several reasons are estimated. The count on the total number of returns made by consumers for any particular product which is referred with the product_id is estimated. The product of the K-Y manufacturer is returned nine times with the specific product_id:AV16khLE-jtxr-f38VFn. The total number of returned count is 9 for the particular product which is marked as $\tau_{\text{p\_id}}$.

Return Percentage of the Manufacturer ($\mathbf{A_{manuf\_return}}$). Therefore, by applying the results obtained we can determine the return percentage value of the manufacturer (A). Just for illustration we will take the sample output values of $\sigma_{\text{p\_id}}$ and $\tau_{\text{p\_id}}$ for the K-Y manufacturer and apply it in Eq. (3). To find the resulting return percentage of the manufacturer. For example,

$$A_{K-Y} = \frac{9_{\text{AV16khLE−jtxr−f38VFn}}}{27_{\text{AV16khLE−jtxr−f38VFn}}} \times 100 \tag{3}$$

$$A_{K-Y} = 33.33\% \tag{4}$$

In Eq. (3). The return percentage of K-Y manufacturer is calculated and the resulting score of A K-Y is found to be 33.33% as shown in Eq. (4). Is obtained in the results as shown in Fig. 4.

```
product.id             manufacturer
AV14LG0R-jtxr-f38QfS   Lundberg                  100.00
AV16khLE-jtxr-f38VFn   K-Y                        33.33
AV1YGDqsGV-KLJ3adc-O   Windex                     25.57
AV1YlENIglJLPUi8IHsX   Kind Fruit & Nut Bars      41.18
AV1YmBrdGV-KLJ3adewb   Pantene                    27.78
                                                    ...
AVpjd_6lilAPnD_xNsA8   Maybelline                 33.05
AVq56bxmv8e3D1O-llXA   Paramount                   7.01
AVqVHr30nnc1JgDc3jVp   WarnerBrothers              7.15
AVzRE7KvGV-KLJ3aatQ9   Kiss Products, Inc          9.79
AVzRGDlw-jtxr-f3yfFH   Opti-Free                  36.16
Length: 311, dtype: float64
```

Fig. 4. Count on the return rate of the manufacturers in percentage

5.2 Feedback Review Percentage ($\mathbf{B_{prod\_feedback}}$)

The return and feedback are related to each other, so whenever a product return is made from unsatisfied customers that will also surely reflect with a poor feedback from the consumers. The feedback percentage ($B_{prod\_feedback}$) is calculated using the sum obtained from the negative feedbacks received for individual products (υ_{p_id}) by the total number of feedbacks received for each product (φ_{p_id}) as shown in Eq. (5).

$$B_{Prod\_feedback} = \frac{\upsilon_{p\_id}}{\varphi_{p\_id}} \times 100 \tag{5}$$

Sum of the Negative Feedbacks Received for Any Product (υ_{p_id}). The feedback for the delivered products are received from the customers by the E-Commerce firm. The feedbacks are received in the form of texts and star ratings. In this work we consider only the feedback received through star ratings and the textual feedback will be included in the future work. The star ratings with less than three stars for any product are considered and summed together. The star value <3 stars for any product is considered as negative feedback in this work, which is denoted as (υ_{p_id}) and the results are obtained.

Total Number of Feedbacks Received for Each Product (φ_{p_id}). The total number of feedbacks received for any particular product, is the summation of all the feedback with star values ranging anything between 0 to 5 is taken for consideration and is denoted as (φ_{p_id}). 0 is the null value representing no star rating, which can be omitted since it's not going to make any difference. The results for the total feedback count is extracted.

Overall Feedback Percentage Received for Any Product ($B_{prod\_feedback}$). The final overall feedback percentage of any product is retrieved by applying the resulting values of (υ_{p_id}) and (φ_{p_id}) in Eq. (5). The output values are shown in Fig. 5. For illustration we will take the sample output values of p_id and (φ_{p_id}) for the K-Y manufacturer and apply it in Eq. (5) to find the overall feedback percentage of a particular product. For example,

$$B_{K-Y} = \frac{11_{\text{AV16khLE–jtxr–f38VFn}}}{27_{\text{AV16khLE–jtxr–f38VFn}}} \times 100 \tag{6}$$

$$B_{K-Y} = 40.74 \tag{7}$$

In Eq. (6) the feedback percentage of the K-Y manufacturer is calculated and the resulting score of B K-Y is found to be 40.74% as shown in Eq. (7). The results can be viewed in implementation as shown in Fig. 5.

5.3 Final Return Prediction Percentage of Any Product (Δ_{tpf})

The final return prediction is calculated by substituting the scores of the results obtained from Eqs. (2) and (5) and applied in Eq. (1). The final resulting value Δ_{tpf} for a particular product from manufacturer k-y is obtained as shown in Eqs. (8) and (9),

$$\Delta_{tpf} = 33.33 + 40.74/2 \tag{8}$$

$$\Delta_{tpf} = 37.035 \tag{9}$$

```
product.id            manufacturer
AV16khLE-jtxr-f38VFn  K-Y                      40.74
AV1YGDqsGV-KLJ3adc-O  Windex                   18.39
AV1YlENIglJLPUi8IHsX  Kind Fruit & Nut Bars     5.88
AV1YmDL9vKc47QAVgr7_  Aussie                    4.21
AV1Yn94nvKc47QAVgtst  CeraVe                    4.00
                                                 ...
AVqkGdQ6v8e3D1O-leAl  Sony                      9.75
AVqkSpW-v8e3D1O-lgn1  Karaoke USA              60.00
AVsRL7C9U2_QcyX9PFuJ  Skittles                 50.00
AVzRE7KvGV-KLJ3aatQ9  Kiss Products, Inc        2.49
AVzRGDlw-jtxr-f3yfFH  Opti-Free                 7.59
Length: 236, dtype: float64
```

Fig. 5. Feedback percentage for each product

For simplicity, we convert the value of Δ tpf ranging between 0 to 1 and based on the range value, we categorize Δ tpf into four groups as shown in Table 1. The description for the four groups Type I, Type II, Type III and Type IV are illustrated in Table 1. In our example the range value is obtained as 37.035% as shown in Eq. (9) which is divided by 100 for conversion to range value and we get the result as 0.37035 which falls in the range converted so the conclusion is made based on the description given in Table 1.

Table 1. Prediction based on categorization of Δ_{tpf}

Groups	Range	Description on prediction
Type I	Δ tpf $= 0$	Strongly very less chances of return
Type II	$0 > \Delta$ tpf ≤ 0.25	Return chances are moderate
Type III	$0.25 > \Delta$ tpf ≤ 0.5	Return chances are high
Type IV	$0.5 > \Delta$ tpf ≤ 0.75	Chances of return is very high
Type V	$0.75 > \Delta$ tpf ≤ 1	Chances of return is extreme

Table 2. Comparison of performance based on Machine Learning Algorithms

ML	Test data	Training data
Linear regression	Mean Absolute Error: 0.01452 Mean Squared Error: 0.00846 Root Mean Squared Error: 0.09200	Mean Absolute Error: 0.01451 Mean Squared Error: 0.00901 Root Mean Squared Error: 0.09492
Random forest	Mean Absolute Error: 0.02002 Mean Squared Error: 0.00903 Root Mean Squared Error: 0.09504	Mean Absolute Error: 0.02031 Mean Squared Error: 0.00947 Root Mean Squared Error: 0.09736
Gradient boosting	Mean Absolute Error: 0.02016 Mean Squared Error: 0.00888 Root Mean Squared Error: 0.09427	Mean Absolute Error: 0.02124 Mean Squared Error: 0.00940 Root Mean Squared Error: 0.09695

The results for the comparison of prediction performances are shown in Table 2. Based on the results, it's found that Gradient Boosting Algorithm correlates well to our approach when compared to Linear Regression and Random Search Algorithms.

6 Conclusion and Future Work

There are no proper mechanisms for predicting the return volume in e-commerce systems with more accuracy. In this work we consider the feedback received through star ratings and the textual feedback are to be included in the future work. The current research work shows the best suited machine learning algorithm for the prediction of return rates. The proposed hybrid approach shows improved accuracy compared to the existing methods. The return rate prediction is not only limited to feedback and products, several other factors shall be included in the future work. The function approximation will be carried out using classification techniques. The future research will include the methodologies on reducing the operational and financial costs for the e-commerce firm in their return policies.

References

1. Cui, H., Rajagopalan, S., Ward, A.R.: Predicting product return volume using machine learning methods. Eur. J. Oper. Res. (2019). https://doi.org/10.1016/j.ejor.2019.05.046
2. E-Commerce Product Return Rate – Statistics and Trends [Infographic] (invespcro.com) (2020)
3. E-CommerceReturns: Stats and Trends (2020), Ecommerce Returns: 2020 Stats and Trends – SaleCycle
4. Walsh, G., Möhring, M.: Effectiveness of product return-prevention instruments: empirical evidence. Electron. Mark. **27**, 341–350 (2017). https://doi.org/10.1007/s12525-017-0259-0
5. Vlachos, D., Dekker, R.: Return handling options and order quantities for single period products. Eur. J. Oper. Res. **151**(1), 38–52 (2003). https://doi.org/10.1016/S0377-2217(02)00596-9
6. Grifs, S.E., Rao, S., Goldsby, T.J., Niranjan, T.T.: The customer consequences of returns in online retailing: an empirical analysis. J. Oper. Manag. **30**(4), 282–294 (2012)
7. Stock, J.R., Mulki, J.P.: Product returns processing: an examination of practices of manufacturers, wholesalers/distributors, and retailers. J. Bus. Logist. **30**(1), 33–62 (2009)
8. Ma, F.: The study on reverse logistics for e-commerce. In: 2010 International Conference on Management and Service Science, Wuhan, China, pp. 1–4 (2010). https://doi.org/10.1109/ICMSS.2010.5575577
9. Shivakumar, S.K., Suresh, P.V.: Maximizing knowledge management returns in e-commerce. In: 2014 International Conference on Computing for Sustainable Global Development (INDIACom), New Delhi, India, pp. 545–550 (2014). https://doi.org/10.1109/IndiaCom.2014.6828018.
10. Yang, H., Wang, J., He, M., Kuang, B.: Research of B2C e-commerce return strategies based on return price. In: 2010 International Conference on Management and Service Science, Wuhan, China, pp. 1–4 (2010). https://doi.org/10.1109/ICMSS.2010.5576679
11. Morganti, E., Seidel, S., Blanquart, C., Dablanc, L., Lenz, B.: The impact of e-commerce on final deliveries: alternative parcel delivery services in France and Germany. Transport. Res. Proc. **4**, 178–190 (2014). https://doi.org/10.1016/j.trpro.2014.11.014

12. Scott Matthews, H., Hendrickson, C.T., Soh, D.L.: Environmental and economic effects of e-commerce: a case study of book publishing and retail logistics. Transp. Res. Rec. **1763**(1), 6–12 (2001). https://doi.org/10.3141/1763-02
13. Amin, S.H., Zhang, G.: A multi-objective facility location model for closed-loop supply chain network under uncertain demand and return. Appl. Math. Model. **37**(6), 4165–4176 (2013). https://doi.org/10.1016/j.apm.2012.09.039
14. Ivanov, D., Pavlov, A., Pavlov, D., Sokolov, B.: Minimization of disruption-related return flows in the supply chain. Int. J. Prod. Econ. **183**, 503–513 (2017). https://doi.org/10.1016/j.ijpe.2016.03.012
15. Ramanathan, R.: An empirical analysis on the influence of risk on relationships between handling of product returns and customer loyalty in e-commerce. Int. J. Prod. Econ. **130**(2), 255–261 (2011). https://doi.org/10.1016/j.ijpe.2011.01.005

Crop Yield Prediction for India Using Regression Algorithms

Devansh Hiren Timbadia(✉), Sughosh Sudhanvan, Parin Jigishu Shah, and Supriya Agrawal

Computer Engineering Department, NMIMS University, Mumbai, India
supriya.agrawal@nmims.edu

Abstract. Agriculture is the backbone of any developing country. Currently, there is no single model that can provide accurate yield predictions at a pan-India level. Thus, in this paper, a yield prediction model has been proposed, which can predict the annual yield for 36 crops, grown in 542 districts of India. The proposed method makes use of various machine learning algorithms, linear regression, support vector machines, and artificial neural networks, to achieve an average Root Mean Square Error of 1.065 (quintals per 10 acres). The proposed model, in addition to predicting yield for all the major districts with an average accuracy of over 90%, also covers more crops as compared to existing works.

Keywords: Machine learning · Agriculture · Yield prediction · SVM · MLR · ANN

1 Introduction

Agriculture and its allied industries act as a basis for most of the industries in a country, especially in a developing country, like India. Agricultural produce not only provides food for people but also for the cattle in the form of fodder and forage. Agricultural produce can be determined by identifying the yield produced from all the different kinds of crops.

Yield prediction is the method of estimating the production of a crop, for a given farm area. The predictions can be seasonal or annual and can be performed for a farm, a district, a state, or the whole nation. A timely and accurate estimation of crop yield can help policymakers to make appropriate decisions on strengthening the national food security [1] and also allow the farmers to make informed financial and management decisions at the time of sowing itself.

The yield of a crop is highly dependent on weather conditions. Extreme weather conditions can affect the crop yield in a major way and at times can completely destroy the crop too, but such events can be predicted and prevented, well in advance. Other factors that affect the yield of a crop are pest infestations or crop diseases, nutrient content of the soil, or improper scheduling of the harvest cycle.

Various aspects of weather play a vital role in optimal crop yield. Rain plays a crucial part in achieving appropriate crop yield, but, untimely or insufficient rains can produce

M. Singh et al. (Eds.): ICACDS 2021, CCIS 1441, pp. 241–251, 2021.
https://doi.org/10.1007/978-3-030-88244-0_23

poor quality crops, causing huge losses to the farmers. Each crop has a specific range of temperature too, within which the yield of the crop is maximized, any deviation from the specific temperature range can affect the crop's yield.

Traditionally, the estimation of crop production was based on the experience of the farmer or an expert. Today, the estimation of crop yield in India is done by the Directorate of Economics & Statistics, Ministry of Agriculture (DESMOA). They release the yield forecast 4 times a year, the first in the middle of September, the second in January, the third towards the end of March, and the fourth by the end of May [2]. At present, various methods are being used to estimate the crop yield like statistical methods [3], stochastic methods [4], various machine learning algorithms, or IoT devices using sensors to estimate crop yield [5].

When machine learning methods are applied for estimating the crop yield, a large amount of data can be analyzed simultaneously and at a much faster rate as compared to traditional or numerical approaches, and as a corollary provides better accuracy in the estimated values due to reduced human involvement.

Section 2 contains the summary of various existing literature works. Section 3 provides a list of the various datasets used in this project. Section 4 explains in detail the components and working of the proposed model. Section 5 provides analysis on the results obtained. Section 7 gives a conclusing remarks, limitations, and future work.

2 Literature Review

The authors, Manjula & Djodiltachoumy [6], have used data mining in crop yield analysis. This paper focuses on only one region, i.e., Tamil Nadu (in India), and uses Association Rules of data mining to predict appropriate value efficiently. The overall accuracy for data from 2000 to 2012 for Tamil Nadu is about 90%.

The authors, Gandge & Sandhya [7], have tried to predict crop yield by mining climatic factors like temperature, rainfall, and agronomical parameters like soil nutrients content. After narrowing down the region and crops, all the data is fetched, and noise is removed. Then the best parameters from all the features are selected from data mining, then the yield is predicted using appropriate machine learning algorithms. The paper presents a list of algorithms used, crops selected, and the accuracies for some main crops. For any crop, they suggest Support Vector Regression.

The authors, Ramesh, D, and Vardhan [8], have used Multiple Linear Regression (MLR) and Density-based Clustering algorithms to predict the crop yield for a single region, i.e., East Godavari district, Andhra Pradesh, India. They have used data of the location dated from 1955 to 2009. The data consists of Rainfall, Area, Yield, Fertilizers, and Production. In MLR, the predicted values varied between -14% to $+13\%$. For the same data, Density-based clustering gave results ranging between -13% and $+8\%$ for 6-clusters approximation.

The authors, Kumar et al. [9], have proposed an algorithm where, if at a time there can be more than one crop that can be grown then the crop with maximum net-yield over the season can be selected. The crops are first categorized as Seasonal Crops, Whole Year Crops, Short Time Plantation Crops, and Long Time Plantation Crops. A combination of crops can be selected based on the timeline of the given period. A dataset of the sowing period, harvesting period, growing days, and yield is created using machine learning algorithms. Then the best crop can be selected at a time.

The authors, Medar et al. [10], are relying on advanced machines and technologies to gather useful and accurate information to be fed as inputs to the machine learning algorithms to predict crop yield. They have used two machine learning techniques, Naïve Bayes and the K-Nearest Neighbour algorithm, and have compared their performances in prediction. The Naïve Bayes algorithm gives a very good accuracy of 91.11% as compared to KNN which gives 75.55% and hence Naïve Bayes is a superior algorithm out of the two.

The authors, Veenadhari et al. [11], have developed a software tool named 'Crop Advisor', which is a webpage for predicting the influence of climatic parameters on crop yield for selected districts of Madhya Pradesh, India. They have created rules for each selected district for soybean, paddy, maize, and wheat, based on main climatic parameters like cloud cover days, wet day frequency, evapotranspiration, rainfall, and minimum temperature. Based on historical values of yield, conditions of climatic parameters are checked against the selected crop and district, and yield is predicted. The average accuracies vary from 76% to 87%.

The authors, Guruprasad et al. [12], have tried to estimate crop yield based on the weather and soil data for paddy crop at different district and taluk levels. They have also performed dis-aggregation of district-level yield data by applying machine learning models trained using district-level data to predict taluk level yield data. The selected district is Siddharth-Nagar (of Uttar Pradesh) and its five Taluks are Bansi, Domariaganj, Etwa, Naugarh, and Shoratgarh. They collected data for these regions from 2011 to 2017 and taluk data yield average error is 3.14% and by dis-aggregation is 6%.

The authors, Suresh et al. [13], have used K-Means and K-Nearest Neighbour algorithms and used MATLAB and WEKA as tools for clustering and classification in crop yield prediction. They have used data of major crops (rice, maize, ragi, sugarcane, and paddy) production, rainfall, groundwater, and cultivation area in the state of Tamil Nadu. Then they have used K-Means to create 5 clusters (very low, low, moderate, high, very high) of supervised data. This data is given to KNN to predict the yield. They have used modified KNN to predict yield which gives an accuracy of about 96%.

The authors, Kantanantha et al. [4], have presented a weather-based regression model with time-dependent varying coefficients to predict yearly crop yield and money earned. For accurate prediction of yield, weekly data of climatic parameters are used which generates a large number of correlated predictors. To overcome this Functional Principal Component Analysis (FPCA) is implemented to reduce the space of predictors. They have used their methods to corn yield and price forecasting for Hancock County in Illinois.

The authors, Rale et al. [14], have tried to create a model that can be accurate as a traditional model or even more accurate. They have used various linear and non-linear 5-fold cross-validation models and have compared their results. They have used two-year data of winter wheat data (2013, 2014) which is specific geolocations.

The authors, Doshi et al. [15], have used a combination of Big Data Analytics and Machine Learning to find the best crop that can be grown in a region having particular soil and climatic characteristics. They have used a dataset containing thirty-year historical records of soil and meteorological parameters of places in India. They provide the data of five major (bajra, jowar, maize, rice, and wheat) and fifteen minor (barley, cotton, groundnut, gram, jute, other pulses, potato, ragi, tur, rapeseed and mustard, sesame, soybean, sugarcane, sunflower, tobacco) crops. They used four machine learning algorithms out of which neural network gave the best accuracy (91%).

The authors, Kulkarni et al. [16], have used the ensembling technique which is to use a combination of machine learning algorithms to make predictions of the right crop to be selected for a specific soil condition. After getting prediction classes of Random Forest, Naïve Bayes, and SVM individually, the correct classes are selected by using the majority voting method. Once we take the combination of climatic parameters too, like rainfall and temperature and soil physical and chemical characteristics, they achieved an accuracy of 99.91%.

The authors, Ratkal et al. [3], have presented a method to predict crop yield and price a farmer could expect from his field by using historical data. They have used a sliding window nonlinear regression model approach to predict the required values. They have done their studies for several districts of the state of Karnataka. They have found the dependency of rainfall, previous price, and temperature values for prediction on production, and the weights are 50%, 35%, and 15% respectively. When they compared their values with the 2014 actual data, they found an average error of 10%.

The authors, Jain et al. [17], have used the WEKA tool to carry out the Machine Learning. They have done two types of prediction, crop selection method, and crop sequencing method. In the crop selection method, crops are selected over a specific season depending upon various environmental as well as economic factors (precipitation levels, average temperature, soil type, market prices, and demand) for the maximum benefit. The Crop Sequencing Method uses a sequencing algorithm to suggest the sequence of the crop(s) based on yield rate and market prices.

Currently, amongst all the existing works, there is no prediction model that can provide accurate yield predictions at a pan-India level. Also, most of the existing works on yield prediction, cover a handful of crops and for a small region (single district or state) only.

3 Datasets Used

The first dataset for this study was taken from the *Open Government Data Platform* [18]. It has complete monthly data ranging from 1997 to 2019 for 542 districts of India. The dataset contains 36 different kinds of crops, horse gram, groundnut, gram, jowar, maize, rice, ragi, green gram, cotton (lint), bajra, castor seed, sesamum, urad, wheat, small millets, oilseeds, turmeric, rapeseed, linseed, onion, potato, jute, mesta, arhar (tur), coriander, tapioca, coconut, cashew nut, arecanut, black pepper, sugarcane, barley, masoor, soyabean, khesari, and other kharif pulses. The dataset required some preprocessing as there were many missing values, such rows had to be deleted, and also there were different spellings for the same districts, which created inconsistency and needed correction. The main parameters extracted from the dataset were production and farm size.

The dataset for temperature was obtained from two different sources, from *India Water Portal* [19] for 1901 to 2008 and from *World Weather Online* [20] for 2009 to 2019. Similarly, the dataset for rainfall was also merged from two different sources, *India Water Portal* [21] and *Environics India* [22]. The combined rainfall dataset was from 1901 to 2019.

4 Proposed Model

The algorithm for yield prediction has been divided into 3 stages:

1. Data Pre-processing
2. Model Comparison
3. Optimized Yield Prediction

4.1 Stage I: Data Pre-processing

In this stage, the primary aim is to merge all the datasets and clean them, i.e., to make them usable for the prediction of yield. In the first step, A location is selected by specifying the state and district, and the required datasets of rainfall, temperature, and yield for the specified location are collected.

To achieve a standardized output, the production data was normalized to "quintals per 10 acres". Based on the selected season, the range of months is decided, for example, *October to March* for *Rabi.* To calculate the annual yield, the average temperature and the total rainfall for the selected months are considered. The flowchart for the same is as shown in Fig. 1.

To prevent unwanted bias in the final prediction, outlier analysis was performed. It was observed that a lot of outliers were found in the 4–6 percentile range, so the top and bottom 5 percentile of data was removed. After performing all the above tasks, a final *'processed dataset'* was obtained.

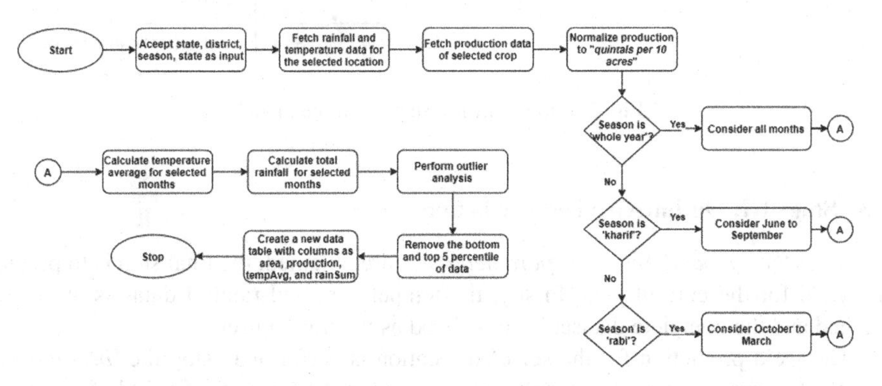

Fig. 1. Flowchart for Stage I of the model

4.2 Stage II: Model Comparison

After reviewing various existing research works and understanding their models used for yield prediction, the authors decided to implement the models, Support Vector Machine (SVM), Multiple Linear Regression (MLR), and Artificial Neural Networks (ANN) in form of an ensemble approach to achieve the best results. The chosen attributes were converted into parameter sets, as (i) Rainfall and Temperature, (ii) Rainfall, and (iii) Temperature. The training dataset was created using data until the 2nd last year and the testing set was for the last year.

Next, for each model, predictions were made using all three parameter sets and the results were saved in the form of a list. For all the results in the list, the error in each production data was calculated and whichever pair of (model, parameter set) gave the least error, was selected as the '*best model*' and '*best parameter set*'. The flowchart for the same is as shown in Fig. 2.

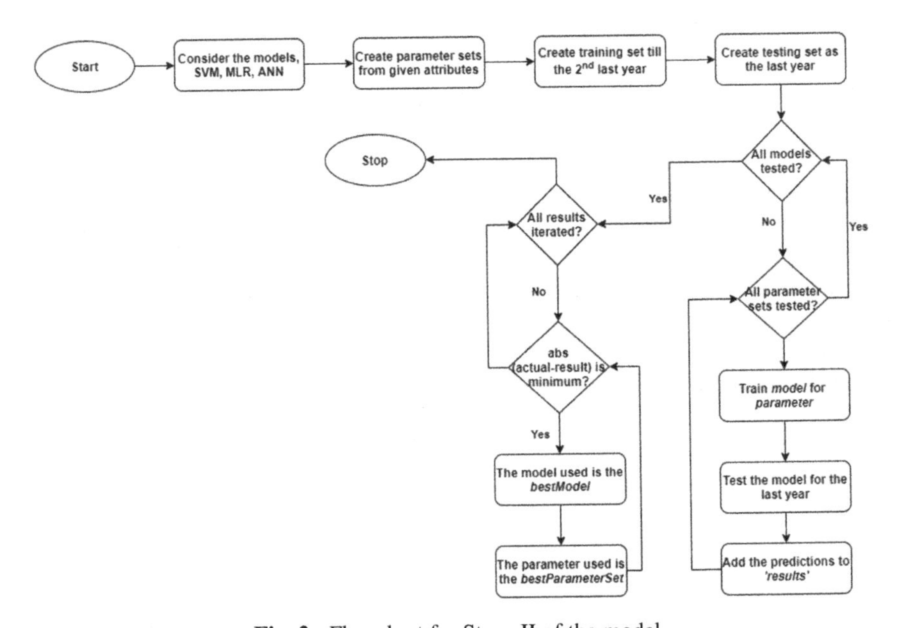

Fig. 2. Flowchart for Stage II of the model

4.3 Stage III: Optimized Yield Prediction

After the '*best model*' and '*best parameter set*' are obtained, the final step is to predict the yield for the current year. Firstly, the temperature and rainfall datasets are to be fetched and the complete dataset is considered as the training set.

The yield prediction for the selected location is performed using the '*best model*' and '*best parameter set*'. The result is then stored in the form of a *CSV* file for ease of conversion while visualizing the results. The flowchart is as shown in Fig. 3.

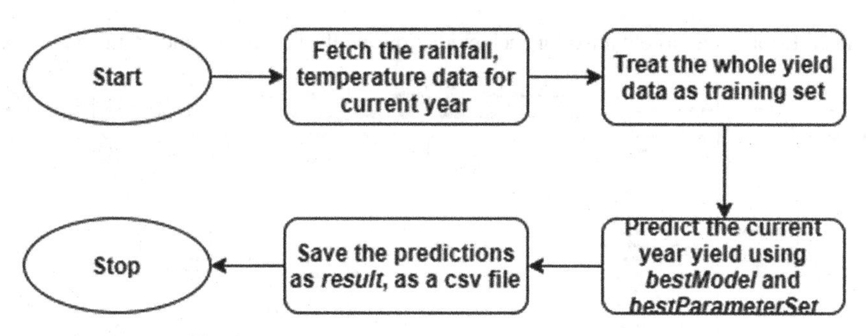

Fig. 3. Flowchart for Stage III of the general model

5 Result Analysis

For testing purpose, yield predictions were performed for two districts of Andhra Pradesh, Anantapur, and Chittoor, for both Rabi and Kharif seasons, consisting of crops like maize, groundnut, rice, ragi, moong, bajra, castor seed, and horse gram.

For each crop, all the three models and three parameter sets (as mentioned in Sect. 4.2) were implemented for each model, as shown in Table 1. All the predicted values of yield were calculated as '*quintals per 10 acres*'.

Table 1. Yield predictions for each crop, using all models and parameter sets

State	District	Season	Parameter Set ->	Temperature, Rainfall			Rainfall			Temperature			Actual
			Model -> / Crop	SVM	MLR	ANN	SVM	MLR	ANN	SVM	MLR	ANN	
andhra pradesh	anantapur	Rabi	Horse-gram	5.25	5.12	10.02	6.35	4.86	1.71	5.38	5.07	1.87	*4.58*
		Kharif	Groundnut	4.71	4.75	1.53	3.30	5.09	1.95	2.26	4.37	16.66	*2.70*
		Rabi	Gram	6.49	7.43	1.46	7.32	7.29	2.63	8.47	7.07	15.57	*5.68*
			Jowar	6.47	9.25	4.13	10.81	9.30	1.82	8.36	8.02	2.41	*4.75*
			Maize	71.33	66.14	36.92	71.24	67.74	26.66	71.57	68.24	39.77	*70.67*
			Rice	24.31	25.06	16.42	24.26	24.91	9.29	23.50	24.72	0.68	*21.94*
		Kharif	Rice	27.31	28.12	12.99	32.78	28.12	10.44	30.55	27.31	2.70	*30.08*
			Ragi	17.35	16.90	13.77	16.96	19.31	6.59	15.64	17.16	24.53	*15.08*
			Moong(Green Gram)	4.46	3.74	2.49	4.22	4.35	1.96	0.84	4.41	11.72	*11.10*
			Maize	28.63	30.99	1.37	30.93	30.22	11.17	22.41	29.68	9.20	*25.35*
			Jowar	13.68	10.31	5.80	10.18	12.09	4.33	5.30	10.49	19.83	*6.73*
		Rabi	Groundnut	13.56	14.07	9.79	13.65	13.45	5.03	15.40	14.10	18.68	*16.50*
		Kharif	Cotton(lint)	11.04	10.35	0.84	15.74	10.66	2.91	4.59	8.96	1.56	*14.07*
			Bajra	2.44	7.79	15.62	2.21	7.09	3.42	12.33	9.25	1.70	*5.87*
			Castor seed	5.37	3.45	3.58	7.31	3.77	2.76	4.79	3.58	19.25	*5.72*
	chittoor	Rabi	Groundnut	24.11	27.04	19.32	24.67	26.96	12.96	28.01	26.92	2.63	*26.05*
		Kharif	Maize	25.84	42.09	19.68	29.29	32.58	17.56	40.72	41.74	20.51	*32.75*
		Rabi	Rice	24.66	27.24	8.84	30.22	27.20	13.88	25.19	28.31	26.33	*31.80*
			Ragi	13.51	15.57	10.43	12.28	15.58	7.41	12.49	15.89	1.96	*15.29*
			Horse-gram	1.27	5.29	5.48	2.02	5.24	2.43	6.29	5.22	3.86	*4.41*
		Kharif	Bajra	8.86	18.83	10.30	17.88	13.22	8.09	40.33	20.66	23.24	*22.32*
			Rice	24.69	30.37	18.06	29.17	24.20	13.61	43.52	32.11	16.59	*33.42*
			Ragi	9.82	10.38	11.41	20.52	9.52	5.51	8.41	10.83	1.98	*8.09*
			Groundnut	0.02	8.44	5.35	-0.22	7.35	3.74	5.16	8.27	1.19	*5.94*
			Horse-gram	5.19	6.48	2.93	6.69	5.52	2.51	12.15	6.16	17.15	*4.49*

In the next step, the error values for each crop were calculated by comparing the predictions with the actual yield values and the difference was noted down. For each crop (each row in Table 2), the model and parameter set pair that gave the least error was noted down. This model and parameter set were tagged as the '*best model*' and '*best parameter set*', respectively.

Table 2. Error-values as obtained for each prediction and the best model and best parameter set

State	District	Season	Parameter Set -> Model -> Crop	Temperature, Rainfall			Rainfall			Temperature			Actual	Minimum Error	Accuracy	Best Model	Best Parameter Set
				SVM	MLR	ANN	SVM	MLR	ANN	SVM	MLR	ANN					
andhra pradesh	anantapur	Rabi	Horse-gram	0.67	0.54	5.44	1.77	0.28	2.87	0.80	0.49	2.71	4.58	*0.28*	*93.91*	MLR	2
		Kharif	Groundnut	2.01	2.05	1.17	0.60	2.39	0.75	0.44	1.67	13.96	2.70	*0.44*	*83.58*	SVM	3
		Rabi	Gram	0.81	1.75	4.22	1.64	1.61	3.05	2.79	1.39	9.89	5.68	*0.81*	*85.73*	SVM	1
			Jowar	1.72	4.50	0.62	6.06	4.55	2.93	3.61	3.27	2.34	4.75	*0.62*	*86.87*	ANN	1
			Maize	0.66	4.53	33.75	0.57	2.93	44.01	0.90	2.43	30.90	70.67	*0.57*	*99.19*	SVM	2
			Rice	2.37	3.12	5.52	2.32	2.97	12.65	1.56	2.78	21.26	21.94	*1.56*	*92.88*	SVM	3
		Kharif	Rice	2.77	1.96	17.09	2.70	1.96	19.64	0.47	2.77	27.38	30.08	*0.47*	*98.42*	SVM	3
			Ragi	2.27	1.82	1.31	1.88	4.23	8.49	0.56	2.08	9.45	15.08	*0.56*	*96.26*	SVM	3
			Moong(Green Gram)	6.64	7.37	8.61	6.88	6.75	9.14	10.26	6.69	0.62	11.10	*0.62*	*94.42*	ANN	3
			Maize	3.28	5.64	23.98	5.58	4.87	14.18	2.94	4.33	16.15	25.35	*2.94*	*88.38*	SVM	3
			Jowar	6.95	3.58	0.93	3.45	5.36	2.40	1.43	3.76	13.10	6.73	*0.93*	*86.22*	ANN	1
		Rabi	Groundnut	2.94	2.43	6.71	2.85	3.05	11.47	1.10	2.40	2.18	16.50	*1.10*	*93.31*	SVM	3
		Kharif	Cotton(lint)	3.04	3.73	13.23	1.66	3.41	11.16	9.49	5.11	12.51	14.07	*1.66*	*88.17*	SVM	2
			Bajra	3.43	1.92	9.76	3.65	1.23	2.45	6.46	3.39	4.17	5.87	*1.23*	*79.09*	MLR	2
			Castor seed	0.35	2.27	2.14	1.59	1.95	2.96	0.93	2.14	13.53	5.72	*0.35*	*93.88*	SVM	1
	chittoor	Rabi	Groundnut	1.94	0.99	6.73	1.38	0.91	13.09	1.96	0.87	23.42	26.05	*0.87*	*96.67*	MLR	3
		Kharif	Maize	6.91	9.34	13.07	3.46	0.18	15.19	7.97	8.99	12.25	32.75	*0.18*	*99.46*	MLR	2
		Rabi	Rice	7.14	4.56	22.96	1.58	4.60	17.92	6.61	3.49	5.47	31.80	*1.58*	*95.03*	SVM	2
			Ragi	1.78	0.28	4.86	3.01	0.29	7.88	2.80	0.60	13.33	15.29	*0.28*	*98.19*	MLR	1
			Horse-gram	3.14	0.88	1.07	2.39	0.83	1.98	1.88	0.82	0.55	4.41	*0.55*	*87.52*	ANN	3
		Kharif	Bajra	13.46	3.49	12.01	4.44	9.09	14.23	18.01	1.66	0.92	22.32	*0.92*	*95.86*	ANN	3
			Rice	8.73	3.05	15.36	4.25	9.22	19.81	10.10	1.31	16.83	33.42	*1.31*	*96.08*	MLR	3
			Ragi	1.73	2.29	3.32	12.43	1.43	2.59	0.32	2.74	6.11	8.09	*0.32*	*96.02*	SVM	3
			Groundnut	5.92	2.50	0.59	6.16	1.41	2.20	0.78	2.33	4.75	5.94	*0.59*	*90.09*	ANN	1
			Horse-gram	0.70	1.99	1.56	2.20	1.03	1.98	7.66	1.67	12.66	4.49	*0.70*	*84.48*	SVM	1
			Jowar	8.36	1.66	0.10	5.23	2.35	2.40	4.58	0.52	2.62	6.96	*0.10*	*98.53*	MLR	1
			Moong(Green Gram)	6.63	7.09	5.40	5.20	6.62	7.22	9.31	6.51	1.83	11.09	*1.83*	*83.49*	ANN	3

Using the *'best model'* and *'best parameter set'*, predictions were made for the selected crops, for the year 2019. The predicted results were then compared with the actual values and an average accuracy of 91.91% was obtained. Also, predictions for 2020 were performed for those selected crops and the results obtained are portrayed in Table 3. Similarly, predictions for all the 542 districts of India can be performed.

Table 3. Yield predictions for 2019 and 2020, for Anantapur and Chittoor

State	District	Season	Crop	2019 Prediction	2019 Actual Value	Accuracy	2020 Prediction
Andhra Pradesh	anantapur	Rabi	Horse-gram	4.86	*4.58*	93.88	3.82
		Kharif	Groundnut	2.26	*2.70*	83.70	1.96
		Rabi	Gram	6.49	*5.68*	85.74	7.42
			Jowar	4.13	*4.75*	86.96	4.59
			Maize	71.24	*70.67*	99.19	73.25
			Rice	23.50	*21.94*	92.89	25.36
		Kharif	Rice	30.55	*30.08*	98.44	28.46
			Ragi	15.64	*15.08*	96.28	19.23
			Moong(Green Gram)	11.72	*11.10*	94.43	13.58
			Maize	22.41	*25.35*	88.40	24.96
			Jowar	5.80	*6.73*	86.18	4.62
		Rabi	Groundnut	15.40	*16.50*	93.33	16.32
		Kharif	Cotton(lint)	15.74	*14.07*	88.17	13.74
			Bajra	7.09	*5.87*	79.17	5.91
			Castor seed	5.37	*5.72*	93.88	6.25
	chittoor	Rabi	Groundnut	26.92	*26.05*	96.66	19.60
		Kharif	Maize	32.58	*32.75*	99.48	29.51
		Rabi	Rice	30.22	*31.80*	95.03	17.26
			Ragi	15.57	*15.29*	98.17	19.05
			Horse-gram	3.86	*4.41*	87.57	5.37
		Kharif	Bajra	23.24	*22.32*	95.87	21.45
			Rice	32.11	*33.42*	96.08	65.37
			Ragi	8.41	*8.09*	96.05	6.48
			Groundnut	5.32	*5.94*	89.56	3.58
			Horse-gram	5.19	*4.49*	84.42	6.49
			Jowar	7.06	*6.96*	98.51	8.59
			Moong(Green Gram)	9.26	*11.09*	83.48	11.46

6 Comparative Analysis

After performing result analysis on selected districts, the authors decided to compare the result of the proposed model with existing works, for the same location as that of the selected model. It was observed that for all three existing models compared, the proposed model fared much better than all the exitng models, as seen in Table 4.

Table 4. Comparative analysis of previous studies

Author	Algorithm used	Crops	Location	Parameters	Dataset size of authors	Dataset of proposed model	Accuracy claimed by authors	Accuracy achieved by proposed model
E. Manjula and S. Djodiltachoumy	Association Rule Mining	Paddy, Cholum, Cumbu, Tanks, Bore wells, Open wells, Production, Yield	Tamil Nadu	Year, District, Crop, Area, Tanks, Bore Wells, Open Wells, Production, Yield	2000 to 2012	1997 to 2019	90.00%	94.65%
B. V. Ramesh, D, and Vardhan	Multiple Linear Regression (MLR) and Density-based Clustering	-	Andhra Pradesh	Year, Rainfall, Area of Sowing, Fertilizers, Production	1955 to 2009	1997 to 2019	88.00%	93.46%
S. Veenadhari, B. Misra, and C. D. Singh	C 4.5 algorithm	Soybean, Maize, Paddy, Wheat	Madhya Pradesh	Rainfall, Max and Min Temp, Potential Evapotranspiration, Cloud Cover, Wetday Frequency	20 years of data	1997 to 2019	81.50%	95.03%

7 Conclusion

Agriculture is a crucial factor in the development of a nation. The primary factor that determines the output of agriculture is the yield produced by the crops. Yield prediction helps the farmers obtain the estimated amount of produce, well in advance. Using machine learning for yield prediction helps the farmers obtain better accuracy and reduced errors as compared to traditional or numerical approaches.

In this study, multiple machine learning models were implemented using an ensemble approach that compares different models and identifies the best model and best parameters for each crop of each state. The yield predictions were performed for selected districts and their crops, achieving an accuracy of 91.91%.

Currently, the proposed model can predict the annual yields for 36 crops, grown in 512 districts of India, with an RMSE of 1.065. To further improve the proposed model, in the future, the authors will try to inculcate intermediary crops, different varieties of each crop, and yield predictions for the multi-cropping method. The authors also plan to cover all the 741 districts of India.

References

1. Horie, T., Yajima, M., Nakagawa, H.: Yield forecasting. Agric. Syst. (1992). https://doi.org/10.1016/0308-521X(92)90022-G
2. Government of India: "Crop Forecasts". Ministry of Statistics & Programme Implementation (2020). http://mospi.nic.in/44-crop-forecasts
3. Ratkal, A.G., Akalwadi, G., Patil, V.N., Mahesh, K.: Farmer's analytical assistant. In: Proc. 2016 IEEE Int. Conf. Cloud Comput. Emerg. Mark, CCEM 2016, pp. 84–89 (2017). https://doi.org/10.1109/CCEM.2016.023
4. Kantanantha, N., Serban, N., Griffin, P.: Yield and price forecasting for stochastic crop decision planning. J. Agric. Biol. Environ. Stat. **15**(3), 362–380 (2010). https://doi.org/10.1007/s13253-010-0025-7
5. Gayatri, M.K., Jayasakthi, J., Mala, G.S.A.: Providing smart agricultural solutions to farmers for better yielding using IoT. In: Proc. 2015 IEEE Int. Conf. Technol. Innov. ICT Agric. Rural Dev. TIAR 2015, Tiar, pp. 40–43 (2015). https://doi.org/10.1109/TIAR.2015.7358528
6. Manjula, E., Djodiltachoumy, S.: A Model for prediction of crop yield. Int. J. Comput. Intell. Inform. **6**(4), 298–305 (2017)
7. Gandge, Y., Sandhya: A study on various data mining techniques for crop yield prediction. In: Int. Conf. Electr. Electron. Commun. Comput. Technol. Optim. Tech. ICEECCOT 2017, vol. 2018, pp. 420–423 (2018). https://doi.org/10.1109/ICEECCOT.2017.8284541
8. Ramesh, B.V., Vardhan, D.: Analysis of crop yield prediction using data mining techniques. Int. J. Res. Eng. Technol. **4**(1), 470–473 (2015). https://doi.org/10.23956/ijarcsse.v7i11.468
9. Kumar, R., Singh, M.P., Kumar, P., Singh, J.P.: Crop selection method to maximize crop yield rate using machine learning technique. In: 2015 Int. Conf. Smart Technol. Manag. Comput. Commun. Control. Energy Mater. ICSTM 2015 – Proc., May, pp. 138–145 (2015). https://doi.org/10.1109/ICSTM.2015.7225403
10. Medar, R., Rajpurohit, V.S., Shweta, S.: Crop yield prediction using machine learning techniques. In: 2019 IEEE 5th International Conference for Convergence in Technology (I2CT), March, pp. 1–5 (2019). https://doi.org/10.1109/I2CT45611.2019.9033611
11. Veenadhari, S., Misra, B., Singh, C.D.: Machine learning approach for forecasting crop yield based on climatic parameters. In: 2014 Int. Conf. Comput. Commun. Informatics Ushering Technol. Tomorrow, Today, ICCCI 2014, pp. 1–5 (2014). https://doi.org/10.1109/ICCCI.2014.6921718
12. Guruprasad, R.B., Saurav, K., Randhawa, S.: Machine learning methodologies for paddy yield estimation in India: a case study, pp. 7254–7257 (2019). https://doi.org/10.1109/igarss.2019.8900339
13. Suresh, A., Ganesh Kumar, P., Ramalatha, M.: Prediction of major crop yields of Tamilnadu using K-means and modified KNN. In: Proc. 3rd Int. Conf. Commun. Electron. Syst. ICCES 2018, ICCES, pp. 88–93 (2018). https://doi.org/10.1109/CESYS.2018.8723956
14. Rale, N., Solanki, R., Bein, D., Andro-Vasko, J., Bein, W.: Prediction of crop cultivation. In: 2019 IEEE 9th Annu. Comput. Commun. Work. Conf. CCWC 2019, pp. 227–232 (2019).https://doi.org/10.1109/CCWC.2019.8666445
15. Doshi, Z., Nadkarni, S., Agrawal, R., Shah, N.: AgroConsultant: intelligent crop recommendation system using machine learning algorithms. In: Proc. 2018 4th Int. Conf. Comput. Commun. Control Autom. ICCUBEA 2018 (2018).https://doi.org/10.1109/ICCUBEA.2018.8697349

16. Kulkarni, N.H., Srinivasan, G.N., Sagar, B.M., Cauvery, N.K.: Improving crop productivity through a crop recommendation system using ensembling technique. In: Proc. 2018 3rd Int. Conf. Comput. Syst. Inf. Technol. Sustain. Solut. CSITSS 2018, pp. 114–119 (2018). https://doi.org/10.1109/CSITSS.2018.8768790
17. Jain, N., Kumar, A., Garud, S., Pradhan, V., Kulkarni, P.: Crop selection method based on various environmental factors using machine learning. Int. Res. J. Eng. Technol. **4**(2) 1530–1533 (2017). https://irjet.net/archives/V4/i2/IRJET-V4I2299.pdf
18. Government of India: "Production Dataset," Open Government Data (OGD) Platform India. https://data.gov.in/catalog/
19. India Water Portal: "Temperature Data 1901 to 2008." https://www.indiawaterportal.org/met data/
20. World Weather Online: "Temperature Data 2009 to 2019." https://www.worldweatheronline.com/developer/api/
21. India Water Portal: "Rainfall Data 1901 to 2010." https://www.indiawaterportal.org/sites/indiawaterportal.org/files
22. Environics India: "Rainfall Data 2011 to 2019." http://environicsindia.in/wp-content/uploads/2018/09/

A Novel Framework for Multimodal Twitter Sentiment Analysis Using Feature Learning

Jamuna S. Murthy(✉), Amulya C. Shekar, Drishti Bhattacharya, R. Namratha, and D. Sripriya

BNM Institute of Technology, Bangalore, India

Abstract. Over the years there has been a lot of speculation with respect to single modal sentiment analysis of twitter (which is one of the world's largest micro blogging platforms) data i.e. either text or image mining. But unfortunately most of the researchers didn't use the non-trivial elements such as memes (i.e. combination of image and text data) and GIFs (i.e. combination of video and audio data) which dominate the twitter world today. Hence looking at the limitations of the existing systems we proposes a novel framework called "Multimodal Twitter Sentiment Analysis using Feature Learning" which defines the polarity of the tweets by considering all types of data such as text, image and GIFs. The framework consist of three main modules i.e. Data Collection for gathering real-time tweets using twitter streaming API, Data Processing module which has context-aware hybrid algorithm used for text sentiment analysis and 'Fast R-CNN' for image sentiment analysis. GIFs are handled using an optical character recognizer which separate texts from images for defining the polarity and finally multimodal sentiment scoring is done by aggregating polarity scores of images are texts. Evaluation results of proposed framework shows accuracy of 96.7% against SVM and Naïve Bayes which outperforms the single modal sentiment analysis models.

Keywords: Lexicon · Machine learning · Multimodal · Twitter sentiment · Context-aware · Optical character recognizer

1 Introduction

Researches in the last few years show that a lot of preference was given to text driven Sentiment Analysis (SA) of twitter data (i.e. both random and benchmark tweets) and very few studies related to visual analysis of image sentiment prediction were reported. Also much of these research works focus only on analysis of single modality of twitter data i.e. either text or image or videos. But today the non-trivial elements such as 'memes' and GIF videos are dominating the twitter world. Most of the people are interested in expressing their views and opinions in the form of visual content which is a combination of text and image data as they are more engaging, interesting and effective to human minds. Hence it is crucial to study the relationship of text-image data which is also called as multimodal data that can effectively modify the semantics and ultimately the sentiment [1–7]. Over the years there have been a lot of classifiers used for SA of tweets.

M. Singh et al. (Eds.): ICACDS 2021, CCIS 1441, pp. 252–261, 2021.
https://doi.org/10.1007/978-3-030-88244-0_24

But most of them include traditional approach of single modal classification of text or image or video data in particular through lexicon based methods or Machine Learning algorithms. Currently there has been a lot of prominence given to Visual Analysis and Multimodal Sentiment of social media data wherein we analyze the topographic and infographic representations of data as a combination of text data which give light to most interesting problems related to text-image relationships [8, 9].

The most prominent researches in the field of visual analysis used techniques and methods from deep learning approach for classification. Also analyzing the topographic and infographic data imposes new challenges on the researchers as they are not easy to analyze. Hence motivated by this problem of content modality we propose a novel framework called "Multimodal Twitter sentiment Analysis using Feature Learning" which facilitate both single modal and multimodal sentiment analysis of twitter data [10].

1.1 Contribution

The proposed framework consists of five different modules such as Data Collection Module, Multimodal Module, Image Module, Text Module and Aggregate Module for sentiment analysis. Data Collection module is used to poll the tweets from the twitter through the Twitter Streaming API. Related tweets are pre-processed and analyzed using the other modules such as Multimodal Module which is used to separated the text from image though Computer Vision API i.e. Optical Character recognizer and send images and texts to respective modules for sentiment score calculations, Image Module is used to calculate image sentiment through SentiBank and SentiStrength scoring for Faster R-CNN or Regions with convolution neural network (R-CNN), Text Module is used to calculate the text sentiment though context-aware hybrid algorithm which is combination of both lexicon and machine learning approach and finally Aggregate Module will combine both the image and text scores to calculate the final sentiment score.

The novelty of the framework includes two main highlights. Firstly the framework analyzes any kind of incoming tweet irrespective of content modality and secondly the framework introduces visual analysis of tweets by analyzing even the associated textual contents with the image and produces the score for image-text sentiment to give better accuracy of 92% than the models which relied only on single modal data for classification such as image or text or video. Thus the framework addresses the problem of data sparsity and sarcasm for classification of tweets. The framework is evaluated for the robustness through some multimodal tweets collected as benchmark datasets based on the three recent topics such as Impact of COVID-19 in India, LGBT verdict of Indian Penal Court (IPC) Section 377 in India (#Section377), Netizens reaction on Kareena Kapoor Khan's delivery of second baby boy. Each of the three datasets is used to evaluate the performance metric accuracy of the framework.

2 Literature Review

Sentiment Analysis is considered as one of the ongoing field of research which determines the polarity of user generated data as positive or negative or neutral. This helps the researcher ad IT Industry to build recommendation systems for the users. Over the years there has been sufficient amount of work carried on SA and few of the recent works are discussed here with their limitations and how we have overcome them in our proposed system. Single modal sentiment analysis covers both texts mining an image classification which helped to build very good prediction models [11–15].

Ankit et al. (2018) in their work "An Ensemble Classification System for Twitter Sentiment" used a ensemble classifier which is a combination of baseline classifiers like Naïve Bayes, SVM. The overall accuracy of this model turned out to be 73.33% which is comparatively less to the proposed hybrid text mining algorithm of our system which outperforms the existing algorithm by increasing the accuracy by 22%. Also this existing model had a limitation of single model sentiment analysis which is over come in our proposed system which is multimodal and produces the accuracy of 96.7% [16].

Zhao Jianqiang et al. (2018) proposed a system called "Deep Convolution Neural Networks for Twitter Sentiment Analysis" where in they used the word embeddings integrated with DCNN for text and emoticons sentiment analysis which produced an accuracy of 87.62% [17]. Also Usman Naseem et al. (2020) in their research work "Transformer based Deep Intelligent Contextual Embedding for Twitter sentiment analysis" developed a model for text sentiment analysis of twitter data using deep learning approach bidirectional LSTM (**BiLSTM**) which gave accuracy of 94%. But our proposed model includes text module which uses context-aware text mining algorithm (i.e. a combination of both lexicon and machine learning approach) which outperforms the existing systems by giving the accuracy of 95% over real-time. Using deep learning models for text sentiment analysis requires very huge amount of data for training the classifier and also makes the system slow [18].

R. Nagamanjula et al. (2020) in their research work "A novel framework based on bi-objective optimization and LAN^{2}FIS for Twitter sentiment analysis" proposed a algorithm LAN^{2}FIS (logistic adaptive network based on neuro-fuzzy inference system) for text mining of tweets which gave an accuracy of 89%. But our proposed model includes context aware hybrid algorithm for text mining of tweets which produced the accuracy 95%. Moreover the existing system was built using Fuzzy logic technique which makes the system complex to understand and cannot be reusable [19]. But proposed system is build using component based approach and the modules are easily reusable.

Ahmed Sulaiman M. Alharbi et al. (2019) used Convolutional Neural Network (CNN) for twitter sentiment analysis in their research work "Twitter sentiment analysis with a deep neural network: An enhanced approach using user behavioral information" which produced an accuracy of 88.46%. But in the proposed system we use a context aware hybrid text mining algorithm which increased the overall accuracy of the existing system by 7% [20]. Also the multimodal nature of our system gives better classification accuracy of 96.7% which was lacking in existing system.

Ashima Yadav et al. (2020) in their research work "A deep learning architecture of RA-DLNet for visual sentiment analysis" used CNN based algorithm residual attention-based deep learning network (RA-DLNet) for addressing the problem of visual sentiment analysis which gave accuracy of 83%. But in the proposed system we have image module built through SentiBank and SentiStrength scoring for 'Faster R-CNN' which takes very less time for training than other CNN models and gives the accuracy of 91% [21]. Also the existing system considered only the image data purely as reference but in the proposed multimodal system we use the image data extracted from memes to calculate the aggregate polarity of images which gave us better accuracy.

Akshi Kumar et al. (2020) used Support vector machine (SVM) classifier trained using bag-of-visual-words (BoVW) for predicting the visual content sentiment in their research work called "Hybrid context enriched deep learning model for fine-grained sentiment analysis in textual and visual semiotic modality social data". The overall accuracy of image sentiment turned out to be 75.4% and info-graphic sentiment was 90.9%. But in the proposed system with the help of 'Faster R-CNN' and OCR methods the overall sentiment of image and info-graphic information turns out to be 91% and 96.7% respectively [22] which clearly outperformed the existing system.

Feiran Huang et al. (2019) in their research work "Image–text sentiment analysis via deep multimodal attentive fusion" used a novel algorithm for image text sentiment analysis called Deep Multimodal Attentive Fusion (DMAF) which produced an accuracy of 87.6% [23]. Also Ziyuan Zhao et al. (2020) proposed a multimodal sentiment analysis system called "An image-text consistency driven multimodal sentiment analysis approach for social media" wherein they used conventional SentiBank approach which is a extension to SentiBank approach for measuring the sentiment of image-text data. The model turned out to be 87% accurate when evaluated for Flickr benchmark dataset. But our proposed model uses combination of SentiBank, SentiStrength scoring for 'Faster R-CNN' for image sentiment and context aware text mining algorithm for text sentiment analysis when aggregated gives a final info-graphic sentiment accuracy of 96.7% on three different real-time datasets collected using twitter Streaming API [24].

Jian Weng et al. (2020) proposed a model called "Attention-Based Modality-Gated Networks for Image-Text Sentiment Analysis" wherein they used Attention-Based Modality-Gated Networks (AMGN) to exploit the correlation between the modalities of images and texts. This model produced sentiment accuracy result of 84%. But our proposed model uses Computer Vision API, OCR which uses 'Faster R-CNN' for image sentiment analysis and Context aware hybrid text mining algorithm which outperformed the existing model by producing the accuracy of 96.7%. Also our proposed system uses real-time data for analysis of sentiment which was lacking in the existing system [25].

3 Proposed Work

The proposed framework consists of five different modules which can handle multimedia data such as text, images. Figure 1 depicts the overall architecture of the system with Data Collection Module, Image Module, Text Module, Multimodal Module and Aggregate Module for twitter sentiment analysis.

3.1 Data Collection

The main Objective of the data collection module is to collect the related tweets based on the particular topic using Twitter Streaming API with the help of python library tweepy. Tweepy is absolutely fast and can capture thousands of messages over real-time and these messages are consumed by Apache Kafka, a distributed message queuing system with Publisher-Subscriber pattern. The benchmark datasets is built for evaluation of the classification techniques considering the tweets of three recent topics such as Impact of COVID-19 in India(#COVID19), LGBT verdict of Indian Penal Court (IPC) Section 377 in India (#Section377), Netizens reaction on Kareena Kapoor Khan's delivery of second baby boy(#saifeena).

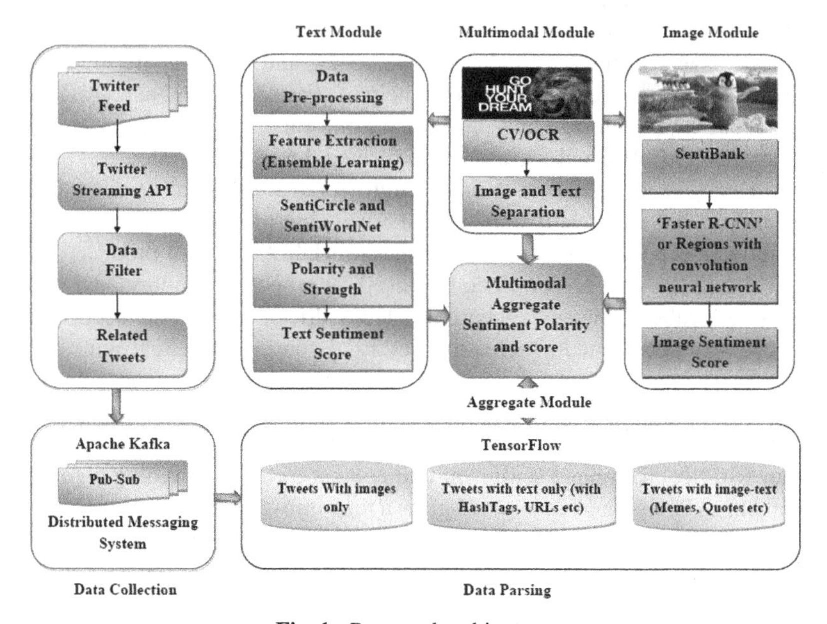

Fig. 1. Proposed architecture

3.2 Image Module

The main objective of this module is to process the tweets with only images and also the image-text tweets sent from Multimodal Module for parsing. This module is built using two major libraries SentiBank, SentiStrength and algorithm R-CNN. SentiBank is a library built using Visual Sentiment Ontology (VSO) consisting of 3244 adjective noun pairs (ANP) and 1200 trained visual concept detectors providing a mid-level representation of sentiment with respect to specific object or scene in noun.

SentiStrength is a library for estimating the strength of positive and negative sentiment in the short web texts. The strength is rated as 5 for 'very positive', –5 for 'very negative', 1 for 'not positive' and –1 for 'not negative'. Region Based Convolutional Neural Networks or 'Faster R-CNN' is one of the popular algorithms under Machine Learning for Computer Vision used in Object detection. It is absolutely fast compared to other CNN natives and reduces the number of bounding boxes fed for classifiers during selective search. The Proposed Algorithm for Image Sentiment Analysis is given below:

Step 1: Separate the Nouns for 1200 APNs.
Step 2: Separate the Objects from 200 'Faster R-CNN' Classes
Step 3: Using Radial basis function (RBF) find the distance between the Nouns and Objects of 200 'Faster R-CNN'.

Step 4: For each APN new APN weights are multiplied with the adjectives of SentiStrength
Step 5: Final Sentiment Score with range (−2, 2) is calculated by summing up all the APNs.

3.3 Text Module

The main Objective of this module is to process the tweets with only texts and also the text information extracted from the images through CV/OCR. This module is built using both lexicon-based and Machine Learning approach for acquiring best classification accuracy. NLTK libraries are used for pre-processing and feature extraction and Ensemble Learning, SentiWordNet and SentiCircle dictionaries are used to determine the sentiment of the tweets. The proposed Algorithm for Text Sentiment Analysis is as Follows:

Step 1: Detection and analysis of slangs and abbreviations are done using JSpell or SpellCheck libraries.
Step 2: Using JSpell or Jazzy Spell Checker or Snow Ball libraries the stemming and lemmatization is carried out.
Step 3: After stemming, lemmatization and correction of words the stop words are removed using Texifier library.
Step 4: Finally the special characters are removed and the most prominent features of tweets such as HashTags and URLs are retained for calculation of sentiment.
Step 5: Manually tagged set of Bag of Words is built as a part of novelty of the framework with 5000 words frequently used in twitter. Feature vector is built for feature extraction by considering features such as Unigrams, POS, Negation, No. of Emoticons, No. of elongated words, Length of tweet, No. of Capitalized words.
Step 6: The ensemble learning technique, gradient boost is used for training the text module and for overall sentiment prediction. SentiWordNet is used to calculate the sentiment score for predicted polarity.
Step 7: SentiCircle considers the contextual co-occurrence patterns in the tweets to capture conceptual information and update strength and polarity in lexicons. Negation terms in the tweets are handled by negating the sentiment scores in the list of negative lexicons. Term Context vector is built based on the words in the particular context of tweets and with the help of Weiszfeld's algorithm the context-aware texts are calculated for sentiment polarity-strength and scores are calculated using SentiWordNet.
Step 8: The final text sentiment score with range (−3, 3) is calculated by taking the aggregation of polarities calculated using gradient boost algorithm and the polarity-strength from SentiCircle.

3.4 Multimodal Module and Aggregate Module

The main objective of this module is to process the tweets with typographic and infographic information to extract the text from images. This module is built using Computer vision API with OCR (CV/OCR) for text detection, text extraction and text recognition. The proposed algorithm for Multimodal sentiment analysis is as follows:

Step 1: Text localization and detection is done using 'Tesseract-OCR' a python library and it is sent for the next step i.e. text extraction.
Step 2: Text extraction is carried out by chopping characters of the words using the pitch of text and is sent for text recognition.
Step 3: In text recognition the chopped characters are converted to machine-encoded text through adaptive recognition.
Step 4: Final accuracy is calculated by aggregating the image and text sentiment scores obtained from image and text module for each tweet.

4 Evaluation Results

The framework is evaluated in three different ways by comparing the 'accuracy' metric of the proposed algorithms for Image Sentiment Analysis, Text Sentiment Analysis and Multimodal Sentiment Analysis for three different dataset with existing algorithms. Image Sentiment is analyzed using proposed algorithm built using 'SentiBank' and 'Faster R-CNN'.

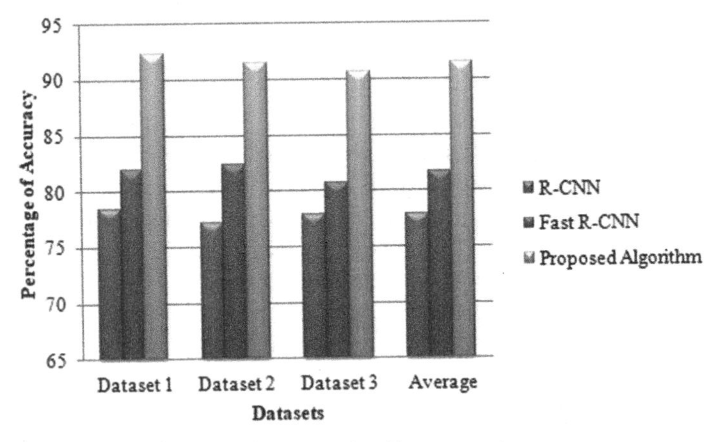

Fig. 2. Comparative analysis of image sentiment accuracy

The accuracy of the proposed algorithm turns out to be on an average of 91% when run on our three benchmark datasets collected using Data Collection Module i.e. Impact of COVID-19 in India (#COVID19), LGBT verdict of Indian Penal Court (IPC) Section 377 in India (#Section377), Netizens reaction on Kareena Kapoor Khan's delivery of second baby boy (#saifeena). Figure 2 depicts the comparative analysis of proposed algorithm with two existing object detection algorithms R-CNN, Fast R-CNN out of which proposed algorithm outperforms the existing ones. Text Sentiment Analysis is done in the 'Text Module' of the proposed framework which is built using context-aware hybrid algorithm. The proposed algorithm is compared with two existing algorithms Naive Bayes and SVM (Support Vector Machine) which outperforms to provide the average accuracy of 95% over three different datasets collected using Data Collection Module

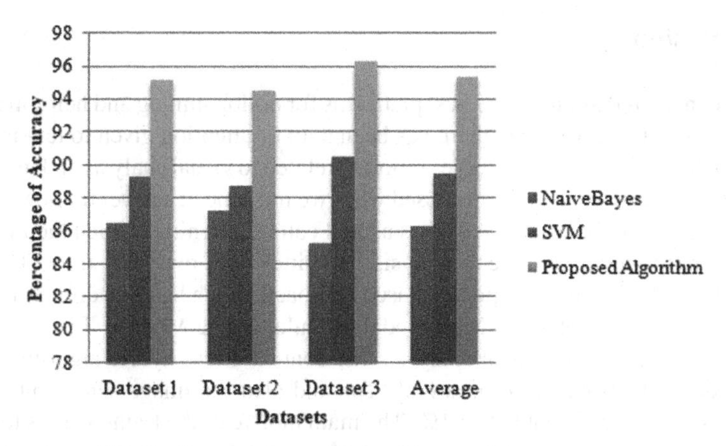

Fig. 3. Comparative analysis of text sentiment accuracy

over real-time. Figure 3 depicts the comparative analysis of accuracy obtained for three algorithms.

Multimodal Sentiment Analysis is a crucial part of our framework. Infographic and typographic data extraction here was done using CV/OCR and hence image-text relationship was highlighted very well. The evaluation results for final multimodal sentiment analysis after aggregating the image and text sentiment score provides best accuracy results of 96.7% over real-time which outperforms the single modal models which gave accuracy average accuracy of 91% for image and 95% for text sentiment analysis. Figure 4 depicts the average modality of tweets distributed over three benchmark datasets. Where image count is 1546, text is 3719 and multimodal is 3201.

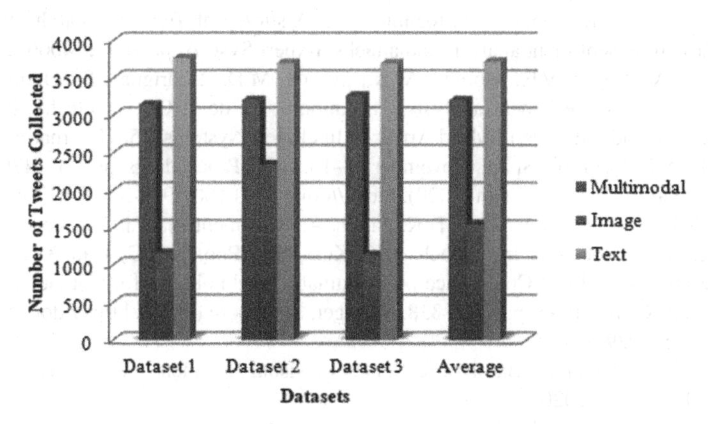

Fig. 4. Distribution of modality of tweets collected

5 Conclusion

Twitter is considered as one of the best platforms for option mining and helps in decision making quickly. Over the years there has been a lot of attention given to text mining of tweets and there were very less studies reported related to visual analysis and multimodal sentiment analysis. Hence in the proposed work we introduced a novel framework called "Multimodal Twitter Sentiment Analysis using Feature Learning" which facilitates text, image and multimodal sentiment analysis of twitter data over real-time. The whole framework was built using component based approach which had five different modules called Data Collection Module, Multimodal Module, Image Module, Text Module and Aggregate Module for sentiment analysis. The context aware hybrid algorithm for text mining produced an overall accuracy of 95% and also the image sentiment accuracy using Fast-RCNN turned out to be 91%. The main motive of the framework is to address the image-text relationship of tweets through multimodal sentiment analysis which was handled well by considering the memes in tweets which outperformed the single modal twitter sentiment analysis models like text sentiment and image sentiment models by giving the overall accuracy of 96.7%. Also the proposed framework was built using component based approach which makes the whole system reusable for R&D. But the scope of our research work is limited to texts, images, infographic and typographic tweets, but GIFs also provide a very useful insight for sentiment analysis as most of the current generation tweeters use fancy ways of texting and expressing their emotions. Hence in the future work the proposed system can be extended to support audio and video data which might increase the overall accuracy of the system.

References

1. Antonakaki, D., Fragopoulou, P., Ioannidis, S.: A survey of Twitter research: data model, graph structure, sentiment analysis and attacks. Expert Syst. Appl. **164**, 114006 (2021)
2. Aquino, P.A., López, V.F., Moreno, M.N., Muñoz, M.D., Rodríguez, S.: Opinion mining system for Twitter sentiment analysis. In: Antonio, E., de la Cal, J., Flecha, R.V., Quintián, H., Corchado, E. (eds.) Hybrid Artificial Intelligent Systems: 15th International Conference, HAIS 2020, Gijón, Spain, November 11–13, 2020, Proceedings, pp. 465–476. Springer International Publishing, Cham (2020). https://doi.org/10.1007/978-3-030-61705-9_38
3. Mehta, R.P., Sanghvi, M.A., Shah, D.K., Singh, A.: Sentiment analysis of tweets using supervised learning algorithms. In: Luhach, A.K., Kosa, J.A., Poonia, R.C., Gao, X.-Z., Singh, D. (eds.) First International Conference on Sustainable Technologies for Computational Intelligence. AISC, vol. 1045, pp. 323–338. Springer, Singapore (2020). https://doi.org/10.1007/978-981-15-0029-9_26
4. Ortis, A., Farinella, G.M., Battiato, S.: Survey on visual sentiment analysis. IET Image Proc. **14**(8), 1440–1456 (2020)
5. Jianqiang, Z., Xiaolin, G.: Comparison research on text pre-processing methods on twitter sentiment analysis. IEEE Access **5**, 2870–2879 (2017)
6. Zimbra, D., Abbasi, A., Zeng, D., Chen, H.: The state-of-the-art in Twitter sentiment analysis: a review and benchmark evaluation. ACM Trans. Manage. Inf. Syst. **9**(2), 1–29 (2018)
7. Symeonidis, S., Effrosynidis, D., Arampatzis, A.: A comparative evaluation of pre-processing techniques and their interactions for twitter sentiment analysis. Expert Syst. Appl. **110**, 298–310 (2018)

8. Chaturvedi, I., Cambria, E., Welsch, R.E., Herrera, F.: Distinguishing between facts and opinions for sentiment analysis: survey and challenges. Inf. Fusion **44**, 65–77 (2018)
9. Giachanou, A., Crestani, F.: Like it or not: a survey of twitter sentiment analysis methods. ACM Comput. Surveys **49**(2), 1–41 (2016)
10. Zou, P., Yang, S.: Multimodal tweet sentiment classification algorithm based on attention mechanism. In: Monreale, A., Alzate, C., Kamp, M., Krishnamurthy, Y., Paurat, D., Sayed-Mouchaweh, M., Bifet, A., Gama, J., Ribeiro, R.P. (eds.) ECML PKDD 2018 Workshops: DMLE 2018 and IoTStream 2018, Dublin, Ireland, September 10–14, 2018, Revised Selected Papers, pp. 68–79. Springer International Publishing, Cham (2019). https://doi.org/10.1007/978-3-030-14880-5_6
11. Mittal, N., Sharma, D., Joshi, M.L.: Image sentiment analysis using deep learning. In: 2018 IEEE/WIC/ACM International Conference on Web Intelligence (WI), pp. 684–687. IEEE (2018)
12. Ahsan, U., De Choudhury, M., Essa, I.: Towards using visual attributes to infer image sentiment of social events. In: 2017 International Joint Conference on Neural Networks (IJCNN), pp. 1372–1379. IEEE (2017)
13. Zhang, L., Wang, S., Liu, B.: Deep learning for sentiment analysis: a survey. Wiley Interdiscipl. Rev. Data Mining Knowl. Discov. **8**(4), e1253 (2018)
14. Soleymani, M., Garcia, D., Jou, B., Schuller, B., Chang, S.F., Pantic, M.: A survey of multimodal sentiment analysis. Image Vis. Comput. **65**, 3–14 (2017)
15. Yue, L., Chen, W., Li, X., Zuo, W., Yin, M.: A survey of sentiment analysis in social media. Knowl. Inf. Syst. **60**(2), 617–663 (2018). https://doi.org/10.1007/s10115-018-1236-4
16. Saleena, N.: An ensemble classification system for twitter sentiment analysis. Procedia Comput. Sci. **132**, 937–946 (2018)
17. Jianqiang, Z., Xiaolin, G., Xuejun, Z.: Deep convolution neural networks for twitter sentiment analysis. IEEE Access **6**, 23253–23260 (2018)
18. Naseem, U., Razzak, I., Musial, K., Imran, M.: Transformer based deep intelligent contextual embedding for twitter sentiment analysis. Futur. Gener. Comput. Syst. **113**, 58–69 (2020)
19. Nagamanjula, R., Pethalakshmi, A.: A novel framework based on bi-objective optimization and LAN 2 FIS for Twitter sentiment analysis. Soc. Netw. Anal. Min. **10**, 1–16 (2020)
20. Alharbi, A.S.M., de Doncker, E.: Twitter sentiment analysis with a deep neural network: an enhanced approach using user behavioral information. Cogn. Syst. Res. **54**, 50–61 (2019)
21. Yadav, A., Vishwakarma, D.K.: A deep learning architecture of RA-DLNet for visual sentiment analysis. Multimedia Syst. **26**(4), 431–451 (2020). https://doi.org/10.1007/s00530-020-00656-7
22. Kumar, A., Srinivasan, K., Cheng, W.H., Zomaya, A.Y.: Hybrid context enriched deep learning model for fine-grained sentiment analysis in textual and visual semiotic modality social data. Inf. Process. Manage. **57**(1), 102141 (2020)
23. Huang, F., Zhang, X., Zhao, Z., Xu, J., Li, Z.: Image–text sentiment analysis via deep multimodal attentive fusion. Knowl. Based Syst. **167**, 26–37 (2019)
24. Zhao, Z., Zhu, H., Xue, Z., Liu, Z., Tian, J., Chua, M.C.H., et al.: An image-text consistency driven multimodal sentiment analysis approach for social media. Inf. Process. Manage. **56**(6), 102097 (2019)
25. Huang, F., Wei, K., Weng, J., Li, Z.: Attention-based modality-gated networks for image-text sentiment analysis. ACM Trans. Multimedia Comput. Commun. Appl. **16**(3), 1–19 (2020)

An Iterative Approach Based Reversible Data Hiding with Weight Update for Dual Stego Images

C. Shaji(✉) and I. Shatheesh Sam

Nesamony Memorial Christian College Affiliated to Manonmaniam Sundaranar University, Abishekapatti, Tirunelveli 627 012, Tamil Nadu, India

Abstract. The paper proposes an iterative approach based reversible data hiding with weight update used for dual stego images. Proposed method initially estimates the error between the adjacent pixels from which the embedding strength $\propto$ is estimated. Proposed method then estimates a weight using the error and the data to be hidden. This weight is updated iteratively by using the subsequent data. Further, to improve the security, the random binary sequence is mixed with the weight. To acquire the two stego images, the weight is entrenched on the original image pixels. In the extraction phase, the data is extracted in a reverse order of embedding, where the inverse weight updation is performed, from which the data is extracted. By averaging the two stego images, the original image can be restored. According to the results of the experimental study, the proposed scheme for embedding data on dual stego images outperforms recent data hiding algorithms in terms of embedding rate and visual efficiency.

Keywords: Embedding rate · Reversible data hiding (RDH) · Dual stego images · Visual quality · Extraction

1 Introduction

Image [1] data hiding is the technique of preserving the confidential or secret content inside a cover image. Data hiding [2] has recently become common in the fields of medicine, military, and copyright protection. In particular, in the field of medicine. [3] patient records such as medical history can be reversibly obscured in scan images. This allows the practitioner to quickly view the medical records without having to re-study them. In military applications [4], the confidential data can be shared between the soldiers of a country by hiding it in a cover image. Data hiding is also essential for maintaining the image content's [5] ownership. The two broad area of data hiding are steganography and cryptography [6, 7]. The steganography algorithm [8] aims to recover the hidden data completely when the stego image is not subject to attacks. When the stego image is attacked, these algorithms typically fail to recover the data. The [7–9] embedding capacity is highly limited, if the steganography algorithm produces a single stego image. As a result, dual stegano image-based algorithms are implemented, which

M. Singh et al. (Eds.): ICACDS 2021, CCIS 1441, pp. 262–270, 2021.
https://doi.org/10.1007/978-3-030-88244-0_25

can carry a large number of data without losing the stego image's visual quality. Stego photos are secret data contained in the original cover image.

Histogram shifting [10], difference expansion [11], LSB substitution [12], and prediction error expansion [13] are some of the most basic data hiding algorithms. The aim of the histogram shifting algorithm is to estimate the cover image's histogram of pixel values; the data is then embedded by creating a vacancy bin. The visual quality of these algorithms is excellent, but the embedding rate is extremely poor. By changing the LSB bit of the cover image pixels, the LSB substitution [12] algorithm hides the details. This scheme has a low degree of distortion, but it is insecure. By shuffling the LSB bit, an unauthorised user can easily extract the secret data. The prediction error expansion [13] and difference expansion scheme provides a very high embedding rate, but it creates a high distortion, since the prediction error or difference value is high near the edges [22].

Several dual data hiding techniques are derived from the prediction error expansion and difference expansion. The algorithm includes exploiting modification direction (EMD) [14], encoding based schemes [15] and location based scheme [16]. The EMD scheme [15] uses a modulus function to embed the data. The centre folding strategy folds the decimal data and then embeds it on the cover image. The modified centre folding strategy [17] uses an encoding scheme to encode the decimal data and then embedded the encoded indices in the cover image. The location based technique uses a pair of pixels $X_{i,j}$ and $X_{i,j+1}$ to embed the data. The cover pixels $X_{i,j}$ and $X_{i,j+1}$ is modified as $X'_{i,j}$ and $X'_{i,j+1}$ after embedding a 2-bit message.

The shiftable position [18] scheme is proposed where it uses an optimal parameter to estimate the code length. Y.L Wang et al. [19] used last 3 bits to embed the data which is derived from the LSB substitution method. The method K.H Jung et al. [20] used modulus function in vertical and horizontal direction instead of estimating in diagonal direction. The binary data is converted to base 5 values for embedding. Chang et al. [14, 21] extended the EMD scheme using a 3×3 and 5×5 modulus function.

2 The Proposed Algorithm

The data embedding scheme and data extraction scheme are the two phases of the proposed process, as shown in the following section.

2.1 Proposed Embedding Scheme

The block diagram of proposed embedding scheme is depicted in Fig. 1.

Let $X_1, X_2 \ldots\ldots\ldots X_K$ be the K number of pixels where the data is to be embedded. By calculating the error between adjacent pixels, the data is embedded. Enable eqn to be used to estimate the error between adjacent pixels

$$e_i = X_i - X_{i-1} \qquad i = 0, 1 \ldots\ldots K \tag{1}$$

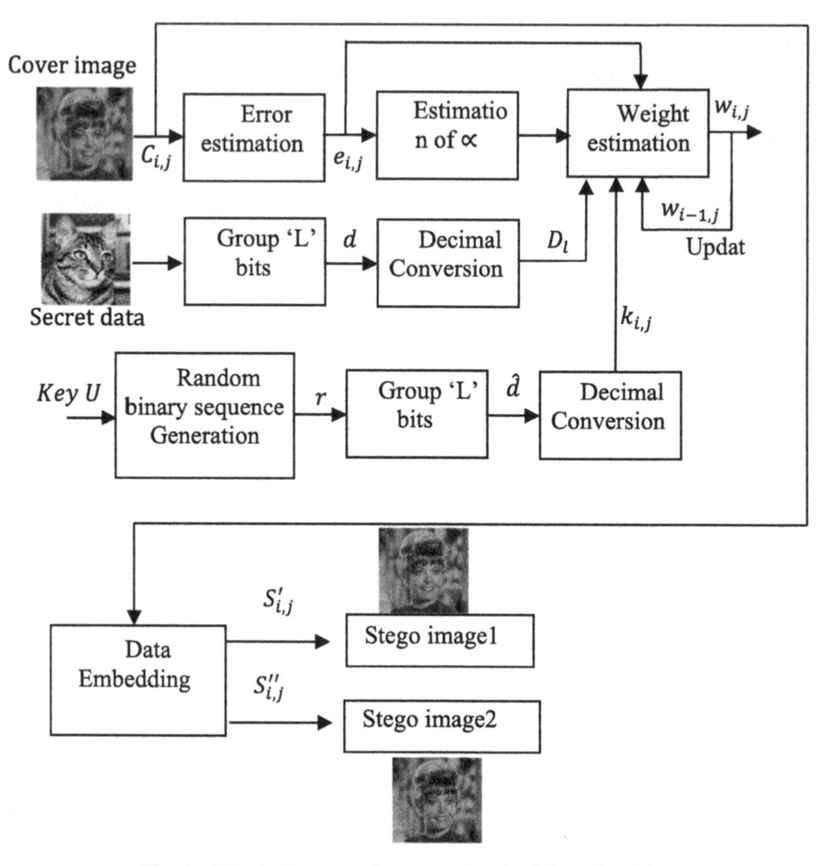

Fig. 1. Block diagram of proposed embedding algorithm

The proposed method estimate an adaptive weight using the error e_i. The adaptive weight is estimated using the relation,

$$w_i = w_{i-1} + e_i \times d \tag{2}$$

e where d is the data to be embedded. The problem with Eq. (2) is, if the error e_i is 0, then the weight becomes $w_i = w_{i-1}$. If the data d is either 0 or 1 (in case of binary), the Eq. (2) loss the property of reversibility. To solve this issue, the Eq. (2) can be modified as,

$$w_i = w_{i-1} + e_i \times d + \alpha \times d \tag{3}$$

$$w_i = w_{i-1} + d \times [e_i + \alpha] \tag{4}$$

where α is the factor that decides the embedding strength. Lower the value of α creates less distortion. The value of α is selected such that using Eq. (5)

$$\alpha = |\min(e_i) - 1| \tag{5}$$

Improve the security, the Eq. (4) is modified as Eq. (6)

$$w_i = w_{i-1} + d \times [e_i + \alpha] - k_i \tag{6}$$

where k_i is the K binary random sequence generated using key 'U'. The Eq. (4) can be modified in two cases based on the value of α,

$$w_i = \begin{cases} w_{i-1} + d \times [e_i + \alpha] - k_i \; if \; \alpha = |\min(e_i) - 1| \\ w_{i-1} + d \times [e_i + \alpha] + k_i \; if \; \alpha = |\max(e_i) + 1| \end{cases} \tag{7}$$

In this paper, we use the weight update Eq. (7) as $w_i = w_{i-1} + d \times [e_i + \alpha] - k_i$ and $\alpha = |\min(e_i) - 1|$. If the data is binary $d \in [0, 1]$, the weight update Eq. (8) is expressed as,

$$w_i = \begin{cases} w_{i-1} + k_i & if \; d = 0 \\ w_{i-1} + [e_i + \alpha] - k_i & if \; d = 1 \end{cases} \tag{8}$$

Instead of using the binary data 0 or 1, the K number of binary data can be grouped together and converted to decimal to form a decimal data D. Therefore the weight update equation can be expressed as using Eq. (9)

$$w_i = w_{i-1} + D \times [e_i + \alpha] - k_i \tag{9}$$

The weight updation in a two dimensional image is expressed as,

$$w_{i,j} = w_{i-1,j} + D_l \times \left[e_{i,j} + \alpha\right] - k_{i,j} \tag{10}$$

$i = 1, 2 \ldots\ldots\ldots. M - 1$ and $j = 1, 2 \ldots\ldots\ldots. N - 1$, where $M \times N$ denotes the cover image size and l denotes the index decimal data.

Let $C_{i,j}$ denote the cover image with coordinates (i, j), and r denote the random binary sequence generated with the key 'U'. To get the random decimal sequence $k_{i,j}$, the L bits of the random binary sequence are grouped together and converted to decimal. To obtain the decimal data D_l, the secret data is grouped in L bits and converted to decimal. The error is calculated using the cover image $C_{i,j}$, which is also used to calculate the parameter α. The cover image $C_{i,j}$ is arranged in a single row for continuous weight updating, and the weight can be modified on that single row.

Using Eqs. (11) and (12), the adaptive weight $w_{i,j}$ is embedded on the cover image $C_{i,j}$ to produce the stego images $S'_{i,j}$ and $S''_{i,j}$.

$$S'_{i,j} = C_{i,j} - \left\lceil \frac{w_{i,j}}{2} \right\rceil \tag{11}$$

$$S''_{i,j} = C_{i,j} + \left\lfloor \frac{w_{i,j}}{2} \right\rfloor \tag{12}$$

Substituting the value of $w_{i,j}$, the stego images can be estimated using the relation,

$$S'_{i,j} = C_{i,j} - \left\lceil \frac{\left(w_{i-1,j} + D \times \left[e_{i,j} + \alpha\right] - k_{i,j}\right)}{2} \right\rceil \tag{13}$$

$$S''_{i,j} = C_{i,j} + \left\lfloor \frac{\left(w_{i-1,j} + D \times \left[e_{i,j} + \alpha\right] - k_{i,j}\right)}{2} \right\rfloor \tag{14}$$

2.2 Proposed Extraction Scheme

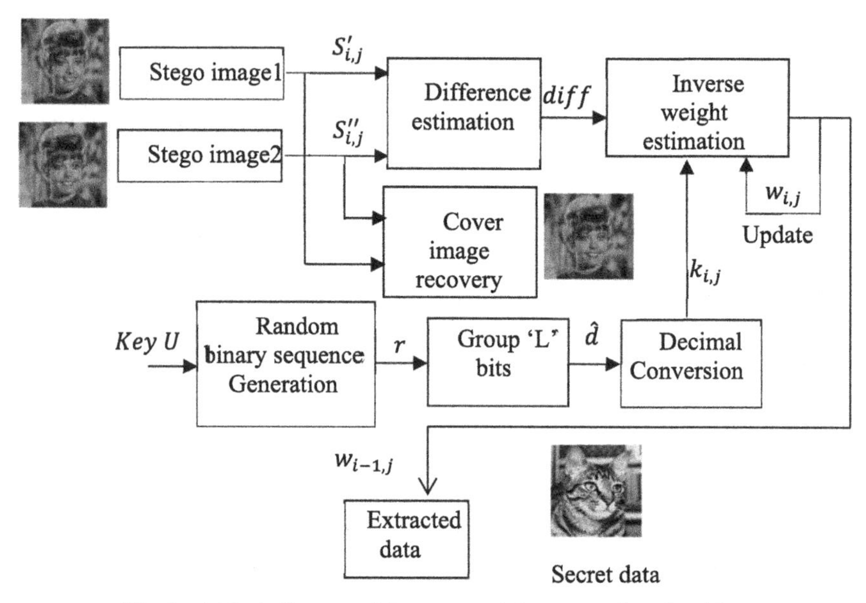

Fig. 2. A block diagram of the proposed data extraction algorithm

The original cover image $C_{i,j}$ is retrieved from the two stego images $S'_{i,j}$ and $S''_{i,j}$ in the extraction process using the relation Eq. (15) (Fig. 2).

$$C_{i,j} = \left\lceil \frac{S'_{i,j} + S''_{i,j}}{2} \right\rceil \tag{15}$$

The Eq. (14) is subtracted from eqn and (15), to get Eq. 16 as

$$\left(w_{i-1,j} + D_l \times \left[e_{i,j} + \alpha\right] - k_{i,j}\right) = S''_{i,j} - S'_{i,j} \tag{16}$$

The decimal data D_l can be estimated using Eq. (17) from the above Eq. (16)

$$D_{l-1} = \frac{1}{\left(e_{i,j} + \alpha\right)}\left(S''_{i,j} - S'_{i,j} - w_{i-1,j} + k_{i,j}\right) = \frac{1}{\left(e_{i,j} + \alpha\right)}\left(diff - w_{i-1,j} - k_{i,j}\right) \tag{17}$$

The data must be extracted in the reverse order from last to first. The last data and the last weight must be known to the extracting module for the complete recovery of data. The weight for preceding pixel pair can be estimated as using Eq. (18)

$$w_{i-1,j} = w_{i,j} - D_l \times \left[e_{i,j} + \alpha\right] + k_{i,j} \tag{18}$$

3 Experimental Results

MATLAB 2018a is used to apply the proposed reversible data hiding scheme, which uses four original images (Boat, Barbara, Cameraman, and Lena) as well as three confidential images (Cat, Pepper, Fruit,). As seen in Fig. 3, the cover image and the secret image are both 8 bit grayscale.

Fig. 3. Sample test images (a)–(d) test cover images (e)–(g) Secret images

The proposed scheme's performance is measured using image performance parameters such as embedding capability and visual consistency (PSNR).

The embedding rate of dual stegano images 'R' was calculated using Eq. (19) to evaluate the proposed method's efficiency.

$$R = \frac{S}{2 \times M \times N}. \tag{19}$$

where S denotes the number of bits embedded in the image and MN denotes the cover image's dimension. The PSNR between the cover image and the stego image is used to determine the visual consistency of the stego image. The dual stegano system will have two PSNR values because there are two stego images (PSNR1 and PSNR2). As eqn, the PSNR of the stego image is calculated using the equation as (20)

$$PSNR = 10\log_{10}\frac{255^2}{MSE}dB \tag{20}$$

where MSE stands for mean square error, which is determined using Eqs. (21) and (22)

$$MSE = \frac{1}{M \times N}\sum\nolimits_{i=1}^{M}\sum\nolimits_{j=1}^{N}\left(C_{i,j} - S_{i,j}^{'}\right)^2 \tag{21}$$

$$MSE = \frac{1}{M \times N}\sum\nolimits_{i=1}^{M}\sum\nolimits_{j=1}^{N}\left(C_{i,j} - S_{i,j}^{''}\right)^2 \tag{22}$$

where $C_{i,j}$ denotes the cover image and $S'_{i,j}$ and $S''_{i,j}$ denote the stego image. Using Eq. (23), the average PSNR is determined from PSNR1 and PSNR2

$$PSNR_{avg} = \frac{1}{2}(PSNR1 + PSNR2) \tag{23}$$

The embedding capacity for L = 2 is greater than 5,23,000 bits, with a visual quality of more than 51.9 dB. Similarly, the embedding capacity for =2 is greater than 7,77,000 bits, with a visual quality of 47.7 dB.

Table 1. Embedding capacity and PSNR comparison for the proposed scheme with traditional methods

Schemes	Embedding capacity (bits)	$PSNR_{av}$ (dB)
Chang et al. [24]	5,24,196	45.12
Chang et al. [23]	802,535	39.89
Qin et al. [22]	5,57,129	46.82
Ki-Hyun Jung [21]	7,86,432	40.84
Wang et al. [20]	629146	47.69
Heng Yao et al. [19]	8,32,124	46.52
Proposed $L = 2$	5,23,878	51.91
Proposed $L = 3$	7,84,191	47.40
Proposed $L = 4$	9,50,372	46.46

The graphical comparison of embedding rate between different schemes and the proposed method is shown in Fig. 4.The graphical comparison for proposed method is depicted for $L = 3$ and $L = 4$. From Fig. 4, it is clear that the visual quality reduces as the embedding rate increases. Also, the lower values of L provides a higher PSNR throughtout different embedding rate when compared to higher values of L. The performance of proposed framework is compared with other latest schemes and state-of-the art methods such as Ki-Hyun Jung [19], Wang et al. [20], Qin et al. [21] Chang et al. [22] and Chang et al. [15] as depicted in Table 1. The proposed scheme outperforms conventional schemes for L = 3 and L = 4 and over a range of embedding rates, as shown in Table 1 and Fig. 4.

4 Conclusion

Using an iterative weight update, this paper proposes a reversible data hiding approach on dual stego images. Iteratively, the weight is updated based on the error and the resulting data. To create dual stego images, the weight is then inserted in the cover pixels. The data is collected in the extraction process by calculating the inverse weight in reverse order. Metrics like embedding rate and PSNR are used to assess the algorithm's

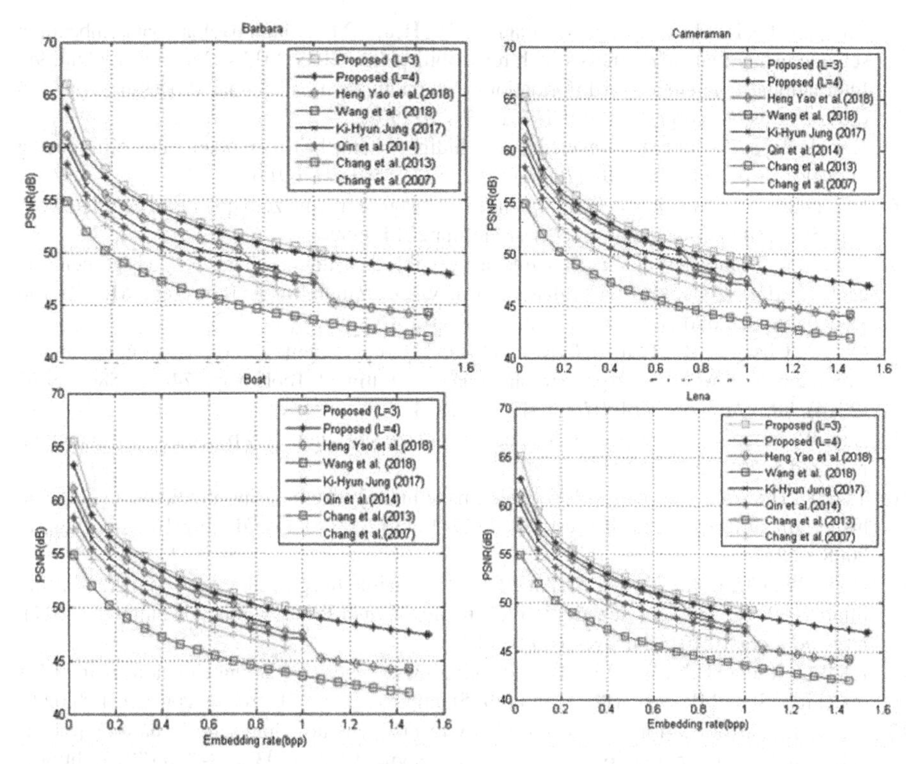

Fig. 4. Performance comparison of proposed method with the traditional methods for different cover images

success. The embedding capacity increases as the value of L increases, but the visual quality decreases. For L = 2 and L = 3, the proposed approach has an embedding capability of about 5,23,000 bits and 7,77,000 bits, respectively. In addition, for L = 2 and L = 3, the proposed approach offers an average PSNR of about 51.9 dB and 47.7 dB, respectively. When compared to other schemes, the proposed scheme offers a high embedding rate with high visual quality. In the future, we will apply our knowledge and skills to the development of a data embedding algorithm that is highly secure and robust and modifications.

References

1. Lu, T., Chang, T., Shen, J.: An Effective Maximum Distortion Controlling Technology in the Dual-image-based Reversible Data Hiding Scheme (2020). https://doi.org/10.1109/ACCESS.2020.2994244
2. Yao, H., Mao, F., Tang, Z., Qin, C.: High-fidelity dual-image reversible data hiding via prediction-error shift. Signal Process. **170**, 107447 (2020). https://doi.org/10.1016/j.sigpro.2019.107447
3. Chang, C.-C., Kieu, T.D., Chou, Y.-C.: Reversible data hiding scheme using two steganographic images. TENCON 2007 - 2007 IEEE Reg. 10 Conference, pp. 1–4 (2007). https://doi.org/10.1109/TENCON.2007.4483783

4. Chang, C.C., Lu, T.C., Horng, G., Huang, Y.H., Hsu, Y.M.: A high payload data embedding scheme using dual stego-images with reversibility. In: ICICS 2013 - Conference Guid. 9th International Conference on Information, Communication and Signal Processing, pp. 1–5 (2013). https://doi.org/10.1109/ICICS.2013.6782790.
5. Kim, P., Ryu, K., Jung, K.: Reversible data hiding scheme based on pixel-value differencing in dual images **16** (2020). https://doi.org/10.1177/1550147720911006
6. Nikolaidis, A.: Reversible data hiding in JPEG images utilising zero quantised coefficients **9**, 560–568 (2015). https://doi.org/10.1049/iet-ipr.2014.0689
7. Shastri, S., Thanikaiselvan, V.: Dual image reversible data hiding using trinary assignment and centre folding strategy with low distortion. J. Vis. Commun. Image Represent. **61**, 130–140 (2019). https://doi.org/10.1016/j.jvcir.2019.03.022
8. Qin, C., Chang, C.-C., Hsu, T.-J.: Reversible data hiding scheme based on exploiting modification direction with two steganographic images. Multimed. Tools Appl. **74**(15), 5861–5872 (2014). https://doi.org/10.1007/s11042-014-1894-5
9. Luo, W., Huang, F., Huang, J.: Edge adaptive image steganography based on LSB matching revisited **5**, 201–214 (2010)
10. Tsai, P., Hu, Y.C., Yeh, H.L.: Reversible image hiding scheme using predictive coding and histogram shifting. Signal Process. **89**, 1129–1143 (2009). https://doi.org/10.1016/j.sigpro.2008.12.017
11. Hong, W., Chen, M., Chen, T.S.: An efficient reversible image authentication method using improved PVO and LSB substitution techniques. Signal Process. Image Commun. (2017). https://doi.org/10.1016/j.image.2017.07.001
12. Lu, T.C., Tseng, C.Y., Wu, J.H.: Dual imaging-based reversible hiding technique using LSB matching. Signal Process. **108**, 77–89 (2015). https://doi.org/10.1016/j.sigpro.2014.08.022
13. Li, X., Li, J., Li, B., Yang, B.: High-fidelity reversible data hiding scheme based on pixel-value-ordering and prediction-error expansion. Signal Process. **93**, 198–205 (2013). https://doi.org/10.1016/j.sigpro.2012.07.025
14. Zhang, X., Wang, S.: Efficient steganographic embedding by exploiting modification direction **10**, 781–783 (2006)
15. Yao, H., Qin, C., Tang, Z., Zhang, X.: A general framework for shiftable position-based dual-image reversible data hiding. EURASIP J. Image Video Process. **41** (2018). https://doi.org/10.1186/s13640-018-0281-y
16. Hu, Y., Lee, H.K., Li, J.: DE-based reversible data hiding with improved overflow location map. IEEE Trans. Circuits Syst. Video Technol. **19**, 250–260 (2009). https://doi.org/10.1109/TCSVT.2008.2009252
17. Chi, L.-P., Wu, C.-H., Chang, H.-P.: Reversible data hiding in dual Stegano-image using an improved center folding strategy. Multimed. Tools Appl. **77**(7), 8785–8803 (2017). https://doi.org/10.1007/s11042-017-4774-y
18. Weng, S., Pan, J., Li, L., Zhou, L.: Reversible data hiding based on an adaptive pixel-embedding strategy and two-layer embedding. Inf. Sci. (Ny). (2016). https://doi.org/10.1016/j.ins.2016.05.030
19. Li, M., Li, Y.: Histogram shifting in encrypted images with public key cryptosystem for reversible data hiding. Signal Process. (2016). https://doi.org/10.1016/j.sigpro.2016.07.002
20. Li, X., Zhang, W., Gui, X., Yang, B.: Efficient reversible data hiding based on Multiple Histograms Modification **10**, 2016–2027 (2016)
21. Lo, C., Hu, Y.: A novel reversible image authentication scheme for digital images. Signal Process. **98**, 174–185 (2014). https://doi.org/10.1016/j.sigpro.2013.11.028
22. Tyagi, V.: Understanding Digital Image Processing. CRC Press, Boca Raton (2018). https://doi.org/10.1201/9781315123905.

Lower and Upper Bounds for 'Useful' Renyi Information Rate

Pankaj Prasad Dwivedi(✉) and D. K. Sharma

Jaypee University of Engineering and Technology, A.B. Road, Raghogarh, District, Guna, M.P., India

Abstract. Choosing the stochastic process with the entropy is attributable to increase the least amount of information to the problem under consideration. As a result, the entropy rate for stochastic processes must be determined. In the current correspondence, we will describe the rate of 'useful' conditional Renyi information measure, as well as, show that rate for 'useful' Renyi Information, the chain rule holds. Therefore, for the rate of 'useful' conditional Renyi information and also for an ergodic Markov chain, we will present a relationship and use it to extract the rate of 'useful' Renyi information. At last, we will show that Shannon's information rate is the rate for 'useful' Renyi information.

Keywords: Utility distribution · Shannon information · 'Useful' information measure · 'Useful' Renyi information · 'Useful' conditional Renyi information · Shannon entropy rate · Renyi information rate · Inequalities

1 Introduction

Suppose $\Delta_m^+ = \left\{p = (p_1, p_2, \ldots\ldots p_m); p_k \geq 0, \sum_k^m p_k = 1\right\}$, be a set of all possible discrete probability distributions of a arbitrary variable Y with utility distribution $U = \{(u_1, u_2, \ldots\ldots u_m); u_k > 0\forall k\}$ attached to each $P \in \Delta_m^+$ such that $u_k > 0$ is utility of the occasion having likelihood of event $p_k > 0$.

Suppose $U = (u_1, u_2, \ldots\ldots u_m)$ collections of positive real numbers, with u_k representing the utility of result y_k. commonly, the usefulness is free of the likelihood of encoding of the source symbol y_k i.e., p_k.

The source of information is thus given by

$$\begin{bmatrix} y_1, y_2 \ldots\ldots\ldots y_m \\ p_1, p_2 \ldots\ldots\ldots p_m \\ u_1, u_2 \ldots\ldots\ldots u_m \end{bmatrix}, u_k > 0, 0 < p_k \leq 1, \sum_{k=1}^m p_k = 1 \quad (1)$$

this is a calculation of the average amount of 'useful' information given by the source of information (1). Belis and Guiasu [2] found that without the qualitative character of a source, the probability distribution P over the source symbols Y does not fully define a source. Thus, in view of the experimenter, it's also possible that the source letters or symbols are given weights based on their importance or usefulness. The following

M. Singh et al. (Eds.): ICACDS 2021, CCIS 1441, pp. 271–280, 2021.
https://doi.org/10.1007/978-3-030-88244-0_26

qualitative-quantitative measure of information was thus introduced by Belis and Guisau [2]:

$$H(P; U) = -\sum_{k=1}^{m} u_k p_k \log p_k \tag{2}$$

It is clear that when utilities are ignored, (2) reduces to Shannon's information measure [36] which is given below:

$$H(P) = -\sum_{k=1}^{m} p_k \log p_k \tag{3}$$

Many authors someone defined the Shannon's information by utilizing distinct postulates. Khinchin [21] made Shannon's evidence much appropriate by utilizing presumptions which were conclude by Fadeev [11]. Shannon's information was additionally defined by Chandy and Mcliod [6], Kendall [20], etc. by considering contrasting sets of postulates.

In 1948 order to incorporate Shannon information, Shannon [36] used certain axioms. Dongdong et al. [22] and Guiasu [14] subsequently proposed a number of sets of rules. In 1961, Shannon entropy was generalized by Li et al. [33] to a one-parameter family of information are characterized by the information of order α called the information of Renyi. The axioms associated with Renyi information [14] have been revised by some authors. Jizba and Arimitsu [18] recently used the existing rules for Shannon and Renyi information to add a set of new axioms from which both Shannon and Renyi information can be derived.

There are a variety of applications in coding theory for the concept of Renyi entropy Csiszar [7], Farhadi and Charalambous [12], statistical mechanics Degregorio and Lacus [9] and also Kirchanov [23], statistics and associated areas [1, 13, 17, 22, 34] and other fields (see for example by Bercher [4] and the references in that). We recognize that conditional entropy for random variables can be obtained in the case of Shannon information. In addition, there is a relation between the random variables of the joint information of Shannon and the conditional information of Shannon. This relationship is referred to as the chain rule given by Cover and Thomas [8]. However, there is no known meaning of this quantity in the case of random variables of conditional Renyi information. Upon this premise of conditional Shannon Information, Cachin [5] has given a description. However, the chain rule doesn't hold for this definition. But by Jizba and Arimitsu [18], the axioms presented can be used to extract conditional Renyi Information, demonstrating the chain method's validity.

The utilization of conditional Renyi information can be applied to various aspects, for instance quantum systems cryptography Cachin [5], clinical engineering Lake [26], statistical fields Yasaei et al. [19], economics by Bentes [3] as well as other fields Dukkipati et al. [10] and also by Paris and Rad [29].

Information and stochastic processes were connected to stochastic processes via the implementation of information in probability theory and the information rate were established.

Shannon proved [36] that for a stationary random procedure with fixed state space, the Shannon Information rate occurs. He also obtained the information charge

$$\overline{K}_1(P; U) = -\frac{\sum_k \pi_k \sum_j u_{kj} p_{kj} \log p_{kj}}{\sum u_j \sum v_j} \tag{4}$$

Above measure reduces to the measure given by Golshani and Pasha [13], when utilities are ignored. Here $\Pi = (\pi_k), k = 1, 2, 3, \ldots\ldots\ldots, m$ and $u_{kj}, p_{kj}, k, j = 1, 2, \ldots\ldots m$ are transition chances and stable distribution of the chain. therein, $\Pi = \Pi P$, Where $\sum_k^m \pi_k = 1$ and $P = (p_{kj})$.

In the case of a finite-state ergodic Markov chain, Rached et al. [31] first described the Renyi information rate. This method's Information rate is expressed like

$$\overline{K}_\alpha(Y) = \frac{1}{1-\alpha} \log \lambda, \alpha \geq 0, \alpha \neq 1 \tag{5}$$

here λ is the matrix's largest positive eigenvalue $(u_{kj}p_{kj})^\alpha k, j = 1, 2, 3, \ldots\ldots m$ and $u_{kj}p_{kj}$ are chain transitional likelihoods. The Renyi entropy rate for an irreducible Markov chain with infinite state space has been demonstrated by Pasha and Golshani [30].

$$\overline{K}_\alpha(Y) = \frac{1}{1-\alpha} \log R^{-1}, \alpha \geq 0, \alpha \neq 1 \tag{6}$$

The chain's transition probabilities are $u_{kj}p_{kj}$ whereas convergence radius is R, of the matrix $(u_{kj}p_{kj})^\alpha k, j = 1, 2, 3, \ldots\ldots m$. Many operational characteristics in error exponents and coding theory have been revealed by the Renyi entropy rate [7, 28, 31, 32, 35]. The rate of Renyi information is given by Jacquet et al. [16] for stochastic processes and for fields related to statistics given by Harvey et al. [15].

This article is structured as follows. The conditional Renyi information is acquired in Sect. 2 as well as certain characteristics of the Renyi information are described. We show that for the Renyi information, In Sect. 3, the chain rule holds, and introduce a relationship to obtain the Renyi information rate. Then, this relationship is used to measure the Renyi information rate for an ergodic Markov chain of finite and infinite state spaces. Ultimately, we demonstrate in Sect. 4 that the Shannon information rate for an ergodic Markov chain is the bound rate of the Renyi information rate. Conclusion is given in Sect. 5.

2 'Useful' Conditional Renyi Information Measure

We know that a generalization form of Shannon information is Renyi information. The conditional information has already been derived for the information of Shannon and its merits are known. But there is still no concept of conditional information in the case of Renyi information. Cachin [5] provided a description close to the concept for the conditional information of Shannon, but he was unable to derive the chain rule for Shannon information that had been obtained for it. In this section, we present Renyi's suitable description of conditional information of random variables, complete with discrete distribution, and discuss its properties.

The Shannon information of a random variable Y or of a probability distribution $P = (p_1, p_2, \ldots\ldots p_m)$, with $P(Y = y_k) = p_k, k = 1, 2, \ldots\ldots m$, is defined by (3) can be valid after attaching the utility distribution in the following form:

$$K_1(Y) \equiv K_1(P; U) = -\frac{\sum_k u_k p_k \log p_k}{\sum_k u_k p_k} = \frac{\sum_k u_k p_k (-\log p_k)}{\sum_k u_k p_k}, \sum_k p_k = 1 \tag{7}$$

It may be noted that when utilities are neglected then above reduces to the value given by Golshani et al. [13]. This relation gives the information $-\log p_k$, with the mean value of its corresponding probability p_k.

The general description of the mean $g^{-1}\left(\sum_k w_k g(y_k)\right)$ (where w_k is the weight of y_k so that $0 \leq w_k \leq 1, \sum_k w_k = 1$) is a particular case. This is the probability theory description of the mean based on Kolmogorov's axioms. It is referred to an arithmetic mean or as a quasi-linear mean which is given by Kolmogorov [25] and Nagumo [27]. In this relationship, g is a continuous, invertible function, and strictly monotonic. We can also write the relation (7) in the following form, $g(y) = y$ based on this description:

$$g^{-1}\left(\frac{\sum_k u_k p_k g(-\log p_k)}{\sum_k u_k p_k}\right) \tag{8}$$

Renyi [34] showed in 1970 that for the function g in (8) two possibilities occur for information which have the additive property: $(a) g(y) = y \quad (b) g(y) = 2^{(1-\alpha)y}$.

In case (a), the relation (8) decreases to the arithmetical mean of the $-\log p_k$ information, and in the case (b), the relation (8) decreases to the $-\log p_k$ exponential mean. The first case leads to Shannon's information, while the second leads to Renyi's information. Renyi information is defined as a random variable obtained through an exponential mean, as follows.

Definition 2.1. The 'useful' Renyi information of random variable Y or 'useful' Renyi information of a chance distribution $P = (p_1, p_2, \ldots\ldots p_m)$, with probability distribution $p_k, k = 1, 2, \ldots, m$, is defined as

$$K_\alpha(Y) \equiv K_\alpha(P; U) = \frac{1}{1-\alpha} \log \frac{\sum_k^m (u_k p_k)^\alpha}{\sum_k^m u_k p_k} \qquad \alpha > 0, \alpha \neq 1 \tag{9}$$

If we expand this definition with probability distribution $P\left(Y = y_k, Z = z_j\right) = p_{kj}$ and the utility distribution $U = \{(u_1, u_2, \ldots\ldots u_m); u_k > 0 \forall k\}$, $k = 1, 2, \ldots\ldots, m$ and $j = 1, 2, \ldots\ldots, n$ to the random vector case (Y, Z), then the joint Renyi information with utility is:

$$K_\alpha(Y : Z; P : U) = \frac{1}{1-\alpha} \log \frac{\sum_{k,j} \left(u_{kj} p_{kj}\right)^\alpha}{\sum_k (u_k p_k)} \qquad \alpha > 0, \alpha \neq 1 \tag{10}$$

Now, for conditional Renyi information, we are finding a suitable description. We will describe the conditional Shannon information to begin with. The conditional Shannon information with $U = \{(u_1, u_2, \ldots\ldots u_m); u_k > 0 \forall k\}$ and $P\left(Z = z_j | Y = y_k\right) = p_{j|k}$ which is the conditional probability distribution for random variable Z, given Y, is by Cover and Thomas [8]:

$$K_1(Z|Y; P : U) = \frac{\sum_k u_k p_k K(Z|Y = y_k)}{\sum_k u_k p_k} = -\frac{\sum_{k,j} u_k p_k p_{j|k} \log p_{j|k}}{\sum_k u_k p_k} \tag{11}$$

where $u_k p_k = P(Y = y_k)$ is chance distribution of Y and $k = 1, 2, \ldots\ldots, m$.

by using above explanation of Shannon's conditional information, subsequent definition for conditional Renyi information has given by Cachin [5],

$$K_\alpha(Z|Y;P:U) = \frac{\sum_k u_k p_k K_\alpha(Z|Y=y_k)}{\sum_k u_k p_k} = \frac{1}{1-\alpha}\frac{\sum_k u_k p_k \log \sum_j (u_k p_k)^\alpha_{j|k}}{\sum_k u_k p_k} \tag{12}$$

For the conditional Renyi information we argue here that the concept described above is not suitable. In 2004, certain axioms for the Renyi and Shannon information's were presented by Jizba and Arimitsu [18], according to which we have utility distribution $U = (u_1, u_2, \ldots\ldots u_m)$ and $V = (v_1, v_2, \ldots\ldots v_n)$ and the probability distributions $P = (p_1, p_2, \ldots\ldots p_m)$ and $Q = (q_1, q_2, \ldots\ldots q_n)$ and $P * Q = (p_1q_1, p_2q_2, \ldots\ldots p_mq_n)$ are the joint probability distribution and $U * V = (u_1v_1, u_2v_2, \ldots\ldots u_mv_m)$, given to the following relationship:

$$K_\alpha(P * Q, U * V) = K_\alpha(P; U) + K_\alpha(Q * V|P * U) \tag{13}$$

With

$$K_\alpha(Q|P;U) = g^{-1}\left(\sum_k \pi_k(\alpha) g(K_\alpha(Q|u_k p_k))\right), \qquad \pi_k(\alpha) = \frac{(u_k p_k)^\alpha}{\sum_k (u_k p_k)^\alpha} \tag{14}$$

Where g is positive and an invertible function on [0,1). These writers have also shown that there are only two possibilities: *(a)* $g(y) = y$ and *(b)* $g(y) = 2^{(1-\alpha)y}$. For $\alpha = 1$, the information of Shannon leads to first case and information of Renyi is gained in the second case. They also demonstrated that a special case of relation (14), namely, by, defines the general mean of relation (8), by the definition:

$$g^{-1}\left(\sum_k \pi_k(1) g(K_1(u_k p_k))\right)$$

Using the function $g(y) = 2^{(1-\alpha)y}$ and the relation (14), the conditional information of Renyi is obtained as

$$K_\alpha(Q|P;U) = \frac{1}{1-\alpha}\log_2^{\sum\limits_k \pi_k(\alpha) 2^{(1-\alpha)K_\alpha(Q|u_k p_k)}}$$

Where $K_\alpha(Q|u_k p_k) = \frac{1}{1-\alpha}\log \sum_j (u_k p_k)^\alpha_{j|k}$ and $\pi_k(\alpha) = \frac{(u_k p_k)^\alpha}{\sum_k (u_k p_k)^\alpha}$ and we have

$$K_\alpha(Q|P;U) = \frac{1}{1-\alpha}\log^{\sum_k \frac{(u_k p_k)^\alpha}{\sum_k (u_k p_k)^\alpha} 2^{\log \sum_j (u_k p_k)^\alpha_{j|k}}} = \frac{1}{1-\alpha}\log\frac{\sum_{kj}(u_{kj} p_{kj})^\alpha}{\sum_k (u_k p_k)^\alpha}$$

Next, for random variables, we describe the conditional 'useful' Renyi information.

Definition 2.2. The random variable Z's conditional Renyi information, given Y, is defined as

$$K_\alpha(Z|Y;P:U) = \frac{1}{1-\alpha}\log\frac{\sum_{kj}(u_{kj} p_{kj})^\alpha}{\sum_k (u_k p_k)^\alpha} \tag{15}$$

From this equation we get:

$$\begin{aligned}K_\alpha(Z|Y;P:U) &= \tfrac{1}{1-\alpha}\log\tfrac{\sum_{kj}(u_{kj}p_{kj})^\alpha}{\sum_k(u_kp_k)^\alpha}\\ &= \tfrac{1}{1-\alpha}\left[\log\sum_k(u_{kj}p_{kj})^\alpha - \log\sum_k(u_kp_k)^\alpha\right]\\ &= K_\alpha(Y,Z;P:U) - K_\alpha(Y;P:U)\end{aligned} \tag{16}$$

To obtain a further description for conditional information. Using this definition, we can demonstrate that when utilities are ignored as $\alpha \to 1$, the conditional Shannon information reduces by the conditional Renyi information.

The conditional 'useful' Renyi information is obtained from the exponential mean by considering the relation (15) and (11) also general definition of the mean $g^{-1}\left(\sum_k w_k g(y_k)\right)$, Whereas, the Shannon conditional information is derived from the arithmetical mean. The definition given by Cachin, however, is based on the arithmetical mean used for the entropy of Shannon and does not seem to be sufficient. In addition, one example shows that the identity $K_\alpha(Z|Y;P:U) = K_\alpha(Y,Z;P:U) - K_\alpha(Y;P:U)$, which holds both the conditional information of Renyi (16) and conditional information of Shannon [8], The relation obtained by Cachin does not hold for. This is another reason for not using the description of Cachin for conditional Renyi information.

3 'Useful' Renyi Information Rates

Suppose that the random vector $(Y_1, \ldots\ldots Y_m)$, and the discrete-time process is $(y_m)_{m\geq 1}$. Then the following likelihood distribution is

$$p(k_1, k_2, \ldots\ldots k_m) = P(Y_1 = k_1, Y_2 = k_2, \ldots\ldots Y_m = k_m) \tag{17}$$

Then, we demonstrate that for the 'useful' Renyi information, the chain rule applies by the following theorem.

Theorem 3.1. Suppose $K_\alpha(Y_1, \ldots\ldots, Y_m)$ be the Renyi information and let $(Y_1, \ldots\ldots, Y_m)$ be a random vector and $u_k > 0$ are the utilities attached to probabilities with the probability distribution. Then for utility distribution $U = (u_1, u_2, \ldots\ldots u_m)$:

$$K_\alpha(Y_1, \ldots\ldots, Y_m) = \sum\nolimits_k^m K_\alpha(Y_k, \ldots\ldots, Y_{m-1}) \tag{18}$$

Proof. We have by (10) and (17), for the random vector $(Y_1, \ldots\ldots, Y_m)$:

$$K_\alpha(Y_1, \ldots\ldots, Y_m) = \frac{1}{1-\alpha}\log\frac{\sum_{k_1,\ldots,k_m}(u_kp_k)^\alpha(k_1, \ldots\ldots k_m)}{\sum_k u_kp_k} \tag{19}$$

We can write:

$$\begin{aligned}&\frac{\sum_{k_1,\ldots,k_m}(u_kp_k)^\alpha(k_1,\ldots,k_m)}{\sum_k u_kp_k}\\ &= \sum\nolimits_{k_1}(u_kp_k)^\alpha(k_1)\frac{\sum_{k_1,k_2}(u_kp_k)^\alpha(k_1,k_2)}{\sum_{k_1}(u_kp_k)^\alpha(k_1)}\cdots\\ &\cdots\frac{\sum_{k_1,\ldots,k_{m-1}}(u_kp_k)^\alpha(k_1,\ldots,k_{m-1})}{\sum_{k_1,\ldots,k_{m-2}}(u_kp_k)^\alpha(k_1,\ldots,k_{m-2})}\frac{\sum_{k_1,\ldots k_m}(u_kp_k)^\alpha(k_1,\ldots,k_m)}{\sum_{k_1,\ldots,k_{m-1}}(u_kp_k)^\alpha(k_1,\ldots,k_{m-1})}\end{aligned}$$

we get the following when this equation is incorporated into (19):

$$K_\alpha(Y_1, \ldots\ldots, Y_m) = \frac{1}{1-\alpha}\left[\log \sum_{k_1} (u_k p_k)^\alpha(k_1) + \log \frac{\sum_{k_1,k_2} (u_k p_k)^\alpha(k_1,k_2)}{\sum_{k_1} (u_k p_k)^\alpha(k_1)} + \cdots + \log \frac{\sum_{k_1,\ldots,k_m} (u_k p_k)^\alpha(k_1,\ldots\ldots,k_m)}{\sum_{k_1,\ldots,k_{m-1}} (u_k p_k)^\alpha(k_1,\ldots\ldots,k_{m-1})}\right]$$

The desired result is obtained, by using (9) and (15).

For all random vectors $Y_1, \ldots\ldots Y_m$, the likelihood distribution is invariant under the time change, if a discrete-time process is static, i.e.

$$p(k_1, k_2, \ldots\ldots k_m) = p\big(k_{1+i}, \ldots\ldots k_{m+i}\big), \quad i \in x \tag{20}$$

We use Eq. (15) and Eq. (20) for this case:

$$K_\alpha(Y_m|Y_1, \ldots\ldots, Y_{m-1}) \le K_\alpha(Y_{m+i}|Y_{1+i}, \ldots\ldots, Y_{m-1+i}) \tag{21}$$

And for stationary processes, the conditional 'useful' Renyi information $K_\alpha(Y_m|Y_1, \ldots\ldots, Y_{m-1})$ is bound:

$$0 \le K_\alpha(Y_m|Y_1, \ldots\ldots, Y_{m-1}) \le K_\alpha(Y_1) + i, i \in R^+ \tag{22}$$

The regular relation when the limit exists for the Renyi information rate is for a discrete-time process $Y = (Y_m)_{m \ge 1}$

$$\overline{K}_\alpha(Y) = \lim_{m\to\infty} \frac{1}{m} K_\alpha(Y_1, \ldots\ldots, Y_m) \tag{23}$$

Multiplying both sides by $\left(\frac{1}{m}\right)$ and taking the limit $m \to \infty$, of the relation (18) we're getting:

$$\lim_{m\to\infty} \frac{1}{m} K_\alpha(Y_1, \ldots\ldots, Y_m) = \lim_{m\to\infty} \frac{1}{m} \sum_{k=1}^{m} K_\alpha(Y_k|Y_1, \ldots\ldots, Y_{k-1})$$

Therefore, by relation (23) we have

$$\overline{K}_\alpha(Y) = \lim_{m\to\infty} \frac{1}{m} \sum_{k=1}^{m} K_\alpha(Y_k|Y_1, \ldots\ldots, Y_{k-1})$$

If there is a limit of conditional Renyi information, $\lim_{m\to\infty} K_\alpha(Y_k|Y_1, \ldots\ldots, Y_{m-1})$ exists, then Cesaro implies that it is equal to $\overline{K}_\alpha(Y)$, and there is a Renyi information rate exists, and we also have another method for obtaining the Renyi information rate.

4 Lower and Upper Bounds for the 'Useful' Renyi Information Rate

We have the following inequalities using the fact that the 'useful' Renyi information is a decreasing function of α:

$$1 - \text{For}\, \alpha < 1, \quad K_1(\cdot) < K_\alpha(\cdot) \tag{24}$$

$$2 - \text{For}\, \alpha > 1, \quad K_1(\cdot) < K_\alpha(\cdot) \tag{25}$$

Here K_1 represent the Shannon information.

Now, by using (24) and (25), we get the bounds of an ergodic Markov chain 'useful' Renyi information rate. Inequality (24) for a random vector $(Y_1, \ldots\ldots, Y_m)$ becomes:

$$K_1(Y_1, \ldots\ldots, Y_m) < K_\alpha(Y_1, \ldots\ldots, Y_m)$$

and therefore

$$\frac{1}{m}K_1(Y_1, \ldots\ldots, Y_m) < \frac{1}{m}K_\alpha(Y_1, \ldots\ldots, Y_m)$$

Taking $m \to \infty$ the entropy limit and considering that the 'useful' Renyi information rate which is given by Pasha and Golshani [30] and Rached et al. [31] and the Shannon information by Shannon [36] and by Bezuglyi et al. [24] exist for state space the Markov ergodic chain with finite and infinite, Therefore: $\overline{K}_1(Y) \leq \overline{K}_\alpha(Y)$.

from (4) we get

$$\frac{\sum_k \pi_k \sum_j u_{kj} p_{kj} \log p_{kj}}{\sum u_k \sum v_j} \leq \overline{K}_\alpha(Y) \tag{26}$$

For the 'useful' Renyi information rate a lower bound can be obtained for the case $\alpha < 1$. We derive from inequality in a similar way (25): $\overline{K}_\alpha(Y) \leq \overline{K}_1(Y)$.

And we have

$$\overline{K}_\alpha(Y) \leq \frac{\sum_k \pi_k \sum_j u_{kj} p_{kj} \log p_{kj}}{\sum u_k \sum v_j} \tag{27}$$

For the case $\alpha > 1$, For the case α > 1, an upper bound rate is obtained for the 'useful' Renyi information rate.

Note that if a system has a finite state space, the rate of 'useful' Renyi information is calculated from (5), and the rate of 'useful' Renyi information is calculated from (6), if the state space is infinite.

5 Conclusion

We presented a new concept for conditional Useful Renyi information in this paper, focusing primarily on axioms proposed by some authors, and demonstrated that the chain rule applies to this definition. Then, for an ergodic Markov chain, we derived a relationship to get the rate of Renyi's information and we used this relationship to obtain the 'useful' Renyi information rate. In addition, it was shown that the Shannon information rate for the aforementioned processes is a bound rate for 'useful' Renyi information.

Acknowledgments. Authors are thankful for Jaypee University of Engineering and Technology, Guna (M.P.), India for its support from time to time.

References

1. Andai, A.: On the geometry of generalized Gaussian distributions. J. Multivar. Anal. **100**, 777–793 (2009)
2. Belis, M., Guiasu, S.: A quantitative-qualitative measure of information in cybernetics system. IEEE Trans. Inform. Theory IT **14**, 593–594 (1968)
3. Bentes, S.R., Menezes, R., Mendes, D.A.: Long memory and volatility clustering is the empirical evidence consistent across stock markets. Phys. A **387**, 3826–3830 (2008)
4. Bercher, J.F.: On some information functionals derived from Renyi information divergence. Inf. Sci. **178**, 2489–2506 (2008)
5. Cachin, C.: Information measures and unconditional security in cryptography. Ph.D. Thesis, Swiss Federal Institute of Technology Zurich (1997)
6. Chandy, T.W., Mcliod, J.B.: On a functional equation. Proc. Edinburgh Maths **43**, 7–8 (1960)
7. Csiszar, I.: Generalized cut off rates and Renyi's information measures. IEEE Trans. Inf. Theory **41**, 26–34 (1995)
8. Cover, T.M., Thomas, J.: The Elements of Information Theory. John Wiley and Sons (1991)
9. Degregorio, A., Lacus, S.M.: On Renyi information for ergodic diffusion processes. Inform. Sci. 279–291 (2009)
10. Dukkipati, A., Bhatnagar, S., Murty, M.N.: Gelfand–Yaglom–Perez, theorem for generalized relative information functionals. Inf. Sci. **177**, 5707–5714 (2007)
11. Fadeev, D.K.: On the concept of entropies of finite probabilistic scheme (Russian). Uspchi Math. Nauk **11**, 227–231 (1956)
12. Farhadi, A., Charalambous, C.D.: Robust codind for a class of sources applications in control and reliable communication over limited capacity channels. Syst. Control Lett. **57**, 1005–1012 (2008)
13. Golshani, L., Pasha, E., Yari, G.: Some properties of Renyi information and Renyi information rate. Inf. Sci. **179**(14), 2426–2433 (2009)
14. Guiasu, S.: Information Theory with Applications. McGraw-Hill Inc (1977)
15. Harvey, N.J.A., Onak, K., Nelson, J.: Streaming algorithms for estimating information. In: IEEE Information Theory Workshop, pp. 227–231 (2008)
16. Jacquet, P.G., Seroussi, W.: Szpankowski: on the information of hidden Markov process. Theor. Comput. Sci. **395**, 203–219 (2008)
17. Jenssen, R., Eltoft, T.: A new information theoretic analysis of sum-of-squared-error kernel clustering. Neurocomputing **72**, 23–31 (2008)
18. Jizba, P., Arimitsu, T.: The world according to Renyi thermodynamics of multifractal systems. Ann. Phys. **312**, 17–59 (2004)
19. Yasaei, S.S., Oselio, B.L., Hero, A.O.: Learning to bound the multi-class Bayes error. IEEE Trans. Signal Process. 1 (2020)
20. Kendall, D.G.: Functional equations in information theory. Z. Wahrs Verw. Geb **2**, 225–229 (1964)
21. Khinchin, A.I.: Mathematical Foundations of Information Theory. Dover Publications, New York (1957)
22. Dongdong, W., Liu, Z., Tang, Y.: A new classification method based on the negation of a basic probability assignment in the evidence theory. Eng. Appl. Artif. Intell. **96**, 103985 (2020). https://doi.org/10.1016/j.engappai.2020.103985
23. Kirchanov, V.S.: Using the Renyi information to describe quantum dissipative systems in statistical mechanics. Theor. Math. Phys. **156**, 1347–1355 (2008)
24. Bezuglyi, S., Karpel, O., Kwiatkowski, J.: Exact number of ergodic invariant measures for Bratteli diagrams. J. Math. Anal. Appl. **480**(2), 123431 (2019)

25. Kolmogorov, A.N.: Sur la notion de la moyenne. Atti della Accademia Nazionale dei Lincei Rend **12**(6), 388–391 (1930)
26. Lake, D.E.: Renyi information measures of heart rate gaussianity. IEEE Trans. Biomed. Eng. **53**, 21–27 (2006)
27. Nagumo, M.: Uber eine klasse der mittelwerte. Jpn. J. Math. **7**, 71–79 (1930)
28. Nilsson, M., Kleijn, W.B.: On the estimation of differential information from data located on embedded manifolds. IEEE Trans. Inf. Theory **53**, 2330–2341 (2007)
29. Paris, J.B., Rad, S.R.: Inference processes for quantified predicate knowledge. Logic Lang. Inform. Comput. **5110**, 249–259 (2008)
30. Pasha, E., Golshani, L.: The Renyi information rate for Markov chains with countable state space. Bull. Iran. Math. Soc. (submitted for publication)
31. Rached, Z., Fady, A., Campbell, L.L.: Renyi's information rate for discrete Markov sources. In: Proceedings of the CISS 1999, Baltimore, MD, pp. 17–19 (1999)
32. Rached, Z., Fady, A., Campbell, L.L.: Renyi's divergence and information rate for finite alphabet Markov sources. IEEE Trans. Inf. Theory **47**, 1553–1560 (2001)
33. Li, W., Li, Y.: Entropy, mutual information, and systematic measures of structured spiking neural networks. J. Theor. Biol. **501**, 110310 (2020). https://doi.org/10.1016/j.jtbi.2020.110310
34. Renyi, A.: Probability Theory. North-Holland, Amsterdam (1970)
35. Seneta, E.: Nonnegative Matrix and Markov Chains. Springer-Verlag, New York (1981)
36. Shannon, C.E.: A mathematical theory of communication. Bell Syst. Tech. J. **27**(3), 379–423 (1948). https://doi.org/10.1002/j.1538-7305.1948.tb01338.x

Sign Language Recognition Using Convolutional Neural Network

Mihir Gandhi(✉), Priyam Shah, Devansh Solanki, and Prasanna Shete

Department of Computer Engineering, K. J. Somaiya College of Engineering, Vidyavihar, Mumbai, India
{mihir.mg,priyam.ds,devansh.solanki,prasannashete}@somaiya.edu

Abstract. A condition in which an individual has trouble forming sounds is called speech impairment, and a condition in which an individual cannot completely receive sounds through their ears is called hearing impairment. Both of these impairments affect an individual's ability to communicate with others. Those affected by these impairments use alternative forms of communication, such as sign language. However, it is difficult for non-sign language speakers to communicate with sign language speakers. This is because most of the sign language recognition solutions are not lightweight or portable and require considerable hardware like a computer for usage. This work focuses on developing a computer vision-based application that translates sign language into text using a Convolutional Neural Network, thus enabling signers and non-signers to communicate. This application will support mobile use and work without an internet connection as well, thus serving as a ubiquitous communication aid.

Keywords: Deep learning · Neural network · Convolutional Neural Network · Computer vision · Hand gesture recognition · American Sign Language

1 Introduction

Speech and Hearing-impaired people use sign language as a primary form of communication. They use hand gestures representing sign language alphabets in order to express their emotions and thoughts. However, it is not possible for them to communicate with people who do not understand sign language. Currently, the communication between them mostly relies on human-based translation. However, due to the involvement of human expertise, this is both expensive and inconvenient. According to the World Health Organization, over 5% of the world's population (or 466 million people) has disabling hearing loss. It is estimated that by 2050, over 900 million people – or 1 in every 10 people – will suffer from disabling hearing loss [1].

For this reason, recent studies have been accelerated in order to help these people communicate more easily and efficiently. Though many solutions have been proposed for sign language recognition, especially using neural networks, most of them involve dense networks that require considerable computing power and space for usage. As a result, these cannot be deployed effectively in portable devices having less processing

M. Singh et al. (Eds.): ICACDS 2021, CCIS 1441, pp. 281–291, 2021.
https://doi.org/10.1007/978-3-030-88244-0_27

power, such as mobile phones. This prevents the technology from being used in the real world as a tool for real-time sign language translation.

Our proposed system aims to overcome this very challenge. To assist people in communicating with the speech and hearing impaired, we propose to develop a Convolutional Neural Network that can identify hand gestures that represent the sign language alphabets in real-time. The American Sign Language Dataset created by B. Kang et al. [2] was used to train the neural network. This dataset was then preprocessed and used to train the CNN designed by us. This CNN achieved an accuracy of 89.5%. This CNN is deployed in an Android application to enable mobile and offline use.

The remainder of this paper is structured as follows: Sect. 2 summarizes the literature survey performed. Section 3 states the dataset used. The architecture of the proposed CNN is discussed in Sect. 4. Section 5 highlights the implementation details, and Sect. 6 summarizes the results of the proposed model. Finally, Sect. 7 presents the conclusion and future works possible.

2 Related Work

Nikam and Ambekar [3] proposed a scale and shape independent method to recognize hand gestures using contour analysis. The algorithm was able to work in offline mode, learn signs, and recognize them with 100% results for fixed sets of signs. However, there can be cases where amputated fingers can cause issues in calculating boundaries and further result in false output. To overcome these issues, advanced techniques like the Residual Neural Network model were used. M. Xie et al. [8] propose an RNN that achieves 99.4% accuracy.

Since Action Recognition has some similarities with Sign Language Recognition, Gunawan et al. [5] attempt to use the i3d Inception model with transfer learning method to recognize sign language. The tests performed showed a 100% accuracy for 100 classes (10 signers and 10 words). However, the model was overfit, and as a result, the validation accuracy was low. Thiracitta et al. [6] compare the performance of two Hidden Markov models – Gaussian and Multinomial – for Sign Language Recognition. It was found that Gaussian MM achieved higher accuracy than Multinomial HMM. However, this research does not reflect real-world scenarios, as the signers are using gloves in the dataset used for training.

The proposed model by K. Bantupalli et al. [9] take a different approach. This model extracts spatial and temporal features from video sequences. It then uses Inception for recognizing spatial features and RNN for training temporal features. The accuracy was greater than 90% for varying sample sizes. The model faced accuracy loss with different skin tones, clothing, and inclusion of various body parts within the video. Soodtoetong et al. [4] and Das et al. [7] use an advanced algorithm for 3D-CNN on recognized processes. The images were captured using Kinect sensors. The algorithm was trained on five gestures only, with a maximum recognition accuracy of 91.23%. It required heavy processing for training, as the stream of data received from Kinect sensors for every single gesture was large.

Rao et al. [10] use a convolutional neural network with 4 convolutional layers to classify sign language gestures. In order to improve accuracy and speed of recognition, varying filter window size is used with each layer. The average recognition rate was found to be 92.88% with this network. Suresh et al. [11] proposed two layered CNN for classification of sign language images. Two different optimizers were used with this model, Adam and SGD, which gave an accuracy of 99.51% and 99.12%, respectively. This CNN showed good accuracy even with images with varying lighting and images containing noise.

3 Dataset

The American Sign Language Dataset created by B. Kang et al. [2] was used for training the neural network. It consists of 31,000 images of hand gestures for alphabets (A-Z except for J and Z) and numbers (1–9), as shown in Fig. 1. The alphabets J and Z were not included as they require movements of the hand and thus cannot be captured in the form of an image. 2/V and 6/W have similar signs and are distinguished depending on the context.

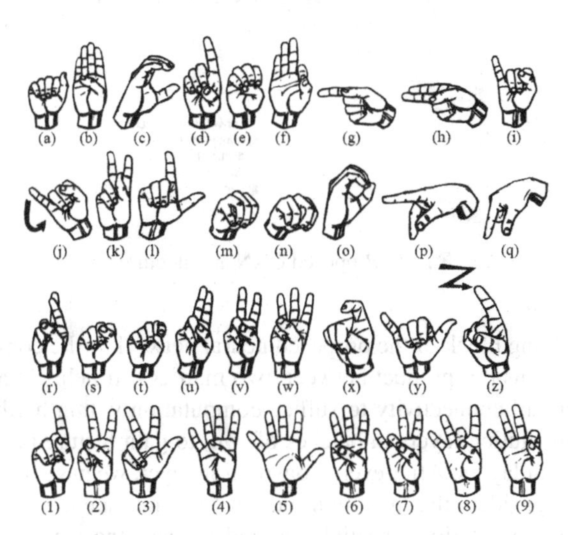

Fig. 1. American Sign Language

4 Architecture

Our proposed CNN consists of the following layers:

- 2 Convolution layers: 64 channel, 3×3 kernel, 3×3 padding
- 1 Maxpool layer: 2×2 pool size, 2×2 stride
- 1 Dropout layer: rate = 0.25
- 2 Convolution layers: 128 channel, 3×3 kernel, 3×3 padding

- 1 Maxpool layer: 2 × 2 pool size, 2 × 2 stride
- 1 Dropout layer: rate = 0.25
- 2 Convolution layers: 256 channel, 3 × 3 kernel, 3 × 3 padding
- 1 Maxpool layer: 2 × 2 pool size, 2 × 2 stride
- 1 Dropout layer: rate 0.25
- 1 Flatten layer
- 1 Dense layer: 31 units

The proposed architecture of our convolutional neural network is shown in Fig. 2.

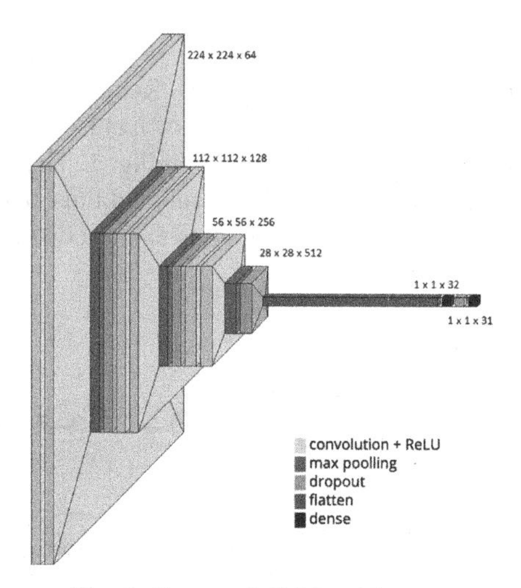

Fig. 2. Proposed CNN architecture

The main building block of the proposed neural network is the convolution layer. It convolves or performs dot product between two matrices. It helps in achieving sparse connectivity over full connectivity to suffice computation within hardware limits. We have used six convolution layers in batches of two, to gain feature maps by increasing parameter sharing at layers. After every convolution layer, we have used a max pooling layer. This layer considers the maximum output from the neighborhood, which helps in reducing the representation's spatial size and reduces the amount of computation required. After pooling, the output is randomly subsampled; this helps in thinning the neural network while training. Finally, we flatten the feature map to a single long feature vector for classification from the output of signs. The dense layer is a fully-connected layer, which helps in understanding the most related features while classifying the output sign.

5 Implementation

5.1 Data Pre-processing

The American Sign Language Dataset created by B. Kang et al. [2] was used to train the CNN. We perform some transformations on the images in the pre-processing stage, such as resizing them to the same size to make them have a uniform size. Then, we used ImageDataGenerator to generate even more images, as this would help to train the model better and improve the accuracy of detection. This was done by varying the rotation, width shift, height shift, zoom, and brightness.

5.2 Designing the CNN

Once our dataset was ready, we designed our proposed neural network architecture, as shown in Fig. 2, inspired by the VGG16 architecture. The architecture of VGG16 is depicted below in Fig. 3. VGG16 is a CNN model proposed in the paper "Very Deep Convolutional Networks for Large-Scale Image Recognition" by K. Simonyan and A. Zisserman. VGG16 achieved a top-5 test accuracy of 92.7% in ImageNet [12].

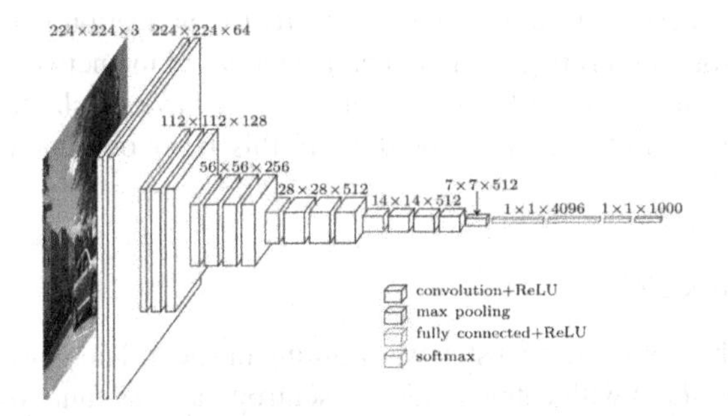

Fig. 3. VGG16 architecture

However, VGG16 has two major disadvantages:

1. The network takes a lot of time for training.
2. Weights of the network are comparatively larger.

VGG16 is larger than 530 MB, primarily due to the number of fully connected nodes and the depth of the network [13]. This large size makes deployment of VGG difficult. As our objective was to develop a sign language recognition application for mobile use, we decided to build our own CNN, which has fewer layers than VGG16, and thus is easy to train, lightweight, and easy to deploy on devices with limited computational power such as mobile phones, while ensuring quick and accurate detection.

The proposed CNN architecture can be seen in Fig. 2. It consists of three types of layers – Convolution, Dense, and Maxpool. The significance of the various layers used is as follows:

1. *Convolutional*: The convolutional layer produces a tensor of outputs by creating a kernel and convolving it with the input to the layer [15]. It helps in summarizing the presence of features in the input image [16].
2. *MaxPooling*: The MaxPooling layer takes the maximum value over a 'pool_size' sized window along every dimension along the feature axis and helps in downsampling the input representation [16].
3. *Dense*: Dense layer is a fully connected layer. It performs the operation: output = activation_function(bias + dot(input, kernel)).

This network was implemented in Python using Keras, a neural network library. Keras is open source, modular, and extensible. It helps to achieve fast experimentation with deep neural networks [14].

We begin by initializing the model as a sequential model, forming a linear stack of layers. Then, we add the aforementioned layers one by one. After the convolution layers, this data is passed to the dense layer. The vector that comes out of the convolution layers is flattened before passing it to the dense layer. RELU activation is used for the dense layer to stop the forward propagation of negative values in the network. Finally, a dense layer of 31 units is used with softmax activation as we have 31 classes to predict from. Depending on the confidence of the prediction, this layer outputs a value between 0 and 1.

5.3 Training the CNN

After designing the CNN, the next step was training the neural network. This step begins by compiling the model with categorical_crossentropy as loss function and accuracy as the metric. We used Adam optimizer to find the global minima while training our model and set the learning rate to 0.001. We then use two modules from Keras to build a good model:

1. *ModelCheckpoint*: This is used to monitor some parameters of the model and to save it. We monitored the val_acc parameter, which represents the validation accuracy.
2. *EarlyStopping*: This is used to stop the training of the model if there is no change in the parameter being monitored. In our case, this parameter was the validation accuracy.

5.4 Deploying the CNN in an Android Application

The output after the training is stored in the form of a.h5 file, which can be used for recognizing the sign language alphabets. However, the.h5 file cannot be directly deployed in an Android application. A.tflite file is required for deployment of the CNN in an Android app. So we used TFLiteConverter to convert the.h5 file to.tflite file. We created an Android application that consisted of a Camera view where the input can be captured, a button to toggle the camera from back to front camera or vice versa, and a button to detect objects, i.e., recognize the alphabets. The detected alphabets are displayed on the screen along with the confidence of the detection. The softmax layer outputs the confidence as a value between 0 and 1. This was converted into a percentage when displaying in the app for a more intuitive understanding.

On clicking the button for detection of the alphabet, an image is captured, which is sent as input to the model. This process is run as a separate thread because the processing may take a second or two, and the application should not appear unresponsive as if it has crashed during this time period. The model, in the form of a tflite file and a file containing all possible labels, needs to be stored on the device with the app to support the offline operation.

6 Results

To analyze the performance of the model, two parameters were considered: Recognition Rate (Accuracy) and Time for Recognition. The experimental setup was as follows: three different testers performed the testing by making hand gestures of various sign language alphabets. Each tester tested the application by making the hand sign and capturing 5 images for each alphabet from A-Z except J and Z, and numbers from 1 to 9 except 2 and 6 (as these are already tested with V and W, respectively). Four such tests are shown in Fig. 4 below. For each test, we record whether the alphabet was recognized correctly as well as the time taken for recognition.

The results were recorded in the form of a table as shown below in Table 1. The first column contains the alphabet being tested. The second and third column contains the number of times the sign was recognized correctly and incorrectly (along with misrecognition result). These false results are captured for further analysis and improvement of the model. The fourth column contains the accuracy of recognition for that alphabet. This is calculated by using Formula 1 shown below.

$$Accuracy = \frac{True\,Result}{True\,Result + False\,Result} \tag{1}$$

The overall accuracy or average accuracy is calculated by using Formula 2.

$$Average\,Accuracy = \frac{\sum True\,Result}{\sum True\,Result + \sum False\,Result} \tag{2}$$

The proposed CNN achieved an average accuracy of 89.25%. Considering its shallow layer architecture and small size, which was necessary for easy deployment and operation in devices with lower computing power and space constraints such as mobile phones, this accuracy is quite good. The average recognition time over the 465 tests was found to be 1.2 s.

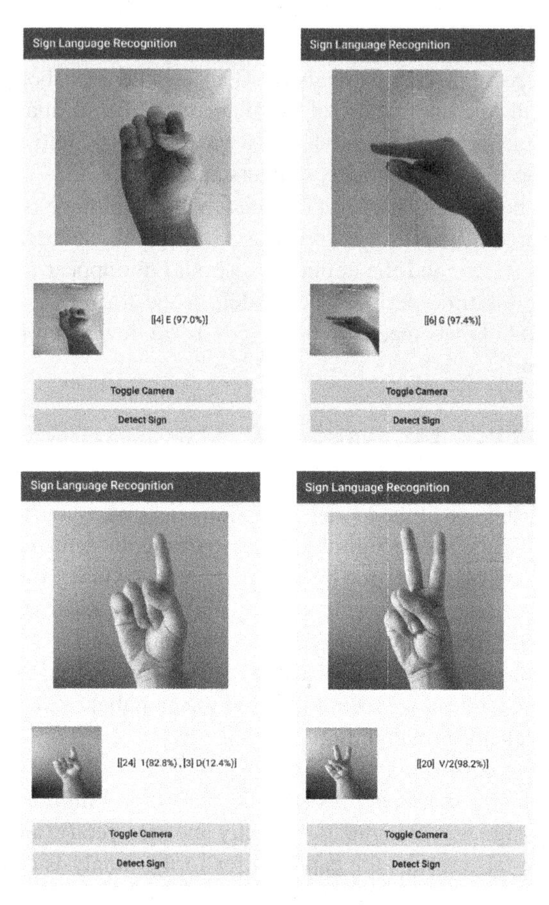

Fig. 4. Recognition results of the CNN deployed in Android app

Table 1. Character-wise results of the tests performed.

Character	True result	False result	Accuracy (in %)
A	10	5 (E,S)	66.67
B	14	1 (4)	93.33
C	15	0	100
D	10	5 (1)	66.67
E	12	3 (A,T)	80
F	15	0	100
G	14	1 (H)	93.33

(continued)

Table 1. (*continued*)

Character	True result	False result	Accuracy (in %)
H	12	3 (G)	80
I	13	2 (Y)	86.67
K	10	5 (V/2)	66.67
L	15	0	100
M	10	5 (N)	66.67
N	13	2 (M)	86.67
O	14	1 (C)	93.33
P	15	0	100
Q	15	0	100
R	15	0	100
S	13	2 (A,E)	86.67
T	13	2 (E)	86.67
U	10	5 (V/2,K)	66.67
V/2	12	3 (K)	80
W/6	15	0	100
X	15	0	100
Y	15	0	100
1	12	3 (D)	80
3	15	0	100
4	13	2 (B)	86.67
5	15	0	100
7	15	0	100
8	15	0	100
9	15	0	100
Total	415	50	89.25

7 Conclusion and Future Work

As a result of communication barriers among the speech and hearing impaired and other individuals in the community, they often experience many difficulties. The proposed Convolutional Neural Network can identify hand gestures representing the alphabets of the American Sign Language to assist people in communicating with the speech and hearing impaired. The CNN was deployed in an android application. Consequently, it could be easily installed on mobile phones, enabling it to be used practically anywhere, even offline. The CNN achieved an accuracy of 89.25% with an average detection time of 1.2 s. Thus, this application can serve as a ubiquitous communication tool and help in the betterment of the speech and hearing-impaired community. Currently, this application supports only American Sign Language. In the future, it can be modified to support other sign languages such as British Sign Language and Indian Sign Language. This will help to develop a consolidated application that can recognize sign language alphabets of multiple sign languages, serving as a global communication tool for people to interact with the speech and hearing impaired.

References

1. World Health Organization: Deafness and hearing loss. https://www.who.int/news-room/fact-sheets/detail/deafness-and-hearing-loss (2021). Accessed 15 Feb 2021
2. Kang, B., Tripathi, S., Nguyen, T.Q.: Real-time sign language fingerspelling recognition using convolutional neural networks from depth map. In: 2015 3rd IAPR Asian Conference on Pattern Recognition (ACPR), pp. 136–140. IEEE (2015)
3. Nikam, A.S., Ambekar, A.G.: Bilingual sign recognition using image based hand gesture technique for hearing and speech impaired people. In: 2016 International Conference on Computing Communication Control and automation (ICCUBEA), pp. 1–6. IEEE (2016)
4. Soodtoetong, N., Gedkhaw, E.: The efficiency of sign language recognition using 3D convolutional neural networks. In: 2018 15th International Conference on Electrical Engineering/Electronics, Computer, Telecommunications and Information Technology (ECTI-CON), pp. 70–73. IEEE (2018)
5. Gunawan, H., Thiracitta, N., Nugroho, A.: Sign language recognition using modified convolutional neural network model. In: 2018 Indonesian Association for Pattern Recognition International Conference (INAPR), pp. 1–5. IEEE (2018)
6. Thiracitta, N., Gunawan, H., Witjaksono, G.: The comparison of some hidden markov models for sign language recognition. In: 2018 Indonesian Association for Pattern Recognition International Conference (INAPR), pp. 6–10. IEEE (2018)
7. Das, P., Ahmed, T., Ali, M.F.: Static hand gesture recognition for American Sign Language using deep convolutional neural network. In: 2020 IEEE Region 10 Symposium (TENSYMP), pp. 1762–1765. IEEE (2018)
8. Xie, M., Ma, X.: End-to-end residual neural network with data augmentation for Sign Language Recognition. In: 2019 IEEE 4th Advanced Information Technology, Electronic and Automation Control Conference (IAEAC), vol. 1, pp. 1629–1633. IEEE (2018)
9. Bantupalli, K., Xie, Y.: American sign language recognition using deep learning and computer vision. In: 2018 IEEE International Conference on Big Data (Big Data), pp. 4896–4899. IEEE (2018)
10. Rao, G.A., Syamala, K., Kishore, P.V.V., Sastry, A.S.C.S.: Deep convolutional neural networks for sign language recognition. In: 2018 Conference on Signal Processing and Communication Engineering Systems (SPACES), pp. 194–197. IEEE (2018)
11. Suresh, S., Mithun, H.T., Supriya, M.H.: Sign language recognition system using deep neural network. In: 2019 5th International Conference on Advanced Computing & Communication Systems (ICACCS), pp. 614–618. IEEE (2019)
12. Neurohive: VGG16–Convolutional Network for Classification and Detection. https://neurohive.io/en/popular-networks/vgg16/ (2018). Accessed 15 Feb 2021
13. Thakur, R.: VGG16 implementation in Keras. In: Towards Data Science. https://towardsdatascience.com/step-by-step-vgg16-implementation-in-keras-for-beginners-a833c686ae6c (2019). Accessed 15 Feb 2021
14. Wikipedia: Keras. https://en.wikipedia.org/wiki/Keras (2021). Accessed 15 Feb 2021

15. Keras: 2D Convolution layer. https://keras.rstudio.com/reference/layer_conv_2d.html (2020). Accessed 15 Feb 2021
16. Brown, J.: Introduction to pooling layers. In: Machine Learning Mastery. https://machinelearningmastery.com/pooling-layers-for-convolutional-neural-networks/ (2019). Accessed 15 Feb 2021

Prediction of Stock Price for Indian Stock Market: A Comparative Study Using LSTM and GRU

Shwetha Salimath, Triparna Chatterjee, Titty Mathai, Pooja Kamble, and Megha Kolhekar(✉)

Fr. C. Rodrigues Institute of Technology, Navi Mumbai, India
megha.kolhekar@fcrit.ac.in

Abstract. The dynamic nature of stock markets makes the task of generalizing stock prediction very challenging. The dependence of stock prices upon many parameters which directly or indirectly affect them; just adds to the challenge. In recent years there is an increased interest in stock prediction using neural networks. Most of the work-related are either based upon global exchange parameters or favor one particular architecture. To the best of our knowledge, a dedicated study of Indian stocks and the prediction of their prices are not reported. This is the main motivation of this work.

In this paper, we describe the implementation of the Long Short-Term Memory (LSTM) network, Gated Recurrent Unit (GRU) networks for stock prediction. We present a comparative study as well as an analysis for four different network architectures. We use the NSEpy 0.8 Python package to extract historic data which is made publicly available by NSE India. We have used three years of training data and one-year of testing data. After an in-depth study of the impact of the parameters on the stock prices, we have chosen ten parameters for training. The metric used is the Loss Function. The prediction is performed for the closing prices of stocks of twenty-five Indian companies. The results indicate that a two-layer GRU outperforms all other networks as far as these twenty-five companies are concerned.

Keywords: Stock price prediction · LSTM · GRU · RNN

1 Introduction

Stock prediction using machine learning algorithms has been a well-studied field in computational finance since the 1990's. However, due to the dynamic nature of the market, it has been a challenging task. India has two stock exchange markets, the Bombay Stock Exchange Ltd (BSE) and the National Stock Exchange of India Ltd (NSE). The BSE is Asia's oldest stock exchange and currently the seventh-largest in the world consisting of 5439 listed companies [1]. While the NSE, which was incorporated in the year 1992 is the eleventh largest has 1952 listed companies [2].

M. Singh et al. (Eds.): ICACDS 2021, CCIS 1441, pp. 292–302, 2021.
https://doi.org/10.1007/978-3-030-88244-0_28

The Indian market has been growing at a very fast pace and has created a lot of wealth, but very few people have invested in the stocks. It is not considered as a primary method for wealth creation, job creation, and channelizing savings in India. Due to the dynamic nature, there is a high risk associated, thus the newcomers who lose money, tend to leave the market forever [2]. A lot of research is occurring in this field, but very few have been on the Indian stock exchange. The state-of-art in this area is restricted to Foreign Exchanges and Foreign company stocks only, limited to a particular architecture. We try in this work to address this gap of work in the area of Indian Stock Market movement with reference to Indian companies. The objective of the paper is to implement stock prediction on twenty-five companies listed in NSE by using LSTM, GRU, and a hybrid network consisting of LSTM and GRU. The paper gives a comparative analysis of these networks. The analysis is performed on a simple computer without any high system requirements. The training dataset is taken for a short duration in consideration with the Indian market as some of the companies have been listed for only a few years. The application of trying to find the best model is to develop a system that predicts the stock price in immediate future while giving deeper insights to the investors with assistance for terminologies and to help them in stock market trading.

The paper is organized as follows, Sect. 2 consists of relevant work done and Sect. 3 contain the methodology explaining the dataset and networks used along with the implementation details. Result, limitation, and conclusion are discussed in Sect. 4.

2 Related Work

This section presents the gist of some seminal and some recent papers which work around predicting stock prices using neural and deep neural networks.

Hossain et al. [3] propose a hybrid model that consists of LSTM and GRU. This model has been implemented on S & P 500 which is trained on historic data of 66 years. The open and volume are used as input to get output as the closing price. The proposed hybrid network outperforms only LSTM and GRU networks over the same dataset. Selvin et al. [4] propose to predict for a single company its stock movement using the closing price. A sliding window approach on RNN, LSTM, and CNN models are used to predict values. For window sizes of 100 min with a 90-min overlap to predict future 10 min, the CNN gives the best accuracy.

Chung et al. [5] propose a hybrid approach which is achieved by integrating the LSTM neural network and Genetic algorithm (GA) which is used to find optimum size of the time window and number of LSTM units. Historic data consisting of open, high, low, close, and volume on a daily basis, and derived quantities like moving average and relative strength index are used. The best window size achieved for Korean Stock Index (KOSPI) is 10 days with LSTM units for the 2 hidden layers being 15 and 7 respectively. Liu et al. [6] propose a hybrid model which consists of two GRU and one LSTM layer where the GRU is used to increase the computational speed. The model is trained and tested on stocks from the Chinese stock exchange to predict the future closing price. The training is performed using the past 10 years of historic data. While the GRU models are faster due to fewer parameters but are less accurate than LSTM.

Nabipour et al. [7], provide a comparison between RNN, LSTM, random forest, decision tree, adaptive boost, and XGboost models to predict future stock market values of which LSTM is found to show the most accurate results. Arif et al. [8] propose a combined model of LSTM and Bi-LSTM to predict future trends. The highest accuracy is obtained when a dataset of 15 years is trained over 100 epochs. Adil et al. [9] use a four-layer LSTM model, where maximum accuracy is obtained for 100 epochs when a dataset for 15 years was used and 50 epochs when a dataset of 10 years was used.

The study of recent literature the state-of-the-art in this area, that LSTM is the most popularly used architecture for stock prediction. We observe that the recent LSTM and GRU models have not been combined with any other factor other than the historic data to get the predicted values. Also, most of the above-mentioned methods have been implemented in foreign exchange market. We have taken twenty-five Indian companies from different fields to test the algorithm so that the resulting model is common and not specific to a particular company performance.

3 Methodology

In this section, we explain the system architecture and the different types of networks used. We have used recurrent units such as LSTM and GRU. Both LSTM and GRU have the ability to retain memory which allows them to remember features for a longer duration and also allows backpropagation to occur through multiple bounded nonlinearities, reducing the possibility of facing a vanishing gradient problem [10].

3.1 Dataset

We have used the NSEpy 0.8 python package from the National Stock Exchange of India to extract the historic stock prices, formed to get publicly available data on the NSE website such as historical data, live indices which can be used to process data for research and projects [11]. For training, we have used 3-year data, from 2016 to 2018 consisting of 724 trading days and for testing, we have used data set for the year 2019 consisting of 229 days. The input paraments used are previous closing price, average of opening, low and high price, trading volume, 10-day simple moving average (SMA), 10-day weighted moving average (WMA), relative strength index (RSI), stochastic K%, stochastic D%. The next day's closing price is the output. The high, low, opening, closing prices, and trading volume were directly obtained from the datasheet which we get from the packages mentioned above, and the other technical indicators have been derived from them. Since the open, high, and low parameters are highly correlated, thus reducing the dimension of dataset and also reducing the total parameters in the model to decrease the computational time, we have taken only one parameter that is the average of the three.

Moving Average (MA) is a parameter used to identify the tread. SMA is the average value over a time frame. In the WMA method the most recent data points are assigned a larger weight. Rather than distant past points, it is the recent data points that carry more relevant information. RSI captures the trend in the price of the security. The scale over which it ranges is from 0 to 100. A 14-day time frame is used. A stochastic oscillator is an indicator of the position of a closing price of a stock over a varying range of the price and time frame. Stochastic %K and stochastic %D are the two types of stochastic oscillators, where %D is the 3-day average of %K [5] (Table 1).

Table 1. Expressions for the parameters

Method	Indicator formula
10-day Simple moving average	$\frac{C_t+C_{t-1}+\cdots+C_{t-10}}{10}$
10-day Weighted moving average	$\frac{(n)C_t+(n-1)C_{t-1}+\cdots+C_{t-10}}{(n+(n-1)+\ldots+1)}$
Relative strength index	$100-\left(\frac{100}{1+\left(\sum_{t=0}^{n-1} Up_{t-1}/n\right)/\left(\sum_{t=0}^{n-1} D\omega_{t-1}/n\right)}\right)$
Stochastic K%	$\frac{C_t-LL_{t-n}}{HH_{t-n}-LL_{t-n}}\times 100$
Stochastic D%	$\frac{\sum_{t=0}^{n-1} K_{t-i}\%}{n}$

Where is C_t closing price, L_t is low price, H_t is high price, LL_t is lowest low price and HH_t is highest high price in the last t days, respectively.

3.2 LSTM and GRU Unit

LSTM [9] is used to overcome the problem of short-term memory of recurrent neural network. An LSTM unit consists of a cell state vector which is the explicit memory and gating unit that help in regulating the information flow in and out of the unit. The functioning of LSTM can be found in [13].

GRU [14] is similar to LSTM, with less complexity and faster computational speed. GRU controls the flow of information through the use of a gating unit and a hidden state. It consists of an update and reset gate, considered as vectors whose combination are used to decide how the hidden state information should be processed. The update gate of GRU is the combination of input and forget gates of LSTM. More details on GRU can be found in [15].

3.3 System Architecture

As shown in Fig. 1, the dataset consists of data from one of the selected twenty-five companies. A normalised dataset will give better results as the input time series data parameters have different scales [3]. As both LSTM and GRU require 3-Dimnetional data, we format the data into a 3-D array, where the third dimension is the number of past days to consider in the prediction of the next day. The pre-processed data is then used to train the RNN model. The Euclidean loss is calculated between the predicted and target value. This loss is backpropagated to the model for it to learn and update the model parameters.

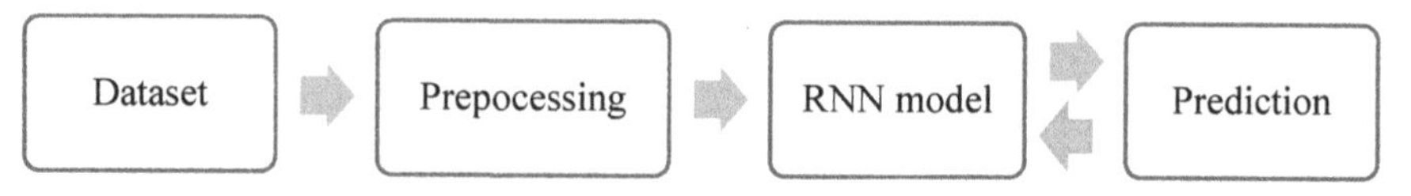

Fig. 1. Proposed system architecture

3.4 Loss Function

Mean square error loss is used to train the network. It is the Euclidean distance between the target value and predicted value given as:

$$L(\theta) = \frac{1}{N}\sum\nolimits_{i=1}^{N} (F(X_i;\theta) - Y_i)^2 \tag{1}$$

where for the proposed network θ is a set of learnable parameters, N is the training data vector, X_i is the ith input and Y_i is the actual value for X_i. $F(X_i;\theta)$ is the predicted value generated for θ parameters over X_i samples.

3.5 Networks Used

The four different networks, which have been compared are shown in Fig. 2. Irrespective of the RNN unit used, the first layer of the model has return sequence true so that the output is three-dimensional for the next RNN layer; which means the output of the first layer consists of the final output given by the units and also the value of hidden layer of each unit. Due to complex architecture and a lot of data, a deep learning network often overfits the training dataset. To overcome this problem dropout layer is added. Dropout regularises the network by randomly dropping out nodes during the training to reduce the overfitting and also improves the generalisation error. The last dense layer is used to convert the dimension of the data into the type required, by performing a simple activation function. All the output nodes of the previous layer are connected to the dense layer.

3.6 Implementation Details

The networks are implemented in Jupyter Notebook and the language used is python. The LSTM and GRU networks are imported from the Keras library [16]. We use the Adam optimizer [17], with a default learning rate of 0.001. Fifteen epochs are used to train the model. The training time required by an Intel Core i5 processor with 1.8 GHz speed processing and 8 GB RAM (Random access memory) is around 1 s for all the networks.

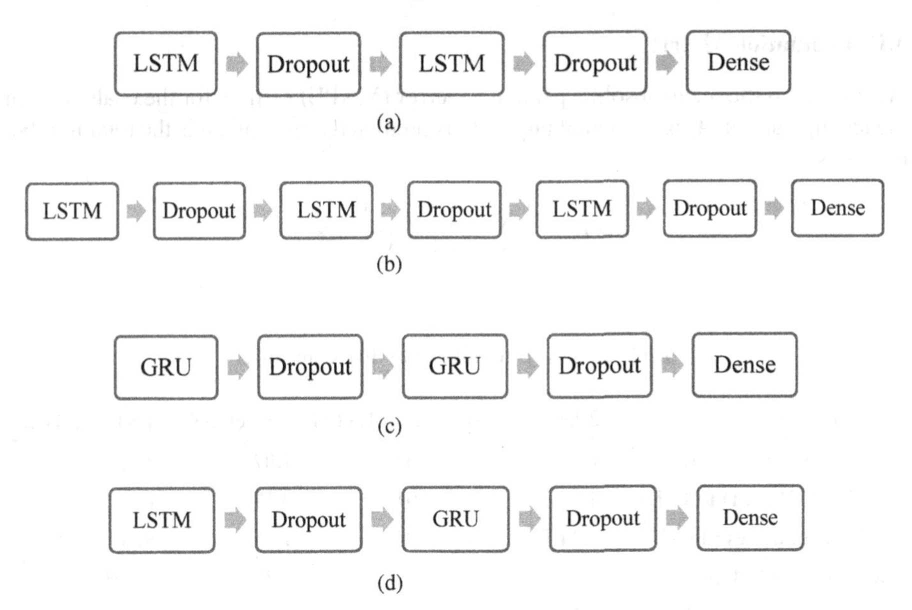

Fig. 2. **a.** Two-layer LSTM network **b.** Three-layer LSTM network **c.** Two-layer GRU network **d.** LSTM-GRU hybrid network

4 Results and Discussion

As the size of training data is less the increase in the number of parameters in the model does not have a large effect on training speed. Figure 3, shows the decrease in the loss as the number of epochs increases.

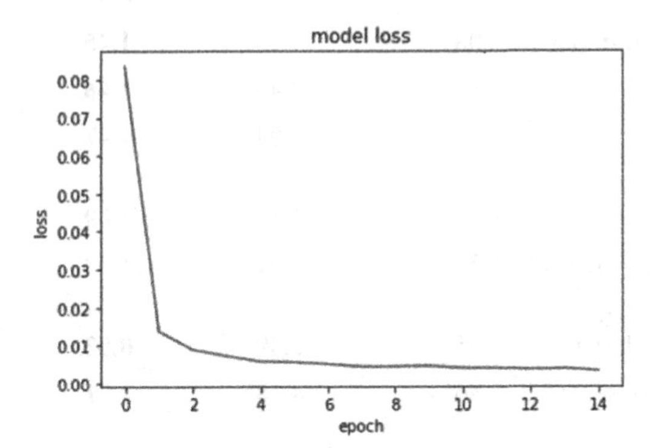

Fig. 3. Training loss

4.1 Evaluation Metric

We have used the mean absolute percentage error (MAPE) metrics for the evaluation of the testing dataset. A_t is the actual target, F_t is predicted target, and n is the total number of points.

$$MAPE = \frac{1}{n}\sum_{t=0}^{n}\left|\frac{A_t - F_t}{A_t}\right| \times 100 \quad (2)$$

Table 2. MAPE for all twenty-five companies

Company name	2-layer LSTM	3-layer LSTM	2-layer GRU	LSTM + GRU
Hindustan Zinc Ltd. (L)	1.69	1.63	**1.07**	1.75
Aurobindo Pharma Ltd. (L)	4.6	13.01	**3.17**	8.31
Divis Laboratories Ltd. (L)	4.24	5.92	**1.55**	4.58
Bajaj Finance Ltd. (L)	1.82	4.14	**1.48**	2.493
HCL Technologies Ltd. (L)	2.12	2.39	**1.82**	2.28
GAIL (India) Ltd. (L)	4.09	7.26	**2.87**	3.73
Hindustan Petroleum Corporation Ltd. (L)	4.84	5.22	**2.02**	3.52
Oil & Natural Gas Corporation Ltd. (L)	1.72	1.62	**1.28**	2.77
HDFC Bank Ltd. (L)	1.69	1.97	**1.33**	1.96
Hindustan Unilever Ltd. (L)	2.64	2.84	**1.27**	1.8
Nestle India Ltd. (L)	1.03	1.66	**0.97**	1.11
Berger Paints India Ltd. (L)	3.08	4.3	**1.45**	2.89
Infosys Ltd. (L)	1.81	2.46	**1.48**	1.57
Sun Pharmaceutical Industries Ltd. (L)	2.26	1.64	**1.47**	1.57
Reliance Industries Ltd. (L)	1.51	2.16	**1.09**	1.91
Procter & Gamble Hygiene & Health Care Ltd. (L)	1.54	1.91	**1.34**	1.59
Larsen & Toubro Ltd. (L)	1.35	2.49	**0.92**	1.74
Colgate-Palmolive (India) Ltd. (L)	2.86	4.78	**1.71**	2.18
Abbott India Ltd. (L)	2.98	3.45	**1.97**	2.81
Info Edge (India) Ltd.(L)	6.04	10.88	**2.17**	2.13
Punjab National Bank (L)	4.71	2.44	**1.85**	2.38

(continued)

Table 2. (*continued*)

Company name	2-layer LSTM	3-layer LSTM	2-layer GRU	LSTM + GRU
Avenue Supermarts Ltd. (L)	3.62	4.29	**2.58**	2.81
DLF Ltd. (L)	3.78	5.15	**3.11**	3.42
Power Grid Corporation of India Ltd. (L)	1.18	1.28	**1.09**	1.15
State Bank of India	2.06	3.68	**1.76**	1.82

Table 2, presents the comparative results of the four networks on all twenty-five companies. Of all the other methods, it can be observed that the 2-layer GRU has performed the best and the 3-layer LSTM has performed the worst. In GRU there is no controlled exposure of memory unlike the LSTM, where memory content seen by the other units is controlled by the output gate. The LSTM units also control the amount of new memory being added to the memory cell from forget gate, but GRU does not [10]. As the dataset taken is small, thus a controlled exposure may have led to dropping of important information during the processing. Thus, the 2-layer GRU model has outperformed all.

Graphs are used to visualize the difference in the accuracy of predicted values by the networks. The graphs plot the predicted and actual value-line for the year 2019, the testing dataset.

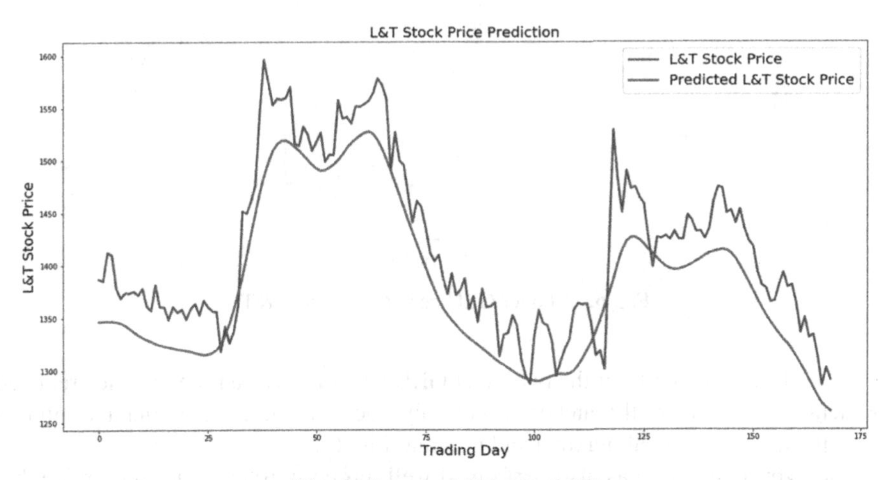

Fig. 4. 2-Layer LSTM for L&T

For two-layer LSTM it can be observed from Fig. 4, that the normal trend of the market has been captured, but it fails to capture the small details. The predicted line is below the actual line; thus, it is under-priced.

In a three-layer LSTM network, the dataset gets overfitted, thus is not able to predict the value to the maximum accuracy. As shown in Fig. 5, the predicted value line is below the actual value line.

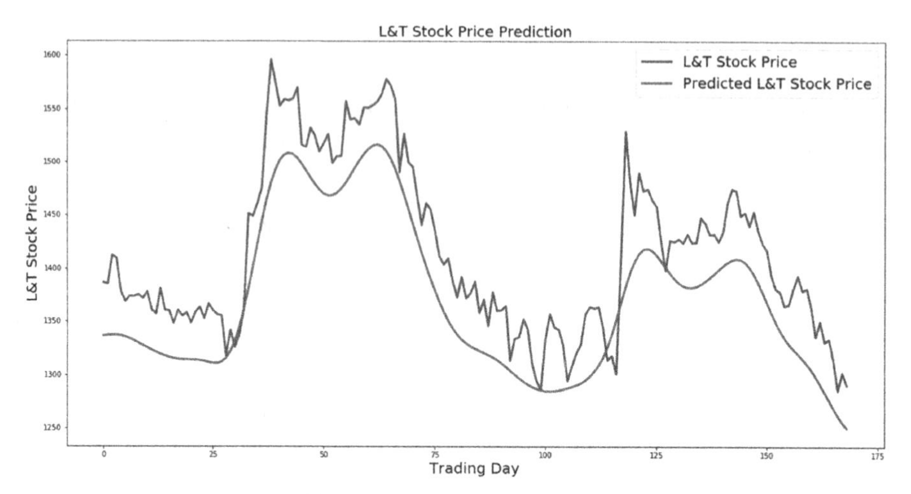

Fig. 5. 3-Layer LSTM for L&T

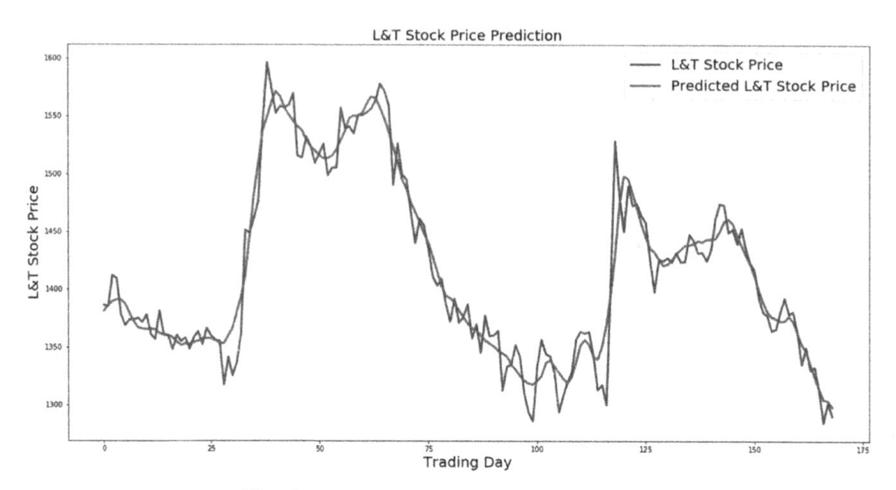

Fig. 6. 2-Layer GRU performance on L&T

From the above graph for the two-layer GRU network, we can see that the predicted and actual values are well matched. It not only traces the global maxima and minima properly but also the local maxima and minima (Fig. 6).

The hybrid network has also performed well, properly tracing the stock price, but the small peaks and valleys have not been predicted (Fig. 7).

4.2 Limitations

A deeper insight into the data provided by companies on public platforms is required for ensuring scalability of the reported architecture on a larger volume of companies. There is high data dependence on performance due to the dynamic nature of the market. Therefore, it is challenging to generalise as the model is prone to overfitting.

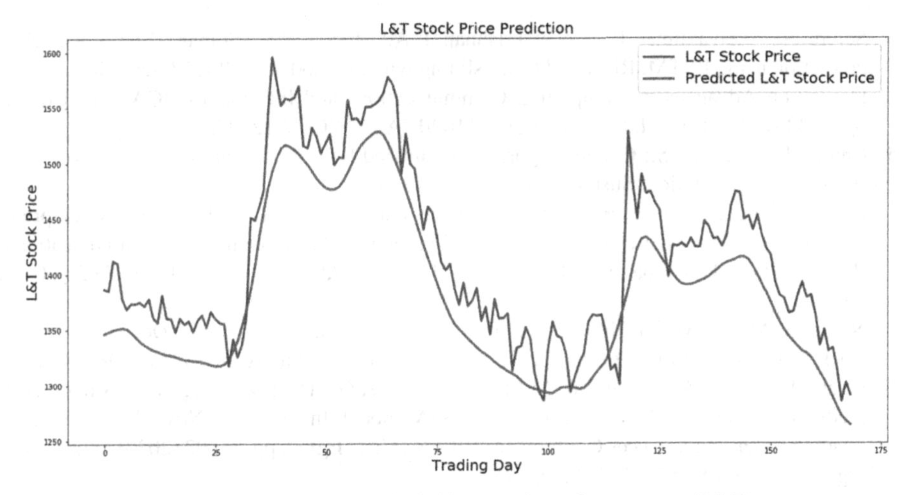

Fig. 7. LSTM + GRU hybrid network performance on L&T

It was observed that all four models provide greater accuracy when a dataset of 3 years is trained for 15 epochs, if we increase the size of the dataset, the accuracy decreases for a few companies.

4.3 Conclusion and Future Work

This paper gives a comparison of performance of four different networks, regarding Indian stocks, for predicting opening price. We use windowing method for training and predicting, with short duration of training data. The multi-layer LSTM networks, GRU network, and hybrid network performance have been studied. Hyperparameter tuning is done, and we find that GRU outperforms the other architectures, and generalizes prediction very well.

In future, we plan to develop a use-friendly website for a deeper understanding of the investor. Also, the work on exploring connections between Global Stock Exchange parameters and Twitter sentiment with Indian stocks is in progress.

Acknowledgment. The authors, thank the department of Electronics and Telecommunication at Fr. Conceicao Rodrigues Institute of Technology for extending support and caring in every aspect for carrying out this work. We would also like to thank the anonymous reviewers whose comments were instrumental in improving the content of this paper.

References

1. https://en.wikipedia.org/wiki/Bombay_Stock_Exchange. Accessed 21 Feb 2020
2. https://en.wikipedia.org/wiki/National_Stock_Exchange_of_India. Accessed 21 Feb 2020
3. Hossain, M.A., Karim, R., Thulasiram, R., Bruce, N.D.B., Wang, Y.: Hybrid deep learning model for stock price prediction. In: 2018 IEEE Symposium Series on Computational Intelligence (SSCI), Bangalore, India, 2018, pp. 1837–1844. https://doi.org/10.1109/SSCI.2018.8628641

4. Selvin, S., Vinayakumar, R., Gopalakrishnan, E.A., Menon, V.K., Soman, K.P.: Stock price prediction using LSTM, RNN and CNN-sliding window model. In: 2017 International Conference on Advances in Computing, Communications and Informatics (ICACCI), Udupi, pp. 1643–1647 (2017). https://doi.org/10.1109/ICACCI.2017.8126078.
5. Chung, H., Shin, K.-S.: Genetic algorithm-optimized long short-term memory network for stock market prediction. Sustain. MDPI Open Access J. **10**(10), 1–18 (2018)
6. Liu, Y., Wang, Z., Zheng, B.: Application of regularized GRU-LSTM model in stock price prediction. In: 2019 IEEE 5th International Conference on Computer and Communications (ICCC), Chengdu, China, pp. 1886–1890 (2019). https://doi.org/10.1109/ICCC47050.2019.9064035.
7. Nabipour, M., Nayyeri, P., Jabani, H., Mosavi, A., Salwana, E., Shahab, S.: Deep learning for stock market prediction. Entropy **22**(8), 840 (2020). https://doi.org/10.3390/e220808408.
8. Istiake Sunny, M.A., Maswood, M.M.S., Alharbi, A.G.: Deep learning-based stock price prediction using LSTM and bi-directional LSTM model. In: 2020 2nd Novel Intelligent and Leading Emerging Sciences Conference (NILES), Giza, Egypt, pp. 87–92 (2020). https://doi.org/10.1109/NILES50944.2020.9257950
9. Moghar, A., Hamiche, M.: Stock market prediction using LSTM recurrent neural network. Procedia Comput. Sci. **170**, 1168–1173 (2020), ISSN 1877-0509. https://doi.org/10.1016/j.procs.2020.03.049.
10. Chung, J., Gulcehre, C., Cho, K., Bengio, Y.: Empirical evaluation of grated recurrent neural networks on sequence modelling. arXiv:1412.355
11. Jariwala, S.: Nsepy Documentation (2020). https://nsepy.xyz/. Accessed October 2020
12. Hochreiter, S., Schmidhuber, J.: Long short-term memory. Neural Comput. **9**(8), 1735–1780 (1997)
13. Understanding LSTM Networks. https://colah.github.io/posts/2015-08-Understanding-LSTMs/. Accessed October 2020
14. Cho, K., et al.: Learning phrase representations using rnn encoder-decoder for statistical machine translation. arXiv preprint arXiv:1406.1078 (2014)
15. Deshpande, M.: Advance Recurrent Neural Network. https://pythonmachinelearning.pro/advanced-recurrent-neuralnetworks/. Accessed Dec 2020
16. Chollet, F., et al.: Keras: deep learning library for Theano and TensorFlow, vol. 7, p. 8 (2015). https://keras.io/k
17. Kingma, D.P., Ba, J.: Adam: a method for stochastic optimization. arXiv preprint arXiv:1412.6980 (2014)

Early Prediction of Cardiovascular Disease Among Young Adults Through Coronary Artery Calcium Score Technique

Anurag Bhatt[1(✉)], Sanjay Kumar Dubey[1], and Ashutosh Kumar Bhatt[2]

[1] Amity University Uttar Pradesh, Noida, India
skdubey1@amity.edu

[2] Birla Institute of Applied Sciences Bhimtal, Bhimtal, India

Abstract. The diseased heart cases are rapidly increasing in lower age groups in addition to the older people. The paper focuses on the early prediction of CVD (Cardiovascular Disease) among young adults. CAC (Coronary Artery Calcium) score or simply the calcium score technique is used to analyze Cardiac Health across different age intervals among young adults. Cardiac Ailments once used to be considered quite often among old-aged people. But, the research study emphasizes that Cardiovascular Disease no longer develops in older persons only, but is tremendously increasing in young adults too. Research outcomes express that young people are also facing Sudden Cardiac Arrest (SCA). Calcium Score is the key indicator to predict cardiac risk at early stage and is obtained through a non-invasive scan of heart i.e. Computerized Tomography (CT). It is efficient to estimate the heart-blockage due to calcification of coronary arteries resulting to plaque formation. This paper systematically analyzes the calcium scores of the young patients suffering from different cardiac health issues. It also focuses to motivate youth towards keeping track of their calcium score, timely monitoring their physical activities, proper nutrition and maintaining a healthy and quality lifestyle.

Keywords: Cardiovascular Disease · Calcium score · Sudden Cardiac Arrest · Computerized Tomography Scan · Cardiac Health · Young adults · Cardiac risk

1 Introduction

Cardiovascular disease is rapidly increasing day by day. It is very alarming that people are frequently facing cardiac health disorders due to the modern lifestyle full of stress, lack of physical activity, improper diet and ignorance towards appropriate nutrition. Therefore early prediction of such CVD is much needed. Calcium Score is one such key indicator to predict cardiac risk at early stage. It is obtained through a non-invasive scan of heart i.e. Computerized Tomography (CT) and used as a metric to calculate the heart-blockage due to calcification of coronary arteries. This calcification will result to plaque formation. People become ignorant towards maintaining good cardiac health until they face severe Cardiac Arrest or Sudden cardiac Arrest (SCA). Cardiovascular Disease

M. Singh et al. (Eds.): ICACDS 2021, CCIS 1441, pp. 303–312, 2021.
https://doi.org/10.1007/978-3-030-88244-0_29

(CVD), Atherosclerosis, Sudden Cardiac Arrest (SCA), Sudden Cardiac Death (SCD) and Coronary Artery Disease (CAD) etc. are majorly affecting human health worldwide. Though modern lifestyle has made our lives comfortable but at the same time takes us away from physical hard work, exercises and day-to-day necessary activities. Urban residents don't devote sufficient time for maintaining physical fitness. They totally ignore its importance. Modern lifestyle is full of stress, lack of physical activity, improper diet and ignorance towards appropriate nutrition. Junk food is preferred more over natural food items which later become the major cause of a number of health issues.

Heart diseases, high cholesterol, high blood pressure, hypertension etc. are very common among people due to modern work-culture, stressed life and less exposure to nature. We all are facing cardiac disorders, high cholesterol, diabetes, hypertension etc. in old patients and young adults too. The reason is our modern approach being very casual towards our health, diet and nutrition until we face some severe health issue or ailment. Therefore, lifestyle moderation, proper dietary practices, physical activities and exposure to nature must be inculcated in our routine [1]. In general, CVD (Cardiovascular Disease) causes death of an individual after every thirty six seconds or approximately half a minute [2]. It is the need of the hour to explore the applicability of various data mining techniques more and more and deploy these methodologies towards healthcare for early prediction of severe and critical diseases like CVD.

In this paper we focus towards one of the most effective predictors i.e. the Coronary Artery Calcium (CAC) Score. It is performed on young adults for early prediction of CVD. If the CAC score is not found in normal range, it is advised to consult heart specialist. The paper focuses on the early prediction of Cardiovascular Disease (CVD) among young adults. Coronary Artery Calcium (CAC) score or calcium score technique is used to analyze Cardiac Health of young adults across different age intervals. Cardiac Ailments once used to be considered quite often among old-aged people.

2 Related Work

Langara et al. [3] unveiled the fact that now-a-days, cardiovascular disease is not restricted to affect older people only. It is no longer depends on old age-factor. Young adults are also getting affected by such CVD. There is a rapid increase in the number of diseased heart cases in lower age groups in addition to the older people. Atherosclerosis leads to the elevation in cardiac damage and artery wall stains. It causes major risks of CVD [4]. Yoon et al. [5] elaborated the higher risk rate of Cardiac Health issues in the patients undergoing chemotherapy or radiotherapy and survived of lung cancer. Yang et al. [6] stated that vascular calcification and severe cardiovascular health issues are deeply related and associated with each other. Sudden cardiac arrest (SCA) majorly causes a number of deaths.

Masethe deployed various data mining techniques (Viz. CART, Naïve Bayes, REP-TREE and J48 etc.) in order to predict Cardiovascular Diseases leading to Cardiac Arrest or Heart Attack with maximum accuracy of 99% [7]. A cardiovascular study proposed a classification based prediction system in order to establish a decision support system. The classification methods deployed various supervised learning techniques like Neural Network, Decision Tree and Bayesian Classification [8]. Code Blue (a prime indicator to predict heart attack) is used to develop a system in order to benefit hospitals in its administrative functioning. The system keeps track of the previously associated heart patients through their ECG reports analysis [9]. Panahi et al. [10] concluded in its study that plasma protein convertase (PCSK9) concentrations, coronary artery disease, acute coronary syndrome and calcification of coronary arteries are deeply associated and affect one another.

Wilkins et al. [11] suggested a deep dose-response relationship between the CAC (Coronary Artery calcium) independent CVD risk factors and apolipoprotein B (apoB) in young adults. Lacking concordance between LDL (Low Density Lipoprotein)-cholesterol and apoB predicts the calcification of coronary artery in young adults. Arnold et al. [12] suggested a Polygenic Risk Score as a clinically beneficial tool to predict CVD, benefiting from Lipid-Lowering Therapy. A Korean research study established the relationship among the categories of blood pressure before the age of 40 years and cardiac risk in later life [13].

Saydah et al. [14] determined the Cardio metabolic risk association with the Mortality risk among young adults and adolescents. C-reactive protein is suggested to be a very efficient a bio-indicator for inflammation and it is explored to be an independent risk factor among clinical practices towards as for cardiovascular ailments and other related events [15]. Over past two decades, young adults (between 18–45 years) have become more vulnerable towards cardiovascular disease as compared to older adults. Heart failure cases have been increasing rapidly in case of young adults [16]. Age is highlighted as a modifiable risk factor towards CVD, which is not considered modifiable factor previously. It proves to be more efficient than other factors viz. blood pressure, cholesterol, smoking, alcohol intake etc. [17].

Age is a key indicator of survival and recovery in type 1 diabetes. It also affects the CVD risk factor. Rawshani et al. [18] focused to identify how age play salient role to diagnose type 1 diabetes, access cardiac risks. Thornburg [19] emphasized that cardiovascular disease depends on the growth pattern before birth that could lead to various functional and structural changes. Cunningham et al. [20] proposed to include mental illness severity towards the key indicators or predictors of Cardiac risks. There is a need to update the already existing equations for CVD risk prediction. Mental illness and its severity should be included as a prominent predictor. All the researches discussed above emphasizes towards the recent trends related to heart disease prediction and analysis.

3 Proposed Work

Cardiac tests results through various key indicators viz. ECG (Electro Cardiograph), LDL (Low Density Lipoprotein), HDL (High Density Lipoprotein), Heart Scan, CT (Computerized Tomography) are collected through clinicians and medical institutions. Their proper monitoring indicates the status of heart's functioning in the individuals. Its proper follow up may alarm us timely in case of any cardiac irregularities well in time at an early stage.

Real time patients' reports are analyzed across different age groups, gender, region and lifestyle. Data gathered is categorized for young adults based on certain age groups and gender. Data is pre-processed and converted into numerical form and its interpretation is compared to the normal human range as referred by the concerned medical practitioners. All the personal details of the patients are not at all disclosed by the team of researchers and are kept highly confidential. It is totally devoted for the benefit of the mankind through this research study. Research work emphasizes the Coronary Artery Calcium (CAC) Score is one of the efficient key indicators for early prediction of CVD and other cardiac ailments among young adults of various age-groups and gender. Coronary Artery Angiography is performed on patients using 80 mL Nonionic Intravenous Contrast (with injection rate of 5 mL/s) with 64 slice scanner.

There are 6 factors on which the CAC score depends and are responsible for its overall estimation viz. LCX, PLV, LAD, RCA, PDA, and OM abbreviated as Left Circumflex Artery, Posterior Left Ventricular, Left Anterior Descending Artery, Right Coronary Artery, Posterior Descending Artery and Obtuse Marginals respectively.

WebMD [21] established the reference range for CAC Score. It states the following points:

- If CAC score value is '0', it expresses 'less than 5%' or very low chances of developing a heart disease.
- If CAC score value is 'between 1 to 10', it expresses 'less than 10%' or low chances of developing a heart disease and confirms less amount of plaque present in coronary arteries.
- If CAC score value is 'between 11 to 100', it expresses mild chances of occurrence of cardiac health issues and confirms mild/fair presence of plaque in coronary arteries.
- If CAC score value is 'between 101 to 400', it expresses approximately 90% or higher chances of heart attack. It confirms blocked arteries due to large amount of plaque present in coronary arteries [21].

Research Questions

Based on today's lifestyle, health practices and current scenario of cardiac ailments and health issues, the following questions need to be answered:

RQ1: Does CVD affect young adults too?
RQ2: Is CAC score age-dependent?
RQ3: Can lifestyle moderation and proper monitoring of CAC score help to reduce the risk of CVD?
RQ4: Is CAC score an efficient predictor for CVD?
RQ5: Are males at higher risk of CVD than women of similar age group?

4 Experimentation Results

Cardiac Tests results of the Coronary Artery Calcium (CAC) score are analyzed through Weka and Microsoft Excel tools. The research results established the relationship between calcium Score, Age, Gender and Lifestyle. It is also explored that Calcium Score increases with advancement of life. The research study comprises of 215 records of patients data i.e. N = 215, taken randomly from the clinicians and medical institutions… Total male and female patients are found to be around 71% (i.e. $N_{mal} = 153$) and 29% (i.e. $N_{fem} = 62$) respectively of the total patients. Average Calcium Score (ACS) is calculated for both male and female patients across various age groups. The age groups are categorized into 4 sub-groups i.e. 1, 2, 3 and 4 corresponding to their age-intervals 0–10, 11–20, 21–30 and 31–40 years respectively.

When the age crosses 31–40 years the likelihood of heart disease severity among adults also increases rapidly. It signifies the initiation of a significant amount of building up of plaque in coronary arteries, which may result to CVD. Overall male adults are more susceptible than females towards cardiac ailments as expressed by their increased Calcium Scores.

Table 1 and 2 represents Age group-wise ACS in female and male adults respectively. It is evident from Table 1 and 2 that no heart patients are visible among 0–10 years of age among both females and males. Females possess ACS value of 8.91, 37.83 and 92.38 among the age groups 0–10, 11–20 and 31–40 years respectively, whereas, males possess a comparatively much higher ACS among almost every age interval, indicating more vulnerability of male adults towards heart disease.

Table 1. Age group-wise ACS in female adults ($N_{fem} = 62$).

Age group	Age interval (years)	No. of patients	Average calcium score	Max average calcium score	Min average calcium score
1	0–10	0	0	NA	NA
2	11–20	5	8.91	76.69	5.87
3	21–30	14	37.83	176.23	6.23
4	31–40	43	92.38	213.87	8.92

Note: NA-Not Applicable

Table 2. Age group-wise ACS in male adults ($N_{mal} = 153$).

Age group	Age interval (years)	No. of patients	Average calcium score	Max average calcium score	Min average calcium score
1	0–10	0	0	NA	NA
2	11–20	11	12.14	81.9	7.34
3	21–30	29	118.34	289.48	8.32
4	31–40	113	188.37	302.28	16

Note: NA-Not Applicable

Figure 1 and 2 represent the ACS vs. Age-Group Plot for female and male adults respectively. In the age interval 0–10 years, no heart disease cases are reported among females and males. As evident from the graphs, ACS increases gradually and comparably in both females and males (0 to 8.91 and (0 to 12.14) respectively in the age-interval of 11–20 years. It signifies lesser chances of developing a Heart Disease. There is tremendous increase in ACS for males (i.e. 118.34) than females (i.e. 37.83) for the age interval 21–30 years. ACS again increases so swiftly attaining higher values for males (i.e. 188.37) and females (i.e. 92.38) in the next age interval of 31–40 years. It expresses clearly that young adults are developing cardiac ailments including CVD among both males and females, males being more vulnerable towards heart disease.

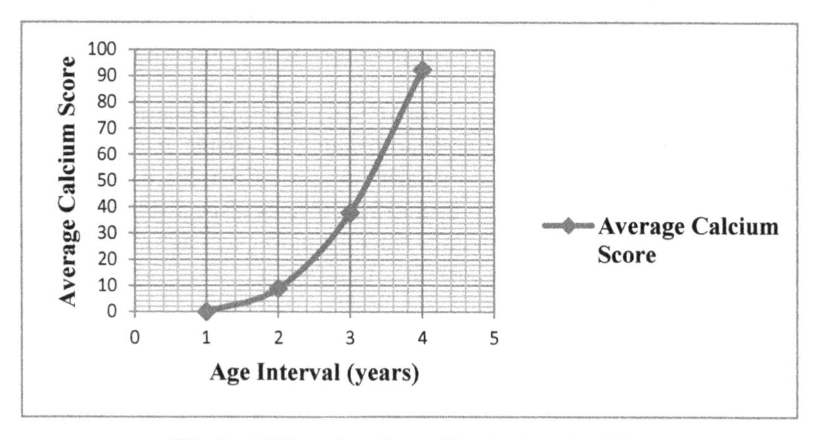

Fig. 1. ACS vs. Age Group Plot for female adults

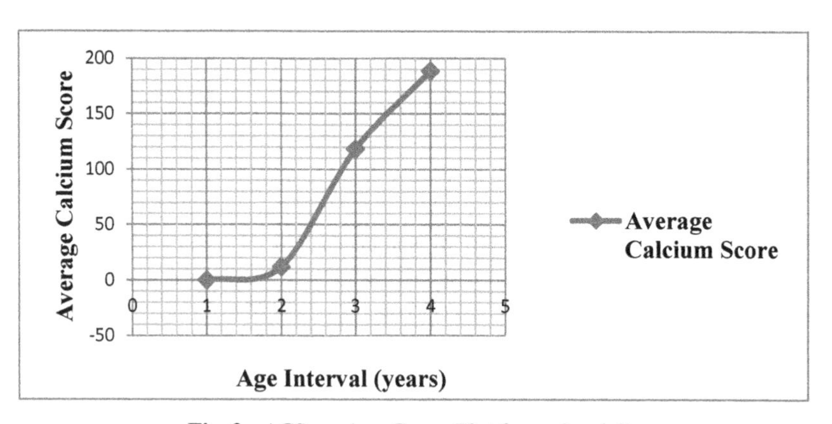

Fig. 2. ACS vs. Age Group Plot for male adults

Figure 3 displays the Age-interval wise ACS Distribution in female adults. A negligible ACS is recorded in females of 0–10 years of age. As age increases the ACS increases lesser in females than males. ACS increase approximately 'four' times in female patients from age interval 11–20 to 21–30 years (i.e. 8.91 to 37.83). ACS shoots up about 'two and a half' times in females from age interval 21–30 to 31–40 years (i.e. 37.83 to 92.38).

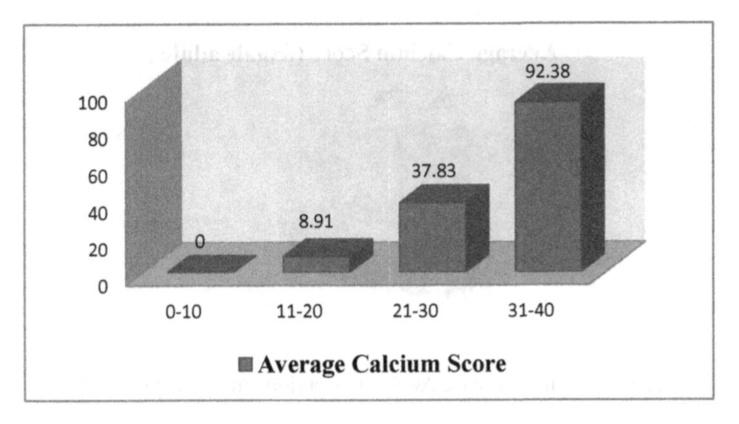

Fig. 3. Age-interval wise ACS Distribution in female adults

Figure 4 displays the Age-interval wise ACS Distribution in male adults. A negligible ACS is reported in males of 0–10 years of age. ACS increases with age more swiftly in males than females. ACS increase approximately 'ten' times in male patients from age interval 11–20 to 21–30 years (i.e. 12.14 to 118.34). Then, ACS shoots up about 'one and a half' times in males from age interval 21–30 to 31–40 years (i.e. 118.34 to 188.37).

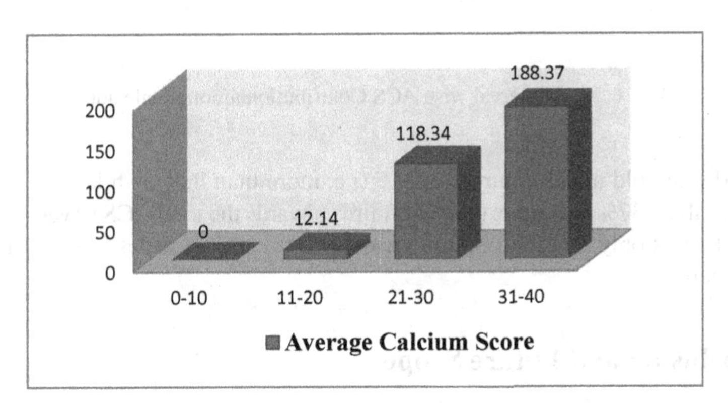

Fig. 4. Age-interval wise ACS Distribution in male adults

Age-interval wise ACS Contribution among female and male adults is displayed by Fig. 5 and 6 respectively. 31–40 years old females contribute 66% (i.e. about two-third), while 21–30 years old females contribute 27% (i.e. about one-fourth) towards the total ACS (Average Calcium Score. There is only 7% contribution form 11–20 years old females towards the overall average score. All males and females belonging to the age group 0–10 years have negligible contribution towards ACS.

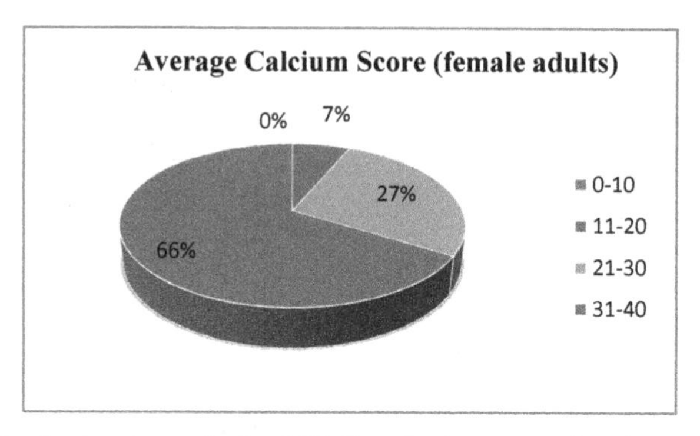

Fig. 5. Age-interval wise ACS Contribution among female adults

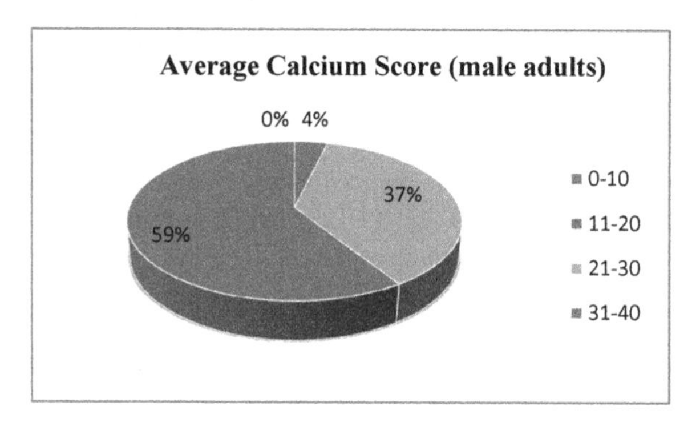

Fig. 6. Age-interval wise ACS Contribution among male adults

31–40 years old males contribute 59% (i.e. more than half), while 21–30 years old males contribute 37% (i.e. more than one-third) towards the total ACS (Average Calcium Score). There is only 4% contribution form 11–20 years old males towards the overall average score.

5 Conclusion and Future Scope

Earlier, there was a perception about Cardiac Issues or Heart Disease that it is an old aged disease or disorder. The research work disproves that Cardiovascular Diseases are associated to old aged individuals only. The study expresses that CAC score is an efficient predictor for cardiac health issues. The research results clearly conclude that male adults are more susceptible than females towards the CVD. They are at more risk than females of same age-group. It is further concluded that CAC score is age-dependent. But lifestyle moderation and proper monitoring of CAC score may be very beneficial to reduce the risk of CVD at early stage. It is advised to avoid smoking, alcohol, and tobacco and get ourselves involved to physical activities, healthy and balanced diet intake. The work emphasizes on the early prediction of CVD (Cardiovascular Disease) among young adults. Coronary Artery Calcium (CAC) score or calcium score technique is used to analyze Cardiac Health of young adults across different age intervals. Cardiac Ailments once used to be considered quite often among old-aged people. Lifestyle moderation and proper monitoring of CAC score may be very helpful to reduce the risk of CVD at primary stage. Thus, research findings mostly answer the research questions framed. Data mining techniques and Data Analytics find a great potential and scope to address the challenges and issues related to Sudden Cardiac Death, Sudden Cardiac Arrest and severe Cardiac Ailments.

Acknowledgment. The authors extend their heartfelt thanks to Dr. Pawan Sharma, Fortis Noida, Dr. S.N. Srivastava, S.N. Hospital Ranikhet, Max Hospital Ghaziabad and its entire staff, Base Hospital Haldwani, Mrs. Mudita Phartiyal, Mr. Priyanshu Belwal., Mr. Manish Joshi, Dr. Vinod Ojha, Dr. Vijay Upadhyay, Earth Ayurveda Center Ranikhet to provide their valuable support throughout the research journey.

References

1. Aljumah, A.A., Ahamad, M.G., Siddiqui, M.K.: Application of data mining: diabetes health care in young and old patients. J. King Saud Univ. Comput. Inf. Sci. **25**(2), 127–136 (2013)
2. Centers for Disease Control and Prevention. Underlying Cause of Death, 1999–2018. CDC WONDER Online Database. Atlanta, GA: Centers for Disease Control and Prevention 2018. Accessed 15 Jan 2021
3. Langara, B., Georgieva, S., Khan, W.A., Bhatia, P., Abdelaziz, M.: Case report: sudden cardiac death in a young man. Breathe **11**(1), 67–70 (2015)
4. Juraschek, S.P., et al.: Subclinical cardiovascular disease and fall risk in older adults: results from the Atherosclerosis risk in communities study. J. Am. Geriatr. Soc. **67**(9), 1795–1802 (2019)
5. Yoon, D.W., et al.: Increased risk of coronary heart disease and stroke in lung cancer survivors: a Korean nationwide study of 20,458 patients. Lung Cancer **136**, 115–121 (2019)
6. Yang, W., Zou, B., Hou, Y., Yan, W., Chen, T., Qu, S.: Extracellular vesicles in vascular calcification. Clin. Chim. Acta **499**, 118–122 (2019)
7. Masethe, H.D., Masethe, M.A.: Prediction of heart disease using classification algorithms. In: Proceedings of the World Congress on Engineering and Computer Science, vol. 2, pp. 22–24, October 2014
8. Taneja, A.: Heart disease prediction system using data mining techniques. Oriental J. Comput. Sci. Technol. **6**(4), 457–466 (2013)
9. Somanchi, S., Adhikari, S., Lin, A., Eneva, E., Ghani, R.: Early prediction of cardiac arrest (code blue) using electronic medical records. In: Proceedings of the 21th ACM SIGKDD International Conference on Knowledge Discovery and Data Mining, pp. 2119–2126, August 2015
10. Panahi, Y., et al.: PCSK9 and atherosclerosis burden in the coronary arteries of patients undergoing coronary angiography. Clin. Biochem. **74**, 12–18 (2019)
11. Wilkins, J.T., Li, R.C., Sniderman, A., Chan, C., Lloyd-Jones, D.M.: Discordance between apolipoprotein B and LDL-cholesterol in young adults predicts coronary artery calcification: the CARDIA study. J. Am. Coll. Cardiol. **67**(2), 193–201 (2016)
12. Arnold, N., Koenig, W.: Polygenic risk score: clinically useful tool for prediction of cardiovascular disease and benefit from lipid-lowering therapy? Cardiovasc. Drugs Ther. **35**(3), 627–635 (2020). https://doi.org/10.1007/s10557-020-07105-7

13. Son, J.S., et al.: Association of blood pressure classification in Korean young adults according to the 2017 American College of Cardiology/American Heart Association guidelines with subsequent cardiovascular disease events. JAMA **320**(17), 1783–1792 (2018)
14. Saydah, S., Bullard, K.M., Imperatore, G., Geiss, L., Gregg, E.W.: Cardiometabolic risk factors among US adolescents and young adults and risk of early mortality. Pediatrics **131**(3), e679–e686 (2013)
15. Avan, A., Tavakoly Sany, S.B., Ghayour-Mobarhan, M., Rahimi, H.R., Tajfard, M., Ferns, G.: Serum C-reactive protein in the prediction of cardiovascular diseases: overview of the latest clinical studies and public health practice. J. Cell. Physiol. **233**(11), 8508–8525 (2018)
16. Andersson, C., Vasan, R.S.: Epidemiology of cardiovascular disease in young individuals. Nat. Rev. Cardiol. **15**(4), 230 (2018)
17. Sniderman, A.D., Furberg, C.D.: Age as a modifiable risk factor for cardiovascular disease. Lancet **371**(9623), 1547–1549 (2008)
18. Rawshani, A., et al.: Excess mortality and cardiovascular disease in young adults with type 1 diabetes in relation to age at onset: a nationwide, register-based cohort study. Lancet **392**(10146), 477–486 (2018)
19. Thornburg, K.L.: The programming of cardiovascular disease. J. Dev. Orig. Health Dis. **6**(5), 366 (2015)
20. Cunningham, R., Poppe, K., Peterson, D., Every-Palmer, S., Soosay, I., Jackson, R.: Prediction of cardiovascular disease risk among people with severe mental illness: a cohort study. PloS One **14**(9), e0221521 (2019)
21. What is a Coronary Calcium Scan? WebMD 16 September 2016. https://www.webmd.com/heart-disease/coronary-calcium-scan

Confidentiality Leakage Analysis of Database-Driven Applications

Angshuman Jana(✉) and Anwesha Kashyap

Indian Institute of Information Technology Guwahati, Guwahati, India
{angshuman,anwesha.kashyap}@iiitg.ac.in

Abstract. In a software system the database technology is one of the most important part that stores external information into the permanent storage and process them accordingly. However, the most prime concern of any information system is to protect the data in the persistent storage from any unauthorised access. Sometimes, the database technology is developed in such a way that sensitive information may be leaked directly or indirectly during information processing along with control structure of a software code. In this paper, we propose a dependency graph based approach to identify possible confidentiality leakage of a software code. We design a framework using the semantic based dependency analysis technique. It refines the existing syntax-based dependency graph by removing false alarms and generates more precise dependency information that leads to improved security analysis results.

Keywords: Structured query language · Security · Data dependency

1 Introduction

To protect a software system and keep the data safe from any unauthorized access i.e. the information security is one of the promising field of research of any information system [1,15]. Over the past, several mechanisms like steganography [16], cryptography [17], access control [18] are used to maintain security of a software system. Sometimes, the database technology is built in such a way that sensitive information may be circumvented directly or indirectly during information processing along with control structure of a software code [9,11]. Therefore, in an information system, one of the major challenge is illegitimate information flow where an application reveals sensitive data to the unauthorised party. For example, given a code fragment that consists of two variables 'x' and 'y'. According to business policies, assume 'x', 'y' are the *public* and *private* variable respectively. If 'x = y' is a statement of a program then data contain by variable 'y' can be circumvented due to assignment which is direct information flow from variable 'y' to 'x'. Other scenarios, if '(y > 0) then x = x + 1 else x = 0' is a program fragment, then the value of 'y' can be inferred by an attacker observing 'x' on the output channel.

M. Singh et al. (Eds.): ICACDS 2021, CCIS 1441, pp. 313–323, 2021.
https://doi.org/10.1007/978-3-030-88244-0_30

Language-based data security analysis has emerged as a brilliant technique to detect illegal data-flow in any software system. In [4] Dennings proposed the work to acknowledge this challenge. Ever since many proposals are published to identify the information leakage in an application for different programming languages [6]. These type of analysis mainly classifies code variables into different security classes. The fundamental security class comprises two types of database attribute: Public (Low) and Private (High). Most of the ideas are published to acknowledge this challenge in different programming paradigm, e.g. imperative, object-oriented, functional, etc. [8]. However, very limited ideas [7,13,15], are proposed to identify possible confidentiality leakage of a software code. A database program needs a different treatment as the values of database attributes differ from that of imperative language variables. To the best of our knowledge, these existing ideas suer from producing high rate of false alarms that leads to an imprecise security analysis.

In this paper, we propose a data-dependency graph based approach to automatically detect all possible confidentiality leakage of an information system. We design a framework using the semantics-based approach that provides a notation of semantic based data dependency analysis and improves the precision of information flow control in a dependency graph by removing the false alarms. This proposal serves as a strong basis to generate the improved security analysis results of a database code.

2 Related Works

Over the past, the variety of works on information leakage have been proposed. Information leakage analysis is one of the most important security challenges and many different approaches are being proposed by different authors for the same. Information flow control (IFC) [8] is an important technique to discover security leaks. However, many IFC can also generate false alarms as it is flow-insensitive, context-insensitive or object-insensitive. The first approach was based on Program Dependence Graph (PDG) [5,8]. PDG's have been developed as a standard device to represent information flow in a program. The PDG's can only decide whether there is a potential information flow or not. It does not give a detailed circumstances of an information flow.

In [14,15] authors have proposed a data-centric approach that captures possible leakage of sensitive database information. Previous existing language-based approaches are coarse-grained and they are based on the assumption that attackers are able to view all the values of insensitive attributes in the database. Here, it performed data-flow analysis on a refined dependency graph to detect all possible information leakage in the case of database applications [7,10,12]. However, it is one of the major challenge for the database statements as attribute indicates values-list rather than just a single value. Many of the approaches are not flow-sensitive which is a prime limitation. In [9,10], different existing approaches like Abstract Interpretation framework, Hoare Logic, etc. are proposed that detects the improper information flow in a software system. Flow-insensitivity is the main shortcoming of these approaches which may produce imprecise outcome.

```
0. public class saleOffer {
1. public static void main(String[] args) throws SQLException {
2.    try { Statement con = DriverManager.getConnection("jdbc sql: ... ", "scot", "tig").createStatement();
        * Flat 5% discount offer on all the low cost products. *
3.      con.executeQuery("UPDATE product SET p_price = p_price − 0.05 * p_price WHERE p_price ⩽ 499 ");
        * Flat 10% discount offer on all the high cost products. *
4.      con.executeQuery("UPDATE product SET p_price = p_price − 0.1 * p_price WHERE p_price ⩾ 500 ");
5.      ResultSet rs1=con.executeQuery("SELECT AVG(p_price) FROM product WHERE p_price ⩽ 2999 ");
        * 10% special discount offer on few products. *
6.      con
      .executeQuery("UPDATE product SET p_price =p_price − 0.1 * p_price WHERE p_price BETWEEN 1000 AND 2999 ");
          ...
          ...
          ...
10.     ResultSet rs2
      =con.executeQuery("SELECT p_id, p_price FROM product WHERE p_price BETWEEN 1500 AND 2500 ");
      } catch (Exception e) { ... }} }
```

Fig. 1. Program fragment "Prog"

3 Running Example

Consider a code fragment "Prog" shown in Fig. 1. This database code implements several functionality of an online shopping system.

The class saleOffer updates product price (stored in attribute p_price) depending on the discount offers decided by the company. For instance, the company decided to include a new policy where 5% discount is given to the customers who purchases products whose product price is less than or equal to 499. This is implemented in our example by statement 3. Similarly, the 10% discount is being given if the product price is more than or equal to 500. This policy is incorporated in this code which is implemented by statement 4. Also, they introduced another policy where the 10% discount is being given for a purchase of product price in between 1000 and 2999 which is implemented by statement 6. From the code "Prog", we can get the several dependencies among statements. Assume that values of the attributes p_price is private and should not be displayed to an unauthorised party. To ensure the confidentiality of the attribute p_price, we analyze the flow of data (stored in the attribute p_price) from the syntax-based DOPDG of "Prog". Observe that the syntax-based DOPDG consists of false dependencies among statements. For example, in the "Prog" although there is a syntax-based dependency from the statements 3 to 4, but there is no semantically dependency exist between them. Because values of an attribute p_price defined by 3 can never be used by 4. Similarly, $3 \rightarrow 6$, $3 \rightarrow 10$ and $4 \rightarrow 10$ are also identified as false dependency. In the subsequent sections, using semantics based analysis our proposed model generates a more precise result of information leakage analysis of "Prog".

4 Proposed Approach

Now, we describe our propose approach, refinement of the syntactic DOPDG for gaining more precise Database-Database (DD) dependency information among statements of an application and generating more precise data leakage analysis

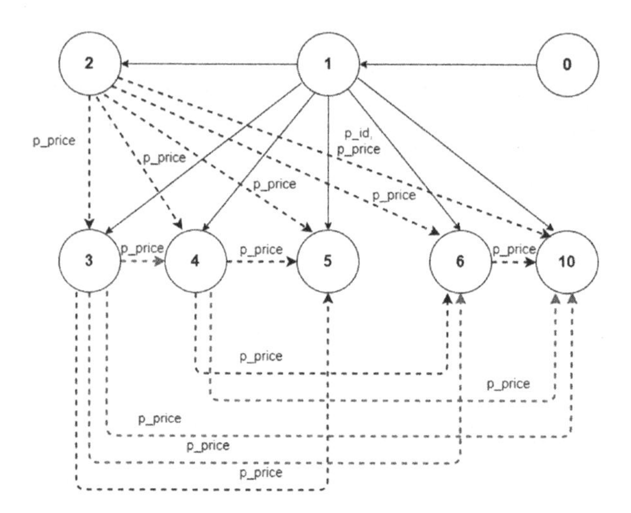

Fig. 2. Syntax-based DOPDG.

result of a database application. We recall from [19] the syntax-based construction of DOPDG. Then, we refine a syntax-based DOPDG by removing the false dependency applying the Condition-Action technique. After that, based on data dependency information, we perform security analysis to identify possible confidentiality leakage of a database application.

4.1 Syntax-Based

In a database program, in place of traditional program dependence graph, DOPDG is constructed. In a DOPDG, there exists additionally two more dependencies:

Definition 1 (Program-Database (PD) dependence [2]**).** *A database statement D is said to be Program-Database dependent on an imperative statement I if it uses a variable y defined by I such that there is no redefinition of y in between I and D.*

Definition 2 (Database-Database (DD) dependence [2]**).** *A database statement D1 is said to be Database-Database dependent on another database statement D2 if D1 uses an attribute y which is defined by D2 such that there is no redefinition of y and no roll-back of D2 between D1 and D2.*

Let $\mathbb{S}$, $\mathbb{V}_d$ and $\mathbb{V}_a$ be the set of statements, database-attributes, and application-variables in a code. Let $\mathbb{V} = \mathbb{V}_d \cup \mathbb{V}_a$ where $\mathbb{V}_d \cap \mathbb{V}_a = \emptyset$. The two following functions are used to construct syntax-based DOPDG. These are:

$$US : \mathbb{S} \rightarrow (\mathbb{V}) \qquad \text{and} \qquad DE : \mathbb{S} \rightarrow (\mathbb{V})$$

This extracts the set of variables used and defined in a statement s $\in \mathbb{S}$.

Example 1. Let us explain the example "Prog" shown in Fig. 1. As in case of traditional PDG, here the control dependencies are computed in a similar way. They are: $\{1 \rightarrow 2,\ 3,\ 4,\ 5,\ 6,\ 10\}$. *Defined(DE)* and *Used(US)* variable are computed to construct the DOPDG for the following program as follows:

DE(2) $= \{p\_id,\ quan,\ p\_price\}$ DE(3) $= \{p\_price\}$ US(3) $= \{p\_price\}$
DE(4) $= \{p\_price\}$ US(4) $= \{p\_price\}$ US(5) $= \{p\_price\}$
DE(6) $= \{p\_price\}$ US(6) $= \{p\_price\}$ US(10) $= \{p\_id,\ p\_price\}$

Following data dependencies are identified from the above information:

- DD-dependencies for attributes *p_id* and *p_price*: $\{2 \rightarrow 3,\ 4,\ 5,\ 6,\ 10\}$, $\{3 \rightarrow 4,\ 5,\ 6,\ 10\}$ and $\{4 \rightarrow 5,\ 6,\ 10\}$,

The syntax-based DOPDG of "Prog" is shown in Fig. 2.

Limitations. It is observed that the DD-dependency from edge $3 \rightarrow 4$ is a false dependency as defined database part by statement 3 will not overlap with the used database part by statement 4. Syntactically we found a dependency between the edge $3 \rightarrow 4$ but semantically there exist no such dependency. Similarly, semantically $3 \rightarrow 6$, $3 \rightarrow 10$ and $4 \rightarrow 10$ are also false dependencies.

4.2 Condition Action Rule Based Approach

To overcome the problem of false dependency, we define the semantic based analysis following the Condition Action rule based approach. The Condition Action rule based approach proposed in [3,12] for expert database system. The rules are expressed in the form $E_{cond} \rightarrow E_{act}$ where E_{cond} represents a condition and E_{act} represents an action as information updating operation. The analysis is being performed by propagation algorithm to predicts an action of one statement affect the condition of another statement. A attribute extension operator ϵ is defined as $\epsilon[\mathrm{Z} = \mathrm{exp}]$, where the expression exp is computed over each tuple t of e under a new schema schema(e) $\cup$ Z. Condition-Action rule based approach can be defined for four Database statements (Select, Update, Delete and Insert).

Consider a Select statement "SELECT marks FROM Student WHERE marks $\geqslant$ 50". The condition E_{cond} and action E_{act} parts are:

$$E_{cond} = \pi_{marks}(\sigma_{marks \geqslant 50} Student) \quad \text{and} \quad E_{act} = \phi$$

Again, consider an Insert statement "INSERT INTO Student (id, marks, roll no) VALUES ('111', '97', '255')". The condition E_{cond} and action E_{act} parts are:

$$E_{cond} = null \quad \text{and} \quad E_{act} = \langle\text{'111', '97', '255'}\rangle$$

Next, an Update statement is taken "UPDATE Student SET marks = marks + 5 WHERE marks $\leqslant$ 50". The condition E_{cond} and action E_{act} parts are:

$$E_{cond} = \pi_{marks}(\sigma_{marks \leqslant 50} Student) \quad \text{and}$$
$$E_{act} = \epsilon[\ \mathrm{Marks} = \mathrm{marks} + 5\ ](\sigma_{marks \leqslant 50} Student)$$

Now we consider a Delete statement "DELETE FROM Student WHERE marks=0". The condition E_{cond} and action E_{act} parts are:

$$E_{cond} = \pi_{marks}(\sigma_{marks=0}Student) \quad \text{and} \quad E_{act} = \pi_{marks}(\sigma_{marks=0}Student)$$

The symbol π is a relational algebra operator, represents for an attribute projection and the symbol σ is also a relational algebra operators, represents for the attribute selection.

Example 2. Now we define the semantics of the running example "Prog" (Fig. 1) using Condition Action rule based approach as follows:

$$E^3_{cond} = \pi_{p\_price}(\sigma_{p\_price \leqslant 499}\ product)$$
$$E^3_{act} = \epsilon[\text{p\_price'} = \text{p\_price} - 0.05 * \text{p\_price}](\sigma_{p\_price \leqslant 499}\ product)$$
$$E^4_{cond} = \pi_{p\_price}(\sigma_{p\_price \geqslant 500}\ product)$$
$$E^4_{act} = \epsilon[\text{p\_price'} = \text{p\_price} - 0.1 * \text{p\_price}](\sigma_{p\_price \geqslant 500}\ product)$$
$$E^5_{cond} = \pi_{p\_price}(\sigma_{p\_price \leqslant 2999}\ product) \qquad E^5_{act} = \phi$$
$$E^6_{cond} = \pi_{p\_price}(\sigma_{p\_price \geqslant 1000 \wedge p\_price \leqslant 2999}\ product)$$
$$E^6_{act} = \epsilon[\text{p\_price'} = \text{p\_price} - 0.1 * \text{p\_price}](\sigma_{p\_price \geqslant 1000 \wedge p\_price \leqslant 2999}\ product)$$
$$E^{10}_{cond} = \pi_{p\_price,p\_id}(\sigma_{p\_price \geqslant 1500 \wedge p\_price \leqslant 2500}\ product) \qquad E^{10}_{act} = \phi$$

4.3 Computation of Data Dependency

Now, we describe Database Database (DD) dependency computations among the database statements. At first we compute Defined- and Used-parts of a database statement and then using this information we compute the semantics based DD-dependency.

Used and Defined Part Computation: Given a database statement S, which is represented in the form $<E_{cond}, E_{act}>$ in the Condition Action rule based approach. To identify the dependency between statements, we first compute the used and defined database parts of each statement. Therefore we divide the semantics of S into three parts $\Delta = \langle T_\circ, F_\circ, T_\bullet \rangle$. Here, $T_\circ$ represents true part, $F_\circ$ represents false part and $T_\bullet$ represents modified part.

Let us define two functions F_d and F_u. Suppose D_f and U_s denotes the *Defined*- and the *Used* database-parts by database statement Q respectively. Therefore -

$$D_f = F_d(Q,\ \Delta) = \langle T_\circ, T_\bullet \rangle \quad \text{and} \quad U_s = F_u(Q,\ \Delta) = \langle T_\circ \rangle$$

Now, we compute the Defined- and Used-part of the running example based on the Condition Action rule based semantics.

Example 3. The Defined- and Used-part of the running example of statements 3, 4, 5, 6 and 10 are illustrated below:

D^3 = <(p_price⩽499), {(p_price' = p_price - 0.05 * p_price) ∧ (p_price⩽ 499)}>
U^3 = <(p_price⩽ 499)>
D^4 = <(p_price⩾500), {(p_price' = p_price - 0.1 * p_price) ∧ (p_price⩾500)}>
U^4 = <(p_price⩾500)>
D^5 = ϕ
U^5 = <(p_price⩽ 2999)>
D^6 = <(p_price⩾ 1000), {(p_price' = p_price - 0.1 * p_price) ∧ (p_price⩾ 1000)}>
U^6 = <(p_price⩾ 1000)>
D^{10} = ϕ
U^{10} = <(p_price⩾ 1500) ∧ (p_price⩽ 2500)>

Dependency: Two statements S_1 and S_2 are dependent to each other iff $D_f^{S_1} \cap U_s^{S_2} \neq \emptyset$. First we will generate the pair-wise dependency for two statements. There will be four possibilities to compute the independency and semantic dependency. $D_f^{S_1}$ can be represented by two components: $D_f^{S_1} = <E_c^{S_1}, E_a^{S_1}>$ where $E_c^{S_1}$ represents the condition part before the action and $E_a^{S_1}$ represents the part after the action and $U_s^{S_1}$ can be represented by $<X_c^{S_1}>$. Similarly for S_2, $D_f^{S_2}$ can be represented by $D_f^{S_2} = <E_c^{S_2}, E_a^{S_2}>$ and $U_s^{S_2}$ can be represented by $<X_c^{S_2}>$. Now, the four conditions are:

1. $E_c^{S_1} \cap X_c^{S_2} \neq \emptyset \wedge E_a^{S_1} \cap X_c^{S_2} = \emptyset$ **2.** $E_c^{S_1} \cap X_c^{S_2} = \emptyset \wedge E_a^{S_1} \cap X_c^{S_2} \neq \emptyset$
3. $E_c^{S_1} \cap X_c^{S_2} \neq \emptyset \wedge E_a^{S_1} \cap X_c^{S_2} \neq \emptyset$ **4.** $E_c^{S_1} \cap X_c^{S_2} = \emptyset \wedge E_a^{S_1} \cap X_c^{S_2} = \emptyset$

A semantic independency is observed between S1 and S2 only in case 4. And the other three cases indicates semantic dependency between them.

Algorithm to Compute Semantic Based DD-Dependency: The algorithm **SEMDEP** considers the *Defined* and *Used* parts (D_f and U_s) for every statement S_i of the database code. The edges (false alarm) between nodes S_i and S_j are then removed based on satisfiability checking of a formula of the *Defined*-part by S_i and *Used*-part by S_j.

Example 4. Let us illustrate our running example by using the Condition Action rule based approach. It is seen that E_{act}^3 operates on a part of data which is not accessed by E_{cond}^4. Therefore $3 \rightarrow 4$ is a false dependency. Similarly there is a false dependency between $3 \rightarrow 6$, $3 \rightarrow 10$ and $4 \rightarrow 10$.

- DD-dependencies for attributes *p_id* and *p_price*: $\{2 \rightarrow 3, 4, 5, 6, 10\}$, $\{3 \rightarrow 5\}$, $\{4 \rightarrow 5, 6\}$ and $\{6 \rightarrow 10\}$.

Algorithm 1: SEMDEP

Input: *Defined-* and *Used*-part (D_f, U_s)
Output: Precise DOPDG

```
for i =1 to n-1 do
    for j=i+1 to n do
        if D_f^i ∧ U_s^j = UNSAT then
            Delete the edge between the nodes i^th and j^th (i → j) ;
            // Satisfiability checking of the formula of the Defined-
            and Used-part of the database statement.
            False alarm reported;
        else
            ...nothing ...
End
```

4.4 Security Analysis

Given the nodes x and y in a dependency graph, information is flowing from x to y, if and only if x $\longrightarrow$ y, which means there is a path. There is no information flow if there is no path. Confidentiality leakage is depicted by assigning truth values to every database attribute depending upon their sensitivity level (public or private) and then we verify satisfiability of a formulae [9,10]. The fundamental target is to identify the illegal data flow in the code because of the private database attributes involvement in the code. Depending on the database attribute type security levels can be allocated for them. For simplicity only two security levels are taken. Therefore, all the database attributes in an application are considered with two security levels that is public (low) and private (high).

Let $\vartheta \longrightarrow$ {L, H} be a function and it appoints security classes to every database attribute, either L (public/low) or H (private/high). A program preserve the privacy of any information, iff information-flow is missing from high to low attributes. To check this property, the function $\hat{\Gamma}$ is defined below:

$$\hat{\Gamma}(\mathrm{x}) = \begin{cases} T \text{ if } \Gamma(\mathrm{x}) = H \\ F \text{ if } \Gamma(\mathrm{x}) = L \end{cases}$$

If $\hat{\Gamma}$ does not satisfy the formula in a path, analysis alerts confidentiality leakage.

5 Propose Framework Architecture

Now, we describe an architecture of our propose framework. It accepts a database code as input and checks whether this code is Safe or Unsafe as output. Here, Safe indicates that there is no information leakage and Unsafe indicates that there is an information leakage. Figure 3 portrays the general architecture of the framework. Now we illustrate various modules below.

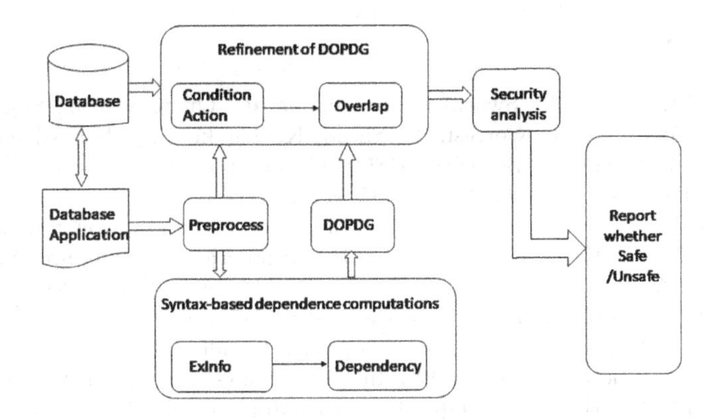

Fig. 3. Architecture of our proposed framework.

- **Preprocess:** The module "Preprocess" annotates the input database code and add numbers to every statement of the database code.
- **ExInfo:** Extraction of detailed information about defined and used variables at every statement in the input code is perform by this module. Information like defined variables, used variables, control statements, etc. are extracted for all the statements in the program.
- **Dependence:** In this module, using the information computed by the "ExInfo" module, syntax-based dependencies are computed among statements.
- **Condition Action:** This module defines the semantics of database code based on the Condition Action rule based approach. Using this semantics we compute Used- and Defined part of every statement.
- **Overlap:** This module checks the overlapping condition to identify the DD-dependency based on the Used- and Defined-part information. If overlap happens then there is a dependency, if not there is no dependency.
- **Security Analysis:** Using the dependency information, this module detects possible illegitimate information flow of the database codes by checking the satisfiability of the formulae based on their security levels.

6 Conclusions and Future Works

Confidential data that is stored in the underlying database may be bypassed to an unauthorized party due to improper development of an information system. One of the most prime concern of a database application is to protect the data from illegitimate flow. In this work, we proposed a new model to detect the illegitimate information flow of an application. This model uses semantic based analysis technique to refine the existing syntax-based dependency graph by removing false alarms. This proposed model serves as a powerful basis to identify possible data leakage of an application. Other stubs *e.g.* sub-queries, JOIN queries and string data-type will be considered as a future aim.

References

1. Ahuja, B.K., Jana, A., Swarnkar, A., Halder, R.: On preventing SQL injection attacks. In: Chaki, R., Cortesi, A., Saeed, K., Chaki, N. (eds.) Advanced Computing and Systems for Security. AISC, vol. 395, pp. 49–64. Springer, New Delhi (2016). https://doi.org/10.1007/978-81-322-2650-5_4
2. Alam, M.I., Halder, R.: Data-centric refinement of information flow analysis of database applications. In: Abawajy, J.H., Mukherjea, S., Thampi, S.M., Ruiz-Martínez, A. (eds.) SSCC 2015. CCIS, vol. 536, pp. 506–518. Springer, Cham (2015). https://doi.org/10.1007/978-3-319-22915-7_46
3. Baralis, E., Widom, J.: An algebraic approach to rule analysis in expert database systems. In: Proceedings of the 20th International Conference on Very Large Data Bases, VLDB 1994, pp. 475–486. Morgan Kaufmann Publishers Inc. (1994)
4. Denning, D.E.: A lattice model of secure information flow. Commun. ACM **19**, 236–243 (1976)
5. Ferrante, J., Ottenstein, K.J., Warren, J.D.: The program dependence graph and its use in optimization. ACM Trans. Program. Lang. Syst. (TOPLAS) **9**(3), 319–349 (1987)
6. Halder, R.: Language-based security analysis of database applications. In: 3rd International Conference on Computer, Communication, Control and Information Technology (C3IT 2015), pp. 1–4. IEEE Press (2015)
7. Halder, R., Zanioli, M., Cortesi, A.: Information leakage analysis of database query languages. In: Proceedings of the 29th Annual ACM Symposium on Applied Computing (SAC 2014), Gyeongju, Korea, 24–28 March 2014, pp. 813–820. ACM Press (2014)
8. Hammer, C., Snelting, G.: Flow-sensitive, context-sensitive, and object-sensitive information flow control based on program dependence graphs. IJIS **8**, 399–422 (2009). https://doi.org/10.1007/s10207-009-0086-1
9. Jana, A.: A static analysis approach to detect confidentiality leakage of database query languages. In: Abraham, A., Piuri, V., Gandhi, N., Siarry, P., Kaklauskas, A., Madureira, A. (eds.) ISDA 2020. AISC, vol. 1351, pp. 794–804. Springer, Cham (2021). https://doi.org/10.1007/978-3-030-71187-0_73
10. Jana, A., Alam, M.I., Halder, R.: A symbolic model checker for database programs. In: ICSOFT, pp. 381–388 (2018)
11. Jana, A., Bordoloi, P., Maity, D.: Input-based analysis approach to prevent SQL injection attacks. In: 2020 IEEE Region 10 Symposium (TENSYMP), pp. 1290–1293. IEEE (2020)
12. Jana, A., Halder, R.: Defining abstract semantics for static dependence analysis of relational database applications. In: Ray, I., Gaur, M.S., Conti, M., Sanghi, D., Kamakoti, V. (eds.) ICISS 2016. LNCS, vol. 10063, pp. 151–171. Springer, Cham (2016). https://doi.org/10.1007/978-3-319-49806-5_8
13. Jana, A., Halder, R., Chaki, N., Cortesi, A.: Policy-based slicing of hibernate query language. In: Saeed, K., Homenda, W. (eds.) CISIM 2015. LNCS, vol. 9339, pp. 267–281. Springer, Cham (2015). https://doi.org/10.1007/978-3-319-24369-6_22
14. Jana, A., Halder, R., Kalahasti, A., Ganni, S., Cortesi, A.: Extending abstract interpretation to dependency analysis of database applications. IEEE Trans. SE **46**(5), 463–494 (2018)
15. Jana, A., Maity, D.: Code-based analysis approach to detect and prevent SQL injection attacks. In: 2020 11th International Conference on Computing, Communication and Networking Technologies (ICCCNT), pp. 1–6. IEEE (2020)

16. Mandal, K.K., Jana, A., Agarwal, V.: A new approach of text steganography based on mathematical model of number system. In: International Conference on Circuits, Power and Computing Technologies, pp. 1737–1741. IEEE (2014)
17. Mohammed, A., Varol, N.: A review paper on cryptography, pp. 1–6, June 2019. https://doi.org/10.1109/ISDFS.2019.8757514
18. Mudarri, T., Al-Rabeei, S., Abdo, S.: Security fundamentals: access control models. Interdisc. Theory Pract. (2015)
19. Willmor, D., Embury, S.M., Shao, J.: Program slicing in the presence of a database state. In: Proceedings of the IEEE ICSM (2004)

Comparative Learning of on Request Direction-Finding Procedures in WSNs

Rakesh Kumar Saini[1(✉)], Mayank Singh[2], and Nishant Gupta[3]

[1] School of Computing, DIT University, Dehradun, Uttarakhand, India
[2] Consilio Research Lab, Tallinn, Estonia
dr.mayank.singh@ieee.org
[3] MGM College of Engineering and Technology, Noida, UP, India

Abstract. In Wireless Sensor networks, on demand direction-finding procedures show an actual significant part for announcement among sensor knots using WSNs Protocol stack. This paper presents substantial study and analysis of WSNs protocol Stack using Bellman-Ford direction-finding procedure to pattern the flow of statistics among the layers. In this broadside we compare the existing on-demand direction-finding protocols like AODV, DSR, DYMO, FSR, IARP are made on the base of concert environments- throughput (bits/s) and normal end to end delay(s) in order to find the Quality of Service routing which can meet requirement of users in WSNs. The result shows that AODV satisfies QoS requirement as the numbers of packet transmitted in a given time is more as compared to others routing protocols.

Keywords: Sensor nodes · Wireless Sensor Network · AODV · DSR · Bellman Ford · Qualnet 5.0.2

1 Introduction

In Wireless Sensor Networks various course discovering conventions exertion for correspondence among sensor hubs. These caption discovering conventions must be energy-proficient and should likewise content Quality of Service (QoS) requests of divergent sales. On the off chance that a sensor hub needs to detect and sends the information to approximately alternative device hub formerly this convention investigation the way in an on-request way and form the connecting to convey and gets the information. On-request steering convention [1, 2] don't support course discovering data at the sensor hubs if there is no correspondence. The course disclosure for the most part occurs by overflowing the course demand packages all through the organizations. On request steering conventions utilize two unique systems to discover and continue courses: first is course recognition measure and the second is course upkeep measure. Various on-request directing conventions have been created for the WSN till today. Remote Sensor Networks are to discover ways for development of energy effectiveness and responsible transmission of detected information to the Sink. Since a sensor hub has restricted detecting and calculation limits, communication execution furthermore, power, an massive number of sensor gadgets are appropriated finished a district of interest for congregation data (temperature, moistness,

M. Singh et al. (Eds.): ICACDS 2021, CCIS 1441, pp. 324–332, 2021.
https://doi.org/10.1007/978-3-030-88244-0_31

undertaking location, and so on) These hubs can speak with each other for sending or getting data either straightforwardly or through other halfway hubs and in this manner structure an organization, so every hub in a sensor network goes about as a switch [4] inside the organization. In direct correspondence steering conventions (single bounce), every sensor hub discusses straightforwardly with a control focus called Base Station (BS) and sends assembled data. The base station is fixed and found far away from the sensors. Base station(s) can speak with the end client either straightforwardly or through some current wired organization. The geography of the sensor network changes habitually. Hubs might not have worldwide recognizable proof. Since the distance between the sensor hubs and base station if there should be an occurrence of direct correspondence is huge, they burn-through energy rapidly. A Wireless Sensor Network means to accumulate natural information and the hub gadgets arrangement might be known or obscure deduced. Organization hubs can have genuine or sensible correspondence with all gadgets; such a correspondence characterizes a topography as designated by the submission. In this paper examination of current on-request leading resolutions DSR, AODV, DYMO, FSR and IARP are completed. On the off chance that a hub needs to send a parcel to some another hub, at that point directing convention investigations for the route in an on-request way and shape the development to communicate and obtains the partition [3, 4].

2 Existing Routing Protocols in Wireless Sensor Networks

1. AODV (Ad-Hoc On-Demand Distance Vector Routing)

An Ad Hoc On-Demand Distance Vector (AODV) is a comportment ascertaining resolution strong-minded for remote and adaptable impromptu nets. This convention establishes approaches to objections on request and funds together unicast and multicast directing. The AODV convention [5, 6] was together settled by Nokia Research Center, the University of California, Santa Barbara and the University of Cincinnati in 1991.

AODV is an approachable heading determining agreement dependent on DSDV. AODV is considered aimed at nets with tens to thousands of device centers. One component of AODV is the utilization of an objective succession amount for respectively steering table section [5]. The succession number is made by the objective hub. AODV utilizes course discovering tables, with one heading section for every objective where every passage stores next bounces towards objective. It transmission course demand (RREQ) parcels and this RREQ is solely perceived by the correspondent statement, objective location and solicitation ID. After meting out the RREP packet the node ahead it towards the source. AODV is a container routing protocol considered for use in WSNs and intended for networks that may cover thousands of sensor nodes. The way detection instrument is prayed only if a way to a journey's end is not recognized [7]. The knot container advanced inform the situation direction-finding info if it determines an improved track or way. The prime objects of AODV protocol are:

- To transmission detection containers only when important.
- To differentiate among local connectivity administration and over-all topology preservation.
- To distribute info about modifications in local connectivity to those adjoining sensor nodes those are possible to need the information.

AODV [8] usages a transmission way finding way. Path detection development is introduced when a node necessitates in announcement with a node for which it has no way by distribution a way demand container comprising the basis statement, source succession number, communicated ID, objective location, objective arrangement number, jump tally to its outsiders. Jump Count is at first 0 and is augmented by every sensor hub as it straight on the course demand toward the end point.

2. DSR (Dynamic Source Routing)
Dynamic Source Routing (DSR) [9] is an on-request steering convention in WSNs. It is a responsive bearing discovering convention which is brilliant to achieve Wireless Sensor Networks without utilizing intruded on table update message like table decided heading discovering conventions do [10]. It uses source heading discovering, which is a strategy wherein the wellspring of a bundle controls the entire request of sensor hubs finished up which the hub has ventured. It was unequivocally determined for use in multi-bounce WSNs. The DSR convention is unsophisticated, compelling and incredibly responsive steering convention [8, 10]. The Dynamic Source Routing permits any host to enthusiastically decide an establishment course to any travel's end in the net. In Dynamic Source Routing the sender closes the total track from the source to the objective hub and stores the addresses of the middle hubs of the course in the parcels. In Dynamic Source Routing convention, the hubs don't have to trade the steering table data intermittently and in this manner lessens the transfer speed overhead in the organization. In Dynamic Source Routing, each source decides the course to be utilized in communicating its parcels to assigned objections. There are two fundamental segments, called Direction Detection and Way Preservation. Course Detection decides the ideal way for a transmission between a given source and objective. Course Maintenance defends that the transmission way stays ideal and circle free as organization conditions change, regardless of whether this necessities adjusting the course all through a transmission.

3. DYMO (Dynamic MANET On Demand)
Dynamic MANET On-request (DYMO) bearing discovering convention is indistinguishable to AODV directing convention and its benefits are similar to AODV steering convention. Basically DYMO steering convention simple to execute and DYMO is improvement of AODV. DYMO is energy-proficient directing convention that are utilized both as a supportive of dynamic and as a responsive steering convention. In DYMO course can be found by sensor hub when they are required. In DYMO directing convention when sensor hub find course then sensor hub sends a course demand (RREQ) message to the objective and afterward objective send steering reply (RREP) message to source sensor hub. DYMO directing convention contain two tasks like AODV steering convention that are: course revelation and course support [13].

Course revelation is performed at source hub to an objective for which it doesn't have a substantial way. What's more, course upkeep is performed to evade the current annihilated courses from the steering table and furthermore to diminish the bundle dropping if there should arise an occurrence of any course break or hub disappointment. The DYMO steering convention [6] is recipient to the well-known Ad hoc On-Demand Distance Vector (AODV) directing convention and offers a significant number of its advantages. DYMO is a responsive directing convention that registers unicast courses on request or when required. It occupations succession numbers to affirm circle self-rule. It permits on demand, multi-bounce unicast steering among the hubs in a portable specially appointed organizations.

4. FSR (FISHEYE STATE ROUTING)

Fisheye is a framed ordered heading discovering convention. It is a functioning convention and is a connection state based heading discovering convention that has been improved to the WSNs. Fisheye offers course data promptly by keeping a geography map at every hub. It diminishes the extent of info that is direct and gets by instrument hubs in WSNs. Fisheye saves a geography map at every hub. Fisheye is a board-single-minded WSNs and its instruments depend on the Connection Government Direction-finding convention utilized in WSNs. Fisheye course discovering convention is utilized to lessen the steering higher up. Fisheye course discovering convention reduces the heading discovering data at each sensor hub. By burning-through fisheye directing convention, each sensor hub has data around additional instrument hubs and course info.

5. IARP (INTRAZONE ROUTING PROTOCOL)

The Intrazone Routing Protocol (IARP) is utilized to maintain headings to end point inside a neighborhood network, which is referenced to as a course discovering district. Extra decisively, a hub's moving district is very much characterized as a social affair of hubs whose least space in jumps from the hub in inquiry is no more noteworthy than a boundary expressed to as the zone territory [2, 8]. Every hub preserves its individual heading discovering zone. A significant hugeness is that the steering zones of neighboring hubs cover.

In IARP every hub screens the adjustments occurs in neighborhood and breaks the overall route recognition to local excursion's end. IARP's steering gives improved, course preservation after courses have been uncovered [16, 17]. The preemptive preservation of steering zones likewise improves the nature of found courses, by making them more powerful to changes in network geography. Whenever courses have been uncovered, IARP's directing zone recommendations improved, ongoing, course support. Connection disappointments can be dodged by different jump ways inside the directing zone.

3 Performance Analysis of AODV, DSR, DYMO, FSR, IARP

Energy efficiency is large matter in WSNs as Sensor nodes have little liveliness effectiveness. In WSNs, direction-finding protocols connect statistics among device knots and sensor nodes transmission intelligence statistics to the improper position. This paper analyzes the diverse WSNs directing convention for example Specially appointed AODV, Dynamic MANET On request (DYMO), Dynamic Source Routing (DSR), Intrazone Routing Protocol (IARP) and Fisheye Routing Protocol (FSR) utilizing 20 sensor hubs. For contemplating the Quality of Service (QoS) boundaries in WSNs we have investigated the presentation of these directing conventions based on Presentation Conditions - quantity (bits/s) and normal start to finish delay(s). Since organization of huge sensor hubs is troublesome in certifiable climate so we build up a recreation climate to check throughput and normal start to finish defer factor of steering conventions [14, 15]. For this reason, we have utilized QualNet 5.0.2 reproduction demonstrating device. In Simulation arrangement, the underlying situation of the hubs is arbitrary in the 100m x

100m Area. In this effort altogether device hubs corporately pass their information to the base station and from that point information is forward through web or satellite to the errand director hub or end client which need to screen the detecting territory. For same Scenario we check execution of AODV, DYMO, DSR, IARP, and FSR. Boundaries that are utilized in this reproduction are appeared in Table 1.

Table 1. Simulation parameters for on demand routing protocols.

Parameters	Values
Simulator	Qualnet 5.0.2
Routing protocol	AODV, DYMO, DSR, IARP, FSR
Number of nodes	20
Simulation area	100 m * 100 m
Traffic type	CRB
Node placement	Random node placement
Simulation time	120 s
Channel frequency	2.4 GHz
Antena	Omni directional

Figure 1 demonstrations the simulation setup situation in which device knots cooperatively permit their information after single device knot to additional device knot (Fig. 2).

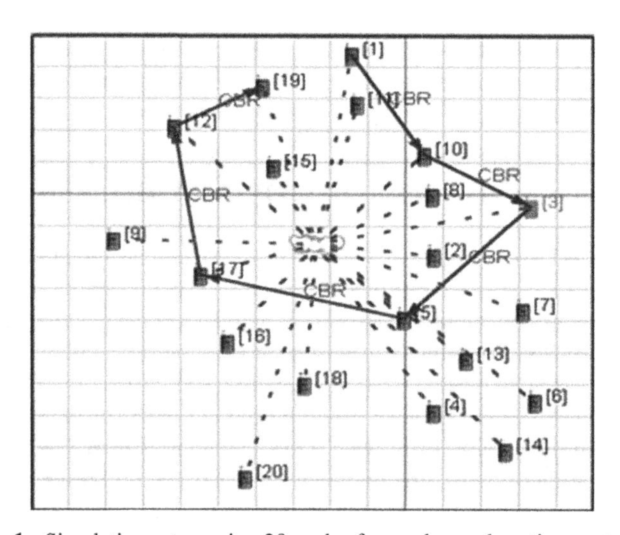

Fig. 1. Simulation setup using 20-nodes for on demand routing protocol.

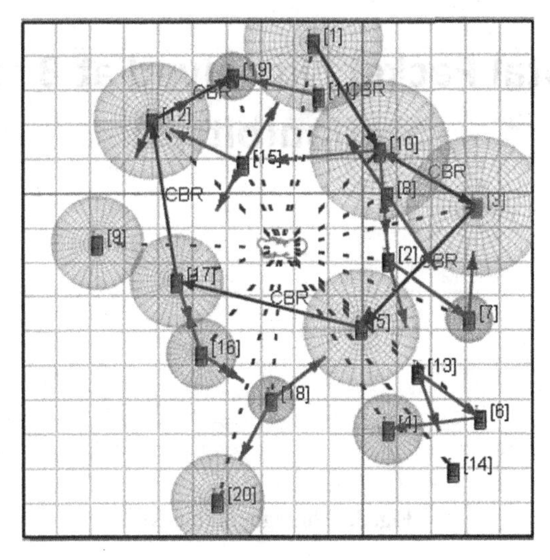

Fig. 2. Running simulation using 20-nodes for on-demand routing protocol.

4 Experimental Results of Routing Protocols

The Presentation of the AODV, DSR, DYMO and IARP are assessed using QualNet 5.0.2 Simulator. Simulation results are investigated for presentation of direction-finding protocols on Quality of Service parameters- throughput and regular completion to completion suspension.

i. Throughput

Quantity [11] is the regular amount of effective communication distribution in an announcement station. Sensor nodes direct containers to base station Therefore base station receives packets from different sensor nodes. This information might be conveyed over a physical or intelligent connection, or pass through a specific organization. Figure 3 shows total number of packets received by base station. Base Station received 1170, 1142, 1095, 1150, 1042 packets (bits/sec) for routing protocols AODV, DSR, DYMO, IARP, FSR respectively.

ii. Average end to end delay

Normal start to finish delay is the normal time where bundles send from sensor hub to the base station. At the point when sensor hubs send information to the base station at that point because of clog in the correspondence networks there might be start to finish delay. By investigating five steering conventions we found that the start to finish delay is most noteworthy in DSR when contrasted with the others. Figure 4 shows normal start to finish deferrals of five steering conventions [11]. The graph is the figure shows the time (in sec) required for sending packets from sensing field to base station using protocols AODV, DSR, DYMO, IARP and FSR are 3.9 s, 4 s, 3.4 s, 5 s, 5.3 s.

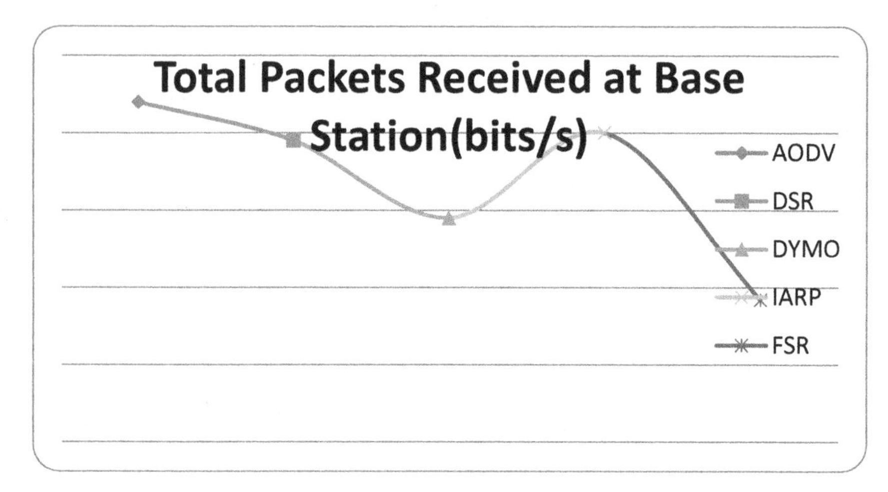

Fig. 3. Throughput (bits/s)

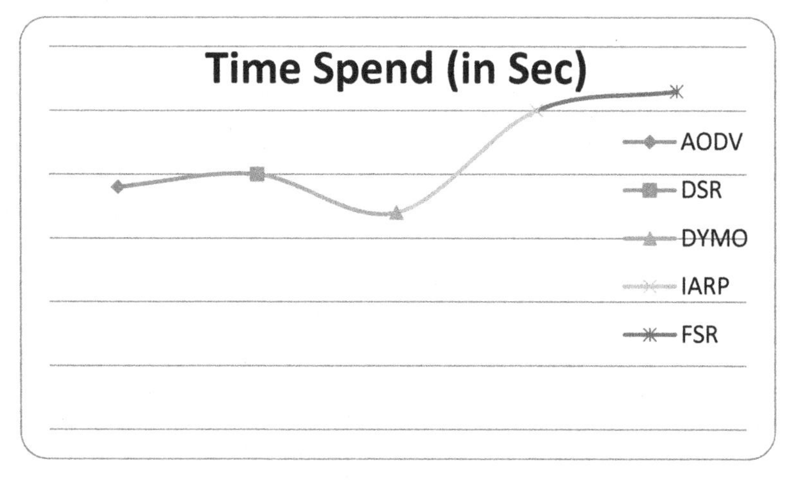

Fig. 4. Average end-to-end delay (in sec).

Directing conventions assumes significant function for correspondence among device hubs in WSNs. Through this examination we discovery that AODV directing convention is more vitality proficient as contrast with others steering convention. Table 2 shows the investigation of aftereffects of the referenced five directing conventions and furthermore graphical portrayal of results are appeared in Fig. 5.

Table 2. Analysis of on-demand routing protocols.

Parameters	AODV	DSR	DYMO	IARP	FSR
Throughput (bits/s)	1170	1142	1095	1150	1042
Average end-to-end delay (in sec)	3.9 s	4 s	3.4 s	5 s	5.3

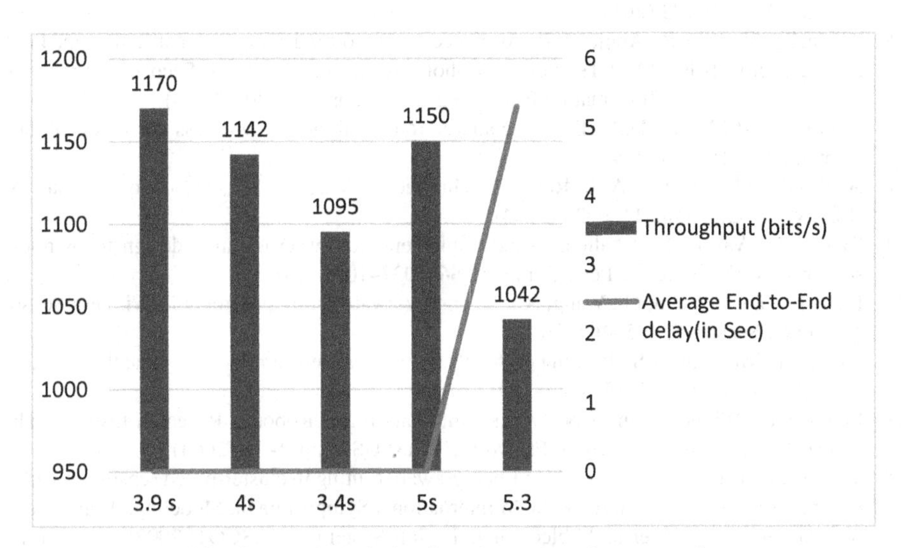

Fig. 5. Result analysis of on-demand routing protocols.

5 Conclusions

In this paper we have carried out the assessment of the five dissimilar on-demand direction-finding procedure consuming similar limitations with 20 sensor nodes. We have analyzed the performance of AODV, DSR, DYMO, FSR, IARP on the basis of presentation environments- quantity (bits/s) and regular completion to completion suspension(s). The result shows that AODV is more energy-efficient routing protocol as compare to others routing protocols as numbers of packet transmitted is more for the given time than the other routing protocols.

References

1. Saleem, M., Ullah, I., Farooq, M.: BeeSensor: an energy-efficient and scalable routing protocol for Wireless Sensor Networks. Inf. Sci. **200**, 38–56 (2012)
2. Ramesh kumar, S.G.: Improving Quality of Service through enhanced node selection technique in Wireless Sensor Networks. Int. J. MC Square Sci. Res. **8**, 133–142 (2016)
3. Xu, N.: A survey of sensor network applications. IEEE Commun. Mag. **40**, 102–114 (2002)
4. Abbasi, A.A., Younis, M.: A survey on clustering algorithms for Wireless Sensor Networks. Computer communication **30**, 2826–28410 (2007)

5. Perkins, C.E., Royer, E.M., Das, S.: Adhoc on demand distance vector routing. In: Proceedings of Second IEEE workshop on Mobile Computing Systems and applications, pp. 207–218 (1999)
6. Arya, S., Nipur, C.A.: Performance analysis of AODV, DSR and DYMO protocols using random waypoint mobility model in MANET. Int. J. Comput. Appl. **67**, 13–17 (2013)
7. Govindasamy, J., Punniakody, S.: A comparative study of reactive, proactive and hybrid routing protocol in wireless sensor networks under wormhole attack. J. Electr. Syst. Inf. Technol. **22**, 167–172 (2017)
8. Ahuja, R., Ahuja, A.B., Ahuja, P.: Performance evaluation and comparison of AODV and DSR routing protocols in MANETs under wormhole attack. In: 2013 IEEE Second International Conference on Image Information Processing (ICIIP), pp. 699–702 (2013)
9. Johnson, D.B., Maltz, D.A.: Dynamic source routing in ad hoc wireless networks. Mobile Comput. **353**, 153–181 (1996)
10. Al-Karaki, J.N., Kamal, A.E.: Routing techniques in Wireless Sensor Networks: a survey. IEEE Wirel. Commun. **11**, 6–28 (2004)
11. Ranjan, R., Varma, S.: Challenges and implementation on cross layer design for wireless sensor networks. Wireless Pers. Commun. **86**, 1037–1060 (2016)
12. Liu, M., Cao, J., Chen, G., Wang, X.: An energy-aware routing protocol in Wireless Sensor Networks. Sensors **9**, 445–462 (2009)
13. Sarkar, A., Murugan, T.S.: Routing protocols for wireless sensor networks: what the literature says? Alexandria Eng. J. **55**, 3173–3183 (2016)
14. Lewis, F.L.: Wireless Sensor Networks. Automation and Robotics Research Institute, The University of Texas at Arlington: Ft. Worth, Texas, USA, pp. 1–18 (2004)
15. Younis, M., Youssef, M., Arisha, K.: Energy-aware routing in cluster-based sensor networks. In: Proceedings of the 10th IEEE/ACM International Symposium on Modeling, Analysis and Simulation of Computer and Telecommunication Systems (MASCOTS2002), Fort Worth, TX, USA, October 2002
16. Liu, J.-S., Lin, C.-H.R.: Cross-layer optimization for performance trade-off in network code-based wireless multi-hop networks. Comput. Commun. **52**, 89–101 (2014)
17. Sahota, H., et al.: An energy-efficient wireless sensor network for precision agriculture. In: 2010 IEEE Symposium on Computers and Communications (ISCC). IEEE (2010)

ECG Based Stress Detection in Automobile Drivers Using Long Short-Term Memory (LSTM) Network

Ramyashri B. Ramteke(✉) and Vijaya R. Thool

Biomedical Instrumentation Laboratory, Shri Guru Gobind Singhji Institute of Engineering and Technology, Nanded, India
vrthool@sggs.ac.in

Abstract. The work presents the driver's acute (short-Time) stress detection during the driving task. Stress is an unavoidable truth of life but it is essential to detect at the earliest to manage effectively and to avoid future health risks. Heightened stress can significantly impede drivers' capacity to respond to dangers. The earlier methods of stress detection use the heart rate variability (HRV) framework. HRV is a plot of the time interval between two successive R peaks derived from an electrocardiogram (ECG). The HRV analysis requires a large length signal usually in minutes, but acute stress appears for short window length. In that case, perceptive deep learning may be a proper solution. The paper utilizes a long short-term memory (LSTM) network for acute stress detection in automobile drivers with short length ECG. In order to detect stressed versus relaxed state, morphology-based and feature-based approaches have been proposed. The proposed algorithm adopted bidirectional-LSTM to enhance the performance of the model. The experimental analysis is carried out on two publicly available Physionet databases *viz.* the driver ECG signal database and the normal ECG signal database. The proposed feature-based approach outperforms existing state-of-art methods by achieving an accuracy of 98.57%.

Keywords: Electrocardiogram · Acute stress · Spectrogram features · Bidirectional long short-term memory network

1 Introduction

Stress can manifest in any situation or thought that causes the individual to feel frustrated, angry, and nervous. Sometimes, a limited amount of stress can be beneficial and can encourage a person to complete tasks before deadlines. However, studies have shown that people who are stressed may aggravate or increase the risk of various diseases. According to the American Psychological Association (APA) research, stress can be divided into acute, and chronic [14]. Acute stress is short-term, described as a response to an immediate threat and is often called a fight or flight response. While chronic stress is defined as a kind of

M. Singh et al. (Eds.): ICACDS 2021, CCIS 1441, pp. 333–342, 2021.
https://doi.org/10.1007/978-3-030-88244-0_32

stress that occurs from internal or external sources of stress over a long period of time, and has been associated with or may be exacerbated by a variety of medical conditions, such as high blood pressure, asthma, arrhythmia, immune system suppression. Both acute and chronic stress may influence other dangers and behaviours, such as cholesterol levels and elevated blood pressure, smoking, physical inactivity, and overeating [14]. The center for disease control reported that 110 million people die every year globally because of mental stress [17]. Electrocardiography is an easily accessible and non-invasive method to diagnose psychological stress, as the variation in ECG waveform is mediated by the sympathetic nervous system and parasympathetic nervous system. The heart rate increases with the increase in sympathetic activation at the ventricular myocardium under stress.

Among the different kinds of stresses driving stress is taken here that can lead to churlish and destructive behaviour on the roadway. The stress during driving is related to certain influences [16], such as impaired decision-making ability, decreased performance, and situational awareness. Stress has been recognized as one of the important factors leading to vehicle crashes, causing huge losses in terms of the loss of life and productivity of the government and society. For road and traffic security, it is essential to detect driving stress early to lift driver performance and awareness.

The rest of the paper is organized into four major sections. The detailed literature is given in the Sect. 2. Section 3 is briefs about the proposed methodology in which database and methods are explained. Section 4 presents the analysis of experimental results, and Sect. 5 concludes the proposed approach.

2 Literature Survey

All the related work is categorized broadly into two main groups based on feature generation as (i) HRV analysis framework, and (ii) Automatic feature generation framework using Deep CNN networks. Existing methods for stress recognition are based on the analysis of heart rate variability (HRV) derived from ECG [6,9,11,15]. Wang et al. [19] implemented a k-nearest neighbor (KNN) classifier for stress detection with HRV feature-based transformation algorithm. The algorithm involves feature generation, selection, and dimension reduction for robust feature generation. To accomplish the study, they used a Physionet driver database and achieved a significant accuracy of 97.78% The authors in [3] investigated HRV based cardiovascular reactivity with short-term psychological stress. The analysis was done in the time domain and the frequency domain. The statistical value of Standard Deviation of RR intervals and Root Mean Square value of RR intervals get reduced under stress state. The Low-frequency component of HRV increased while the high-frequency component of HRV decreased

Voluminous literature exists for stress detection using a traditional approach. This approach uses HRV based features that require a large window length of the signal, usually in minutes [2,12]. However, acute stress detection requires the decision made in a short window [5]. The performance of traditional HRV based methods might be constricted with the decreased window length.

For the past few years, with a great interest in deep neural network architectures, numerous articles were published on mental stress detection for different stressors with a short length of signals. Rastgoo et al. [10] proposed a multimodal fusion CNN-LSTM network for driver stress detection that combines vehicle data, contextual data, and ECG signal data to enhance the classification results and achieved an accuracy of 92.8%. The diversified mental states of pilots [4] were detected using a multimodal approach (ECG, EEG, and respiration rate) with a fusion of CNN and LSTM. Many researchers formulated multimodality-based stress detection study with a variety of stressors using deep learning [5,7,13]. Generally, multimodal input adopted fusion-based models that are computationally complex.

Due to the significant cost of solving the driver's stress, it is essential to build a practical system to detect the driver's stress in real-time with conclusive accuracy. This work proposes a novel spectrogram feature-based approach for mental stress detection to improve the effectiveness of LSTM. To the best of our knowledge, it is the first attempt of the utilization of spectrogram features for stress recognition. The contributions of the proposed work are (1) Filters are designed for removing the unwanted portion of ECG signal (2) Features (Instantaneous frequency and spectral entropy) are extracted from ECG spectrogram (3) Bidirectional LSTM is used to learn the sequence of extracted features effectively.

3 Proposed Methodology

In this study, a Long short-term memory (LSTM) network has employed for the classification of stressed versus normal individuals using the ECG signals, as shown in the Fig. 1. This work presents two different approaches as follows.

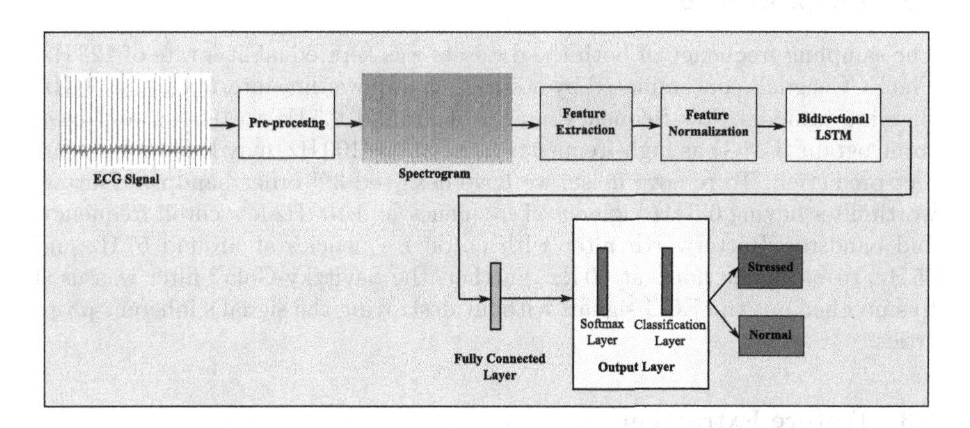

Fig. 1. Proposed driver stress recognition architecture.

Approach-I: Morphology-Based Method. The filtered ECG as a single one-dimensional feature data was fed directly to an input of the LSTM that learns time-dependant morphology of a given ECG to discriminate the classes.

Approach-II: Feature-Based Method. The feature extraction has prepared to improve the performance of a classifier further. The two features were extracted using the spectrogram (time-frequency representation). Hence in the second strategy, the LSTM adopted the knowledge from two highly efficient one-dimensional features. The instantaneous frequency and spectral entropy are distributed over time and vary with signal-to-signal [1,18]. The Fig. 3 shows that instantaneous frequency has the time varying dependency and spectral entropy evaluates the spectral complexity of signal in time. Hence it is essential to learn sequences step-by-step to generate a concrete decision about the signal category. Bidirectional LSTM is used to learn sequences in a forward state and in a backward state for a better understanding of a provided context.

3.1 ECG Dataset

In this paper, two open-source databases were used, for the task of stress identification from drivers, available at Physionet [8]. The first is the driver stress ECG database (drivedb), and secondly, the normal sinus rhythm ECG database (nsrdb). Both the datasets consist of 18 recordings each, so there were only a total of 36 samples. Hence to increase the dataset, we had taken 1-h recordings and then segmented to a 1-min each. Now there are 997 stressed samples and 1098 normal samples.

3.2 Pre-processing

The sampling frequency of both the datasets was kept equal at a rate of 128 Hz. The ECG signals contaminated by noises such as powerline interference at 60 Hz, Baseline wanders (low-frequency noise) of around 0.5 Hz to 0.6 Hz, and electromyogram (EMG) as high-frequency noise above 100 Hz, may lead to an incorrect prediction. To remove noise, we have designed 3rd order bandpass Butterworth filter having 0.6 Hz high cutoff frequency and 100 Hz low cutoff frequency, and bandstop Butterworth filter with cutoff frequencies at around 57 Hz and 65 Hz, to block the noise at 60 Hz. Further, the Savitzky-Golay filter was used to smoothen out the ECG signals without destroying the signal's inherent properties.

3.3 Feature Extraction

Feature extraction improves the performance of the classifier. The time-frequency (TF) moments were extracted in the time domain, inspired by the use of the spectrogram as a time-frequency image to train the convolutional neural network (CNN). The Fig. 2 shows the filtered ECG signal and its spectrogram. We translated a spectrogram approach to a TF moment so it can be used as a 1D feature to the LSTM. Two TF moments were extracted:

I. Instantaneous Frequency
II. Spectral Entropy

Instantaneous Frequency computes the first spectral moment of the time-frequency distribution of the input signal. It is the first moment of a spectrogram obtained using a short-time Fourier transform [1]. Thus, the instantaneous frequency I_{freq} is estimated as,

$$I_{freq}(t) = \frac{\int_0^\infty fS(t,f)df}{\int_0^\infty S(t,f)} \tag{1}$$

where, $S(t, f)$ is the power spectrogram.

Spectral Entropy is defined as a measure of a probability distribution of a signal in the frequency domain [18]. It measures the spikiness or flatness of a signal's spectrum. A spiky spectrum has low spectral entropy, and high spectral entropy is associated with a flat spectrum. It is widely used for feature extraction to detect and diagnose a fault in biomedical signal processing. For a signal $z(n)$ with $Z(k)$ as its discrete Fourier transform, $S(k) = |Z(k)|^2$ is the power spectrum, the probability distribution is,

$$P(k) = \frac{S(k)}{\sum_j S(j)} \tag{2}$$

Then the normalized spectral entropy E is calculated as,

$$E = -\frac{\sum_{k=1}^{M} P(k)\log_2 P(k)}{\log_2 M} \tag{3}$$

where, M are the frequency points. The probability distribution in the presence of power spectrogram $S(t, f)$ is then calculated at time t,

$$P(t,k) = \frac{S(t,k)}{\sum_f S(t,f)} \tag{4}$$

Hence, the spectral entropy becomes,

$$E(t) = -\sum_{k=1}^{M} P(t,k)\log_2 P(t,k) \tag{5}$$

Now the signal with 7500-sample-long each gets transformed into two 129-sample-long features that consisting of instantaneous frequency and spectral entropy shown in the Fig. 3.

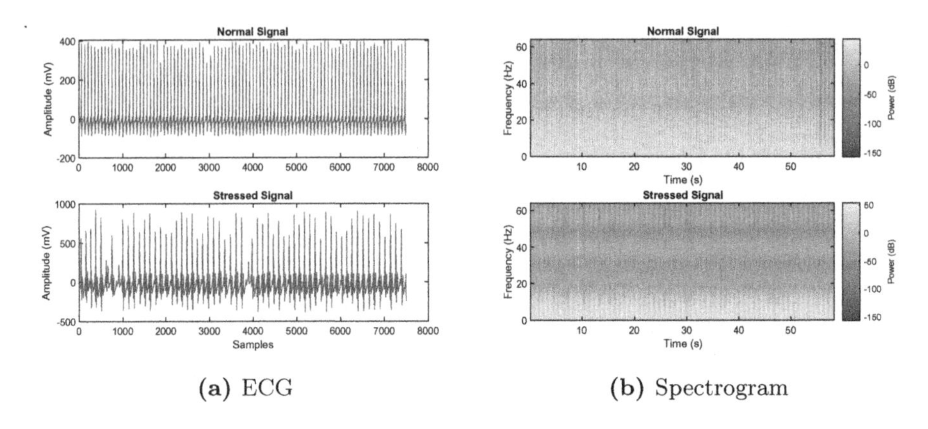

(a) ECG (b) Spectrogram

Fig. 2. Electrocardiogram (ECG) and it's equivalent spectrogram for the normal and stressed individual.

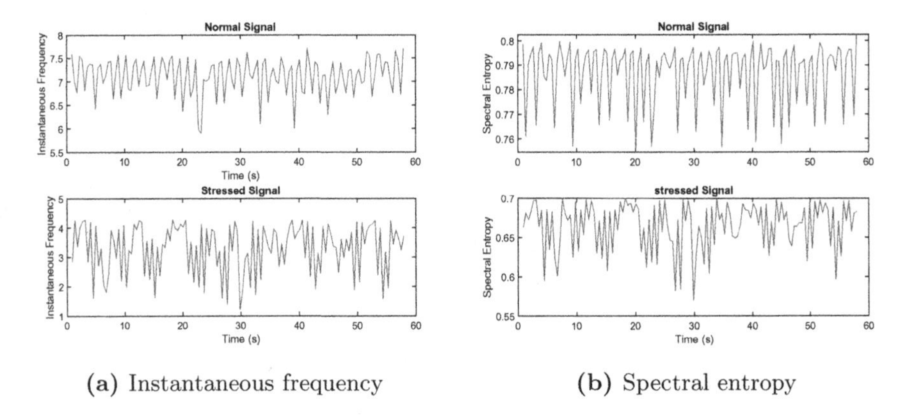

(a) Instantaneous frequency (b) Spectral entropy

Fig. 3. Features extracted from the normal and stressed individual.

3.4 Feature Normalization

The difference in both features is almost an order of magnitude. The instantaneous frequency is a high range value, it could slow down the learning process and hence network also converges slowly. The mean and standard deviation of a training set was calculated to normalize the training set and testing set, which further helps in the improvement of network performance during training.

Z_{train} and Z_{test} was considered as the training and testing sets respectively. The standardized data was obtained then,

$$Z_{trainSD} = \frac{Z_{train} - \mu}{\sigma} \tag{6}$$

$$Z_{testSD} = \frac{Z_{test} - \mu}{\sigma} \tag{7}$$

where, μ and σ are the mean and standard deviation of Z_{train} respectively.

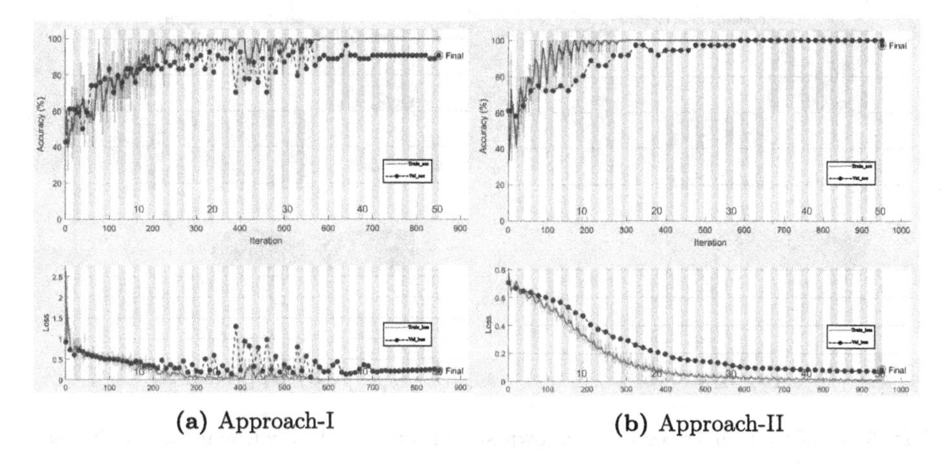

Fig. 4. Training and validation plot for proposed approaches using bi-directional LSTM.

3.5 Hyper-parameter Setting

The proposed deep learning based approach optimizes the categorical crossentropy loss by Adam optimizer for softmax classification. The hidden units of the LSTM and bidirectional LSTM was set to 100. The training process was terminated by 50 number of epochs with fix global learning rate of 0.01 and batch size of 150. The weight and bias learning rate of fully connected layer is kept 5 times the global learning rate. All the experiments are implemented on the system configured with 2 GB NVDIA GeForce MX230 GPU using software MATLAB R2020a.

4 Experimental Results

The ultimate objective of the research work is to detect the severe stress of the driver. To achieve the objective, the dataset has divided into three sets: training set with 1675 samples, while the validation and testing set each with 210 samples.

Table 1. Results obtained using different approaches.

Method	Approach	Classifier	Sensitivity	Specificity	F1 score	Accuracy
Conventional	HRV analysis	k-NN	79%	73.63%	75.96%	76.19%
		SVM	86%	80.91%	83.09%	83.33%
Deep learning	Approach-I	LSTM	85%	80%	82.12%	82.38%
		Bi-directional LSTM	86%	90%	87.31%	88.09%
	Approach-II	LSTM	95%	93.63%	94.05%	**94.28%**
		Bi-directional LSTM	99%	98.18%	98.51%	**98.57%**

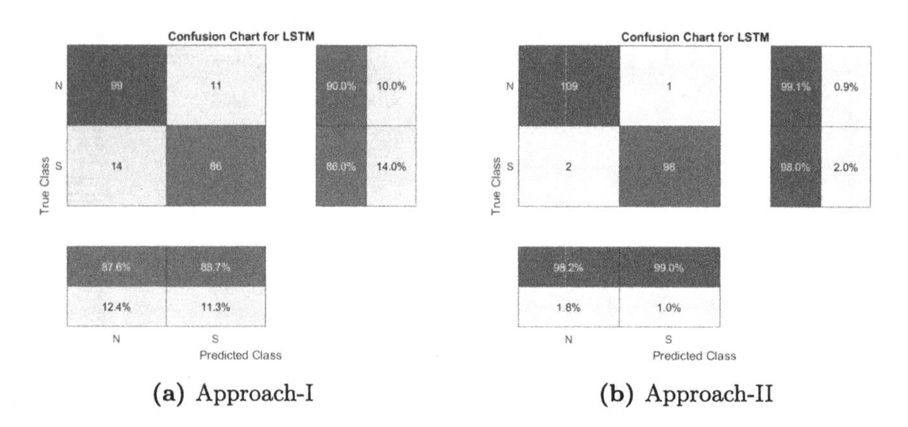

(a) Approach-I **(b)** Approach-II

Fig. 5. Confusion matrices for unknown samples using bi-directional LSTM (N: Normal, S: Stressed).

The overall results comprised of the classification of stress and relax condition of human. Initially HRV based conventional methods are implemented for classification. The time domain features, frequency domain features and Poincare plot features are computed for HRV analysis. Further, the deep learning based method consisting of LSTM and bi-directional LSTM network are utilized for classification. To evaluate the performance of deep learning based model, two approaches are proposed. In first approach, filtered ECG is directly applied to LSTM, while in second approach, the spectrogram features of filtered ECG signal are inputs to the LSTM. The result analysis shows the substantial enhancement of classification using the deep learning based method over the conventional method. The efficiency of the LSTM is improved by incorporating the spectrogram features, which signifies the importance of instantaneous frequency and spectral entropy. The overall classification results are further intensified by replacing the LSTM with bi-directional LSTM for leaning the sequential features in both forward and backward direction. The Table 1 exhibits discrimination of different approaches as stated above in which, first approach achieved 88.09% accuracy and second approach achieved highest accuracy of 98.57% on test set using bi-directional LSTM. The confusion matrices for both approaches are shown in the Fig. 5. The Table 2 shows a comparison of results obtained using various approaches proposed by researchers with the proposed work that indicates our method is excellent in feature extraction and classification. The training and validation performance shown in the Fig. 4 expressed the fast convergence of the proposed approaches by reducing the overfitting.

Table 2. Performance of ECG based studies for mental stress detection.

Sr. no.	Author	Stressor	Approach	Classifier	Accuracy
1.	Rigas, et al. [11] (2012)	Self-created driver stress data	HRV analysis with driving event	Bayesian network	96%
2.	Wang, et al. [19] (2013)	Physionet driver stress data	Trend-based and parameter-based feature analysis	k-NN	97.78%
3.	George, et al. [15] (2014)	Self-created pilot stress data	Linear and non-linear HRV features	Naive Bayes	90%
4.	Rastgoo, et al. [10] (2019)	Self-created driver stress data	ECG, vehicle and contextual features	CNN-LSTM	92.8%
5.	Han, et al. [4] (2020)	Pilot's diversified mental stress data	ECG, EEG and RESP. features	CNN-LSTM	85.2%
6.	**This paper**	Physionet driver stress data	**ECG spectrogram features**	**Bi-directional LSTM**	**98.57%**

5 Conclusions

This paper presents a robust bidirectional long short-term memory network-based system for driver mental stress recognition to avoid accident risks on the roadway. In this study, morphology-based and feature-based approaches were proposed. The morphology-based method uses filtered ECG of each 60-s window and generates 88.09% accuracy for unknown subjects. In a feature-based approach, instantaneous frequency and spectral entropy were extracted from the spectrogram of ECG to generate significant features. The spectrogram feature-based method improved test accuracy from 88.09% to **98.57%**.

Learning ECG signals gives only time-domain information. While learning these two features provides time-dependent spectral information. Therefore LSTM network generated improved accuracy with extracted features.

In the future, the work will emphasize the detection of different levels of stress with a variety of mental stressors.

References

1. Boashash, B.: Time-frequency and instantaneous frequency concepts. In: Time-Frequency Signal Analysis and Processing (2016). https://doi.org/10.1016/B978-0-12-398499-9.00001-7
2. Camm, A.J., Malik, M., et al.: Heart rate variability: standards of measurement, physiological interpretation and clinical use. Task force of the European society of cardiology and the North American Society of pacing and electrophysiology. Circulation **93**(5), 1043–1065 (1996)
3. Delaney, J.P.A., et al.: Effects of short-term psychological stress on the time and frequency domains of heart-rate variability. Percept. Mot. Skills **91**(2), 515–524 (2000)
4. Han, S.-Y., Kwak, N.-S., et al.: Classification of pilots' mental states using a multimodal deep learning network. Biocybern. Biomed. Eng. **40**(1), 324–336 (2020)

5. He, J., Li, K., Liao, X., Zhang, P., Jiang, N.: Real-time detection of acute cognitive stress using a convolutional neural network from electrocardiographic signal. IEEE Access **7**, 42710–42717 (2019)
6. Melillo, P., et al.: Nonlinear heart rate variability features for real-life stress detection. Case study: students under stress due to university examination. Biomed. Eng. Online **10**(1) (2011). https://doi.org/10.1186/1475-925X-10-96
7. Oskooei, A., Chau, S.M., et al.: DeStress: deep learning for unsupervised identification of mental stress in firefighters from heart-rate variability (HRV) data. arXiv preprint arXiv:1911.13213 (2019)
8. PhysioBank. http://www.physionet.org/physiobank/database/drivedb/. http://www.physionet.org/physiobank/database/nsrdb/
9. Ramteke, R., Thool, V.R.: Stress detection of students at academic level from heart rate variability. In: 2017 International Conference on Energy, Communication, Data Analytics and Soft Computing (ICECDS), pp. 2154–2157. IEEE (2017)
10. Rastgoo, M.N., et al.: Automatic driver stress level classification using multimodal deep learning. Expert Syst. Appl. **138**, 112793 (2019)
11. Rigas, G., et al.: Real-time driver's stress event detection. IEEE Trans. Intell. Transp. Syst. **13**(1), 221–234 (2011)
12. Ribeiro, R.T., et al.: A regression approach based on separability maximization for modeling a continuous-valued stress index from electrocardiogram data. Biomed. Sig. Process. Control **46**, 33–45 (2018)
13. Seo, W., Kim, N., Kim, S., et al.: Deep ECG-respiration network (DeepER net) for recognizing mental stress. Sensors **19**(13), 3021 (2019)
14. Stress: The different kinds of stress. American Psychological Association. http://www.apa.org/helpcenter/stress-kinds.aspx. Accessed 11 Feb 2010
15. Tanev, G., et al.: Classification of acute stress using linear and non-linear heart rate variability analysis derived from sternal ECG. In: 2014 36th Annual International Conference of the IEEE Engineering in Medicine and Biology Society, pp. 3386–3389. IEEE (2014)
16. Taylor, A.H., et al.: Stress, fatigue, health, and risk of road traffic accidents among professional drivers: the contribution of physical inactivity. Ann. Rev. Public Health **27**, 371–391 (2006)
17. The Science of Stress. https://www.slma.cc/the-science-of-stress/. Accessed 20 Aug 2020
18. Vakkuri, A., et al.: Time-frequency balanced spectral entropy as a measure of anesthetic drug effect in central nervous system during sevoflurane, propofol, and thiopental anesthesia. Acta Anaesthesiol. Scand. **48**(2), 145–153 (2004)
19. Wang, J.-S., et al.: A k-nearest-neighbor classifier with heart rate variability feature-based transformation algorithm for driving stress recognition. Neurocomputing **116**, 136–143 (2013)

Evaluation of Soil Moisture for Estimation of Irrigation Pattern by Using Machine Learning Methods

Abhishek Khanna(✉) and Sanmeet Kaur(✉)

Computer Science and Engineering Department, Thapar Institute of Engineering and Technology, Patiala 147004, India
abhikhanna@hotmail.com, {abhishek.khanna,sanmeet.bhatia}@thapar.edu

Abstract. The presence of soil moisture is of paramount importance in agricultural domain, as it constitutes largely towards variations in soil texture, and development of crops. Hence, evaluation of this parameter can turn out to be very effective while performing agricultural activities. Mainly, evaluations for the parameter is done with an aim to estimate and reduce water consumption within the fields. In this article, the presence of soil moisture has been evaluated at three different levels, *i.e.*, 10 cm, 45 cm, and 80 cm through Autoregressive Integrated Moving Average (ARIMA) modeling technique based on Time Series Analysis, to predict the future possible values so that precise distribution water can be done within the fields. The intermediate diminution in error rates attained by the modeling technique attained between 74%–77% in comparison to other modeling techniques, depicting its superiority. Based on the results, distribution of water was scheduled in advance as per the minimal requirement, resulting lesser consumption and better crop yields

Keywords: Soil moisture · Smart irrigation · Autoregressive Integrated Moving Average (ARIMA)

1 Introduction

Agricultural domain has witnessed a radical change over the last few years Khanna and Kaur (2019). Over the years, newer concepts like, Internet of Things (IoT), Wireless Sensor & Actuator Network (WSAN), incorporation of drones, *etc.* have been introduced within the domain Khanna and Kaur (2020). All the latest studies conducted during the past few years depict evaluation of agricultural parameters such as soil moisture, luminosity, atmospheric pressure, temperature, humidity, *etc.* Jin and Young (2001). The presence of water plays a significant role while performing agricultural activities. It has been observed that

The authors would like to acknowledge Council of Scientific and Industrial Research (CSIR) for funding grants vide No. 38(1464)/18/EMIR-II for carrying out research work.

M. Singh et al. (Eds.): ICACDS 2021, CCIS 1441, pp. 343–352, 2021.
https://doi.org/10.1007/978-3-030-88244-0_33

farmers blindly irrigate their fields without even estimating the actual requirement. For the same purpose, majority of water is acquired from underground. Due to this, the water level had drastically decreased over the past few years turning into a critical situation Molden (2013). In this situation, the adaptation of precision agricultural practices becomes the only resort to reply upon. One of the most important parameter within this domain is the presence of soil moisture, because consumption of water by the fields is directly associated according to its percentage presence within the soil Khanna (2020). The main objective of irrigating the fields is to maintain the soil moisture level so that the crops can acquire the required amount of water for its growth. Hence, the estimation of soil moisture becomes an important factor, as the process of irrigation can be scheduled in advance to maintain the soil moisture. Incorporation of agricultural sensors can help in providing different parametric values. Furthermore, machine learning or statistical analysis can performed on acquired values to predict the future possible values.

An another advantage to this practice is that it helps in retaining of water from misuse. This process would also help in soil and crop's health. With a similar aspect in mind, this study aims at retaining the water by estimating the future possible values for soil moisture through Autoregressive Integrated Moving Average (ARIMA) modeling technique. The next subsection, highlights some of the research articles published in similar context.

1.1 Existing Studies

Over the years, several techniques have been depicted through various research articles to predict soil moisture within the fields. Atluri et al. (1999) in their study proposed Artificial Neural Network (ANN) was proposed to predict the soil moisture using remotely sensed data. In their study, Levenberg Marquardt and Back Propagation Learning (BPL) were applied. Results of the study proved that Levenberg Marquardt algorithm is faster than the BPL technique. Song et al. (2008) in their study proposed a method to predict soil moisture using Support Vector Machine (SVM). The performance of SVM was compared with back propagation based Artificial Neural Network (ANN). The performance was compared in terms of Root Mean Square of Error (RMSE) and Mean Absolute Error (MAE). Gill et al. (2006) in their study proposed a technique to predict the presence of soil moisture through ANN based Feature Selection Technique. Esmaeelnejad et al. (2015) in their study proposed pedotransfer function to predict soil moisture using the ANN in northern area of Iran. Matei et al. (2017) in their study proposed data mining based soil moisture prediction method. The proposed technique gathered weather data from different stations with an aim to analyze and predict the soil moisture for the next day. Results for the study suggested that the K-nearest neighbour (KNN) provided better result than SVM, ANN, and logistic regression. Chatterjee et al. (2018) in his study proposed ANN based technique to predict soil moisture. ANN was trained by a modified flower pollination algorithm. In this paper, the results of soil moisture prediction of ANN with modified flower pollination algorithm was compared with Particle Swarm

Optimization (PSO) supported ANN and Cuckoo search supported ANN. On similar evaluating methodology, Abdel-Fattah et al. (2021) in their study evaluated upon the quality of water drainage for irrigation purpose was evaluated using ARIMA modeling technique and other integrated approaches and Artificial Neural Network (ANN) model. Kumar et al. (2020) in their study installed IoT embedded sensors in an agricultural habitat to capture soil moisture, humidity, light, and temperature. This experiment was conducted in a farmhouse of Chennai where a special cultivable zone positioned in the central area of the field of vegetarian crops and watered horticulture. This experiment aimed to analyze the patterns of the crop and maintaining the novice quality of the crops such as well-organized blooming stem, well-hygienic plants.

In all the above studies, soil moisture has been the area of prime focus. It has also been concluded that consumption of water is blindly done within the fields, irrespective of estimating or understanding farm's minimal requirement. Hence keeping all the factors, current scenarios, and previous studies under consideration, this article aims at evaluating future possible values for the fields, in order to estimate the precise requirement of water for irrigating the fields by creating a test bed. The next section depicts the methodology adopted for gathering values related to various parameters, to estimate the future possible values and to schedule the distribution of water accordingly.

2 Materials and Methods

Time series analysis is a technique which comprises of various methods that are used for an analyzing time series data in order to extract meaningful statistics and other characteristics for the data. Since time series analysis has natural temporal ordering, hence it makes it an easy technique in context to other computational techniques. An another advantage of Time Series Analysis is that it is distinct from spatial data analysis where the observations typically relate to geographical locations, hence it makes it easy to adapt the theory and analyze the current standings for zone. In order to evaluate the parameters, a test bed was identified the Malwa region of Punjab, *i.e.*, Patiala and parameters acquired by incorporating Libelium's Waspmote Plug & Sense device. Autoregressive Integrated Moving Average (ARIMA) modeling technique was taken in consideration for evaluating the parameters. The experimentation was performed on wheat crop. To estimate the presence of soil moisture within the fields, dughalls at 10 cm, 45 cm, and 8.0 cm were done to obtain values at different levels of soil. Apart from soil moisture, four other parameters were also considered for evaluation. They were humidity, luminosity, soil temperature, and atmospheric pressure. All the sensed data was recorded by the device on its built in memory card and later the recorded data was transmitted to www.smartfasal.in. This was a dedicated portal to precure and access the real-time readings for both prediction and future reference purpose. Recordings were generated by the sensors every $3^r d$ second. The acquired values could be retrieved over .csv format from the portal. Figure 1 depicts the pictorial image for the hardware device mounted within the field for generating values (Fig. 2).

Fig. 1. Libelium's Waspmote Plug & Sense device deployed within the field

A1 | 1573839142

	A	B	C	D	E	F	G	H	I	J
16	1573840267	0	2688.17	3676.47	22.31	63.9	98692.94	0	6	15
17	1573840342	0	2688.17	3676.47	22.27	63.57	98694.5	0	6	16
18	1573840417	0	2688.17	3676.47	22.22	62.7	98696.08	0	6	17
19	1573840492	0	2688.17	3676.47	22.21	63.1	98687.42	0	6	18
20	1573840567	0	2702.7	3649.63	22.19	65.5	98694.15	0	6	19
21	1573840642	0	2688.17	3649.63	22.18	64.88	98697.9	0	6	20
22	1573840717	0	2688.17	3676.47	22.14	65.7	98691.39	0	6	21
23	1573840792	0	2702.7	3649.63	22.09	63.56	98693.22	0	6	22
24	1573840867	0	2702.7	3649.63	22.05	63.56	98692.1	0	6	23
25	1573840942	0	2688.17	3676.47	22.01	66.35	98692.58	0	6	24
26	1573841017	0	2702.7	3649.63	21.98	64.3	98694.98	0	6	25
27	1573841092	0	2702.7	3649.63	21.93	63.34	98695.72	0	6	26
28	1573841167	0	2688.17	3676.47	21.88	63.3	98699.72	0	6	27
29	1573841242	0	2688.17	3649.63	21.86	66.64	98692.38	0	6	28
30	1573841317	0	2688.17	3649.63	21.84	65.09	98698.85	0	6	29
31	1573841392	0	2688.17	3676.47	21.82	63.68	98692.06	0	6	30
32	1573841467	0	2688.17	3649.63	21.79	64.74	98686.06	0	6	31
33	1573841542	0	2702.7	3649.63	21.77	63.36	98686.58	0	6	32
34	1573841617	0	2702.7	3649.63	21.75	63.64	98686.25	0	6	33
35	1573841692	0	2688.17	3649.63	21.7	65.87	98687.81	0	6	34
36	1573841767	0	2688.17	3649.63	21.68	67.27	98691.01	0	6	35
37	1573841842	0	2688.17	3649.63	21.62	64.43	98684.49	0	6	36
38	1573841917	0	2688.17	3676.47	21.59	63.69	98682.52	0	6	37
39	1573841992	0	2702.7	3649.63	21.57	63.06	98675.49	0	6	38
40	1573842067	0	2702.7	3649.63	21.55	65.24	98677.82	0	6	39

S_AgriA (7)

Fig. 2. Readings gathered and stored at the dedicated portal

The next section depicts the conceptual functioning of the modeling technique incorporated within the study to predict the future possible values.

2.1 Autoregressive Integrated Moving Average (ARIMA) Modeling Technique

ARIMA model is composed of three parameters such as Auto regressive (AR) - p, Differencing - d, and Moving Average (MA) - q. Auto-regression is a time series model that uses previous time steps of the data observations as a new input to the data and a regression equation is used to predict the next time step. Hence, Automatic Regression is a simplified term that provides accurate results to forecasts a range of data. A simple formula of autoregressive is depicted in Eq. 1.

$$yhat = B0 + B1 * xi \tag{1}$$

where the *yhat* is the term to be predicted, *B0* and *B1* are coefficients terms. These coefficients terms are founded by optimizing the model using the dataset, and X_i is an input value at the ith term. This technique is mostly used with the lag variables. The lag variable is defined as a dataset that takes input variable as observations of the previous time steps. As an instance, the next observation of the dataset is predicted using two previous observations of the dataset and the regression model is used to predict the next observation. Equation 2 depicts the working of the autoregressive model for three observations.

$$X(t+1) = B0 + B1 * x(t-1) + B2 * x(t-2) \tag{2}$$

This regression model is also called as the auto-regression (regression of self) when it uses data from the same input variable at previous time steps.

The differencing is the number of times that a dataset must be differenced from the original by a lag value to make a dataset in a stationary form, it is also called as the degree of differencing. The differencing of the data depicts the reshaping of the data to the next lag observations to transform it into a stationary format of the data.

The Moving Average is the series of the mean value of the consecutive different subset of datasets. It includes the average value of the n consecutive terms, where n is a value used to include the number of transactions of the last n terms, that represents as a window's width. It depicts the number of n observation is used to calculate the moving average value. The "moving" part of the moving average refers to the ongoing window width of the data that increments the element for the next n observations by one transaction and decrements the last observation by one. An example of the *MA* is depicted in Eq. 3.

$$MA(t) = mean(obs(t-1), obs(t), obs(t+1)) \tag{3}$$

The ARIMA model uses the linear regression model to constructs the ACF and PACF to compute the values for the p(AR) and q(MA) that used to make the dataset stationary by changing the degree of the differencing that helps to remove the trends and the seasonal structure of the data. A formula of the ARIMA is shown in Eq. 4.

$$\triangle Yt = pi\triangle yt - 1 + di\triangle yt - 1 + qi\triangle yt - 1 \tag{4}$$

Equation 4 elaborates the mathematically working of the ARIMA model, here p_i includes all the Autoregressive values, d_i has a degree of differencing that manipulate the dataset by a lag order, while the q_i computes the Moving average for the model Haiges et al. (2017).

3 Results and Discussion

The proposed model has been trained using the data acquired and the predictions from the model have been projected using graphical interface. The required libraries and directories which have been used for the model building such as *numpy, pandas,* and *sklearn* that were used for arrays and dataset processing, *matplotlib* for graph plotting. These predictions were inversely molded to the normal dataset for actual values and saved onto a new variable. In the final step, the predicted values are compared with the testing dataset (of the proposed model) to compute the error rate of the proposed model.

Soil Moisture

Soil moisture content is a value that determines the presence of amount of water within the soil. This presence can either be measured by the weight, volume of soil, or inches of water per foot of soil. Sufficient levels of soil moisture is an important factor for proper nourishment of plant and its formation into high crop yields. The presence of water within soil no only acts as an agent of moisture restoration, but also as a temperature regulator. In addition, the presence of water within soil has following purposes:

- Soil moisture levels affect air content, salinity, and the presence of toxic substances.
- Regulates soil structure, ductility, and its density.
- Influences soil temperature, and heat capacity.
- Prevents soil from weathering.
- Determines the readiness of fields to be worked upon.

Based on the importance of its presence, observations were conducted to acquire the values pertaining to the precise presence soil moisture within the test bed. The hardware device generated a reading after every 3rd second from all the sensors. For one complete month readings were acquired. For 10 cm minimum and maximum temperature reading acquired was 15.41 g/m^3 and 429.92 g/m^3 respectively. Similarly, for 45 cm and 80 cm, readings that were observed were 2380.98 g/m^3 and 3246.75 g/m^3, 2873.56 g/m^3 and 3381.25 g/m^3 respectively. Table 1 depicts the maximum and minimum values that were obtained through sensors. On these acquired readings, ARIMA modeling technique was implemented. The results depicted that the readings and drawings observed for 10 cm was perfectly accurate, however the readings that were observed for 45 cm and 80 cm were slightly onto the higher side than expected, *i.e.*, the presence of soil moisture was relatively more than the one required. Figure 3, Fig. 4, and Fig. 5 depicts the graph for actual readings and the predicted values.

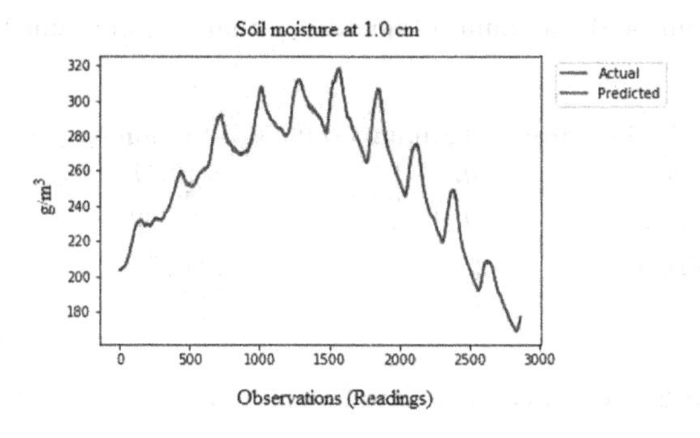

Fig. 3. Soil moisture evaluation at 10 cm

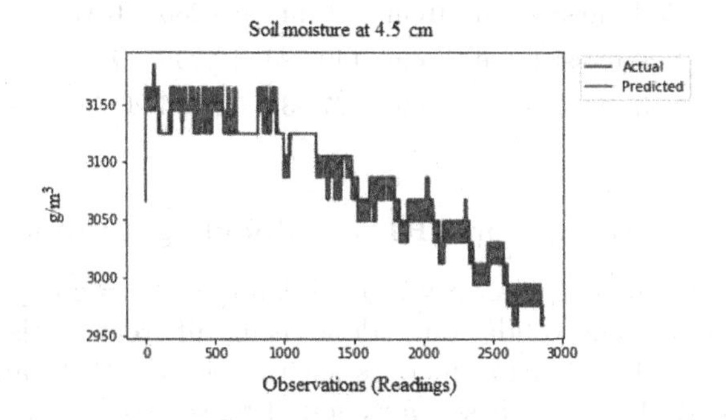

Fig. 4. Soil moisture evaluation at 45 cm

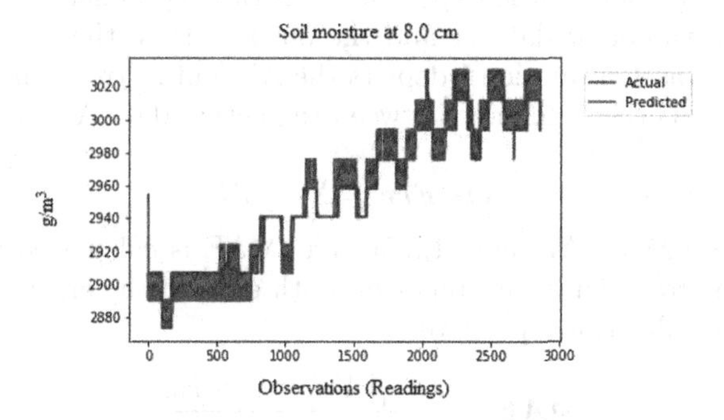

Fig. 5. Soil moisture evaluation at 80 cm

The forecasting model evaluated the data and provided possible error rate, based on which the proficiency of the model was determined. The less the error rate, the more appropriate the results are. The error rates obtained by the model at 10 cm for all the three parameters, *i.e.*, *MAE, MSE,* and *RMSE* were 0.143, 0.285, & 0.378 respectively. Table 2 depicts the evaluated error rate for soil moisture at all three levels.

Table 1. Maximum and minimum observations that were obtained by all the three sensors

Soil moisture level	Maximum value (g/m^3)	Minimum value (g/m^3)
10 cm	151.42	429.92
45 cm	2380.95	3246.75
80 cm	2873.56	3381.25

Table 2. Evaluated error rate for soil moisture (at all depths)

Depth level	MAE	MSE	RMSE
Soil moisture at 10 cm	0.143	0.285	0.378
Soil moisture at 45 cm	149.693	12.235	7.526
Soil moisture at 80 cm	127.833	7.096	11.306

3.1 Validation Strategy for ARIMA Modeling Technique

The validation of a model is a very crucial component to compute the performance of the model. The validation is done using different statistical models of the performance and evaluation metrics such as *MAE, MSE,* and *RMSE* Fischer and Krauss (2018). The reason for computing several metrics is to compare the proposed study to another forecasting model. The lower value of the error depicts the best model for the experiment. In the experiment, the testing data set is used as a validated dataset and the outputs from the model are used as the predicted dataset. Equation 5 depicts the Absolute Error. The *Ab Err* is the absolute error, it is the difference between the actual data *Xi* and the predicted data *Yi.*

$$\boldsymbol{AB\ Err} = Xi - Yi \tag{5}$$

Equation 6 is the Mean Absolute Error. The MAE is calculated as the average of the absolute error. There are pipes on both ends of the error that makes the nature of this result always positive.

$$\boldsymbol{MAE} = \frac{\Sigma_{i=1}^{n}|AbsoluteError|}{n} \tag{6}$$

Equation 7 equated the MSE. MSE is the minimum square error. It is the square mean error of the absolute error.

$$\boldsymbol{MSE} = \frac{\Sigma_{i=1}^{n}(AbsoluteError)^2}{n} \tag{7}$$

Equation 8 is the Root Means Square of an Error value. RMSE is the mean of the square root of differences between actual readings and predicted readings.

$$\boldsymbol{RMSE} = \frac{\sqrt{\Sigma_{i=1}^{n}(AbsoluteError)}}{n} \tag{8}$$

3.2 Comparative Analysis with Other Modeling Techniques

In order to validate the superiority for the proposed model and its efficiency, the acquired dataset was evaluated on two other modeling techniques, *i.e.*, Long Short Term Modeling (LSTM), & Pophet. Results of LSTM depicted a higher error rates in comparative to ARIMA model. Moreover, ARIMA results depicted that they were better suited for short term forecasting, in contrast to the results obtained from LSTM. Similarly, when the results of ARIMA were compared with the results obtained from Prophet modeling technique, it was concluded that this model was best suited for non-linear data trends, whereas the dataset generated by the hardware device was linear. Hence, ARIMA scored better on linear data in comparison to Prophet modeling technique.

For all the three cases, *i.e.* MAE, MSE, & RMSE the error rates generated by ARIMA modeling technique were comparatively less than the other two models. Hence, proving its superiority.

4 Conclusion

Based on acquired results, it has been concluded that the modern day agricultural practices are driven by continuous improvements in digital tools and on the basis of acquired data. In today's date it is very much important to first evaluate provide all the important parameters to have a pragmatic identification of all the aspects that are involved within agricultural practices. Based on the results, it is easy to estimate the future possible values, and conditionally favorable scenario for performing agricultural activities. Furthermore, it has been concluded that the observations are relatively close to the ones that are relatively available, that are required for performing fruitful agricultural activities. The average reduction in error rates obtained by ARIMA were found within a range of 74%–77% giving estimation regarding precise requirement of water distribution within fields.

References

Abdel-Fattah, M.K., Mokhtar, A., Abdo, A.I.: Application of neural network and time series modeling to study the suitability of drain water quality for irrigation: a case study from egypt. Environ. Sci. Pollut. Res. **28**(1), 898–914 (2021)

Atluri, V., Hung, C.-C., Coleman, T.L.: An artificial neural network for classifying and predicting soil moisture and temperature using Levenberg-Marquardt algorithm. In: Proceedings IEEE Southeastcon 1999. Technology on the Brink of 2000 (Cat. No. 99CH36300), pp. 10–13. IEEE (1999)

Chatterjee, S., Dey, N., Sen, S.: Soil moisture quantity prediction using optimized neural supported model for sustainable agricultural applications. Sustain. Comput. Inform. Syst. **28**, 100279 (2018)

Esmaeelnejad, L., Ramezanpour, H., Seyedmohammadi, J., Shabanpour, M.: Selection of a suitable model for the prediction of soil water content in north of Iran. Span. J. Agric. Res. **13**(1), 1202 (2015)

Fischer, T., Krauss, C.: Deep learning with long short-term memory networks for financial market predictions. Eur. J. Oper. Res. **270**(2), 654–669 (2018)

Gill, M.K., Asefa, T., Kemblowski, M.W., McKee, M.: Soil moisture prediction using support vector machines 1. JAWRA J. Am. Water Resour. Assoc. **42**(4), 1033–1046 (2006)

Haiges, R., Wang, Y., Ghoshray, A., Roskilly, A.: Forecasting electricity generation capacity in Malaysia: an auto regressive integrated moving average approach. Energy Procedia **105**, 3471–3478 (2017)

Jin, L., Young, W.: Water use in agriculture in China: importance, challenges, and implications for policy. Water Policy **3**(3), 215–228 (2001)

Khanna, A.: Agro-based sensor's deployment for environmental anticipation: an experimental effort for minimal usage of water within agricultural practices. Culture **4**(3), 219–236 (2020)

Khanna, A., Kaur, S.: Evolution of internet of things (IoT) and its significant impact in the field of precision agriculture. Comput. Electron. Agric. **157**, 218–231 (2019)

Khanna, A., Kaur, S.: Internet of things (IoT), applications and challenges: a comprehensive review. Wireless Pers. Commun. **114**, 1687–1762 (2020)

Suresh Kumar, K., Balakrishnan, S., Janet, J.: A cloud-based prototype for the monitoring and predicting of data in precision agriculture based on internet of everything. J. Ambient. Intell. Humaniz. Comput. **12**(9), 8719–8730 (2020). https://doi.org/10.1007/s12652-020-02632-5

Matei, O., Rusu, T., Petrovan, A., Mihuţ, G.: A data mining system for real time soil moisture prediction. Procedia Eng. **181**, 837–844 (2017)

Molden, D.: Water for Food Water for Life: A Comprehensive Assessment of Water Management in Agriculture. Routledge (2013)

Song, J., Wang, D., Liu, N., Cheng, L., Du, L., Zhang, K.: Soil moisture prediction with feature selection using a neural network. In: 2008 Digital Image Computing: Techniques and Applications, pp. 130–136. IEEE (2008)

Sentiment Analysis in Online Learning Environment: A Systematic Review

Sarika Sharma[1], Vipin Tyagi[2](✉), and Anagha Vaidya[1]

[1] Symbiosis Institute of Computer Studies and Research, Symbiosis International (Deemed University), Atur Centre, Model Colony, Pune 411016, India
sarika.sharma@sicsr.ac.in

[2] Jaypee University of Engineering and Technology, Raghogarh-Guna 473226, MP, India

Abstract. The global pandemic of COVID-19, has impacted various sectors around the globe, including education sector. It has compelled the educators and learners to go for the teaching/learning activities in online mode, rather than traditional face to face teaching. The technology-enabled interactions can be effective only when the student teacher bonding is created and the sentiments of the learners are understood fully. To be prepared for such outbreaks in future is the need of hour. The study imbibes the role of sentiment analysis with the introduction of what it means and how it can help in such outbreaks in an online learning environment. Recently few studies are being contributed for covering the various aspects of this evolving area of sentiment analysis. The literature however is scattered and unorganized, therefore there is a need to conduct a systematic literature review to compile all the relevant studies together and to arrange it in a framework. This paper attempts towards this to provide better insight on the usage of sentiment analysis for education sector. The outcome of this paper is a step towards proposal of future areas of the research in this emerging field.

Keywords: Sentiment analysis · Online learning · Text analytics · Higher education · Virtual learning

1 Introduction

Education sector is evolving with more and more usage of information and communication technologies (ICT) in teaching, learning, and evaluation. Since last one year, because of COVID-19 pandemic, various digital collaborative platforms such as zoom, Microsoft teams, google meet etc. are being adopted by the educational institutes to continue the teaching activities. Most of the institutes are forced to replace the traditional method of classroom teaching in physical mode with this new normal of online, virtual, or e-learning system. For effectiveness of any system it is vital to know the user's acceptance and attitude towards that system. The users here are students who are taking technology enabled education. Teachers often find it difficult to understand the learning issues faced by students in their virtual classroom settings. A solution for this can be adoption of a Sentiment Analysis (SA) methodology for the detection of the classroom

M. Singh et al. (Eds.): ICACDS 2021, CCIS 1441, pp. 353–363, 2021.
https://doi.org/10.1007/978-3-030-88244-0_34

mood during the learning process [1]. Sentiment analysis is proven to be an effective method to do so [2].

In the recent years a swift inclination towards the area of sentiment analysis is being seen. [3]. Sentiment analysis, also called opinion mining, "is the field of study that analyses people's opinions, sentiments, evaluations, appraisals, attitudes, and emotions towards entities such as products, services, organizations, individuals, issues, events, topics, and their attributes". The roots emerges from public opinion analysis in the era of twentieth century. Natural language Processing give rise to Sentiment Analysis, which is the way towards AI based text analysis. Today the buzz word is data analysis and data scientist, who looks for extraction of meaning from unstructured data, sentiment analysis is the first step to provide additional information with lesser cost and time. [4] Describes the sentiment analysis, "which combines rule based classification, supervised learning and machine learning into a new combined method known as hybrid classification". The authors also proposed a semi-automatic complementary approach for achieving the good level of effectiveness wherein each classifier is contributing to the other classifier. There are various levels of sentiment analysis. As per [5] sentiment analysis is the task of classification of target document into positive or negative. It can be classified into three main classifications of document, sentence, and aspect levels. The document level represents about the whole document expressing positive or negative sentiment whereas the sentence level categorizes the sentiments of each sentence and lastly the entity level categorizes the sentiments on entity and level for example what exactly likes and dislikes of people opinion are on a particular aspect.

There has been an increase in interest in exploring the learning experiences of students through the study of emotions, monitoring, expressions, and interactions in context of online learning environments [6–8]. Various researchers have contributed to this imperative area of sentiment analysis. The present study attempts to consider and analyze the relevant studies on SA and in online education domain through a systematic literature review of journal papers from prominent databases. The paper is arranged as follows: Introduction part talks about the definitions and foundation of the topic undertaken. Research gaps are identifies and accordingly research questions are set in next section. The methodology section describes the steps followed for conducting the systematic literature review. The outcome is in form of a framework for SA. The paper culminates with the suggestion of further research avenues.

2 Research Gap and Objective

The sentiment analysis for online education is an emerging area and hence attracted research studies highlighting its various dimension. [9] presented a literature survey on educational research, although not specifically for online education. [10] considered sentiment analysis e-learning for distance education. [11] explained about the techniques for improving sentiment analysis for the same. [12] considered the situation of COVID-19 as emergency online, and put forward an empirical investigation of sentiments from teacher's perspective. [13] conducted a qualitative and quantitative analysis on sentiment analysis of the students in online teaching.

It is observed from the above discussion that although various studies are contributed on the said topic, the research is however scattered among the methods, stakeholders' approach and the techniques of sentiment analysis. The literature reviews conducted are not holistic in nature. Therefore there exists a gap and need of a robust systematic review of literature which can bridge this gap. The presents study aims to fill this gap through a systematic approach so that relevant studies are identified and are included for the review. We posit these research questions, which this study will address:

RQ1: What is the present state of research in field of sentiment analysis for online education?
RQ2: What approach is applied to select the studies included in the Literature review?
RQ3: Can the selected studies be arranged in a framework?

3 Methodology of Systematic Literature Review

A systematic literature review should follow a procedure. There are various guidelines and steps for systematic literature review. We followed the steps suggested by [14] which is widely accepted in scientific studies. It comprises of five steps which are adopted and are followed as given below. The procedure is also represented in a pictorial form for clarity (Fig. 1).

Step 1: Research Objectives
The main objective of this review is to propose a conceptual framework for sentiment analysis in online teaching environment. The three research questions are framed and presented in the previous section.

Step 2: Database Search
This step is about identification of database for further search for relevant articles using a search string formed by the researchers. The present research connects the two areas of Sentiment analysis and online learning environment. The database selected is Scopus as it suits the multi-disciplinary nature of the study. Scopus database provides quality research papers and offers a wide range of research studies, particularly in emerging fields, which makes Scopus a natural choice of databases for most of the literature review based studies.

For the search string formation, first the keywords are identified from both the areas a) Sentiment analysis b) online education/learning/teaching, e-learning, distance learning. Considering keywords from both 'a' and 'b' database is searched using following search string:

"Sentiment Analysis" AND ("education" OR "teaching" OR "e-learning" OR "distance learning")

The search is conducted in March 2021. The initial search for keywords was conducted with this string, which resulted in search resulted in 330 number of articles.

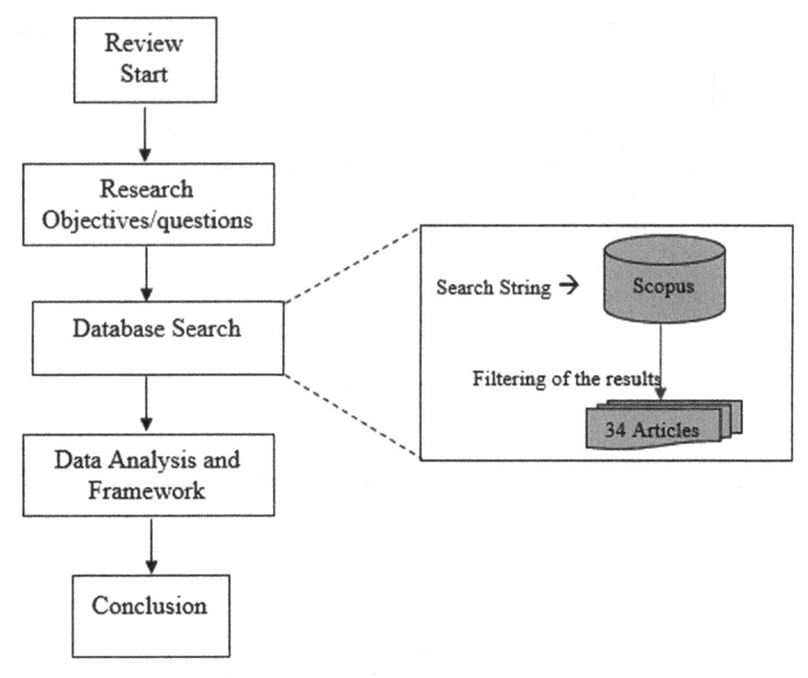

Fig. 1. Systematic review process steps

Step 3: Assessment of Studies for Quality and Further Inclusion
Out of these 330 articles we included articles from journals only and in English languages, which resulted in 88 articles. These 88 articles' abstract were visited to decide on the relevance and suitability of the article for further inclusion. All three researchers individually carried out this activity, and finally all outcomes were compiled together, some difference of opinions were sorted out with further discussions among the researchers and a common conscience was formed. This exercise led to a number of 34 articles found to be in scope of the study. Articles talking about sentiment analysis, educational data mining, machine learning with respect to online teaching/learning were included for further analysis. It can be observed that the field on sentiment analysis in online educational environment is quite resent as all 34 articles are between the time duration of 2014–2021 (see Fig. 2). An excessive expanse of studies can be noted from year 2019 onwards.

Step 4: Analysis and Compilation
The selected 34 research articles are presented with their major findings in Table 1.

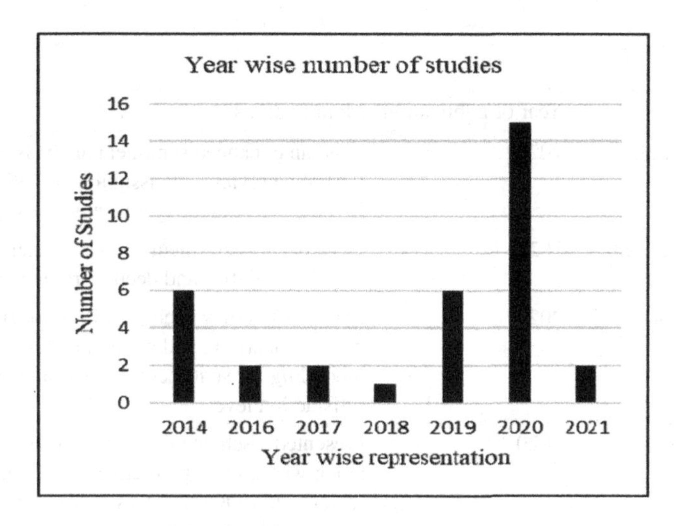

Fig. 2.

Table 1.

Author(s)	Year of publication	Main findings
[15] Martinho et al.	2021	Proposed conceptual model for sentiment analysis for emergency online teaching adopted due to pandemic
[16] Dina et al.	2021	Designed a framework for Massive Online Open courses (MOOCs) for sentiment analysis of users and using text mining for the same
[17] Zhang et al.	2020	To understand the emotional changes and sentiments of the users the authors developed a bidirectional encoder representation from transformers (BERT) using an experimental approach
[18] Madani et al.	2020	For the adaptive e-learning platforms, researchers used social filtering method for sentiment analysis for adoption of certain online courses
[19] Amala Jayanthi and Elizabeth Shanthi	2020	Analyzed the educational data mining for sentiment analysis of students. Adopted various theoretical approaches for education and sentiment analysis
[20] Omar et al.	2020	Extracted sentiments for huge e-learning platforms using machine learning, unsupervised learning, Lexicon based approach using the programming language Python

(continued)

(*continued*)

Author(s)	Year of publication	Main findings
[21] Saeed et al.	2020	Used an enhanced sentiment analysis approach for online reviews by assigning a sentiment score to the positive and negative emotions
[22] Spatiotis et al.	2020	Developed a sentiment analysis system using opinion analytics and deep learning approach
[23] Hew et al.	2020	For MOOCs they used algorithm for supervised machine learning and hierarchical linear modelling for sentiment analysis and users' satisfaction level
[24] Onan	2020	Presented a scheme for sentiment classification along with a great predictive performance for reviews in MOOCs. It was achieved with deep and ensemble learning
[25] Gkontzis et al.	2020	Implemented data mining techniques for learning management systems to analyze emotions of teachers as well as students
[26] Kastrati et al.	2020	Proposed a framework for analysis of students' views based on aspect-level sentiment analysis through deep learning techniques on MOOCs
[27] Okoye	2020	Proposed a model for Educational Process and Data Mining (EPDM). This model influences the views expressed by the learners. The teachers are evaluated by students
[28] Khamparia et al.	2020	Used the techniques classification-support Naïve Bayes and vector machine for sentiment analysis of the students
[29] Ray et al.	2020	A natural language processing based method is applied for the comparison of various MOOCs platforms by using opinion-mining and sentiment analysis for user intentions
[30] Attili et al.	2020	The authors with the implementation of notions of sentiment analysis developed a method. Social and collaborative filtering is also applied
[31] Grljević et al.	2020	Analyzed initially developed a corpus and dictionary. Next for sentiment analysis they applied machine learning-based approaches
[32] Cobos et al.	2019	Used a sentiment Analysis tool called edX-CAS, which also supports polarity detection

(*continued*)

(continued)

Author(s)	Year of publication	Main findings
[33] Sai Tulasi and Deepa	2019	Used sentiment analysis using learning inspect for calculating students various emotions and learning style
[34] Elia et al.	2019	Implemented the concept of learning analytics with big data for reals time insights of sentiments
[35] Huang et al.	2019	Adopted the algorithm of Constructing Base Classifiers based on Product Attributes to combine the sentiment information of each attribute in a review so as to obtain better performance on sentiment classification
[36] Yuan	2019	Used machine learning algorithms (PMI and SVM) for lexicon sentiment analysis and text sentiment orientation
[37] Sahu et al.	2019	The applicability of sentiment analysis and opinion mining is shown in the paper for the evaluation of teaching learning process. The proposed computational model of sentiment analysis is an automated system to analyze the textual feedback of faculty submitted by students
[38] Suwal and Singh	2018	Researchers analyzed the sentiments of students towards online platforms using text analytics methods
[39] Shapiro et al.	2017	Learners' engagement and experience is analyzed employing text analysis in context of MOOCs
[40] Mandal et al.	2017	Using machine learning approach for analysis of emotions in e-learning. A sentiment analysis classifier for polarity scoring and a support vector machine was deployed for the same
[41] Al-Rubaiee et al.	2016	Authors used text classifiers for opinion mining of students with the help of algorithms: Support Vector Machine and Naive Bayes
[42] Zarra et al.	2016	Applied the Latent Semantic Analysis model to establish a relationship between messages published by learners and course content
[43] Sun et al.	2014	Presented 'feature frequency-adaptive on-line training' method for natural language processing systems and identified it as a new method of training

(continued)

(*continued*)

Author(s)	Year of publication	Main findings
[44] Ortigosa et al.	2014	Used 'facebook' application 'SentBuk' for classification method implementation for sentiment analysis of users. 'Sentbuk' is a combination of machine-learning and lexicon techniques implemented for e-learning
[45] Colace et al.	2014	Adopted probabilistic approach based on 'Latent Dirichlet Allocation' for sentiment grabbing in context of e-learning
[46] Slavakis et al.	2014	The researchers implemented 'Stochastic approximation' for online learning for big data analytics
[47] Moreno-Jiménez et al.	2014	Adopted opinion analysis techniques and developed a new tool for extracting arguments
[48] Ravichandran and Kulanthaivel	2014	Suggested a method for sentiment mining for twitter messages to extract learners sentiment polarity and to model it to monitor the emotion changes using Naïve Bayesian approach

Step 5: Interpretation and Proposed Framework
The research articles are first visited for an overall understanding of them. Qualitative analysis is carried out using conductive thematic analysis, based on which a framework is proposed by the researchers. This framework is presented below:

I Tools for sentiment analysis
  I(A) Supervised
  I(B) Unsupervised
II Techniques of sentiment analysis
III Application domain.

4 Conclusion and Future Scope

This research paper makes a valuable contribution to the theoretical foundations of research on sentiment analysis by arranging the literature on its application in the area of online learning environment, in a conceptual framework. Although researchers have tried to compile all relevant literature it cannot be claimed as holistic one. Therefore this is a limitation of the study. Another limitation is that only one database that is Scopus is included for the search of the studies. Other prominent database such as web of science, and EBSCO can also be considered for further insights.

Researchers have identified some future research areas based on the framework and literature review:

- Learner's satisfaction for online courses and evaluation process in online mode is to be done through sentiment analysis.
- A comparative study of various sentiment analysis techniques for their effectiveness can be taken up for future.
- Instructor's sentiments are of importance too, but there are few studies for the same. Further it can be explored with empirical research.
- In-fact the collective study of sentiment analysis of all stakeholders (Learners, instructors, management) will be of great importance.

References

1. Clarizia, F., Colace, F., De Santo, M., Lombardi, M., Pascale, F., Pietrosanto, A.: E-learning and sentiment analysis: a case study. In: Proceedings of the 6th International Conference on Information and Education Technology (ICIET 2018). Association for Computing Machinery. In book: Cyberspace Safety and Security, pp. 291–302 (2018)
2. Kechaou, Z., Mahmoud, A.B., Alimi, A.: Improving e-learning with sentiment analysis of users' opinions. In: Proceedings, 2011 IEEE Global Engineering Education Conference (EDUCON), pp. 1032–1038 (2011)
3. Mäntylä, M.V., Graziotin, D., Kuutila, M.: The Evolution of sentiment analysis. Comput. Rev. **27**, 16–32 (2018)
4. Prabowo, R., Thelwall, M.: Sentiment analysis: a combined approach. J. Inform. **3**(2), 143–157 (2009)
5. Ray, P., Chakrabarti, A.: A mixed approach of deep learning method and rule-based method to improve aspect level sentiment analysis. Appl. Comput. Inform. (2020, Article in press)
6. Arguel, A., Lockyer, L., Lipp, O.V., Lodge, J.M., Kennedy, G.: Inside out: detecting learners' confusion to improve interactive digital learning environments. J. Educ. Comput. Res. **55**(4), 526–551 (2017)
7. Lajoie, S.P., Pekrun, R., Azevedo, R., Leighton, J.P.: Understanding and measuring emotions in technology-rich learning environments. Learn. Instruc. **70**, 101272 (2020)
8. Malekzadeh, M., Mustafa, M.B., Lahsasna, A.: A review of emotion regulation in intelligent tutoring systems. Educ. Technol. Soc. **18**(4), 435–445 (2015)
9. Zhou, J., Jun-min, Y.: Sentiment analysis in education research: a review of journal publications. Interactive Learning Environment. Published online: 01 Oct 2020
10. Clarizia, F., Colace, F., De Santo, M., Lombardi, M., Pascale, F., Pietrosanto, A.: E-learning and sentiment analysis: a case study. In Proceedings of the 6th International Conference on Information and Education Technology (ICIET 2018), pp. 111–118. Association for Computing Machinery, New York (2018)
11. Lin, X.-M., Ho, C.-H., Xia, L.-T., Zhao, R.-Y.: Sentiment analysis of low-carbon travel APP user comments based on deep learning. Sustain. Energy Technol. Assess. **44**, 101014 (2021)
12. Martinho, D., Sobreiro, P., Vardasca, R.: Teaching sentiment in emergency online learning-a conceptual model. Educ. Sci. **11**(53), 2–16 (2021)
13. PraveenKumar, T., Manorselvi, A., Soundarapandiyan, K.: Exploring the students feelings and emotion towards online teaching: sentimental analysis approach. In: Sharma, S.K., Dwivedi, Y.K., Metri, B., Rana, N.P. (eds.) TDIT 2020. IAICT, vol. 617, pp. 137–146. Springer, Cham (2020). https://doi.org/10.1007/978-3-030-64849-7_13
14. Khan, K.S., Kunz, R., Kleijnen, J., Antes, G.: Five steps to conducting a systematic review. J. R. Soc. Med. **96**(3), 118–121 (2003)

15. Martinho, D., Sobreiro, P., Vardasca, R.: Teaching sentiment in emergency online learning—a conceptual model. Educ. Sci. **11**(2), 1–16 (2021)
16. Dina, N.Z., Yunardi, R.T., Firdaus, A.A.: Utilizing text mining and feature-sentiment-pairs to support data-driven design automation massive open online course. Int. J. Emerg. Technol. Learn. **16**(1), 134–151 (2021)
17. Zhang, H., Dong, J., Min, L., Bi, P.: A BERT fine-tuning model for targeted sentiment analysis of chinese online course reviews. Int. J. Artif. Intell. Tools **29**, 7–8 (2020)
18. Madani, Y., Ezzikouri, H., Erritali, M., Hssina, B.: Finding optimal pedagogical content in an adaptive e-learning platform using a new recommendation approach and reinforcement learning. J. Ambient Intell. Hum. Comput. **11**(10), 3921–3936 (2019). https://doi.org/10.1007/s12652-019-01627-1
19. Amala, J.M., Elizabeth, S.I.: Role of educational data mining in student learning processes with sentiment analysis: a survey. Int. J. Knowl. Syst. Sci. **11**(4), 31–44 (2020)
20. Omar, M.A., Makhtar, M., Ibrahim, M.F., Aziz, A.A.: Sentiment analysis of user feedback in e-learning environment. SSRG Int. J. Eng. Trends Technol. **1**, 153–157 (2020)
21. Saeed, N.M.K., Helal, N.A., Badr, N.L., Gharib, T.F.: An enhanced feature-based sentiment analysis approach. Wiley Interdisc. Rev. Data Min. Knowl. Disc. **10**(2), e1347 (2020)
22. Spatiotis, N., Periko, I., Mporas, I., Paraskevas, M.: Sentiment analysis of teachers using social information in educational platform environments. Int. J. Artif. Intell. Tools **29**(1), 2040004 (2020)
23. Hew, K.F., Hu, X., Qiao, C., Tang, Y.: What predicts student satisfaction with MOOCs: a gradient boosting trees supervised machine learning and sentiment analysis approach. Comput. Educ. **145**, 103724 (2020)
24. Onan, A.: Sentiment analysis on massive open online course evaluations: a text mining and deep learning approach. Comput. Appl. Eng. Educ. (2020, Article in press)
25. Gkontzis, A.F., Kotsiantis, S., Kalles, D., Panagiotakopoulos, C.T., Verykios, V.S.: Polarity, emotions and online activity of students and tutors as features in predicting grades. Intell. Decis. Technol. **14**(3), 409–436 (2020)
26. Kastrati, Z., Imran, A.S., Kurti, A.: Weakly supervised framework for aspect-based sentiment analysis on students' reviews of MOOCs. IEEE Access. **8**, 106799–106810 (2020)
27. Okoye, K., et al.: Impact of students evaluation of teaching: a text analysis of the teachers qualities by gender. Int. J. Educ. Technol. High. Educ. **17**(1), 1–27 (2020)
28. Khamparia, A., Singh, S.K., Luhach, A.Kr., Gao, X.-Z.: Classification and analysis of users review using different classification techniques in intelligent e-learning system. Int. J. Intell. Inf. Database Syst. **13**(2–4), 139–149 (2020)
29. Ray, A., Bala, P.K., Kumar, R.: An NLP-SEM approach to examine the gratifications affecting user's choice of different e-learning providers from user tweets. J. Decis. Syst. (2020, Article is press)
30. Attili, V.R., Annaluri, S.R., Gali, S.R., Somula, R.: Behaviour and emotions of working professionals towards online learning systems: sentiment analysis. J. Amb. Intell. Hum. Comput. **11**(10), 3921–3936 (2020)
31. Grljević, O., Bošnjak, Z., Kovačević, A.: Opinion mining in higher education: a corpus-based approach. Enterpr. Inf. Syst. (2020, Article in press)
32. Cobos, R., Jurado, F., Blazquez-Herranz, A.: A content analysis system that supports sentiment analysis for subjectivity and polarity detection in online courses. Revista Iberoamericana de Tecnologias del Aprendizaje **14**(4), 177–187 (2019)
33. Sai, T.K., Deepa, N.: Sentiment exploration system to improve teaching and learning. Test Eng. Manage. **81**(11–12), 5560–5565 (2019)
34. Elia, G., Solazzo, G., Lorenzo, G., Passiante, G.: Assessing learners' satisfaction in collaborative online courses through a big data approach. Comput. Hum. Behav. **92**, 589–599 (2019)

35. Huang, J., Xue, Y., Hu, X., Jin, H., Lu, X., Liu, Z.: Sentiment analysis of Chinese online reviews using ensemble learning framework. Cluster Comput. **22**(2), 3043–3058 (2018). https://doi.org/10.1007/s10586-018-1858-z
36. Yuan, X.: Emotional tendency of online legal course review texts based on SVM algorithm and network data acquisition. J. Intell. Fuzzy Syst. **37**(5), 6253–6263 (2019)
37. Sahu, Y., Thakur, G.S., Dhyani, S.: Dynamic feature based computational model of sentiment analysis to improve teaching learning system. Int. J. Emerg. Technol. **10**(4), 17–23 (2019)
38. Suwal, S., Singh, V.: Assessing students' sentiments towards the use of a Building Information Modelling (BIM) learning platform in a construction project management course. Eur. J. Eng. Educ. **43**(4), 492–5064 (2018)
39. Shapiro, H.B., Lee, C.H., Wyman Roth, N.E., Li, K., Çetinkaya-Rundel, M., Canelas, D.A.: Understanding the massive open online course (MOOC) student experience: an examination of attitudes, motivations, and barriers. Comput. Educ. **110**, 35–50 (2017)
40. Mandal, L., Das, R., Bhattacharya, S., Basu, P.N.: Intellimote: a hybrid classifier for classifying learners' emotion in a distributed e-learning environment. Turk. J. Electr. Eng. Comput. Sci. **25**(3), 2084–2095 (2017)
41. Al-Rubaiee, H., Qiu, R., Alomar, K., Li, D.: Sentiment analysis of Arabic tweets in e-learning. J. Comput. Sci. **12**(11), 553–563 (2016)
42. Zarra, T., Chiheb, R., Faizi, R., El Afia, A.: Using textual similarity and sentiment analysis in discussions forums to enhance learning. Int. J. Softw. Eng. Appl. **10**(1), 191–200 (2016)
43. Sun, X., Li, W., Wang, H., Lu, Q.: Feature-frequency-adaptive on-line training for fast and accurate natural language processing. Comput. Ling. **40**(3), 563–586 (2014)
44. Ortigosa, A., Martín, J.M., Carro, R.M.: Sentiment analysis in Facebook and its application to e-learning. Comput. Hum. Behav. **31**(1), 527–541 (2014)
45. Colace, F., de Santo, M., Greco, L.: Safe: a sentiment analysis framework for e-learning. Int. J. Emerg. Technol. in Learn. **9**(6), 37–41 (2014)
46. Slavakis, K., Kim, S.-J., Mateos, G., Giannakis, G.B.: Stochastic approximation vis-à-vis online learning for big data analytics. IEEE Signal Process. Mag. **31**(6), 124–129 (2014)
47. Moreno-Jiménez, J.M., Cardeñosa, J., Gallardo, C., De La Villa-Moreno, M.A.: A new e-learning tool for cognitive democracies in the Knowledge Society. Comput. Hum. Behav. **30**, 409–418 (2014)
48. Ravichandran, M., Kulanthaivel, G.: Twitter sentiment mining (TSM) framework based learner's emotional state classification and visualization for e-learning system. J. Theoret. Appl. Inf. Technol. **69**(1), 84–90 (2014)

Image Splicing Forgery Detection Techniques: A Review

Kunj Bihari Meena and Vipin Tyagi(✉)

Jaypee University of Engineering and Technology, Raghogarh, Guna, MP, India

Abstract. The authenticity of digital images is openly challenged today due to the easy availability of various advanced image editing software. The semantic meaning of an image can be changed upto any extent with the help of these software. Image splicing forgery is one of the most popular ways to manipulate the content of an image. In image splicing forgery, two or more images or the parts of the images are used to create a spliced (composite) image. Spliced images can be misused in many ways. Therefore, to revive the trustworthiness of digital images, several efforts are made by researchers to develop various methods to detect image splicing forgery in the last few years. The main objective of this study is to review and analyze the recent work in this area. In this paper, first, a generalized workflow to detect image splicing forgery is presented. Second, this paper categorized the existing image splicing detection methods as hand-crafted feature-based and deep learning-based. Third, various publicly available image datasets are also summarized. Finally, future research directions are provided to help the researchers.

Keywords: Tampering detection · Image forgery · Image forgery detection · Copy-move · Image splicing · Deep learning · CNN

1 Introduction

"A picture is worth a thousand words". This proverb says information represented using a picture or image can be more effective than the information conveyed by thousands of words. However, if the image is manipulated using image editing software, then the same image may carry incorrect information and may harm society in different ways. For example, in 2004, to defame Senator John Kerry (US Secretary of State at that time) a spliced image [6, 12] was circulated by showing that John Kerry had shared an anti-Vietnam War rally with Jane Fonda (Hollywood actress). Very soon it was revealed that the image was tampered with. Later, Farid et al. also verified that the image was spliced by proposing a source light direction-based image splicing detection technique. This technique determines that the source light directions of Kerry and Fonda were 123° and 86°, respectively. Recently, in 2020, a forged image [12] was posted on Twitter to demonstrate that Iranian President Hassan Rouhani shaking hands with US President Barack Obama. However, according to Buzzfeed News, the source image was most likely one of Obama's meetings with an Indian Prime Minister in 2011.

M. Singh et al. (Eds.): ICACDS 2021, CCIS 1441, pp. 364–388, 2021.
https://doi.org/10.1007/978-3-030-88244-0_35

According to Machado et al. [55], a number of fake images political agenda are shared on different social media platforms. During the Brazilian presidential elections held in 2018, many fake images comprising misinformation were circulated on different platforms. A survey [55] concluded that 13.1% of WhatsApp posts were fake during this election. Recently, every individual has witnessed that uncountable fake images or videos or messages were shared around the globe that are pertaining to the COVID-19 pandemic [43].

Today's state-of-the-art computer graphics rendering software allow us to create images with desired scenes having high photorealism. Such types of Computer-Generated (CG) images can be easily misused. Hence, a computer-generated image that is created with malicious intent is considered as image forgery [59–62]. Therefore, a digital image can be considered forged if it is created using computer graphics rendering software with malicious intentions or its content is tampered with to distort the semantic meaning.

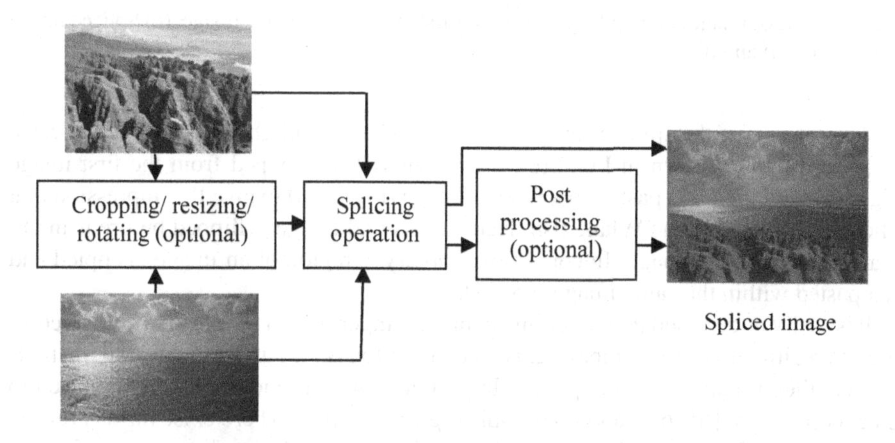

Fig. 1. Different steps to create a spliced image

According to Meena and Tyagi [63], there can be two types of image tampering: legitimate and illegitimate. In legitimate image tampering, the semantic meaning of an image does not change significantly. For example, retouching and resampling can be considered as legitimate image tampering. Generally, the main purpose of applying image retouching is to enhance the quality of an image so that image should look visually more appealing. Retouched images are used mostly to upload on social media or to print as a cover photo of a magazine. Similarly, resampling is mainly used to resize the images for different purposes. For example, while uploading the image as a profile photo on most of the websites we may have to resize it to a certain size. On the contrary, illegitimate tampering is tampering in which the semantic meaning of an image can be changed up to any extent. Copy-move forgery and image splicing forgery are the two main categories of illegitimate image tampering. In image splicing forgery, two or more images can be merged to get a single composite image [64, 87]. As shown in Fig. 1, the whole source images can be considered for creating the composite image; or optionally, particular parts of the source images can be used for creating a spliced image. In some cases, each of the source images may be postprocessed by applying operations such as resizing, rotation, etc. Similarly, some postprocessing operations can be applied to the spliced image to create more convincing forgery or to hide any discernible clues.

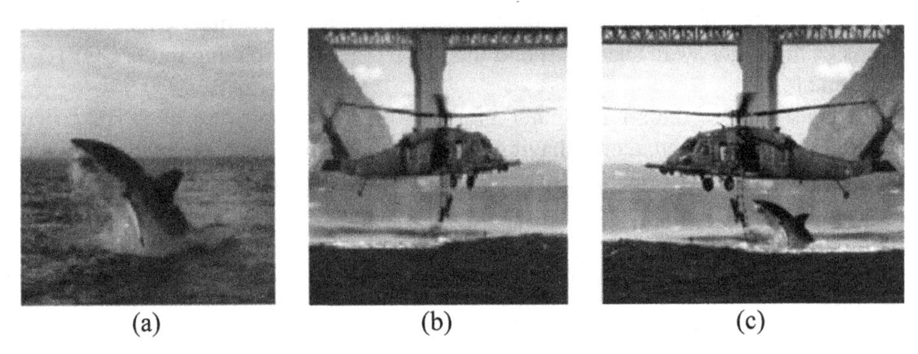

Fig. 2. A case of image splicing forgery [6]: (a) First image, (b) Second image, (c) Spliced image created from (a) and (b)

An instance of the spliced image taken from [6] is highlighted in Fig. 2. To create the spliced image shown in Fig. 2(c), the Whale fish is cropped from the first image (Fig. 2(a)) and then it is pasted in a second image (Fig. 2(b)) to give the impression that helicopter is rescuing the Whale fish. Here the second image is flipped before using to create the composite image. In copy-move forgery, a region of an image is copied and then pasted within the same image [65–67].

It is easily understood that illegitimate image tampering can be more harmful as compared to legitimate image tampering. Hence, more focus was given to develop methods to detect illegitimate image tampering. In past, a number of methods were proposed to detect copy-move [65–67], and image splicing forgery [63]. There exist various review papers [7, 15, 22, 30, 63, 94], that provides an in-depth analysis of copy-move forgery detection methods. However, very few review papers are available on image splicing forgery detection. Table 1 provides a brief description of existing survey papers on image splicing forgery detection. From Table 1 it can be observed that only five review papers are available in the literature and none of the papers provide a detailed survey on image splicing detection methods which are published after 2016. Besides, we have also observed that each of the existing survey papers listed in Table 1 included very few approaches related to image splicing detection. For example, a total of 14 image splicing detection methods are reviewed by Birajdar et al. [15], whereas a total of 11 methods are reviewed by Asghar et al. [8]. Similarly, Zampoglou et al. [104] have also considered 14 methods. The main focus of Birajdar et al. [15] was to survey various image forgery detection methods; therefore this paper provides limited discussion on image splicing detection. However, Meena et al. [63] have provided a detailed survey on image splicing detection methods published between 2000 and 2016. Therefore, in this paper, we review recent image splicing forgery detection methods published between 2017 and 2021.

Table 1. Existing survey papers related to image splicing detection

Authors	Description
Birajdar et al. [15]	Reviewed only 14 image splicing detection techniques published between 2004 and 2011
Asghar et al. [8]	Reviewed only 11 image splicing detection techniques published between 2010 and 2015. Also categorized the existing techniques into six groups: camera response function-based, bi-coherence feature-based, pixel correlation-based, DCT and DWT-based, invariant image moment-based, and other algorithms
Zampoglou et al. [104]	Reviewed and evaluated 14 image splicing detection localization techniques published between 2000 and 2016
Wang et al. [89]	Reviewed various image splicing detection techniques and classified all the existing techniques into five classes: pixel-based, camera-based, format-based, noise-based, special scenario-based. A comparison of six image splicing detection localization techniques is also provided
Meena et al. [63]	Reviewed and compared 19 image splicing detection techniques published between 2000 and 2016

A total of 78 techniques that are proposed to detect image splicing are considered in this paper. The statistical data of year-wise publications is shown in Fig. 3. From Fig. 3, it can be noticed that 53 papers are published in journals, whereas, only 25 papers are conference papers. Figure 3 also suggests that more researchers are showing interest in image splicing detection, as the most number of papers (23 research papers) is published in 2020.

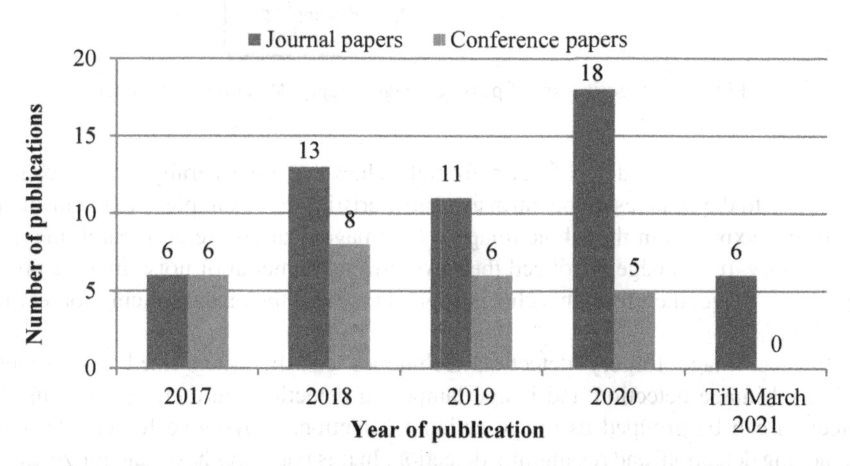

Fig. 3. Number of publications (between 2017 and 2021) for image splicing detection

2 Image Forgery Detection

Active and passive are the two main categories of existing image forgery detection techniques. In active forgery detection techniques, some prior extra information is required about the query image. The extra information may be inserted in an image either during the acquisition process of the image or at a later stage using suitable software or hardware. In active forgery detection, this embedded information is extracted from the query image and then the possibility of image forgery is determined. However, if the query image is obtained from the Internet or any social media platform, then we cannot expect any embedded information. Therefore, active forgery detection methods have limited scope in today's time of Internet and social media [65–67]. Consequently, researchers have started developing forgery detection techniques that can work blindly and require no history or knowledge about the query image [63]. Such type of image forgery detection approaches is commonly referred to as passive approaches. Two main categories under active forgery detection techniques include digital watermarking [48] and digital signature [54].

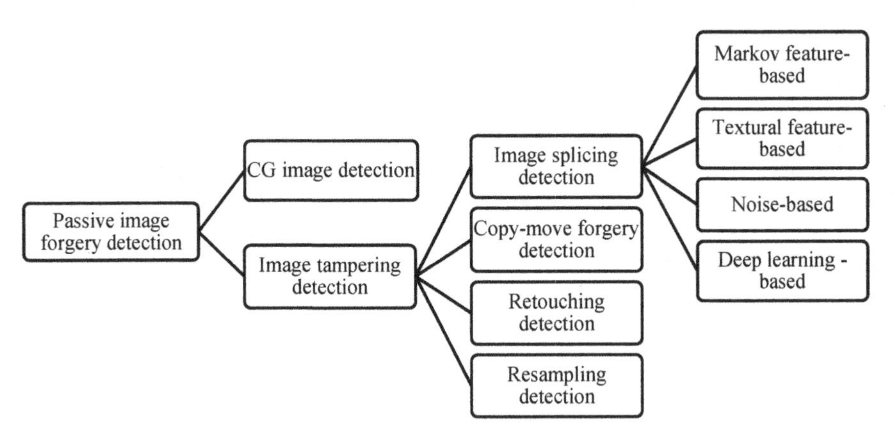

Fig. 4. Categorization of passive image forgery detection approaches

Passive approaches detect forgery from the clues that are generally introduced in the image due to the changes in the intrinsic characteristics. For example, a uniform amount of noise is expected in the whole image if the image is an unaltered natural image; on the contrary, if an image is spliced then two different amount of noise may be present within the image; therefore, this clue is applied to reveal the image splicing forgery [38, 93, 106].

Passive image forgery detection techniques can be categorized as computer-generated image detection and image tampering detection. Further, image tampering detection can be grouped as image splicing detection, copy-move forgery detection, retouching detection, and resampling detection. In this paper, we have categorized image splicing detection techniques into four main categories: Markov feature-based, textural feature-based, noise-based, and deep learning-based techniques. Figure 4 shows a complete categorization of passive approaches.

3 Image Splicing Forgery Detection

The source images used to create a spliced image may have different color temperature, illumination conditions, and noise levels based on various factors. Illumination conditions may vary between the source images as each of the source images may be captured in different lighting conditions such as daytime or nighttime. Similarly, depending on the ISO settings or imperfection in the sensor used in a digital camera, a uniform noise is introduced in the captured image [107]. A spliced image is expected to show two different levels of noise. To create splicing boundaries smooth and less visually different from its surrounding, forger always applies average filtering or some other related image processing operation as postprocessing that may introduce some local discontinuities in the spliced image. Moreover, a forger can resize, crop, rotate, retouch each of the source images to match the visual conditions or shape and size of the target image so that the forged image can look realistic [93]. Each of these postprocessing operations introduces some anomalies in the resulting forged image. For example, cropping, resizing, and rotation operations internally require interpolation operations; hence, interpolation-based artifacts can be applied to detect the image splicing forgery. In a nutshell, there can be various clues that can be used to reveal splicing forgery in an image.

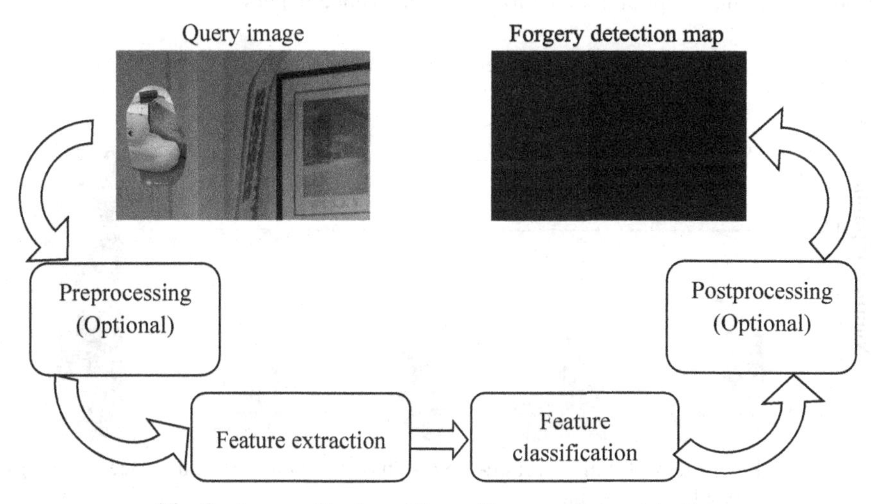

Fig. 5. A generalized workflow of image splicing detection

A generalized workflow to detect or localize image splicing forgery in digital images is shown in Fig. 5. There are four steps in the workflow that include **preprocessing**, **feature extraction**, **feature classification**, and **postprocessing**.

In **preprocessing** step, the query image can be converted from RGB color space to any other suitable color space. Sometimes the input image can be converted to a grayscale image. Several authors [3, 26, 42, 44, 71, 83, 84, 102] have converted RGB image into YCbCr color space in the preprocessing step. Similarly, some authors [20, 79] preferred HSV color space for representing query images in their proposed methods.

The **feature extraction** is the most important phase in the image splicing detection method. Various feature extraction mechanisms are explored in past to represent the image so that the forgery can be detected accurately. Different feature extraction mechanisms focus on different characteristics of the image. For example, local discontinuities introduced by splicing operation can be extracted using Markov features, after that these features can be used to distinguish between authentic and spliced images [47]. Based on the feature extraction mechanisms, the image splicing detection techniques can be categorized as hand-crafted feature-based and deep learning-based. Traditional features used in image processing or computer vision are commonly referred to as hand-crafted features. Common hand-crafted feature extraction mechanisms include Discrete Cosine Transform (DCT) [75, 83], Discrete Wavelet Transform (DWT) [108], Local Binary Pattern (LBP) [5, 88], etc. Most of the existing techniques proposed to detect image splicing employed such hand-crafted features. In this paper, we classify hand-crafted feature-based techniques into four classes as Markov feature-based, textural feature-based, noise-based, and other hand-crafted feature-based techniques. Figure 6 shows category-wise publications on image splicing detection that are published between 2017 and 2021. From Fig. 6, it can be observed that in the last three to four years, most of the researchers have developed deep learning-based image splicing detection methods. Figure 6 also suggests that the number of hand-crafted feature-based methods has reduced over the years. Further, it can also be noticed that more journal papers are published on image splicing detection than conference papers.

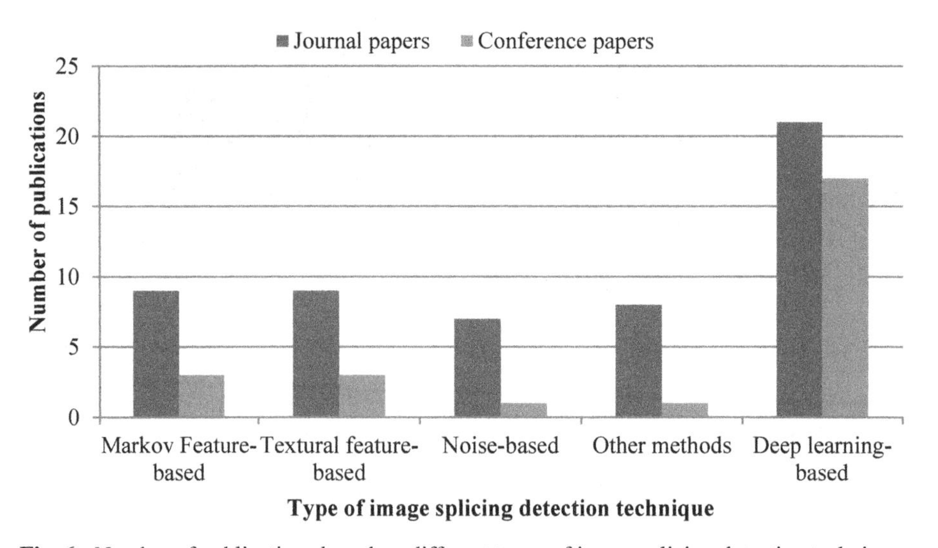

Fig. 6. Number of publications based on different types of image splicing detection techniques

The **feature classification** step also plays a significant role in any image splicing detection method. Some of the popular feature classification techniques are Support Vector Machine (SVM) [3, 47, 73, 92, 108], Linear Discriminant Analysis (LDA) [70], Random Forest [38, 88], etc. From the literature review, we have noticed that SVM is the most popular choice as a feature classifier in image splicing detection techniques. In deep learning-based techniques, some of the authors have performed feature extraction and classification as a single module, whereas many authors use existing Convolutional

Neural Network (CNN) as feature extractor, and then feature classification is applied explicitly.

The **postprocessing** step is also an optional step in the image splicing detection process. Postprocessing operation is mainly applied for the localization of splicing forgery [106]. Discussion on image splicing detection based on hand-crafted features is presented in the next section.

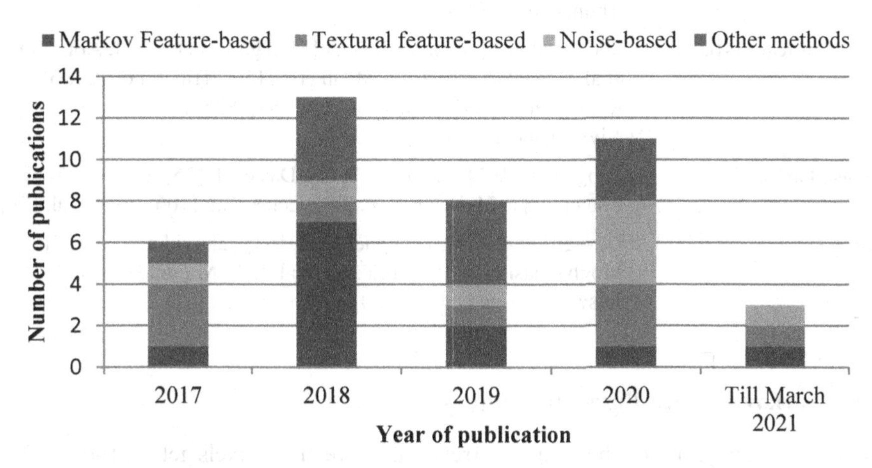

Fig. 7. Year-wise publications on hand-crafted feature-based techniques

4 Hand-Crafted Features-Based Methods

Image splicing detection techniques that utilize traditional feature extraction mechanisms are considered as hand-crafted feature-based methods [64]. In other words, all the techniques that are not based on deep learning fall under the category of hand-crafted feature-based methods. All the existing hand-crafted features-based methods can be grouped into four groups: Markov feature-based, textural feature-based, noise-based, and other hand-crafted feature-based techniques. Figure 7 shows year-wise publications on hand-crafted feature-based techniques. Table 2 lists all the hand-crafted feature-based techniques in each category.

In Table 3 through Table 7, the first column indicates technique; the second column presents the main concept (s) used in the technique, the third column presents the performance of the technique in terms of classification accuracy (Acc) [64] or True Positive Rate (TPR) and False Positive Rate (FPR) [21], F1-measure (F1) [86] or Precision [49] or Area Under Curve (AUC) [57], Block Detection Accuracy (BDA) and Block False Positive (BFP) [106]. Note that, for image splicing localization techniques the performance is calculated at the pixel level, therefore, it is also mentioned in column three. The fourth column shows the evaluation dataset (s) (detail of each dataset is provided in Table 8) used to evaluate the technique. Note that, we considered the best performance of the technique from the different datasets used by the authors; therefore, the performance mentioned in the third column is corresponding to the first dataset (in the case of multiple datasets used) mentioned in the fourth column.

Table 2. Publications on hand-crafted feature-based image splicing detection techniques

Feature extraction	Techniques
Markov feature-based	Li et al. [47], Pham et al. [73], Wang et al. [92], Zhang et al. [108], Han et al. [34], Yildirim et al. [71], Yildirim et al. [102], Chen et al. [21], Kumar et al. [46], Yıldırım et al. [103], Wang et al. [90], Rhhman et al. [75]
Textural feature-based	Shen et al. [83], Alahmadi et al. [5], Agarwal et al. [3], Abrahim et al. [2], Jalab et al. [42], Manu et al. [56], Tripathi et al. [86], Kanwal et al. [44], Srivastava et al. [84], Sharma et al. [82], Niyishaka et al. [70]
Noise-based	Zeng et al. [107], Zhu et al. [111], Das et al. [25], Liu et al. [49], Zou et al. [112], Wang et al. [93], Zeng et al. [106], Itier et al. [38]
Others	Devagiri et al. [26], Hadigheh et al. [33], Iakovidou et al. [37], Moghaddasi et al. [68], Prakash et al. [52], Majumdar et al. [57], Jaiswal et al. [41], Rhee [80], Gokhale et al. [32]

4.1 Markov Feature-Based Techniques

In image splicing forgery, the natural correlation between the pixels gets distorted. These changes in pixel correlation can be highlighted using Markov features effectively [73]. Therefore, Markov-based features are widely studied (Fig. 7) for detecting image splicing forgery and got overwhelming success. From the literature review and the comparative study provided in Table 3, it is observed that Markov features obtain in the frequency domain such as DCT, DWT, etc. achieves better performance than the Markov features obtained in the spatial domain. Therefore, many researchers [34, 73, 92, 108] have used the Markov feature in the frequency domain to reveal image splicing forgery. From the comparison provided in Table 3, it can be observed that the technique proposed by Yildirim et al. [71] obtains the best classification accuracy of 99.98% on the CASIA v1.0 image dataset.

4.2 Textural Feature-Based Techniques

During the image splicing forgery, the local distribution of micro-edge patterns is replaced by some new micro-patterns in the spliced area [5]. Besides, the edges become sharp along the boundary of spliced regions. Due to these reasons, the local frequency distributions in the spliced image significantly differ from the local frequency distributions of an authentic image [5]. Each of the source images used for creating a composite image may have a distinct spatial arrangement of colors or intensities. Various existing methods are surveyed that use such textural anomalies to detect image splicing forgery. LBP [2, 5, 83, 85, 88], Gray Level Co-occurrence Matrices (GLCM) [83], Otsu based Enhanced Local Ternary Pattern (OELTP) [44], Enhanced local ternary pattern [84], Local Directional Pattern (LDP) [82], are some of the descriptors used for representing the textural information of the images for revealing the splicing forgery. Table 4

Table 3. Comparison of various Markov feature-based image splicing detection techniques

Technique	Main concepts used	Performance	Dataset used
Li et al. [47]	Markov features, SVM	Acc = 92.67%	CASIA v2.0
Pham et al. [73]	Block-wise Markov features in DCT domain, SVM	Acc = 96.90%	CASIA v2.0, CASIA v1.0
Wang et al. [92]	Markov features in quaternion DCT domain and quaternion wavelet transform domain, SVM	Acc = 86.91%	CUISDE, CASIA v1.0, CASIA v2.0
Zhang et al. [108]	Markov features in DWT domain, SVM	Acc = 89.88%	CISDE
Han et al. [34]	Markov features in DCT domain, SVM	Acc = 98.50%	CASIA v1.0, CASIA v2.0, CUISDE
Yildirim et al. [71]	Markov features in DCT domain, SVM	Acc = 99.98%	CASIA v1.0, CASIA v2.0, CUISDE
Yildirim et al. [102]	Markov features in DWT domain, SVM	Acc = 99.77%	CASIA v2.0, CASIA v1.0, CUISDE
Chen et al. [21]	Markov features, SLIC, k-means clustering	TPR = 47.6%, FPR = 25.8% (at pixel level)	CUISDE
Kumar et al. [46]	Markov features in block DCT domain Discrete Meyer Wavelet Transform (DMWT) domain, SVM	Acc = 88.43%	CISDE
Yıldırım et al. [103]	Markov features, SVM	Acc = 99.58%	CASIA v2.0, CASIA v1.0, CUISDE
Wang et al. [90]	Markov features in Quaternion DCT, SVM	Acc = 97.52%	CASIA v2.0, CASIA v1.0
Rhhman et al. [75]	Markov features in Quaternion DCT, SVM	Acc = 99.00%	CISDE

provides a comparison of existing textural feature-based image splicing detection techniques. From Table 4, it can be observed that LBP is a mostly used descriptor for image splicing detection. Median Robust Extended Local Binary Pattern (MRELBP)-based technique [56] performs best under this category and achieves a classification accuracy of 98.89% on the CUISDE image dataset.

Table 4. Comparison of various textural feature-based image splicing detection techniques

Technique	Main concepts used	Performance	Dataset used
Vidyadharan et al. [88]	LBP, Local Phase Quantization, Binary Statistical Image Features and Binary Gabor Pattern, Random Forest	Acc = 97.33%	CASIA v2.0, CASIA v1.0
Shen et al. [83]	GLCM, Block-DCT, SVM	Acc = 98.54%	CASIA v1.0, CASIA v2.0
Alahmadi et al. [5]	LBP, SVM	Acc = 97.50%	CASIA v2.0, CASIA v1.0, CUISDE
Agarwal et al. [3]	RICLBP, SVM	Acc = 96.81%	CASIA v1.0, CUISDE, DSO-1
Abrahim et al. [2]	LBP, HOG, ANN classifier	Acc = 98.60%	CASIA v2.0, CASIA v1.0, CUISDE
Jalab et al. [42]	Approximated Machado fractional entropy in DWT domain, SVM	Acc = 98.60%	CASIA v2.0
Manu et al. [56]	Median Robust Extended Local Binary Pattern, SVM	Acc = 98.89%	CUISDE, CASIA v2.0, Wild Web
Tripathi et al. [86]	Grey Level Run Length Matrix, Fuzzy-SVM	F1 = 90.00%	CASIA v1.0
Kanwal et al. [44]	Otsu based Enhanced local ternary pattern (OELTP), SVM	Acc = 97.59%	CASIA v2.0
Srivastava et al. [84]	Enhanced local ternary pattern, SVM	Acc = 98.65%	CASIA v1.0, CASIA v2.0, CUISDE
Sharma et al. [82]	Local Directional Pattern (LDP), SVM	Acc = 98.55%	CUISDE
Niyishaka et al. [70]	LBP, Linear Discriminant Analysis (LDA)	Acc = 94.59%	CASIA v2.0

4.3 Noise-Based Techniques

Noise is introduced in the digital images during the acquisition process. There can be various sources of this noise. Two types of noise in a natural image include; physical noise and hardware noise [107]. Physical noise is mainly related to the nature of the light. Poisson noise is a type of physical noise that can be introduced due to photons arrive irregularly on the photosites. Therefore, the adjacent pixels may get different photon counts. Whereas, hardware noise like Photo Response Non-Uniformity (PRNU) is introduced mainly due to the imperfection in the sensors. Hence, it can be noticed that the images captured using different camera devices tend to have a different level of noise. Thus, it is expected that the spliced image will show two different noise levels. Hence, several noise-based techniques to detect image splicing forgery are introduced in past. The comparative study of existing noise-based techniques is presented in Table 5. The effectiveness of noise-based approaches depends on the robustness of the noise level estimator used. Therefore, various noise level estimators are explored to improve the accuracy [25, 38, 107]. From Table 5 it can be noticed that most of the noise-based techniques are developed for localizing the image splicing forgery.

Table 5. Comparison of various noise-based image splicing detection techniques

Technique	Main concepts used	Performance	Dataset used
Zeng et al. [107]	PCA-based noise level estimation, k-means algorithm	TPR = 47.9%, FPR = 18.5% (at pixel level)	CUISDE
Zhu et al. [111]	Noise level function	Acc = 83.75% (at pixel level)	NIST16
Das et al. [25]	Block-Matching and 3D Collaborative Filtering (BM3D), k-means clustering	F1 = 96.63% (at pixel level)	Four images from CUISDE dataset
Liu et al. [49]	Adaptive-SVD noise estimation and vicinity noise descriptor, SVM	Precision = 99.82%	UM-IPPR
Zou et al. [112]	Signal-Dependent Noise estimation	F1 = 77.23% (at pixel level)	CUISDE, DSO-1
Wang et al. [93]	SLIC, Noise level estimation, Coarse-to-fine grained splicing localization, Fuzzy C-means (FCM) clustering	TPR = 71.80% (at pixel level)	CUISDE
Zeng et al. [106]	Local noise level estimation, k-means clustering	BDA = 88.60%, BFP = 4.50% (at block level)	Self-created dataset of 100 spliced and 100 authentic images
Itier et al. [38]	Noise level estimation, Random Forest	Acc = 98.13%	CUISDE

4.4 Other Techniques

A comparative study on image splicing detection techniques that use miscellaneous features or hybrid features is presented in Table 6. From this Table, it can be noticed that the method proposed by Rhee [80] performs best in this category and achieves a classification accuracy of 99.77%. This method uses Median Filter Residual (MFR) as a feature extractor and an SVM as a feature classifier.

Table 6. Comparison of other hand-crafted feature-based techniques

Technique	Main concepts used	Performance	Dataset used
Devagiri et al. [26]	DWT, One tailed Wilcoxon signed rank test, SVM	Acc = 82.32%	CUISDE
Hadigheh et al. [33]	DWT, GLCM, N-Run Length, SVM	TPR = 89.47%	CISDE
Iakovidou et al. [37]	Content-aware filtering	F1 = 53.7% (at pixel level)	DSO-1, VIPP, Wild Web, Korus
Moghaddasi et al. [68]	SVD-based feature, SVM	Acc = 99.30%	CASIA v1.0, CASIA v2.0, CUISDE
Prakash et al. [52]	Block DCT, SVM	Acc = 99.44%	CASIA v1.0, CASIA v2.0
Majumdar et al. [57]	Lighting environment estimation	AUC = 98.10%	Self-created dataset of 15 spliced and 15 authentic images
Jaiswal et al. [41]	LBP, 36 features of DWT, 15 LTE features, 32 HoG features, Logistic regression	Acc = 99.50%	CASIA v2.0, CASIA v1.0, CUISDE
Rhee [80]	Median Filter Residual (MFR), SVM	Acc = 99.77%	Self created dataset
Gokhale et al. [32]	DCT, Random Forest	Acc = 82.35%	AbhAS dataset

5 Deep Learning-Based Techniques

Traditionally, different filters or descriptors are used for obtaining specific features from an image. For example, Canny, Sobel, Prewitt are some of the filters used to obtain the edges from an image. As there can be a large variety of features in an image, it becomes difficult to design the filters for extracting each type of feature. Besides this, we may need an expert view to choose the particular filter for highlighting specific features. To overcome these limitations of traditional or hand-crafted filters, the concept of deep learning was developed [19]. As deep learning can learn a rich amount of features automatically, hence several methods to detect image splicing forgery are proposed in the last few years based on deep learning. Year-wise publications on deep learning-based methods are shown in Fig. 8. From this figure, it can be noticed that the number of publications on deep learning-based methods is increasing every year.

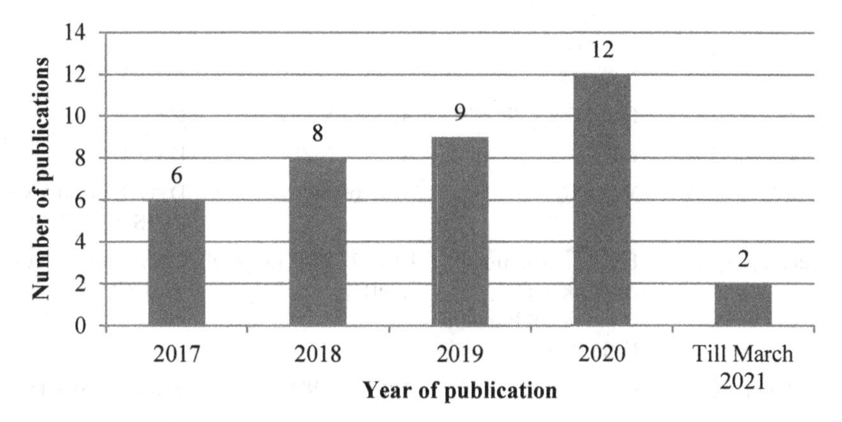

Fig. 8. Year-wise publications on deep learning-based techniques

A comparative study of various deep learning-based image splicing detection methods is presented in Table 7. From this Table, it can be noticed that Quaternion Back-Propagation Neural Network (QBPNN) based technique [18] performs best in this category and achieves a classification accuracy of 100% on a self-created dataset of 60000 grayscale images.

Table 7. Comparison of various deep learning-based image splicing detection techniques

Technique	Main concepts used	Performance	Dataset used
Chen et al. [20]	Camera response function, SVM, CNN	Acc = 97.00%	CUISDE
Bunk et al. [16]	CNN	Acc = 94.86%	NIST16
Rao et al. [77]	CNN, SVM	Acc = 97.83%	CASIA v2.0
Liu et al. [53]	CNN	F1 = 40.63% (at pixel level)	IFC-TC, Korus
Bappy et al. [9]	CNN-LSTM	AUC = 93.90%	NIST16, IFC-TC
Chen et al. [18]	Quaternion Back-Propagation Neural Network (QBPNN)	Acc = 100%	DVMM, CASIA v1.0 CASIA v2.0, Wild Web

(continued)

Table 7. (*continued*)

Technique	Main concepts used	Performance	Dataset used
Cun et al. [24]	Semi-Global Network, Semi-Global Network	Acc = 97.81%	NIST16
Zhang et al. [110]	Fast SCNN, SCNN	Acc = 85.83%	CASIA v2.0
Rezende et al. [79]	ResNet-50, SVM	Acc = 95.00%	DSO-1
Salloum et al. [81]	VGG-16	F1 = 60.40%	DSO-1, CASIA v1.0, CUISDE, NIST16
Chen et al. [19]	Fully convolutional network and Conditional Random Field (CRF)	F1 = 73.88% (at pixel level)	CASIA v1.0, CASIA v2.0, CUISDE
Bayar et al. [11]	CNN	Acc = 99.95%	Self-created dataset of 60000 grayscale images
Peng et al. [72]	Two-stream Faster R-CNN network	F1 = 72.20% (at pixel level)	NIST16, CASIA v2.0, CUISDE
Liu et al. [50]	Fully Convolutional Network, Conditional Random Field	TPR = 82.6%, FPR = 7.20% (at pixel level)	CUISDE, CASIA v2.0 [33]
El-Latif et al. [27]	CNN, SVM	Acc = 96.36%	CASIA v2.0
Cozzolino et al. [23]	Siamese Network, Noiseprint	pixel F1 = 78.0	DSO-1, VIPP, NIST16, Korus
Bi et al. [14]	Ringed residual U-Net (RRU-Net)	F1 = 91.50%	CUISDE,CASIA v2.0
Wu et al. [96]	CNN, SVM	Acc = 95.51%	CISDE
Ye et al. [101]	FPLN, end-to-end network	F1 = 71.10%	NIST16
Xiao et al. [98]	Coarse-to-Refined Convolutional Neural Network (C2RNet), Diluted Adaptive Clustering	F1 = 69.50% (at pixel level)	CASIA v2.0, CUISDE
Elsharkawy et al. [29]	Homomorphic image processing, SVM, Neural Network	Acc = 97.20%	Self-created dataset 200 forged and 800 authentic
Majumdar et al. [58]	Siamese CNN, SVM	Acc = 97.00%	DSO-1, DSI-1

(*continued*)

Table 7. (*continued*)

Technique	Main concepts used	Performance	Dataset used
Wu et al. [97]	VGG-19 network	AUC = 82.40% (at pixel level)	CUISDE, NIST16, CASIA v2.0
Rao et al. [78]	Pre-trained CNN-based local descriptor, SVM	Acc = 96.70%	CASIA v2.0, DSO-1
Liu et al. [51]	Revised version of DenseNet CNN	Acc = 99.00%	CUISDE
Ahmed et al. [4]	ResNet-50, Mask-RCNN	AUC = 96.70%	Self-created dataset from COCO dataset
Yao et al. [100]	CNN, Reliability Fusion Map (RFM)	Acc = 92.20%	Self-created dataset of 500 spliced and 500 authentic images
El-Latif et al. [28]	CNN, SVM	Acc = 97.27%	CASIA v2.0, CASIA v1.0
Bi et al. [13]	CNN	F1 = 87.75%	NIST16, CASIA v2.0, CUISDE
Hussien et al. [36]	CNN, Color Filter Array (CFA), PCA	Acc = 98.19%	CISDE
Bappy et al. [10]	LSTM	AUC = 79.36%	NIST16
Hao et al. [35]	Deep Belief Networks (DBN)	AUC = 89.60%	Self-created dataset of 500 spliced and 193 authentic images
Wei et al. [95]	Neural learning network controlled by multiple scales (MCNL-Net)	F1 = 86.60% (at pixel level)	CASIA v2.0, CUISDE
Wang et al. [91]	CNN, edge features, Photo Response Non-Uniformity (PRNU) features	Acc = 99.45%	CASIA v1.0, CASIA v2.0
Meena et al. [64]	ResNet-50, SVM	Acc = 97.24%	CUISDE
Zhang et al. [109]	CNN, DenseCRFs	Acc = 94.80% (at pixel level)	NIST16, CASIA v2.0
Abhishek et al. [1]	CNN, Deep CNN	Acc = 98.69% (at pixel level)	CUISDE

6 Publicly Available Image Datasets

Table 8 provides details of 14 image datasets that are publicly available for evaluating image splicing detect techniques.

Table 8. Description of various available image datasets used to evaluate image splicing detection techniques

Dataset name	Total images	Number of spliced/ authentic images	Image format	Image size (in pixels)	Ground truth mask available
CISDE (2004) [69]	1845	933/912	BMP grayscale	128 × 128	No
CUISDE (2006) [99]	363	180/183	TIFF	757 × 568 to 1152 × 768	Yes
CASIA v1.0 (2009) [39]	1725	925/800	JPEG, TIFF	324 × 256	No
VIPP (2013) [31]	9600	4800/4800	JPEG	420 × 300 to 3888 × 2592	Yes
DSO-1 (2013) [17]	200	100/100	JPEG	2018 × 1536	Yes
DSI-1 (2013) [17]	50	25/25	JPEG	Different resolutions	Yes
IFS-TC (2013) [104]	1492	442/1050	PNG	922 × 691 to 4752 × 3168	Yes
CASIA v2.0 (2015) [40]	12614	5123/7491	JPEG, TIFF	240 × 160 to 900 × 600	No
Wild Web (2015) [105]	10746	10646/100	JPEG, PNG	122 × 120 to 2560 × 1600	Yes
Korus dataset (2016) [45]	220	220/0	TIFF	Different resolutions	Yes
NIST16 (2016) [113]	1439	564/875	JPEG	Different resolutions	Yes
UM-IPPR (2016) [74]	1530	1530/0	JPEG	1152 × 768	Yes
SMIFD (2019) [76]	500	250/250	JPEG, PNG	200 × 200 to 1080 × 2340	No
AbhAS dataset (2020) [32]	93	48/45	JPEG	278 × 181 to 3216 × 4288	Yes

7 Conclusions and Future Research Directions

A concise review of image splicing detection approaches can help the research community to easily investigate the new challenges in the research area. This paper provides an extensive review and analysis of state-of-the-art techniques proposed between 2017 and 2021 to detect and localize image splicing forgery. Specifically, a generalized workflow to detect image splicing forgery is also presented. This paper, classify image splicing detection techniques as hand-crafted feature-based and deep learning-based. Hand-crafted feature-based techniques are further categorized as Markov feature-based, textural feature-based, noise-based, and other hand-crafted features-based. Detailed analysis of a number of publications in each of the categories is also provided that will help researchers to identify the latest research trend. More specifically, detailed comparative studies on image splicing detection techniques in each of the categories are also presented. These comparative studies will help readers to understand which features or concepts are more conducive to detect image splicing, which is less powerful for this task.

Significant work has been done to detect image splicing forgery; however, the following are the main challenges in this field:

(1) Although there exist several image datasets to evaluate image splicing detection techniques; however, none of the datasets includes the forged images with various geometric transformations and postprocessing operations such as scaling, rotation, additive noise, image blurring, color reduction, JPEG compression with different quality factors, etc. As it becomes an essential requirement to assess the performance of the proposed approach under such challenging situations, there should be a well-designed image dataset comprising images under such attacks.
(2) As the hand-crafted feature-based methods focus on a certain characteristic of an image, these techniques perform poorly, especially under different geometric transformations and postprocessing operations.
(3) The existing techniques show low classification and localization accuracy under different geometric transformations and postprocessing operations; therefore there is still scope for improving the accuracy.

In the future, researchers may create a well-designed image dataset by considering all possible geometric transformations and postprocessing operations. Further, as deep learning techniques can learn rich features from the images, deep learning-based techniques should be explored to develop robust image splicing detectors that can work under different challenging situations. Moreover, we believe that deep learning-based techniques can improve the localization accuracy drastically if the image splicing localization problem is modeled as an image segmentation problem. Further, the exiting techniques can be enhanced to detect the spliced videos that may be created by merging several videos.

References

1. Abhishek, Jindal, N.: Copy move and splicing forgery detection using deep convolution neural network, and semantic segmentation. Multimed. Tools Appl. **80**, 3571–3599 (2021)
2. Abrahim, A.R., Rahim, M.S.M., Sulong, G.: Bin: Splicing image forgery identification based on artificial neural network approach and texture features. Cluster Comput. **22**, 647–660 (2019)
3. Agarwal, S., Chand, S.: Image forgery detection using co-occurrence-based texture operator in frequency domain. In: Sa, P.K., Sahoo, M.N., Murugappan, M., Wu, Y., Majhi, B. (eds.) Progress in Intelligent Computing Techniques: Theory, Practice, and Applications. AISC, vol. 518, pp. 117–122. Springer, Singapore (2018). https://doi.org/10.1007/978-981-10-3373-5_10
4. Ahmed, B., Gulliver, T.A., alZahir, S.: Image splicing detection using mask-RCNN. SIViP **14**(5), 1035–1042 (2020). https://doi.org/10.1007/s11760-020-01636-0
5. Alahmadi, A., Hussain, M., Aboalsamh, H., Muhammad, G., Bebis, G., Mathkour, H.: Passive detection of image forgery using DCT and local binary pattern. SIViP **11**(1), 81–88 (2016). https://doi.org/10.1007/s11760-016-0899-0
6. Ali Qureshi, M., Deriche, M.: A review on copy move image forgery detection techniques. In: 2014 IEEE 11th International Multi-Conference Syst. Signals Devices, SSD 2014, pp. 1–5 (2014)
7. Ansari, M.D., Ghrera, S.P., Tyagi, V.: Pixel-based image forgery detection: a review. IETE J. Educ. **55**, 40–46 (2014)
8. Asghar, K., Habib, Z., Hussain, M.: Copy-move and splicing image forgery detection and localization techniques: a review. Aust. J. Forensic Sci. **49**, 281–307 (2017)
9. Bappy, J.H., Roy-Chowdhury, A.K., Bunk, J., Nataraj, L., Manjunath, B.S.: Exploiting spatial structure for localizing manipulated image regions. In: Proceedings of the IEEE International Conference on Computer Vision, pp. 4980–4989 (2017)
10. Bappy, J.H., Simons, C., Nataraj, L., Manjunath, B.S., Roy-Chowdhury, A.K.: Hybrid LSTM and encoder-decoder architecture for detection of image forgeries. IEEE Trans. Image Process. **28**, 3286–3300 (2019)
11. Bayar, B., Member, S., Stamm, M.C.: Constrained convolutional neural networks : a new approach towards general purpose image image manipulation detection. IEEE Trans. Inf. Forensics Secur. **6013**, 1–17 (2018)
12. Ben-Meir, I.: Anti-Iran Deal TV Ad Uses Fake Image Of Obama Meeting Iranian President. https://www.buzzfeed.com/ilanbenmeir/anti-iran-deal-tv-ad-uses-fake-image-of-obama-meeting-irania?utm_term=.per5RX6Bw#.ftNZW4Ney
13. Bi, X., et al.: D-UNet: A dual-encoder U-Net for image splicing forgery detection and localization, http://arxiv.org/abs/2012.01821 (2020)
14. Bi, X., Wei, Y., Xiao, B., Li, W.: RRU-Net : the ringed residual U-Net for image splicing forgery detection. In: IEEE/CVF Conference on Computer Vision and Pattern Recognition Workshops, pp. 30–39 (2019)
15. Birajdar, G.K., Mankar, V.H.: Digital image forgery detection using passive techniques: a survey. Digit. Investig. **10**, 226–245 (2013)
16. Bunk, J., Bappy, J.H., Mohammed, T.M., Nataraj, L.: Detection and localization of image forgeries using resampling features and deep learning. In: IEEE Computer Society Conference on Computer Vision and Pattern Recognition Workshops, pp. 1881–1889 (2017)
17. Carvalho, T.J.D., Riess, C., Angelopoulou, E., Pedrini, H., Rocha, A.D.R.: Exposing digital image forgeries by illumination color classification. IEEE Trans. Inf. Forensics Secur. **8**, 1182–1194 (2013)

18. Chen, B., Qi, X., Sun, X., Shi, Y.Q.: Quaternion pseudo-Zernike moments combining both of RGB information and depth information for color image splicing detection. J. Vis. Commun. Image Represent. **49**, 283–290 (2017)
19. Chen, B., Qi, X., Wang, Y., Zheng, Y., Shim, H.J., Shi, Y.Q.: An improved splicing localization method by fully convolutional networks. IEEE Access. **6**, 69472–69480 (2018)
20. Chen, C., Mccloskey, S.: Image splicing detection via camera response function analysis. In: IEEE Conference on Computer Vision and Pattern Recognition, pp. 5087–5096 (2017)
21. Chen, H., Zhao, C., Shi, Z., Zhu, F.: An image splicing localization algorithm based on SLIC and image features. In: Hong, R., Cheng, W.-H., Yamasaki, T., Wang, M., Ngo, C.-W. (eds.) PCM 2018. LNCS, vol. 11166, pp. 608–618. Springer, Cham (2018). https://doi.org/10.1007/978-3-030-00764-5_56
22. Christlein, V., Riess, C., Jordan, J., Riess, C., Angelopoulou, E.: An evaluation of popular copy-move forgery detection approaches. IEEE Trans. Inf. Forensics Secur. **7**, 1841–1854 (2012)
23. Cozzolino, D., Verdoliva, L.: Noiseprint: a CNN-based camera model fingerprint. IEEE Trans. Inf. Forensics Secur. 1–13 (2019)
24. Cun, X., Pun, C.-M.: Image splicing localization via semi-global network and fully connected conditional random fields. In: Leal-Taixé, L., Roth, S. (eds.) ECCV 2018. LNCS, vol. 11130, pp. 252–266. Springer, Cham (2019). https://doi.org/10.1007/978-3-030-11012-3_22
25. Das, A., Aji, S.: A fast and efficient method for image splicing localization using BM3D noise estimation. In: Krishna, A.N., Srikantaiah, K.C., Naveena, C. (eds.) Integrated Intelligent Computing, Communication and Security. SCI, vol. 771, pp. 643–650. Springer, Singapore (2019). https://doi.org/10.1007/978-981-10-8797-4_65
26. Devagiri, V.M., Cheddad, A.: Splicing forgery detection and the impact of image resolution. In: Proceedings of the 9th International Conference on Electronics, Computers and Artificial Intelligence, ECAI 2017, pp. 1–6 (2017)
27. El-Latif, E.I.A., Taha, A., Zayed, H.: A passive approach for detecting image splicing using deep learning and haar wavelet transform. Int. J. Comput. Netw. Inf. Secur. **11**, 28–35 (2019)
28. Abd El-Latif, E.I., Taha, A., Zayed, H.H.: A passive approach for detecting image splicing based on deep learning and wavelet transform. Arab. J. Sci. Eng. **45**(4), 3379–3386 (2020). https://doi.org/10.1007/s13369-020-04401-0
29. Elsharkawy, Z., Abdelwahab, S., Abd El-Samie, F., Dessouky, M., Elaraby, S.: New and efficient blind detection algorithm for digital image forgery using homomorphic image processing. Multimed. Tools and Appl. **78**(15), 21585–21611 (2019). https://doi.org/10.1007/s11042-019-7206-3
30. Farid, H.: Image forgery detection a survey. IEEE Signal Process. Mag. **26**, 16–25 (2009)
31. Fontani, M., Bianchi, T., De Rosa, A., Piva, A., Barni, M.: A framework for decision fusion in image forensics based on Dempster-Shafer Theory of Evidence. IEEE Trans. Inf. Forensics Secur. **8**, 593–607 (2013)
32. Gokhale, A.L., et al.: AbhAS : a novel realistic image splicing forensics dataset. J. Appl. Secur. Res. 1–23 (2020)
33. Hadigheh, H.G., Sulong, G.: Bin: splicing forgery detection based on neuro fuzzy fusion. Life Sci. J. **15**, 2017–2019 (2018)
34. Han, J.G., Park, T.H., Moon, Y.H., Eom, I.K.: Quantization-based Markov feature extraction method for image splicing detection. Mach. Vis. Appl. **29**(3), 543–552 (2018). https://doi.org/10.1007/s00138-018-0911-5
35. Hao, H., Delp, E.J., Lafayette, W.: Manipulation detection in satellite images using deep belief networks. In: IEEE/CVF Conference on Computer Vision and Pattern Recognition Workshops (CVPRW), pp. 2832–2840 (2020)
36. Hussien, N.Y., Mahmoud, R.O., Zayed, H.H.: Deep learning on digital image splicing detection using CFA artifacts. Int. J. Sociotechnol. Knowl. Dev. **12**, 31–44 (2020)

37. Iakovidou, C., Zampoglou, M., Papadopoulos, S., Kompatsiaris, Y.: Content-aware detection of JPEG grid inconsistencies for intuitive image forensics. J. Vis. Commun. Image Represent. **54**, 155–170 (2018)
38. Itier, V., Strauss, O., Morel, L., Puech, W.: Color noise correlation-based splicing detection for image forensics. Multimed. Tools Appl. **80**(9), 13215–13233 (2021). https://doi.org/10.1007/s11042-020-10326-5
39. Dong, W.W.: CASIA tampered image detection evaluation database (2013). http://forensics.idealtest.org
40. Dong, W.W.: CASIA2 tampered image detection evaluation (TIDE) database (2015). http://forensics.idealtest.org
41. Jaiswal, A.K., Srivastava, R.: A technique for image splicing detection using hybrid feature set. Multimed. Tools Appl. **79**(17–18), 11837–11860 (2020). https://doi.org/10.1007/s11042-019-08480-6
42. Jalab, H., Subramaniam, T., Ibrahim, R., Kahtan, H., Noor, N.: New texture descriptor based on modified fractional entropy for digital image splicing forgery detection. Entropy **21**, 371 (2019)
43. McDonald, J.: Social Media Posts Spread Bogus Coronavirus Conspiracy Theory. https://www.factcheck.org/2020/01/social-media-posts-spread-bogus-coronavirus-conspiracy-theory/
44. Kanwal, N., Girdhar, A., Kaur, L., Bhullar, J.S.: Digital image splicing detection technique using optimal threshold based local ternary pattern. Multimed. Tools Appl. **79**(19–20), 12829–12846 (2020). https://doi.org/10.1007/s11042-020-08621-2
45. Korus, P., Huang, J.: Multi-scale fusion for improved localization of malicious tampering in digital images. IEEE Trans. Image Process. **25**, 1312–1326 (2016)
46. Kumar, A., Prakash, C.S., Maheshkar, S., Maheshkar, V.: Markov feature extraction using enhanced threshold method for image splicing forgery detection. In: Panigrahi, B.K., Trivedi, M.C., Mishra, K.K., Tiwari, S., Singh, P.K. (eds.) Smart Innovations in Communication and Computational Sciences. AISC, vol. 670, pp. 17–27. Springer, Singapore (2019). https://doi.org/10.1007/978-981-10-8971-8_2
47. Li, C., Ma, Q., Xiao, L., Li, M., Zhang, A.: Image splicing detection based on markov features in QDCT domain. Neurocomputing **1**, 297–303 (2015)
48. Lin, C.Y., Wu, M., Bloom, J.A., Cox, I.J., Miller, M.L., Lui, Y.M.: Rotation, scale, and translation resilient watermaking for images. IEEE Trans. Image Process. **10**, 767–782 (2001)
49. Liu, B., Pun, C.M.: Locating splicing forgery by adaptive-SVD noise estimation and vicinity noise descriptor. Neurocomputing. **387**, 172–187 (2020)
50. Liu, B., Pun, C.M.: Locating splicing forgery by fully convolutional networks and conditional random field. Signal Process. Image Commun. **66**, 103–112 (2018)
51. Liu, B., Pun, C.M.: Exposing splicing forgery in realistic scenes using deep fusion network. Inf. Sci. (Ny) **526**, 133–150 (2020)
52. Liu, B., Pun, C.M., Yuan, X.C.: Digital image forgery detection using JPEG features and local noise discrepancies. Sci. World J. **2014** (2014)
53. Liu, Y., Guan, Q., Zhao, X., Cao, Y.: Image forgery localization based on multi-scale convolutional neural networks. In: ACM Workshop on Information Hiding and Multimedia Security, pp. 85–90 (2017)
54. Lu, C., Liao, H.M., Member, S.: Structural digital signature for image authentication : an incidental distortion resistant scheme. IEEE Trans. Multimed. **5**, 161–173 (2003)
55. Machado, C., Kira, B., Howard, P.N.: A study of misinformation in WhatsApp groups with a focus on the Brazilian Presidential Elections. In: WWW 2019: Companion Proceedings of The 2019 World Wide Web Conference, pp. 1013–1019 (2019)

56. Manu, V.T., Mehtre, B.M.: Tamper detection of social media images using quality artifacts and texture features. Forensic Sci. Int. **295**, 100–112 (2019)
57. Mazumdar, A., Bora, P.K.: Estimation of lighting environment for exposing image splicing forgeries. Multimed. Tools Appl. **78**(14), 19839–19860 (2019). https://doi.org/10.1007/s11042-018-7147-2
58. Mazumdar, A., Bora, P.K.: Deep learning-based classification of illumination maps for exposing face splicing forgeries in images. In: 2019 IEEE International Conference on Image Processing, pp. 116–120 (2019)
59. Meena, K.B., Tyagi, V.: A novel method to distinguish photorealistic computer generated images from photographic images. In: 2019 Fifth International Conference on Image Information Processing (ICIIP), Shimla, India, pp. 385–390 (2019)
60. Meena, K.B., Tyagi, V.: A deep learning based method to discriminate between photorealistic computer generated images and photographic images. In: Singh, M., Gupta, P.K., Tyagi, V., Flusser, J., Ören, T., Valentino, G. (eds.) ICACDS 2020. CCIS, vol. 1244, pp. 212–223. Springer, Singapore (2020). https://doi.org/10.1007/978-981-15-6634-9_20
61. Meena, K., Tyagi, V.: Methods to distinguish photorealistic computer generated images from photographic images: a review. In: Singh, M., Gupta, P.K., Tyagi, V., Flusser, J., Ören, T., Kashyap, R. (eds.) ICACDS 2019. CCIS, vol. 1045, pp. 64–82. Springer, Singapore (2019). https://doi.org/10.1007/978-981-13-9939-8_7
62. Meena, K.B., Tyagi, V.: Distinguishing computer-generated images from photographic images using two-stream convolutional neural network. Appl. Soft Comput. J. **100**, 107025 (2021)
63. Meena, K., Tyagi, V.: Image forgery detection: survey and future directions. In: Shukla, R.K., Agrawal, J., Sharma, S., Singh Tomer, G. (eds.) Data, Engineering and applications, pp. 163–194. Springer, Singapore (2019). https://doi.org/10.1007/978-981-13-6351-1_14
64. Meena, K.B., Tyagi, V.: A deep learning based method for image splicing detection. J. Phys. Conf. Ser. **1714** (2021)
65. Meena, K.B., Tyagi, V.: A copy-move image forgery detection technique based on Gaussian-Hermite moments. Multimed. Tools Appl. **78**(23), 33505–33526 (2019). https://doi.org/10.1007/s11042-019-08082-2
66. Meena, K.B., Tyagi, V.: A copy-move image forgery detection technique based on tetrolet transform. J. Inf. Secur. Appl. **52**, 102481–102490 (2020)
67. Meena, K.B., Tyagi, V.: A hybrid copy-move image forgery detection technique based on Fourier-Mellin and scale invariant feature transforms. Multimed. Tools Appl. **79**, 8197–8212 (2020)
68. Moghaddasi, Z., Jalab, H.A., Noor, R.M.: Image splicing forgery detection based on low-dimensional singular value decomposition of discrete cosine transform coefficients. Neural Comput. Appl. **31**(11), 7867–7877 (2018). https://doi.org/10.1007/s00521-018-3586-y
69. Ng, T.-T., Chang, S.-F.: A data set of authentic and spliced image Blocks. ADVENT Technical Report #203-2004-3, Columbia University, New York (2004)
70. Niyishaka, P., Bhagvati, C.: Image splicing detection technique based on Illumination-Reflectance model and LBP. Multimed. Tools Appl. **80**(2), 2161–2175 (2020). https://doi.org/10.1007/s11042-020-09707-7
71. Odabaş Yildirim, E., Ulutaş, G.: Markov-based image splicing detection in the DCT high frequency region. In: 2018 International Conference on Artificial Intelligence and Data Processing, IDAP 2018 (2019)
72. Peng, Z., Xintong, H., Davis, L.S.: Learning rich features for image manipulation detection. In: IEEE Conference on Computer Vision and Pattern Recognition (CVPR), pp. 1053–1061 (2018)

73. Pham, N.T., Lee, J.-W., Kwon, G.-R., Park, C.-S.: Efficient image splicing detection algorithm based on markov features. Multimed. Tools Appl. **78**(9), 12405–12419 (2018). https://doi.org/10.1007/s11042-018-6792-9
74. Pun, C.M., Liu, B., Yuan, X.C.: Multi-scale noise estimation for image splicing forgery detection. J. Vis. Commun. Image Represent. **38**, 195–206 (2016)
75. ur Rhhman, H., et al.: Comparative analysis of various image splicing algorithms. In: Balas, V.E., Jain, L.C., Balas, M.M., Shahbazova, S.N. (eds.) SOFA 2018. AISC, vol. 1222, pp. 211–228. Springer, Cham (2021). https://doi.org/10.1007/978-3-030-52190-5_15
76. Rahman, M., Tajrin, J., Hasnat, A., Uzzaman, N., Rahaman, G.M.A.: SMIFD: novel social media image forgery detection database. In: 22nd International Conference on Computer and Information Technology (ICCIT), pp. 18–20. IEEE (2019)
77. Rao, Y., Ni, J.: A deep learning approach to detection of splicing and copy-move forgeries in images. In: 8th IEEE International Workshop on Information Forensics Security, WIFS 2016, pp. 1–6 (2017)
78. Rao, Y., Ni, J., Zhao, H.: Deep learning local descriptor for image splicing detection and localization. IEEE Access, 25611–25625 (2020)
79. Rezende, E., Rocha, A., Carvalho, T.: Image splicing detection through illumination inconsistencies and deep learning. In: 25th IEEE International Conference on Image Processing (ICIP), pp. 3788–3792 (2018)
80. Rhee, K.H.: Detection of spliced image forensics using texture analysis of median filter residual. IEEE Access. **8**, 103374–103384 (2020)
81. Salloum, R., Ren, Y., Jay Kuo, C.C.: Image splicing localization using a multi-task fully convolutional network (MFCN). J. Vis. Commun. Image Represent. **51**, 201–209 (2018)
82. Sharma, S., Ghanekar, U.: Spliced image classification and tampered region localization using local directional pattern. Int. J. Image, Graph. Signal Process. **11**, 35–42 (2019)
83. Shen, X., Shi, Z., Chen, H.: Splicing image forgery detection using textural features based on the grey level co-occurrence matrices. IET Image Process. **11**, 44–53 (2017)
84. Srivastava, V., Yadav, S.K.: Texture operator based digital image splicing detection using ELTP technique. In: 9th International Conference System Modeling and Advancement in Research Trends (SMART), pp. 345–348 (2020)
85. Tiwari, D., Tyagi, V.: Dynamic texture recognition using multiresolution edge-weighted local structure pattern. Comput. Electr. Eng. **62**, 485–498 (2017)
86. Tripathi, E., Kumar, U., Tripathi, S.P., Yadav, S.: Automated image splicing detection using texture based feature criterion and fuzzy support vector machine based classifier. In: 2019 International Conference on Cutting-edge Technologies in Engineering, ICon-CuTE 2019, pp. 81–86 (2019)
87. Tyagi, V.: Understanding Digital Image Processing. CRC Press, London (2018)
88. Vidyadharan, D.S., Thampi, S.M.: Digital image forgery detection using compact multi-texture representation. J. Intell. Fuzzy Syst. **32**, 3177–3188 (2017)
89. Wang, J., Li, Y.: Splicing image and its localization: a survey. J. Inf. Hiding Priv. Prot. **1**, 77–86 (2019)
90. Wang, J., Liu, R., Wang, H., Wu, B., Shi, Y.: Quaternion Markov splicing detection for color images based on quaternion discrete cosine transform. KSII Trans. Internet Inf. Syst. **14**, 2981–2996 (2020)
91. Wang, J., Ni, Q., Liu, G., Luo, X., Kr, S.: Image splicing detection based on convolutional neural network with weight combination strategy. J. Inf. Secur. Appl. **54**, 102523 (2020)
92. Wang, R., et al.: Digital image splicing detection based on Markov features in QDCT and QWT domain. Int. J. Digit. Crime Forensics. **10**, 61–79 (2018)
93. Wang, X., Zhang, Q., Jiang, C., Zhang, Y.: Coarse-to-fine grained image splicing localization method based on noise level inconsistency. In: 2020 International Conference on Computing, pp. 79–83 (2020)

94. Warif, N.B.A., et al.: Copy-move forgery detection: survey, challenges and future directions. J. Netw. Comput. Appl. **75**, 259–278 (2016)
95. Wei, Y., Wang, Z., Xiao, B., Liu, X., Yan, Z., Ma, J.: Controlling neural learning network with multiple scales for image splicing forgery detection. ACM Trans. Multimed. Comput. Commun. Appl. **16**, 1–22 (2020)
96. Wu, J., Chang, X., Yang, T., Feng, K.: Blind forensic method based on convolutional neural networks for image splicing detection. In: 2019 IEEE 5th International Conference on Computer Communication, ICCC 2019, pp. 2014–2018 (2019)
97. Wu, Y., Abdalmageed, W., Natarajan, P.: ManTra-Net: manipulation tracing network for detection and localization of image forgeries with anomalous features. In: 2019 IEEE/CVF Conference on Computer Vision and Pattern Recognition (CVPR), Long Beach, CA, USA, pp. 9543–9552 (2019)
98. Xiao, B., Wei, Y., Bi, X., Li, W., Ma, J.: Image splicing forgery detection combining coarse to refined convolutional neural network and adaptive clustering. Inf. Sci. (Ny) **511**, 172–191 (2019)
99. Hsu, Y. F., Chang, S.F.: Detecting image splicing using geometry invariants and camera characteristics consistency. In: International Conference, pp. 549–552 (2006)
100. Yao, H., Xu, M., Qiao, T., Wu, Y., Zheng, N.: Image forgery detection and localization via a reliability fusion map. Sensors (Switzerland) **20**, 1–18 (2020)
101. Ye, K., Dong, J., Wang, W., Peng, B., Tan, T.: Feature pyramid deep matching and localization network for image forensics. In: 2018 Asia-Pacific Signal and Information Processing Association Annual Summit and Conference, APSIPA ASC 2018 – Proceedings, pp. 1796–1802 (2019)
102. Yildirim, E.O.Ş.: Image splicing detection with DWT domain extended markov features. In: 26th Signal Processing and Communications Applications Conference (SIU), pp. 3–6 (2010)
103. Yıldırım, E.O., Uluta, G.: Augmented features to detect image splicing on SWT domain. Expert Syst. Appl. **131**, 81–93 (2019)
104. Zampoglou, M., Papadopoulos, S., Kompatsiaris, Y.: Large-scale evaluation of splicing localization algorithms for web images. Multimed. Tools Appl. **76**(4), 4801–4834 (2016). https://doi.org/10.1007/s11042-016-3795-2
105. Zampoglou, M., Papadopoulos, S., Kompatsiaris, Y.: Detecting image splicing in the wild (Web). In: 2015 IEEE International Conference on Multimedia & Expo Workshops (ICMEW), pp. 1–6 (2015)
106. Zeng, H., Peng, A., Lin, X.: Exposing image splicing with inconsistent sensor noise levels. Multimed. Tools Appl. **79**(35–36), 26139–26154 (2020). https://doi.org/10.1007/s11042-020-09280-z
107. Zeng, H., Zhan, Y., Kang, X., Lin, X.: Image splicing localization using PCA-based noise level estimation. Multimed. Tools Appl. **76**(4), 4783–4799 (2016). https://doi.org/10.1007/s11042-016-3712-8
108. Zhang, Q., Lu, W., Wang, R., Li, G.: Digital image splicing detection based on Markov features in block DWT domain. Multimed. Tools Appl. **77**(23), 31239–31260 (2018). https://doi.org/10.1007/s11042-018-6230-z
109. Zhang, Y., Zhang, J., Xu, S.: A hybrid convolutional architecture for accurate image manipulation localization at the pixel-level. Multimed. Tools Appl. **80**(15), 23377–23392 (2021). https://doi.org/10.1007/s11042-020-10211-1
110. Zhang, Z., Zhang, Y., Zhou, Z., Luo, J.: Boundary-based image forgery detection by fast shallow CNN. arXiv:2658-2663 (2018)
111. Zhu, N., Li, Z.: Blind image splicing detection via noise level function. Signal Process. Image Commun. **68**, 181–192 (2018)

112. Zou, M., Yao, H., Qin, C., Zhang, X.: Statistical analysis of signal-dependent noise : application in blind localization of image splicing forgery. arXiv Prepr. arXiv:2010.16211, pp. 1–16 (2020)
113. Open Media Forensics Challenge. https://www.nist.gov/itl/iad/mig/open-media-forensics-challenge.

A Predictive Model for Classification of Breast Cancer Data Sets

S. Venkata Achuta Rao[1](✉) and Pamarthi Rama Koteswara Rao[2]

[1] Department of CSE, SREYAS Institute of Engineering and Technology, Hyderabad, Telangana, India

[2] ECE, NRI Institute of Technology, Krishna District, Agiripalli, AP, India

Abstract. Medical professionals need a reliable methodology to predict diseases. The process of Machine Learning is used to identify unknown and useful patterns to assist in important tasks of disease prediction and treatment. The techniques that combine multiple classifiers are used for classifying the data sets. Each feature of data sets in the Wisconsin Breast Cancer Dataset (WBCD) collected from fine needle ambitious from human breast tissue. This data set was used to develop a predictive model for the classification and prediction of breast cancer. Support Vector Machine algorithm exhibited good performance when differentiating to other algorithms in such a way that it could be confirmed as the effective classification algorithm with respect to the accuracy, sensitivity, and mean absolute error when applied to diabetes, data sets. Classification and prediction accuracy varied with the quality of the data set.

Keywords: Machine learning · Prediction · Classification and support vector machine and accuracy

1 Introduction

A new field of theory or study has been introduced in the last two decades, which have had a direct impact on the integration of biology, medicine and computer science [1, 2]. Now a day's for finding the solutions to the clinical questions the scientist are utilizing the results of medical and biological research and this process is termed as evidence-based medicine [3–5]. The advanced research findings of biology and medicine depend on the collection, storage, management and analysis of huge medical databases. To extract new knowledge from this data we can use data mining techniques [6–8]. The process of machine learning is dependent on inductive inference. It is a process of examining a phenomenon. After careful examination we have to develop a new model based on that phenomenon. This new model can be used for predictions [9, 10]. The scientists and engineers who are working in various fields have been getting complex experimental data like gigabytes of protein sequences and DNA, and the main objective of data mining is to find out valuable and useful information from large volumes of these medical data sets [11]. The advancement of these technologies was achieved because of the involvement of computer science and technology that expands the multidisciplinary aspects of medicine [12–14]. The biology and medicine can supply the data, and the computer science provides the tools which are required to obtain the knowledge from this data.

M. Singh et al. (Eds.): ICACDS 2021, CCIS 1441, pp. 389–403, 2021.
https://doi.org/10.1007/978-3-030-88244-0_36

This research mainly focuses on the study of effective classification algorithms for prediction of diseases [15, 16]. There are different aspects in terms like accuracy, sensitivity, specificity, precision and efficiency etc. Feature selection techniques were used to the data sets and obtained a reduced representation of the medical data sets which were lesser in volume, yet closely maintains the originality of the data [14]. The Information Gain measure was used to identify the important & unimportant attributes [16, 17].

2 Related Work

Data mining is now seeing broad use in different fields, for example, in computer games, web bots, search engines, scientific applications, and personalized assistants [12]. The process of data mining starts with problem definition, identifying a suitable data set for mining process, and assessing the data quality. The data quality is crucial for data mining and this could have an impact on the quality of the mining results. According to Rygielski, Wang, and Yen [19], the integration of data and transformation of data is most important in knowledge discovery process from the database.

In the traditional database query handling, the data that can satisfy the query can be returned by the database. The returned output can be a subset of the data present in the database and it can also contain aggregations or may be an extracted view. But data mining access the database in a different way from this traditional query access. Compared to traditional query the data mining produces an effective outcome and also it can be implemented in an automated manner in huge data set. It is evident from the existing data mining tools that a wide range of data analysis methods are covered by data mining [14].

The medical practitioners can predict or guess effectively various diseases of patients by applying the classification techniques of data mining on medical data sets. The literature presented here was useful in understanding the main theories which provide the basis of analysis of medical data research. Classification or supervised learning algorithms became most popular and familiar because they are simple to understand and easy to implement. Various applications of classification techniques are classifying financial market trends, medical diagnosis, pattern and image recognition, etc. Prediction and estimation can be considered as kinds of classification [20]. The classification approach needs some knowledge about data. To build the classification model training data set can be used and the model can be validated with the help of test data set.

The data mining process has various techniques. Classification technique is one of the outstanding techniques to predict the target attribute in data set. Classification process of a collection contains separating the items that make up the collection into different classes or categories. In the data mining context, classification is performed by utilizing a model, which is created depending on the historical data and an objective of the classification process is to predict accurately results of the target class value for every latest record in the data set. Various classification methods were available which uses various kinds of techniques for discovering relationship between the values of predictor attributes and the target attribute. The medical practitioners can predict or diagnose patients with different diseases accurately by using the data mining classification techniques on medical data set. This literature review has helped to interpret the important theories that provide the groundwork of analysis for medical data sets research (Fig. 1).

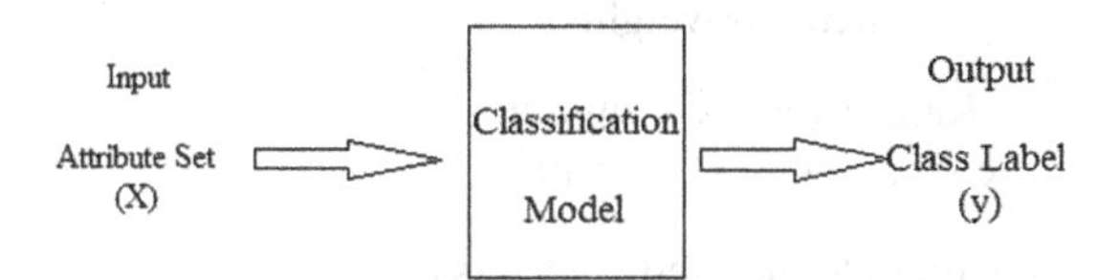

Fig. 1. Classification model [20]

In this study he used the 10-fold cross validation technique to perform that research work to compare and analyze the above mentioned classification techniques on the data set acquired from the Kent Ridge Bio Medical Data set repository. The result of that work shows that among the entire classification methods ensemble achieved high accuracy [18]. This work has disclosed that there exists no single top-quality algorithm which gives better results for the every data set. In this study, concepts and the theorems of prediction and classification or supervised learning techniques were presented, which were applied in the medical database analysis. The main aim of predictive technique is to guess the values of target attribute based on remaining attributes values. Present research is based on the comparative study of different data mining classification algorithms, which were described in the next sections.

2.1 Naive Bayes Model

It takes low computing time and gives high accuracy when applied to the big size data sets. Naïve Bayes was based on the assumption that the consequence of value of an attribute on a given class is not dependent on the values of the other attributes.

2.2 Back Propagation Model

It is a multi-layered Neural Network technique for learning rules, The following pseudo-code algorithm outlines the principle behind training a network:

Algotithm trainingNetwork
(Tr: training set, var R:ANN)begin

// R is initial network with a particular topologyW := a vector of all weights;
$W^{(0)}$:= W with randomly generated weights;

Repeat for each training example t=< xi, ci > in Tr

docompute the predicted class output $c^{(k)}$ for each weight $W^{(k)}$ in the weight vector $W^{(k)}$ do update the weight $W^{(k)}$ to $W_{(k+1)}$

endfor

endfor

until a stopping criterion is met modify R with the updated W;

end;

2.3 Support Vector Machine (SVM) Model

According to SVM literature, a transformed attribute which is used to identify the hyper plane is called a feature and the predictor variable is known as an attribute [19]. Here, selecting the most appropriate representation can be called as feature selection. The support vectors are those which are near the hyper plane. It was shown in Fig. 2.

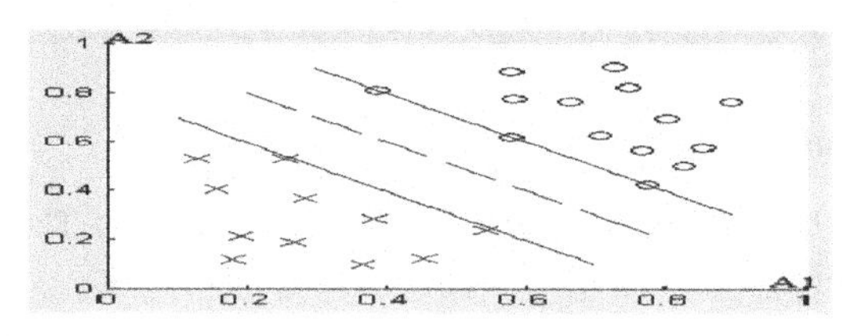

Fig. 2. SVM Hyperplane from two classes, given a labelled training data as data points of the form

2.4 K-Nearest Neighbor Model

K-Nearest Neighbour algorithm (KNN) is also a supervised learning technique that has been applied in several different applications in data mining field, statistical pattern recognition, and many others.

2.5 Model

The technique is the most famous technique in the data mining as it is simple to interpret and easy to understand how it makes the predictions [18]. Different decision tree techniques were available for developing a decision tree including C4.5, ID3, and SLIQ etc.

3 Methodology

Before diagnosing the physician has to examine a number of features which makes physician's job difficult. A medical diagnosis comes under a classification process to predict the disease. This paper presents the results of testing the following data mining algorithms described on data set.

- Feature selection by using Information Gain () filter and Back ward feature elimination method.
- Naïve Bayes Classifier using Weka's Naïve Bayes.
- Back propagation Neural Network using Weka's Multilayer Perceptron () function.
- K-Nearest Neighbor using Weka's IBK function.
- Decision Tree using Weka's J48 tree.

The comparative study of the results obtained from each supervised learning algorithms using the Wisconsin data will follow.

3.1 Data Collection

The data set contains 10 attributes and 699 patient records as presented in the Table 1. All attributes were numeric only. Each and every record contains 10 input attributes and also contains a class attribute. All these 10 attributes were graded on an interval scale between 1 and 10 except sample code number, with 10 considered as the most critical state. To confirm a benign label either a periodic examination or biopsy is used.

3.2 Data Preprocessing

This data set contains no missing values. So we have considered all the instances of data set for classification. The Statistical data of each feature in the data set was displayed in the Table 2.

3.3 Feature Selection

It is an important step of the classification process. A medical data set contains several attributes because, diagnosis of most of the diseases require many tests. So, disease prediction is an expensive process as many tests are required to predict the disease. A medical specialist collects data from several sources and tests to confirm a diagnostic impression but all these tests were not necessary or useful for the prediction of the

Table 1. Wisconsin data set attributes

Sl. no	Name	Data type
1	Sample code number	Numeric
2	Uniformity of cell size	Numeric
3	Clump thickness	Numeric
4	Bare nuclei	Numeric
5	Single decidua cell size	Numeric
6	Marginal adhesion	Numeric
7	Cell shape consistency	Numeric
8	Conventional nucleoli	Numeric
9	Bland chromatin	Numeric
10	Mitoses	Numeric

Table 2. Statistical data set features in WBCD

Sl. no	Name	Min	Max	Mean	Std. dev
1	Breadth of strumble	1	10	4.4185	2.8163
2	Cell size consistency	1	10	3.1341	3.0511
3	Cell shape consistency	1	10	3.2071	2.9723
4	Marginal cohesion	1	10	2.8070	2.8550
5	Single decidua cell size	1	10	3.2163	2.2141
6	Undressed nuclei	1	10	3.5452	3.6440
7	Soft chromatin	1	10	3.4381	2.4382
8	Conventional nucleoli	1	10	2.8671	3.0540
9	Mioses	1	10	1.5890	1.7151

breast cancer. By applying the data mining methods we can cut the cost of diagnosis, by avoiding many tests and by selection of those features which are really required for prediction of disease. Feature selection was performed in the data set to significance of important attributes. The information gain values of features present in this data set were shown in Table 3.

Based on data set Information Gain value only the feature ranks were given. Ordering of the features was shown in the Table 4.

Table 3. Information gain values of features in WBCD

Sl. no	Name	Info. gain
1	Breadth of strumble	0.4591
2	Cell size consistency	0.6750
3	Cell shape consistency	0.6601
4	Marginal cohesion	0.4431
5	Single decidua cell size	0.5052
6	Undressed nuclei	0.5641
7	Soft chromatin	0.5432
8	Conventional nucleoli	0.4661
9	Mioses	0.1980

4 Results

Each of these algorithms was applied on the WBC data set. Confusion Matrix values and the values of the various performance measures of classification algorithms for the different feature set combinations of WBC data set were presented in the following Tables 4, 5 and 6. In every iteration a feature with minimum information gain was left and the rest of the features were sent as inputs. The confusion matrix values for total set of attributes for data set were shown in the Table 4.

Table 4. Confusion matrix values for nine feature set combination of WBCD.

Algorithm	TP	FN	FP	TN
Naïve Bayes	436	22	6	235
Back prop.	442	16	16	225
SVM	446	12	9	232
KNN	443	15	18	223
C4.5	438	20	18	223

The various measure of performance was observed. The measures Sensitivity, Specificity Precision and Accuracy were evaluated based on these confusion matrix values. Performances of Classification Algorithms for combination of the nine feature set of WBC data set were shown in the Table 5.

We get only Eight features after applied the back propagation method. These eight feature data set was supplied as input to the classification methods. The confusion matrix values for the 8 features set of data set for each algorithm used were given in the Table 6.

Table 5. Classifiers performance of nine feature set combination of WBCD.

Algorithm	Sensitivity	Specificity	Precision	Accuracy	MAE	ROC	Time
Naïve Bayes	95.20	97.50	98.60	96.00	4.08	0.98	0.02
Back prop.	96.50	93.40	96.50	95.42	5.14	0.99	2.53
SVM	97.40	96.30	98.00	96.99	3.00	0.97	0.19
KNN	96.70	92.50	96.10	95.30	4.59	0.97	0.01
C4.5	95.60	92.50	96.10	94.60	6.94	0.96	0.03

Table 6. Confusion matrix values for eight feature set combination of WBCD.

Algorithm	TP	FN	FP	TN
Naïve Bayes	439	19	8	233
Back prop.	441	17	19	222
SVM	445	13	10	231
KNN	443	15	18	223
C4.5	439	19	16	225

Performances of Classification Algorithms for the combination of eight feature set WBCD were explained in the Table 7.

Table 7. Classifiers performance of eight feature set of combination of WBCD.

Algorithm	Sensitivity	Specificity	Precision	Accuracy	MAE	ROC	Time
Naïve Bayes	95.90	96.70	98.20	96.10	3.76	0.99	0.01
Back prop.	96.30	92.10	95.90	94.84	5.53	0.98	2.09
SVM	97.20	95.90	97.80	96.70	3.29	0.97	0.08
KNN	96.70	92.50	96.10	95.30	4.59	0.97	0.01
C4.5	95.90	93.40	96.50	95.00	6.73	0.95	0.02

From the Table 7 we can observe that the performance measures of each algorithm was increased or almost remains same after removing the lowest information gain feature from the data set. But the execution speed was increased.By applying backward elimination method we can remove the lowest information gain feature (Marginal Adhesion) from the data set. We get only seven features. These seven feature data set was supplied as input to the classification methods. The confusion matrix values for the 8 features set of data set for each algorithm used were shown in the Table 8.

Performance of the Classifiers for the 7 feature set combination was presented in the Table 9.

Table 8. Confusion matrix values for seven ordered features of WBCD.

Algorithm	TP	FN	FP	TN
Naïve Bayes	441	17	8	233
Back prop.	439	19	14	227
SVM	446	12	10	231
KNN	443	15	17	224
C4.5	438	20	20	221

Table 9. Classifiers performance of seven feature set combination of WBCD.

Algorithm	Sensitivity	Specificity	Precision	Accuracy	MAE	ROC	Time
Naïve Bayes	96.30	96.70	97.20	96.40	3.61	0.99	0.02
Back prop.	95.90	94.20	96.90	95.27	5.53	0.98	2.00
SVM	97.40	95.90	97.80	96.85	3.29	0.97	0.03
KNN	96.70	92.90	96.30	95.42	4.70	0.98	0.02
C4.5	95.60	91.70	95.60	94.27	6.70	0.95	0.05

From the above results we can observe the speed of execution of each algorithm was increased by keeping the other performance measures almost the same after removing the unimportant features from the data set. Figure 3, 4, 5, 6, 7 and 8 shows Accuracy, Sensitivity, Specificity, Precision, MAE, and computing time of the selected classification algorithms applied on WBCD.

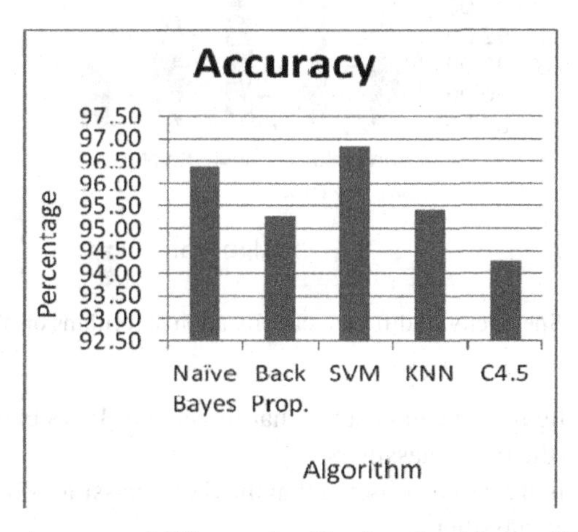

Fig. 3. Accuracy of different classification algorithms on WBCD.

From the above figure we can observe that the accuracy of SVM classifier is high when compared to the other classifiers.

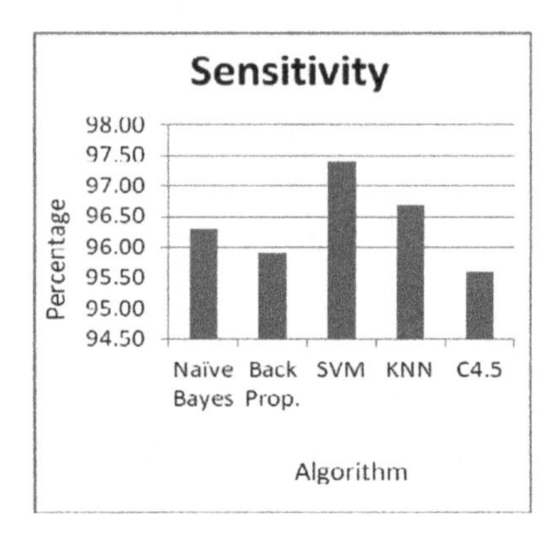

Fig. 4. Sensitivity of different classification algorithms on WBCD.

From the above figure we can observe that the sensitivity of SVM classifier is high when compared to the other classifiers.

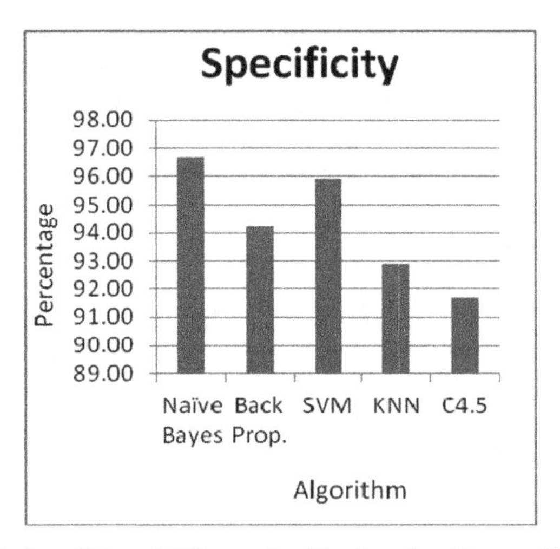

Fig. 5. Specificity of different classification algorithms on WBCD

From the above figure we can observe that the Naïve Bayes classifier specificity is more with respect to the other classifiers.

From the above figure we can observe that the SVM classifier Precision is high when compared to the other classifiers.

From the above figure we can observe that the MAE of SVM classifier is low when compared to the other classifiers.

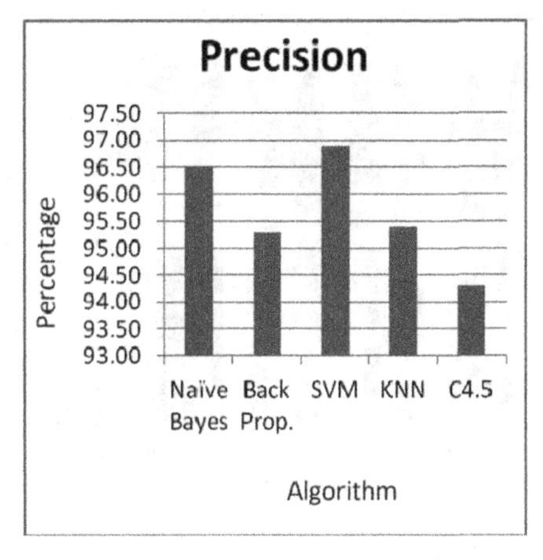

Fig. 6. Precision of different classification algorithm on WBCD.

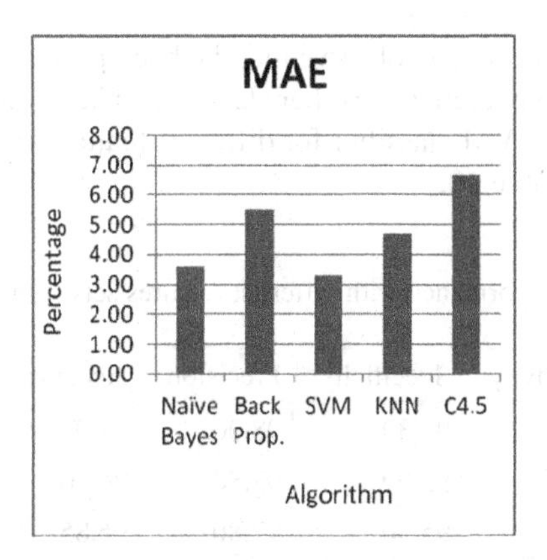

Fig. 7. Mean absolute error of different classification algorithms on WBCD.

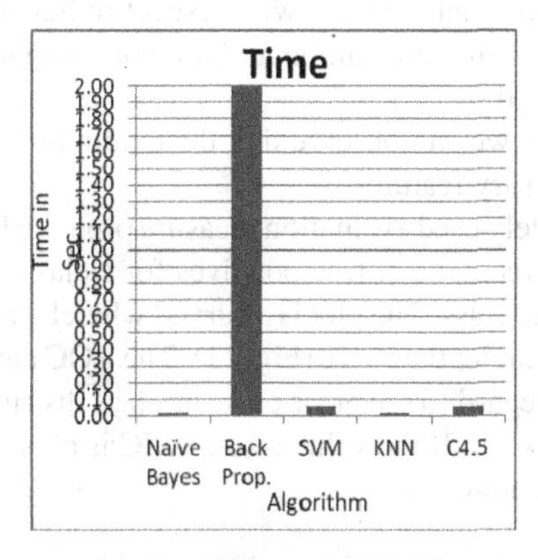

Fig. 8. Time of computing for different classification algorithm on WBCD.

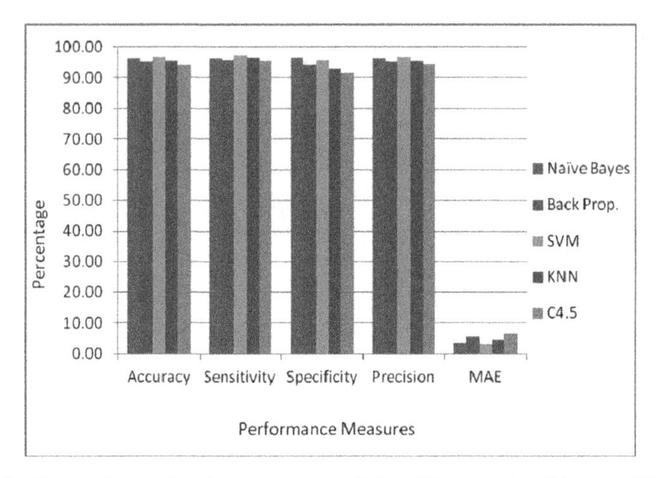

Fig. 9. Comparison of various measures of classification algorithms on WBCD

From the above figure we can observe that the back propagation method takes more computing time when compared to the other classifiers. Clearly it was presented in Fig. 9.

The performance of SVM classifier for different feature set combination of WBCD were explained in the Table 10.

Table 10. SVM performance with different features sets combination on WBCD

No. of attributes	Sensitivity	Specificity	Precision	Accuracy	MAE	ROC	Time
9-attr.	97.40	96.30	98.00	97.00	3.00	0.97	0.19
8-attr.	97.20	95.90	97.80	96.70	3.29	0.97	0.08
7-attr.	97.40	95.90	97.80	96.85	3.29	0.97	0.03

The removal of unimportant attribute with respect to Sensitivity, Specificity, Precision, Accuracy and ROC Values are similar and time of computing is lowered. This was presented in the Fig. 10 and 11.

From the above figure we can observe that the computing time is decreasing when we eliminate the unnecessary features.

The ROC area is a widely used evaluation measure produced using ROC curves. ROC curves are plots of the percentage of true positives for a class against the percentage of false positives for the same class. The plot is a curved whereby the accuracy of predicting a class is equal to the area under the curve (Fig. 11). The ROC curve presents the accuracy of the classifier without regard to the error costs or class distribution. ROC curves have values in between 0.0 and 1.0. If the value of the AUC is closure to 1.0, then the model is considered as more accurate.

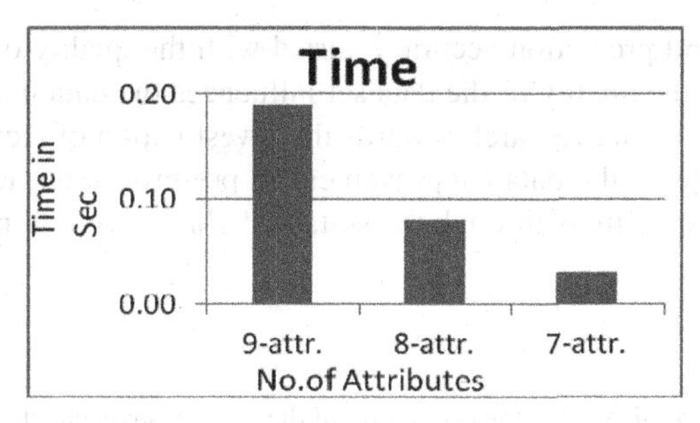

Fig. 10. SVM time comparision with different feature set

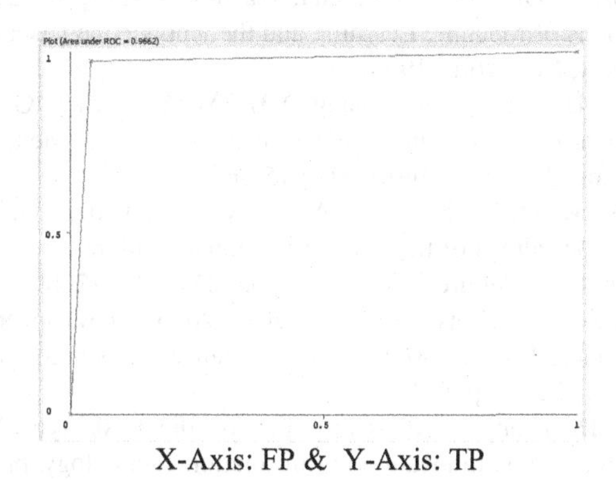

X-Axis: FP & Y-Axis: TP

Fig. 11. The SVM ROC with seven feature set combination of WBCD.

5 Conclusion

Results indicate that, the SVM algorithm exhibited better performance when compared to other algorithms in such a way that it could be confirmed as the effective classification algorithm with respect to accuracy, sensitivity and mean absolute error when applied to the data sets (Table 11).

Table 11. Prediction results of the classifiers for WBCD.

Algorithm	Sensitivity	Specificity	Precision	Accuracy	MAE	ROC	Time
Naïve Bayes	96.30	96.70	97.20	96.40	3.61	0.99	0.02
Back prop.	95.90	94.20	96.90	95.27	5.53	0.98	2.00
SVM	97.40	95.90	97.80	96.85	3.29	0.97	0.03
KNN	96.70	92.90	96.30	95.42	4.70	0.98	0.02
C4.5	95.60	91.70	95.60	94.27	6.70	0.95	0.05

Classification and prediction accuracy varied with the quality of the data set. This confirms again that the quality of the data set influences the data mining results. There is always scope for further research towards the investigation of techniques in enhancing the standard trait of the data (improvement in pre-processing techniques) and the performance of the classifiers through the suitable hybridization of methods.

References

1. Domingos, P., Pazzani, M.: On the optimality of the simple Bayesian classifier under zero-one loss. Mach. Learn. **29**(2), 103–130 (1997). https://doi.org/10.1023/A:1007413511361
2. Kohavi, R., Provost, F.: On applied research in machine learning. In: Editorial for the Special Issue on Applications of Machine Learning and the Knowledge Discovery Process, vol. 30. Columbia University, New York (1998)
3. Rajendra Acharya, U., Ng, E.Y.K., Chang, Y.H., Yang, J., Kaw, G.J.L.: Computer-based identification of breast cancer using digitized mammograms. J. Med. Syst. **32**(6), 499–507 (2008). https://doi.org/10.1007/s10916-008-9156-6
4. Ali, A., Tufail, A., Khan, U., Kim, M.: A survey of prediction models for breast cancer survivability. In: Proceedings of the 2nd International Conference on Interaction Sciences: Information Technology, Culture and Human, pp. 1259–1262 (2009)
5. American Cancer Society: Facts and figures 2015–2016. www.cancer.org/content/dam/cancer-org/research/cancer-facts-and-statistics/breast-cancer-facts-and-figures/breast-cancer-facts-and-figures-2015-2016.pdf
6. Aljarullah, A.A.: Decision tree discovery for the diagnosis type-2 diabetes. In: IEEE International Conference on Innovation in Information Technology, pp. 303–307 (2011)
7. Barnum, S.R.: Biotechnology: An Introduction, Cengage Learning (2006)
8. Batista, G.E., Monard, M.C.: An analysis of four missing data treatment methods for supervised learning. Appl. Artif. Intell. Int. J. **17**(5), 519–533 (2003)
9. Berry, M.J.A., Linoff, G.S.: Data Mining Techniques for Marketing, Sales, and Customer Relationship Management. Wiley, Indianapolis (2004)
10. Ghosh, B.: Using fuzzy classification for chronic disease management. Indian J. Econ. Bus. **11**(1), 231–240 (2012)
11. Cortes, C., Vapnik, V.: Support-vector networks. Mach. Learn. **20**(3), 273–297 (1995). https://doi.org/10.1007/BF00994018
12. Cruz-Ramírez, N., Acosta-Mesa, G.H., Carrillo-Calvet, H., Nava-Fernández, L.A., Barrientos-Martínez, R.E.: Diagnosis of using Bayesian networks: a case study. Comput. Biol. Med. **37**(11), 1553–1564 (2007)
13. European Public Health Alliance. [20]. http://www.epha.org/a/2352
14. Kandwal, R., Garg, P.K., Garg, R.D.: Health GIS and HIV/AIDS studies: perspective and retrospective. J. Biomed. Inform. **42**(4), 748–755 (2009)
15. Kleissner, C.: Data mining for the enterprise. In: Proceeding of the 31st Annual Hawaii International Conference on System Science, CA, US, vol. 7, pp. 295–304. IEEE Computer Society (1998)
16. Liao, S.C., Lee, I.N.: Appropriate medical data categorization for data mining classification techniques. Med. Inform **27**(1), 59–67 (2002)
17. Michie, D., Spiegelhalter, D., Taylor, C.: Machine learning: neural and statistical classification, Ellis Horwood, NJ, USA (1994)
18. Newman, D.J., Hettich, S., Blake, C.L., Merz, C.J.: UCI repository of machine learning databases (1998). University of California, Department of Information and Computer Science, Irvine. http://www.ics.uci.edu/~mlearn/MLRepository.html

19. Ali, P.U.S., Ventakeswaran, C.J.: Improved evidence theoretic kNN classifier based on theory of evidence. Int. J. Comput. Appl. **15**(5), 37–41 (2011)
20. Vijiyarani, S., Sudha, S.: Disease prediction in data mining technique – a survey. Int. J. Comput. Appl. Inf. Technol. **2**(1), 17–21 (2003)

Impact of COVID-19 on the Health of Elderly Person

Ravindra Kumar(✉)

Navvis Healthcare, Saint Louis, USA

Abstract. Health is one of the important indicators of the development of the nation. This study focusses on the impact of COVID-19 on the health of the elderly population. The study was carried in the state of Andhra Pradesh, India and survey questionnaire was administered to two set of people one set having high blood glucose level and another set having both high thyroid and blood glucose level. The findings of the study showed that 70.5% male population and 61.5% -female population have lost their lives due to a complication of blood glucose levels with COVID 19 infection. In another set of people having both thyroid and blood level complication, it was found that 70.9% male and 62.1% female population have lost their lives due to COVID -19 infection. The result clearly shows that thyroid is not a major concern as far as death related to COVID -19 infection is concerned.

Keywords: COVID-19 · Health · Elder population · Blood glucose level and thyroid

1 Introduction

Water, power and health are the three most essential things for any country. The good health of the citizen of the country provides a sufficient potential that can aid in the development of the country [1]. Countries with weak health and health care logistic conditions find it harder to achieve sustained growth. In developing countries, the middle-class people are subjected to the high risk of contracting diseases and lower access to quality health care [2]. In the present circumstances entire world is affected by the global diseases known as Corona virus, which originates from the Wuhan region of China [3]. The first human cases of Corona virus were reported by officials in Wuhan City, China, in the month of December 2019. This Corona virus was designated as COVID-19 where 'CO' means Corona, 'VI' for virus and'D' for diseases.

History of human health has witnessed many viruses, but the spread of these viruses is not as rapid as COVID 19. Due to the rapid transmission nature of this virus between humans, this virus spread all over the world within a short span of time. This is the main reason that the World Health Organization (WHO) announced COVID 19 as pandemic, which is giving alarming signals to the health of human being. World Health Organization (WHO) declared a public health emergency against this pandemic specially for the elder age population [4]. One study in April 2020 reported that the Corona virus affected the

M. Singh et al. (Eds.): ICACDS 2021, CCIS 1441, pp. 404–411, 2021.
https://doi.org/10.1007/978-3-030-88244-0_37

health of a larger section of the society. It claimed that lives of around 1000,000 have been affected by COVID-19 - which lead to mass panic and hysteria [5]. This panic created a negative impact on the mental as well as physical health of the human being. It was observed that the COVID 19 affects the humans with low immune systems more severely than - high immune one. That's why, the persons with co-morbidities are at higher risk compared to the healthy one [6]. In this case an additional precautionary measure needs to be taken against the COVID 19 affected person.

In general, the elderly persons have higher chances of medical conditions as compared to young population. Further, the elderly population is also associated with mental, social, and natural weaknesses. Frailty in the elderly population brings in the risk of various infections and decrease all forms of their immune response which makes them susceptible and an easy target for COVID 19 pandemic [7]. In many studies, it was observed that the COVID 19 has a severe health impact on the elderly population [8, 9]. This is why the elderly population across the globe was suggested to self-isolate for "a very long time" as this virus transmits quickly [10]. Due to this attempts were made to create for the protection of the elder population against this pandemic, which provide the least exposure to the virus.

Even though, self-isolation and country lock down were good strategic steps for the elderly against the direct exposure to the virus, but there were other drawbacks observed during this self-isolation and country lock down; such as lack of socialization, impact on mental health & sense of self-worth, feeling of helplessness etc. [11]. These challenges create health anxiety, panic, adjustment disorders, depression, chronic stress, and insomnia to the elder age population [12]. Therefore, in this research work, a survey-based study was performed to understand the impact of COVID 19 on the elder age population. This study will help in understanding the psychic and medical level of the elder population during this pandemic. The entire manuscript has five major segments where second segments gives the information about the COVID-19 pandemic, which is followed by the Methodology section. The fourth segment of the paper give the results and discussion of the present research work which is followed by the last segment which focusses on conclusions of the present work.

2 Perusal of COVID-19 Pandemic

In the year 2020, whole world has faced a global threat from COVID-19. This virus caused by the SARS-CoV-2, which has symptoms like dry cough, sore throat, fever, malaise, and fatigue. Though, in advance stage pneumonia can lead to acute respiratory distress syndrome (ARDS) and multi-organ failure, eventually leading to death. The novel corona virus was found to be of the same family of SARS (Severe Acute Respiratory Syndrome) 2002-03 [13]. It was observed that the mortality rate is 2% to 3% which is much lower compared to its earlier congeners like SARS [14]. The brutality and casualty of Corona virus has been directly related to age and immune system. In one of the studies it was found that 15% of the first wave of deaths in China were aged above 60 years. As per the Chinese Centre for Disease Control and Prevention

(CCDCP), the 3.6% is the mortality rate for the age group from 60 to 69 years where as the mortality rate of 18% for 80 years and above [15]. This data clearly depicts that the COVID 19 infection has the severe impact on the elderly age population. The outbreak of COVID -19 was announced by WHO as a global disaster. There is no definite proof as to how COVID-19 virus emerged but there has been strong belief that it might have been originated from the Bats [16].

3 Methodology

A community-based survey was conducted among the elderly population over 80 years of age to understand the effect of the COVID 19 pandemic on their health. A total of 499 gender samples were considered in this study, consisting of male (314) and female (185), as shown in Fig. 1. The male gender sample is represented by - 1 and the female sample by 0. The questionnaire prepared for the survey is summarized in Table 1. The survey was conducted amongst the block level corona worriers (Volunteers) across Andhra Pradesh state, as stated in Table 1. The questionnaire was administered through google form. The collected data was analyzed using Python Language The two parameters, infection status and living status, were taken into account during the analysis of the data. In the data analysis, 0 and 1 codes were used to denote the status of non-infected and infected, respectively. Similarly, codes 0 and 1 were used to indicate the status of the collected samples, both alive and dead. The data obtained from the survey was analyzed on the basis of two questionnaire parameters, namely thyroid (TSH) and blood glucose.

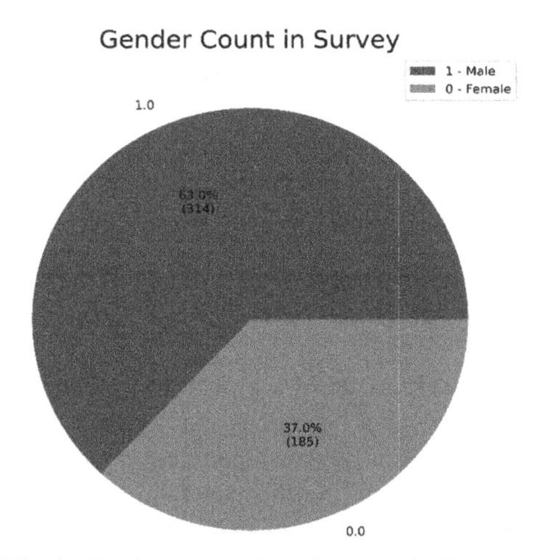

Fig. 1. Total number of gender count in the survey

Table 1. Preparation of questionnaire

Sl. No.	Questions	Questionnaire	Response collection
1	Q1	What is the gender?	Y/N
2	Q2	Is he/she diabetic?	Y/N
3	Q3	What is random blood glucose value?	Value
4	Q4	Is he/she have thyroid (TSH)?	Y/N
5	Q5	What is TSH value?	Value
6	Q6	Is he/she have cardiovascular issues?	Y/N
7	Q7	Is he/she have arthritis patient?	Y/N
8	Q8	Is he/she had TB in his/her life? If yes, how many times?	Y/N
9	Q9	Is he/she have Blood pressure issues?	Y/N
10	Q10	What is the value of BP?	Value
11	Q11	Does she/he have any respiratory system issues?	Y/N
12	Q12	Is he/she have immunity problem?	Y/N
13	Q13	Does she/he is taking nutritious food regularly?	Y/N
14	Q14	Does he/she feel any physiological changes during lock down and isolation?	Y/N
15	Q15	Does he/she receive social respect during lock down and isolation?	Y/N
16	Q16	Does he/she take all prescribed medicines timely?	Y/N
17	Q17	Does he/she do yoga regularly?	Y/N

4 Result and Discussions

The collected response from the survey was analyzed with the help of python language data analysis library. During the analysis the two parameter, Random Blood Glucose level (RGL) and Thyroid (TSH) were considered to study the impact of COVID-19 pandemic on the elderly population. The COVID-19 impact on elder male and female subject of age 80 year or higher with RGL were analyzed separately and the same is plotted in Fig. 2 and 3. The representation format on the pie chart of Fig. 2 follows the pattern of Infection Status (IS), RGL Range and Alive Status which is presented in pie chart as (IS, RGL, Alive). As shown in Fig. 2, the COVID 19 pandemic impact on the age of 80 years or above population is 18.4% for the data format of (1.0, (385.4,460.4], 1.0) which indicates the conversion of infected person into death with average RGL of (385.4, 460.4). The Fig. 2 is basically representing the statistics of conversion from infected to the death stage with varied RGL. Similarly, the Fig. 3 presents the impact of

COVID 19 pandemic on the elder age female population with various random glucose level (RGL). From the Fig. 3, it can be observed that the highest mortality conversion from the infection of 10.9% with RGL of (385.4, 460.4), which was the same in the case of elder age male population. In Fig. 2 and 3, the male (70.5%) and female (61.5%) have lost their life due to complication of blood glucose level with infection.

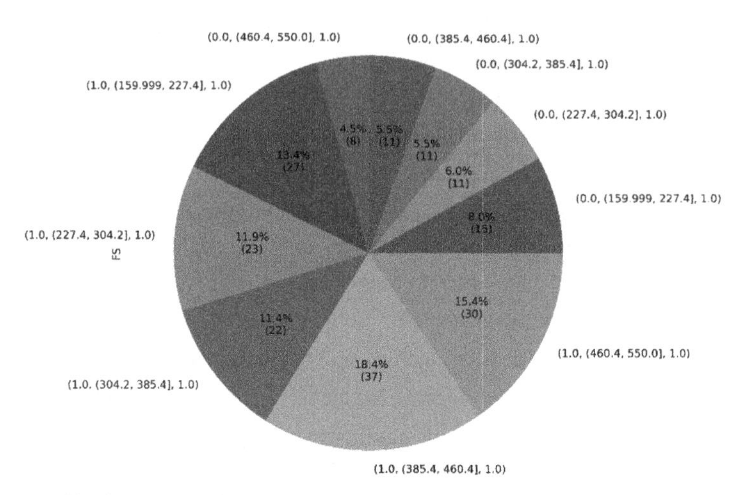

Fig. 2. Impact of COVID-19 pandemic on Elderly male with RGL

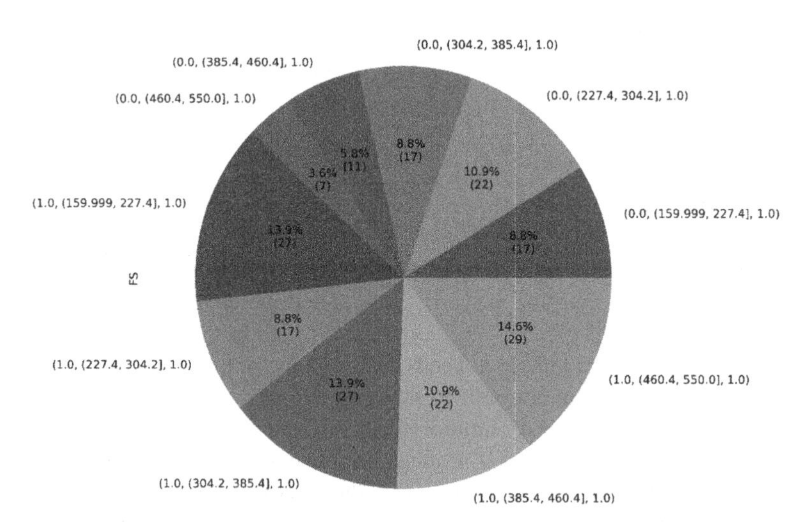

Fig. 3. Impact of COVID-19 pandemic on Elderly female with RGL

Further, the combine effect of two diseases namely thyroid (TSH) and random glucose value (RGL) were considered to analyses the impact of COVID on elder population. Based on the data collected from the survey a pie graph is plotted for analyzing the impact of (RGL) and Thyroid (TSH) on the infected elderly male and female subjects having age more than 80 years with their alive status by considering the infection status (i.e., infected or not infected). The pie chart is shown in Fig. 4 and 5. The data on the pie chart is plotted with a certain format of representation which is like (Infection Status, TSH Range, RGL Range, Alive Status). The alive status is drawn on 0 or 1 scale which indicates the alive or dead, respectively. Similarly, the infection status as 0 or 1 which means not-infected or infected. As depicted in Fig. 4, the highest percentage of 10.9% infected to death conversion is observed for the survey format of (1.0, (8.239, 15.726], (385.4, 460.4], 1.0).

Likewise, in Fig. 5 the highest percentage of 7.3% infected to death conversion is observed for the three different survey formats such as F1(1.0, (8.239, 15.726], (385.4, 460.4], 1.0), F2(1.0, (8.239, 15.726], (304.2, 385.4], 1.0), F3(1.0, (8.239, 15.726], (159.999, 227.4], 1.0). These variation RGL range in F1, F2 and F3 indicates that even though the RGL for all three patterns were different but the conversion percentage is same which means the conversion of infected to dead strongly depends on the TSH. In Fig. 4 and 5, the male (70.9%) and female (62.1%) have lost their live due to the complication of Thyroid (TSH) and Blood Glucose level. Further, the death counts comparison graph of not infected and infected elderly people with Blood glucose range and Thyroid (TSH) range is plotted which is shown in Fig. 6. The Fig. 6 clearly explains the difference between the death of infected people with not-infected people. As shown in Fig. 6, one can clear inference that elderly infected people die more compare to the not infected people.

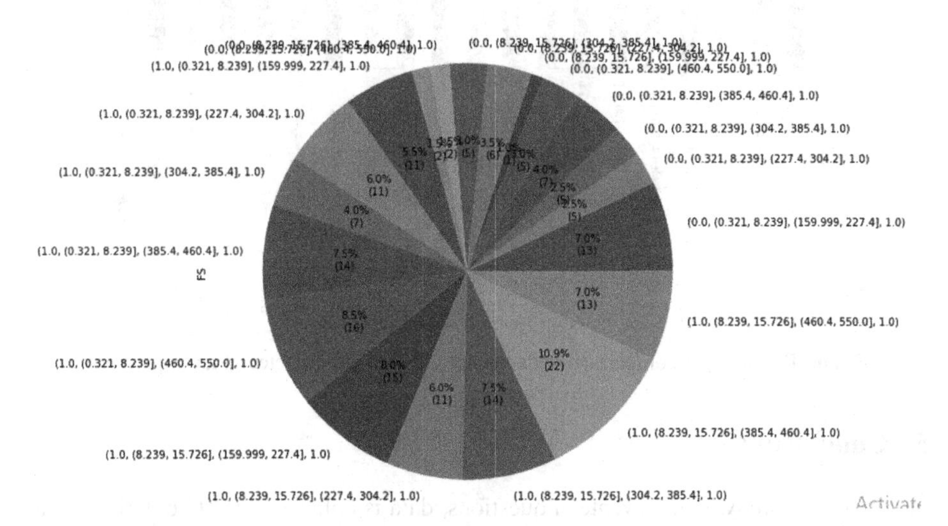

Fig. 4. Impact of COVID-19 pandemic on Elderly male with TSH and RGL

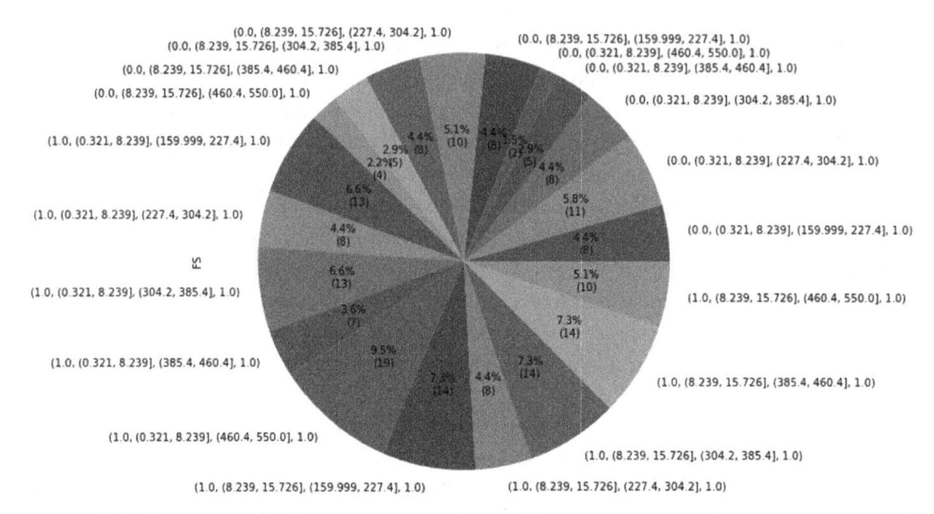

Fig. 5. Impact of COVID-19 pandemic on Elderly Female with TSH and RGL

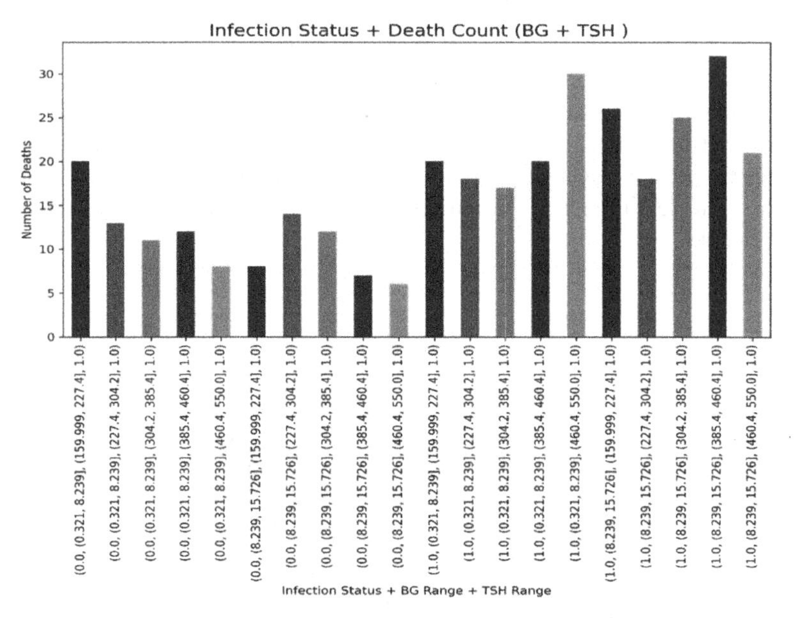

Fig. 6. Death count comparison between not infected and infected elder population

5 Conclusion

In the current study, with seventeen questions, data is collected of 499 elderly people (male-314, female-185), over 80 years, from block-level corona warriors (Volunteers) across the Indian state of Andhra Pradesh. The effect of covid-19 on the elderly population is correctly inferred with different factors that are taken into account in the questionnaire. Two key inferences are drawn in this study, first, male (70.5%) and female

(61.5%) have lost their lives due to infection complications of blood glucose level. Second, due to the complication of Thyroid (TSH) and Blood Glucose level, men (70.9%) and women (62.1%) have lost their lives. In the current research, a minor differential was found between males and females in both cases. It can therefore be easily inferred that thyroid (TSH) is not a significant cause of concern for elderly people if they are affected by covid-19. In future, few other cardiovascular and respiratory system diseases and other medical parameter lines can be studied to observe the effect of covid-19 on elderly people.

References

1. Strauss, J., Thomas, D.: Health, nutrition, and economic development. J. Econ. Lit. **36**(2), 766–817 (1998)
2. Lange, S., Vollmer, S.: The effect of economic development on population health: a review of the empirical evidence. Br. Med. Bull. **121**(1), 47–60 (2017)
3. Wu, P., et al.: The psychological impact of the SARS epidemic on hospital employees in China: exposure, risk perception, and altruistic acceptance of risk. Can. J. Psychiatry **54**(5), 302–311 (2009)
4. Tosepu, R., Gunawan, J., Effendy, D.S., Lestari, H., Bahar, H., Asfian, P.: Correlation between weather and Covid-19 pandemic in Jakarta. Indonesia. Sci. Total Environ. **725**, 138436 (2020)
5. Banerjee, D.: The impact of Covid-19 pandemic on elderly mental health. Int. J. Geriatric Psychiatry **35**(12), 1466–1467 (2020)
6. Stoian, A.P., Banerjee, Y., Rizvi, A.A., Rizzo, M.: Diabetes and the COVID-19 pandemic: how insights from recent experience might guide future management. Metab. Syndr. Relat. Disord. **18**(4), 173–175 (2020)
7. Hubbard, R.E., Maier, A.B., Hilmer, S.N., Naganathan, V., Etherton-Beer, C., Rockwood, K.: Frailty in the face of COVID-19. Age Ageing **49**(4), 499–500 (2020)
8. Kundi, H., et al.: The role of frailty on adverse outcomes among older patients with COVID-19. J. Infect. **81**(6), 944–951 (2020)
9. Ma, Y., et al.: The association between frailty and severe disease among COVID-19 patients aged over 60 years in China: a prospective cohort study. BMC Med. **18**(1), 1–8 (2020)
10. Shinohara, T., Saida, K., Tanaka, S., Murayama, A.: Do lifestyle measures to counter COVID-19 affect frailty rates in elderly community dwelling? Protocol for cross-sectional and cohort study. BMJ Open **10**(10), e040341 (2020)
11. El-Khatib, Z., Al Nsour, M., Khader, Y.S., Abu Khudair, M.: Mental health support in Jordan for the general population and for the refugees in the Zaatari camp during the period of COVID-19 lockdown. Psychol. Trauma Theory Res. Pract. Policy **12**(5), 511 (2020)
12. Bu, F., Steptoe, A., Fancourt, D.: Loneliness during a strict lockdown: trajectories and predictors during the COVID-19 pandemic in 38,217 United Kingdom adults. Soc. Sci. Med. **265**, 113521 (2020)
13. Cohen, B.J.: Mining coronavirus genomes for clues to the outbreak's origins. Science, vol. 31 (2020)
14. Liu, T., Luo, S., Libby, P., Shi, G.P.: Cathepsin L-selective inhibitors: a potentially promising treatment for COVID-19 patients. Pharmacol. Therapeutics **213**, 107587 (2020)
15. Meo, S.A., et al.: Novel coronavirus 2019-nCoV: prevalence, biological and clinical characteristics comparison with SARS-CoV and MERS-CoV. Eur. Rev. Med. Pharmacol. Sci. **24**(4), 2012–2019 (2020)
16. Guo, Y.R., et al.: The origin, transmission and clinical therapies on coronavirus disease 2019 (COVID-19) outbreak–an update on the status. Mil. Med. Res. **7**(1), 1–10 (2020)

The Determinants of Visit Frequency and Buying Intention at Shopping Centers in Vietnam

Dam Tri Cuong(✉) and Nguyen Thanh Long

Industrial University of Ho Chi Minh City, Ho Chi Minh City, Vietnam
damtricuong@iuh.edu.vn

Abstract. In the view' clients, shopping centers are as yet in customers' interest for their purchasing experience and entertainment objectives. Buyers visit shopping centers to look for items; however, they additionally see this visit as a pastime action that gives pleasure from the shopping experience. Thus, this paper aims to analyze the determinants of the frequency of the visit and purchasing intention at malls in Ho Chi Minh City, Vietnam. We proposed the research model from early investigations. We researched 277 purchasers who purchased at shopping centers in Ho Chi Minh City, Vietnam. We applied the partial least squares (PLS) technique to estimate the proposed model. The results uncovered that the determinants (convenience, tenant variety, and internal environment) have decidedly connected with visit frequency and buying intention. Additionally, the outcomes likewise depicted that visit frequency has decidedly connected to buying intentions.

Keywords: Convenience · Tenant variety · Internal environment · Visit frequency · Purchase intention

1 Introduction

Because of the extensive impacts of the COVID-19 epidemic worldwide and its effects on each edge of everyday life, the retail business in Vietnam also faces a few difficulties, particularly with supermarkets or shopping. In recent times, the Vietnam market has witnessed the departure or closure of many branches of many big players in the retail sector, such as Auchan, Parkson, Big C, etc. This failure can come from many different factors: buyers' changing needs, the other competition, etc. Thus, to survive, the shopping mall can create some schemes to pull clients. Moreover, designing, decorating models at the mall, giving an exclusive experience to their clients when they shop at the mall, training staff, increasing promotions, and so forth. These experiences of clients are difficult to find on online channels [1]. Buyers visit shopping centers to look for items; however, they additionally see this visit as a pastime action that gives pleasure from the shopping experience [1]. Thus, it can be noticed and remarked that shopping malls are still in clients' demand for their buying experience and recreational goals [2]. Besides, with a fiercely competitive and changing business context, firms' ability and demand level continuously increase. Developing long-term relationships with consumers is essential

M. Singh et al. (Eds.): ICACDS 2021, CCIS 1441, pp. 412–421, 2021.
https://doi.org/10.1007/978-3-030-88244-0_38

for firms' success and survival [3]. Likewise, the shopping malls' change appeared because the malls did not concentrate only on selling goods/services to their end-users, but the shopping centers have changed the areas where people gather, spent their free time, and met customer demand. Moreover, shopping centers have developed over the years, and factors centered on marketing, administration, attractiveness, or tenant variety have impacted shopping malls' activities [4].

Some studies involved the visit frequency and buy intent at shopping centers (e.g., [2, 5, 6], etc.). Not many investigations have considered the determinants (convenience, tenant variety, and internal environments) of visit frequency and purchasing intents at retail plazas in Vietnam. Accordingly, this examination researches try to fill this gap.

2 Literature review

2.1 Convenience

The convenience concept has attracted scholars recently. An expanded collection of literature has developed (e.g., see scholars' studies such as [2, 5, 6]. In the now's market, convenience is becoming more significant in influencing purchasing behavior and consumer choice. Convenience was the benefit received from the mall's capability to give clients the chance to differentiate shopping duties with the least time and attempt no leaving the shopping malls [7]. Other scholars said that convenience in shopping was easy shopping, the least amount of time spent searching for the place, shop, goods, and information [2]. The shopping center could give time convenience through one-stop buying, spread transacting hours, an enclosed ecosystem, and locations near the place clients live or work [8]. Besides, shopping centers also have supplied convenience store operating hours, large parking areas [9].

Convenience was one of the vital factors influencing shopping malls' choices [10]. Some previous studies said that convenience was an antecedence of visit frequency [2, 5, 6]. Prior empirical researchers revealed that convenience significantly impacted visit frequency [2, 5, 6].

Thus, we suggest the following hypothesis:

H1: Convenience positively affects visit recurrence.

2.2 Tenant Variety

A shopping center that offered tenant variety was expected to pull more customers because of the excitement it has created and because one-stop buying has enabled clients to compare goods offerings easily [11]. Shopping centers consisted of malls for various retailers/tenants. The retail outlets consolidated numerous shops inside a specific region that could draw customers, who might see a broad combination of merchandise prepared inside these stores [7]. In this way, tenant variety, known as the mall's retail inhabitant connecting, has been vital to the customers' pull for malls. It is firmly depicted as the primary benefit of the shopping experience [12]. Sinha and Banerjee [13] questioned that retail plazas, including a general classification of branded shops (e.g., garments stores, departmental stores, basic food item shops, and so on) to acquire the recurrence of visits from the customers [13]. The renter blends could pull more customers and supplement purchasing openings when customers visited the malls [5]. Some previous studies exhibited that tenant variety positively impacted visit frequency [5, 14].

Accordingly, we suggest the accompanying hypothesis:

H2: Tenant diversity emphatically affects visit frequency.

2.3 Internal Environment

The mall environment included five significant aspects: exterior, overall interior, layout, interior display, and personal factors, which were acknowledged to affect result variables, e.g., time used inside the center, general assessment, and rebuying intentions. They likewise suggested that the inside atmosphere includes outside appearance, inside highlights, colors, designs, lighting, and so forth. These points straightforwardly pulled the feelings, and they could have a fundamental impact on buyer joy [15]. Besides, the malls' environment has flawless and enchanting, established a pleasant climate to accumulate and invite partners or associate with different customers [7]. A retail outlet with a reasonable inner temperature could also pull and somewhat stimulate selling at the malls [16]. Moreover, the inside climate has added to shaping a passionate reaction and affecting the shopping center's general judgment [17]. The past study contended that the interior environment had kept visitors around and experience unwinding [2]. Therefore, we recommend the hypothesis as follows:

H3: Internal environment has a significant favorable influence on visit frequency.

2.4 Visit Frequency and Purchase Intention

Customers' frequency of visits to the shopping center was concerned by retail managers and mall managers, who might create their marketing activities to purpose various customer groups. Frequency of visits has been an essential behavioral aspect of customers, and understanding the client behavior of different client segments will support managers in predicting client transfer movements. Besides, scholars might be more focused on the process underlying trips to shopping malls and could describe shopping trips or purchase rates adequately [18]. In this study, clients' visit frequency was considered the number of times that clients visit shopping malls in a specific period.

The notion of buying intention remained an essential part of marketing, though customers' purchasing decision-making remains very complicated [19, 20]. Purchase intention was the purchasers' actual intent towards goods/services [21]. From the same perspective, purchase intention was regarded as the possibility that customers are ready to examine purchasing particular products or brands [22]. Therefore, purchase intention can forecast client purchasing behavior. The clients perform a decision to buy a specific well in a select store; they had been made by their prior intention that has been formed [19].

Linking between clients' visit frequency and purchase intention, some scholars disclosed that clients' visit frequency had a preditor of purchase intention [2, 5, 6]. Prior empirical studies also said that clients' visit frequency significantly impacted purchase intention [2, 5, 6].

Thus, we propose the following hypothesis:

H4: Clients' visit frequency is significantly related to purchase intention.

Figure 1 depicts the recommended research model.

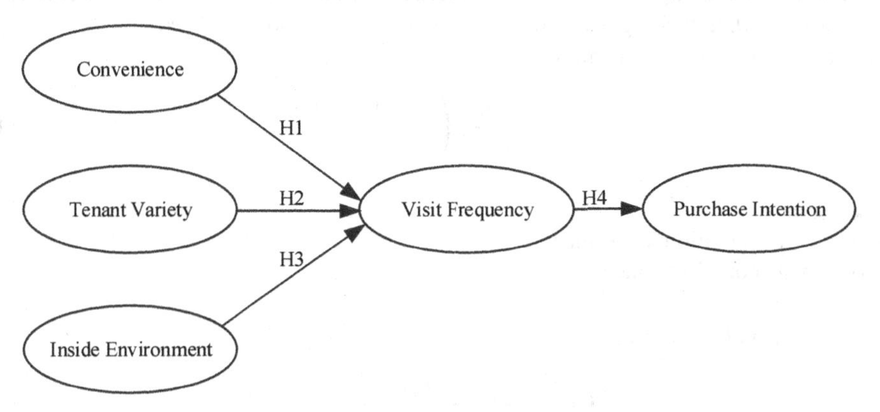

Fig. 1. Research model

3 Research Methodology

Estimation indicators were picked and adjusted from earlier investigations to coordinate with the examination conditions. We take a Likert scale with 1 (emphatically objection) to 5 (absolutely consent) to assess the indicators. There were five main concepts: convenience, tenant variety, internal environment, visit frequency, and buying intention. We change four indicators of convenience and tenant variety from [2, 5], four items of inner atmosphere from [2, 10], two indicators of visit frequency from [2, 10], and four items of purchasing intention from [23].

The population sample information was acquired from purchasers in Ho Chi Minh City, Vietnam, by a convenient sampling procedure. The respondents were chosen who purchased at shopping centers by surveying offline and online. After dropping, the respondents didn't give sufficient information. Two hundred seventy-seven responses were utilized in the last analysis. The sample comprised 122 male users (40.04%) and 155 female users (55.96%). Users were in the age group 18 to 25 made 18.1% of the respondents; 41.5% were in the period group 26 to 35; 26% in the age group 36 to 45; and 14.4% were more extensive than 45.

We estimate the measurement and structural models using the partial least squares (PLS) technique [24].

4 Results and Discussion

4.1 Results

Measurement Model

The reliability of the internal consistency of concepts was measured by Cronbach's alpha (α) and composite reliability (CR). As displayed in Table 1, α coefficients were above 0.60 (from 0.670 to 0.800), and CR criteria were over 0.70 (from 0.813 to 0.870). These results showed evidence of concept reliability [25].

The factor loading of all items and average variance extracted (AVE) of concepts were over 0.50, indicating each construct's convergent validity [24, 26].

Cronbach alpha (α)'s formula:

$$\alpha = \frac{k}{k-1}\left(1 - \frac{\sum \sigma_i^2}{\sigma_x^2}\right) \tag{1}$$

- k: factor
- σ_i^2: the observation's variance
- σ_x^2: total scores' variance

CR and AVE's formula:

$$CR = \frac{\left(\sum \lambda_i\right)^2}{\left[\left(\sum \lambda_i\right)^2 + \sum 1 - \lambda_i^2\right]} \tag{2}$$

$$AVE = \frac{\sum \lambda_i^2}{\left[\sum \lambda_i^2 + \sum 1 - \lambda_i^2\right]} \tag{3}$$

- λ_i: the outer (element) loading to an item
- $1 - \lambda_i$: in case of standardized items

In this study, the Fornell-Larcker standards were applied to estimate discriminant validity [27]. As displayed in Table 2, the AVEs' square root indexes (in bold - from 0.724 to 0.866) were higher than correlations between each construct with any other construct. Therefore, the constructs' discriminant validity was met.

Structural Model

The standard was regarded assessing the structural model, including the coefficient of determination (R-Square), because the weight relations determined the values of the constructs. R^2 coefficients were 0.67, 0.33, 0.19, were equivalent substantial, moderate, and low [28].

As displayed in Fig. 2, the R^2 coefficient is 0.238, below 0.33, considered a moderate-weak effect. In other words, visit frequency explains 23.8% of the variance of purchase intention. Likewise, convenience, tenant variety, and internal environment describe 39.4% of the variance on visit frequency; we also show that the internal environment has the most critical influence (0.312), followed by tenant variety (0.238) and convenience (0.187).

Hypothesis Testing

A bootstrapping process of 5000 resamplings was used for hypothesis testing. The visit frequency of consumers has positive effect by three elements ($\beta = 0.187, 0.238, 0.312$, $p < 0.05$), as well as has positive effect to purchase intention ($\beta = 0.488$, $p < 0.05$) (see Table 3). Thus, hypotheses were approved.

Table 1. Results of measurement items

Constructs	Factor loading	α	CR	AVE
Convenience		0.717	0.824	0.540
1. I like to visit a shopping center with a comfortable and convenient accessibility	0.763			
2. I like to visit a shopping center a have good client service	0.752			
3. I like to visit a shopping center a have free parking	0.653			
4. I like to visit a shopping center I believe is protected by the center's security	0.767			
Tenant variety		0.695	0.813	0.524
1. I like to visit a shopping center having famous retailers and franchises	0.784			
2. I like to visit a shopping center having a supermarket	0.798			
3. I like to visit a shopping center with related services (bank, mobile phone firms, etc.)	0.601			
4. I like to visit a shopping center having a broad of goods presentations	0.694			
Internal environment		0.780	0.859	0.603
1. I like to visit a shopping center having appealing facilities and an attractive internal environment	0.797			
2. I like to visit a shopping center that has rest places	0.723			
3. I like to visit a shopping center having adapted signposting in a commercial area	0.786			
4. I like to visit a shopping center having separate play areas for kids	0.798			
Visit frequency		0.670	0.857	0.750
1. I will plan to visit the shopping center	0.894			
2. I will often visit the mall	0.837			
Purchase intention		0.800	0.870	0.625
1. I would intend to purchase at the shopping mall in the future	0.814			
2. I would plan to buy at the shopping mall	0.785			
3. I would attempt to get to buy at the shopping mall	0.746			
4. I would certainly buy at the shopping mall	0.816			

Table 2. Discriminant validity

	Convenience	Internal environment	Visit frequency	Purchase intention	Tenant variety
Convenience	**0.735**				
Internal environment	0.543	**0.777**			
Visit frequency	0.503	0.553	**0.866**		
Purchase intention	0.491	0.583	0.488	**0.791**	
Tenant variety	0.613	0.585	0.535	0.487	**0.724**

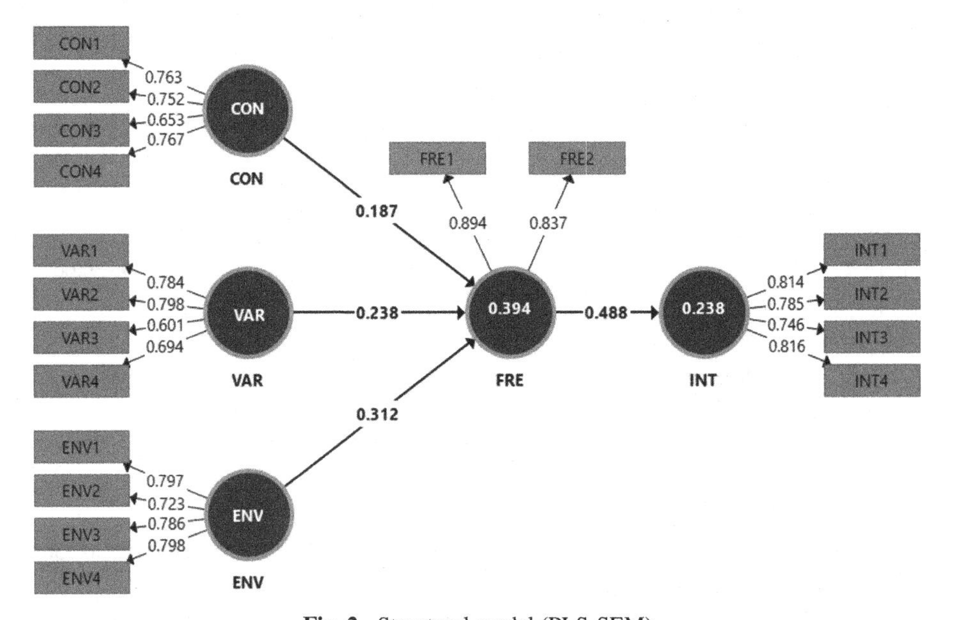

Fig. 2. Structural model (PLS-SEM)

Table 3. Hypothesis testing

Links	Hypotheses	Path coefficients	t-value	p-value
Convenience → Visit frequency	H1	0.187	2.996	0.003
Tenant variety → Visit frequency	H2	0.238	3.297	0.001
Internal environment → Visit frequency	H3	0.312	4.364	0.000
Visit frequency → Purchase intention	H4	0.488	8.484	0.000

4.2 Discussion

The author's outcomes affirmed the inner environment has a significantly positive influence on visiting frequency, and it had the most effect on visit frequency. The internal environment was an antecedent of visit frequency. The prior empirical study supported this study [2]. Likewise, the research results revealed that tenant variety also has a positive relationship with visit frequency. The tenant variety was also a predictor of visit frequency. The previous empirical study indicated this study [5, 14]. Besides, the research finding reinforced that convenience also has a positive correlation with visit frequency. The convenience was an antecedent of visit frequency. The prior empirical study declared this study [2, 6].

Additionally, the study results also verified that visit frequency was positively associated with buying intention. Visit frequency was an antecedent of purchase intention. The prior empirical investigations explained this research outcome [2, 6].

5 Conclusions and Limitations

This research has analyzed the determinants (convenience, tenant variety, and internal environments) of visit frequency and buying intention at shopping centers in Vietnam's circumstances. Therefore, practitioners should have marketing plans that enhance shoppers' perception of convenience, tenant variety, and internal environment to increase the visit frequency and purchase intention at the shopping centers.

This study helps practitioners to know the significance of the internal environment on visit frequency and buying intention. It has the most influence on clients' visit frequency. Therefore, managers should create a charming internal environment to increase clients' visiting frequency. Some of the highlights that managers should mention for the lovely internal environment cover appealing facilities, having rest places, having adapted signposting in a commercial area and displaying products, and having separate play areas for kids. If shoppers recognize charming internal environment positively, they will point to favor buying intention at shopping malls.

Moreover, managers also recognize the importance of the tenant variety and convenience on visit frequency and buying intention. Thus, managers also should have to plan to draw the tenant variety as it drives clients' visiting frequency to malls. When the purchasers recognize the malls have the tenant variety, they will increase visit frequency at malls. Likewise, managers should build a communicating scheme to help clients having convenience when shopping at malls. Some points should be considered: having comfortable and convenient accessibility, having good client service, having free parking, and feeling safe when shopping. When the purchasers realize the malls have a positive convenience, they will tend to enhance visit frequency.

Finally, the research will support practitioners in understanding the essence of visit frequency on buying intent. Thus, practitioners should focus on determinant factors (internal environment, tenant variety, and convenience) to pull customers to malls. When they have a positive recognition of visit frequency, they will tend to increase purchase intention.

This study has limitations. First, this research only explains 23.8% of buying intent's variance by visit frequency and 39.4% of visit frequency's variance by three independent variables (internal environment, tenant variety, and convenience). Therefore, future studies should add independent variables to explain variance in visit frequency and purchase intention. Second, this current examination may not be generalizable to all different fields, so the future should examine various industrials.

References

1. Kim, Y.-H., Lee, M.-Y., Kim, Y.-K.: A new shopper typology: utilitarian and hedonic perspectives. J. Global Acad. Market. Sci. **21**, 102–113 (2011). https://doi.org/10.1080/12297119.2011.9711017
2. Calvo-Porral, C., Lévy-Mangín, J.P.: Pull factors of the shopping malls: an empirical study. Int. J. Retail Distrib. Manage. **46**, 110–124 (2018). https://doi.org/10.1108/IJRDM-02-2017-0027
3. Mirabi, V., Akbariyeh, H., Tahmasebifard, H.: A study of factors affecting on customers purchase intention case study: the agencies of bono brand tile in Tehran. J. Multidisc. Eng. Sci. Technol. (JMEST). **2**, 267–273 (2015)
4. Kunc, J., Reichel, V., Novotná, M.: Modelling frequency of visits to the shopping centres as a part of consumer's preferences: case study from the Czech Republic. Int. J. Retail Distrib. Manage. **48**, 985–1002 (2020). https://doi.org/10.1108/IJRDM-04-2019-0130
5. Yeo, A.C.M., Ong, W.S., Kwek, C.L.: The antecedents influencing shoppers' frequencies of visit and purchase intention in the shopping mall: a study on the pull factors. HOLISTICA J. Bus. Public Admin. **10**, 53–74 (2019). https://doi.org/10.2478/hjbpa-2019-0029
6. Tjandra, C., Muqarrabin, A.M.: Analysis of factors affecting the frequency of visits and their impact on purchase intention at XYZ mall. J. Bus. Entrepr. **7**, 11–18 (2019)
7. El-Adly, M., Eid, R.: Measuring the perceived value of malls in a non-Western context: the case of the UAE. Int. J. Retail Distrib. Manage. **43**, 849–869 (2015)
8. Reimers, V., Clulow, V.: Retail centres: it's time to make them convenient. Int. J. Retail Distrib. Manage. **37**, 541–562 (2009). https://doi.org/10.1108/09590550910964594
9. Yan, R.N., Eckman, M.: Are lifestyle centres unique? Consumers' perceptions across locations. Int. J. Retail Distrib. Manage. **37**, 24–42 (2009). https://doi.org/10.1108/09590550910927144
10. Kushwaha, T., Ubeja, S., Chatterjee, A.S.: Factors influencing selection of shopping malls: an exploratory study of consumer perception. Vision **21**, 274–283 (2017). https://doi.org/10.1177/0972262917716761
11. Berman, B., Evans, J., Chatterjee, P.: Retail Management: A Strategic Approach. Pearson Education Limited, Upper Saddle River (2018)
12. Anselmsson, J.: Sources of customer satisfaction with shopping malls: a comparative study of different customer segments. Int. Rev. Retail Distrib. Consum. Res. **16**, 115–138 (2006). https://doi.org/10.1080/09593960500453641
13. Sinha, P.K., Banerjee, A.: Store choice behaviour in an evolving market. Int. J. Retail Distrib. Manage. **32**, 482–494 (2004). https://doi.org/10.1108/09590550410558626
14. Pan, Y., Zinkhan, G.M.: Determinants of retail patronage: a meta-analytical perspective. J. Retail. **82**, 229–243 (2006). https://doi.org/10.1016/j.jretai.2005.11.008
15. Raajpoot, N.A., Sharma, A., Chebat, J.C.: The role of gender and work status in shopping center patronage. J. Bus. Res. **61**, 825–833 (2008). https://doi.org/10.1016/j.jbusres.2007.09.009
16. Chebat, J.C., Sirgy, M.J., Grzeskowiak, S.: How can shopping mall management best capture mall image? J. Bus. Res. **63**, 735–740 (2010). https://doi.org/10.1016/j.jbusres.2009.05.009

17. Loureiro, S.M.C., Roschk, H.: Differential effects of atmospheric cues on emotions and loyalty intention with respect to age under online/offline environment. J. Retail. Consum. Serv. **21**, 211–219 (2014). https://doi.org/10.1016/j.jretconser.2013.09.001
18. Roy, A.: Correlates of mall visit frequency. J. Retail. **70**, 139–161 (1994). https://doi.org/10.1016/0022-4359(94)90012-4
19. Zahid, W., Dastane, O.: Factors affecting purchase intention of South East Asian (SEA) young adults towards global smartphone brands. Asean Market. J. **8**, 66–84 (2016)
20. Morrison, D.: Purchase intentions and purchase behavior. J. Market. **43**, 65–74 (1979)
21. Fishbein, M., Ajzen, I.: Belief, Attitude, Intention and Behavior: An Introduction to Theory and Research. Addison-Wesley Publishing, Reading (1975)
22. Choi, N.H., Qiao, X., Wang, L.: Effects of multisensory cues, self-enhancing imagery and self goal-achievement emotion on purchase intention. J. Asian Fin. Econ. Bus. **7**, 141–151 (2020). https://doi.org/10.13106/jafeb.2020.vol7.no1.141
23. Mathur, A.: Incorporating Choice into an Attitudinal Framework (1999). https://doi.org/10.1300/j046v10n04_06
24. Hair, J.F., Hult, G.T.M., Ringle, C.M., Sarstedt, M.: A Primer on Partial Least Squares Structural Equation Modeling (PLS-SEM). SAGE Publications Inc, Los Angeles (2017)
25. Hair, J.F., Black, W.C., Babin, B.J., Anderson, R.E.: Multivariate Data Analysis. Pearson Education Limited, London (2014)
26. Bagozzi, R.P., Yi, Y.: On the evaluation of structural equation models. Acad. Market. Sci. **16**, 74–94 (1988)
27. Fornell, C., Larcker, D.F.: Evaluating structural equation models with unobservable variables and measurement error. J. Mark. Res. **18**, 39–50 (1981)
28. Chin, W.W.: The partial least squares approach to structural equation modeling. In: Macoulides, G.A. (ed.) Modern Methods for Business Research, pp. 295–336. Lawrence Erlbaum Associates, Mahwah (1998)

Correction to: Frequency Based Feature Extraction Technique for Text Documents in Tamil Language

M. Mercy Evangeline, K. Shyamala, L. Barathi, and R. Sandhya

Correction to: Chapter "Frequency Based Feature Extraction Technique for Text Documents in Tamil Language" in: M. Singh et al. (Eds.): *Advances in Computing and Data Sciences*, CCIS 1441, https://doi.org/10.1007/978-3-030-88244-0_8

In the originally published chapter 8 the name of the author was incorrect. The author's name has been corrected as "K. Shyamala".

The updated version of this chapter can be found at
https://doi.org/10.1007/978-3-030-88244-0_8

M. Singh et al. (Eds.): ICACDS 2021, CCIS 1441, p. C1, 2021.
https://doi.org/10.1007/978-3-030-88244-0_39

Author Index

Adi, Sourav P. II-1
Agarwal, Vasavi I-162
Agrawal, Supriya II-241
Aiswarya, L. S. I-112
Alam, Mahboob I-290
Al-ghanim, Farah Jawad I-386
Al-juboori, Ali mohsin I-386
Al-Taei, Ali I-373
Amballoor, Renji George II-22
Amjad, Mohammad I-561
Amritha, P. P. I-224
Anjum, Samia II-52
Antonijevic, Milos I-604
Anusha, K. S. I-550
Anusha, M. Geerthana II-208
Anwar, Md. Imtiyaz I-410
Arman, Md. Shohel I-244, II-160
Asim, Mohammad I-290

Baba, Asifa Mehraj I-410
Baba, Usman Ahmad I-317
Bacanin, Nebojsa I-604
Bal, Bal Krishna II-85
Balabanov, Georgi I-528
Barathi, L. II-76
Behera, Gopal I-627
Belkasmi, Mohammed Ghaouth II-97
Bettadapura Adishesha, Vivek II-1
Bezdan, Timea I-604
Bhanu Prakash, M. I-538
Bharadwaj, Keshav V. II-1
Bhatt, Anurag II-303
Bhatt, Ashutosh Kumar II-303
Bhatt, Brijesh II-107
Bhattacharya, Drishti II-252
Bhattacharyya, Balaram I-463
Bhavsar, R. P. I-102
Biplob, Khalid Been Badruzzaman II-160
Biswas, Al Amin I-338
Bouchentouf, Toumi II-97

C, Srikantaiah K. I-617
Chakrabarty, Navoneel I-255
Chakraborty, Partha I-338
Chandra, Varun II-218
Chatterjee, Riju I-56
Chatterjee, Triparna II-292
Chaudhari, Bhushan I-124
Chaurasiya, Rishikesh I-592
Chitale, Pranjal II-33
Choudhury, Hussain Ahmed I-137
Chouhan, Nitesh I-1
Chowdhury, Kounteyo Roy I-90
Cuong, Dam Tri II-412

Dabhi, Dipak I-327
Datta, Alak Kumar I-463
Dawadi, Pankaj Raj II-85
De, Paramita I-204
Deshmukh, Saurabh I-234
Doshi, Pankti II-63
Dubey, Sanjay Kumar I-512, II-303
Dwivedi, Pankaj Prasad II-271
Dwivedi, Vimal I-660

Ettifouri, El Hassane II-97
Evangeline, M. Mercy II-76

Fore, Vivudh II-116

Gala, Jay II-33
Ganatra, Amit II-12
Gandhi, Mihir II-281
Ganesan, Shamika I-475
Ghosh, Archan I-56
Gonde, Anil I-193
Gorty, V. R. Lakshmi II-172
Goswami, Kalporoop I-56
Govind, D. II-208
Gupta, Abhishek I-37
Gupta, Deepti II-136
Gupta, Govind P. I-279
Gupta, Nishant II-324
Gupta, P. K. I-351, I-729
Gupta, Subhash Chandra I-561

Habeeb, Nada Jasim I-373
Haneesh, Annabathuni Chandra I-690

Hasan, Afia II-160
Herle, Harsha I-452
Hossain, Md. Ekram I-244

Ibrahim, Mohammed Fadhil I-373
Islam, Arni I-244
Islam, Majharul II-160
Iyengar, Srivatsan I-81

Jadhav, Mayur Sunil I-124
Jadhav, Santosh P. I-528
Jadhav, Yuvraj Anil I-124
Jahan, Mahmuda Rawnak II-160
Jain, Arush I-204
Jain, Divya I-592
Jain, Kapil I-571
Jain, S. C. I-1
Jain, Sakshi Jitendra I-124
Jaiswal, Garima I-739
Jana, Angshuman II-313
Jayakumar, M. I-550
John, Jacob I-181, II-147
Joshi, Nikhil I-81
Joshi, Puneet I-268
Juneja, Mamta II-136

Kabir, Nasrin I-338
Kamat, R. K. II-183, II-195
Kamble, Pooja II-292
Kanani, Pratik II-33
Kannan, A. I-68
Kashyap, Anwesha II-313
Katkar, S. V. II-183, II-195
Kaur, Harjinder II-136
Kaur, Paramjeet I-26
Kaur, Sanmeet II-343
Kekre, Kaustubh II-33
Kesav, R. Sai I-538, II-42
Khanna, Abhishek II-343
Kharade, K. G. II-183, II-195
Kharade, S. K. II-183, II-195
Khosla, Arun I-410
Kolhekar, Megha II-292
Kumar, Ajay I-290
Kumar, Krishanth I-538
Kumar, Parmalik I-571
Kumar, Pradeep I-290
Kumar, Ravindra II-404
Kumar, S. V. N. Santhosh I-68
Kumar, Sandeep II-116
Kumar, Sanjay I-279
Kumar, Vidit I-701
Kumari, Nandini I-648

Londhe, Alka I-677
Long, Nguyen Thanh II-412
Longe, Olumide I-317

Madhavan, R. I-638
Madkar, Sanat II-63
Mahadevan, Krish I-81
Mahendru, Mansi I-512
Maheshwari, Tanay II-63
Mangaiyarkarasi, P. I-47
Mary Dayana, A. I-401
Mathai, Titty II-292
Meena, Kunj Bihari II-364
Mellah, Youssef II-97
Menon, Vijay Krishna II-208
Merani, Mann II-63
Merchant, Rahil II-63
Mishra, Girish I-711
Mishra, Preeti I-37
Mistree, Kinjal II-107
Mohite, Nilima I-193
Moon, Ayaz Hussain I-410
More, Bhushan Sanjay I-124
Muddana, Akkalakshmi I-214
Murthy, Jamuna S. II-252
Musa, Md. I-244
Musku, Ujwala I-488

Naik, Shankar B. II-22
Nain, Neeta I-627
Nair, Jayashree I-112
Namratha, R. II-252
Nand, Parma I-26
Nisat, Nahid Kawsar I-244
Norta, Alex I-660

Padmaja, K. V. I-452
Paliwal, Deepa I-15
Pant, Bhaskar I-701
Parmar, Khinal II-172
Patel, Manthan I-224
Patel, Vibha II-12
Patil, Kavita. T. I-102
Patil, Manisha I-193
Patil, Mrinmayi I-303
Patil, N. S. II-183, II-195

Author Index

Adi, Sourav P. II-1
Agarwal, Vasavi I-162
Agrawal, Supriya II-241
Aiswarya, L. S. I-112
Alam, Mahboob I-290
Al-ghanim, Farah Jawad I-386
Al-juboori, Ali mohsin I-386
Al-Taei, Ali I-373
Amballoor, Renji George II-22
Amjad, Mohammad I-561
Amritha, P. P. I-224
Anjum, Samia II-52
Antonijevic, Milos I-604
Anusha, K. S. I-550
Anusha, M. Geerthana II-208
Anwar, Md. Imtiyaz I-410
Arman, Md. Shohel I-244, II-160
Asim, Mohammad I-290

Baba, Asifa Mehraj I-410
Baba, Usman Ahmad I-317
Bacanin, Nebojsa I-604
Bal, Bal Krishna II-85
Balabanov, Georgi I-528
Barathi, L. II-76
Behera, Gopal I-627
Belkasmi, Mohammed Ghaouth II-97
Bettadapura Adishesha, Vivek II-1
Bezdan, Timea I-604
Bhanu Prakash, M. I-538
Bharadwaj, Keshav V. II-1
Bhatt, Anurag II-303
Bhatt, Ashutosh Kumar II-303
Bhatt, Brijesh II-107
Bhattacharya, Drishti II-252
Bhattacharyya, Balaram I-463
Bhavsar, R. P. I-102
Biplob, Khalid Been Badruzzaman II-160
Biswas, Al Amin I-338
Bouchentouf, Toumi II-97

C, Srikantaiah K. I-617
Chakrabarty, Navoneel I-255
Chakraborty, Partha I-338
Chandra, Varun II-218
Chatterjee, Riju I-56
Chatterjee, Triparna II-292
Chaudhari, Bhushan I-124
Chaurasiya, Rishikesh I-592
Chitale, Pranjal II-33
Choudhury, Hussain Ahmed I-137
Chouhan, Nitesh I-1
Chowdhury, Kounteyo Roy I-90
Cuong, Dam Tri II-412

Dabhi, Dipak I-327
Datta, Alak Kumar I-463
Dawadi, Pankaj Raj II-85
De, Paramita I-204
Deshmukh, Saurabh I-234
Doshi, Pankti II-63
Dubey, Sanjay Kumar I-512, II-303
Dwivedi, Pankaj Prasad II-271
Dwivedi, Vimal I-660

Ettifouri, El Hassane II-97
Evangeline, M. Mercy II-76

Fore, Vivudh II-116

Gala, Jay II-33
Ganatra, Amit II-12
Gandhi, Mihir II-281
Ganesan, Shamika I-475
Ghosh, Archan I-56
Gonde, Anil I-193
Gorty, V. R. Lakshmi II-172
Goswami, Kalporoop I-56
Govind, D. II-208
Gupta, Abhishek I-37
Gupta, Deepti II-136
Gupta, Govind P. I-279
Gupta, Nishant II-324
Gupta, P. K. I-351, I-729
Gupta, Subhash Chandra I-561

Habeeb, Nada Jasim I-373
Haneesh, Annabathuni Chandra I-690

Hasan, Afia II-160
Herle, Harsha I-452
Hossain, Md. Ekram I-244

Ibrahim, Mohammed Fadhil I-373
Islam, Arni I-244
Islam, Majharul II-160
Iyengar, Srivatsan I-81

Jadhav, Mayur Sunil I-124
Jadhav, Santosh P. I-528
Jadhav, Yuvraj Anil I-124
Jahan, Mahmuda Rawnak II-160
Jain, Arush I-204
Jain, Divya I-592
Jain, Kapil I-571
Jain, S. C. I-1
Jain, Sakshi Jitendra I-124
Jaiswal, Garima I-739
Jana, Angshuman II-313
Jayakumar, M. I-550
John, Jacob I-181, II-147
Joshi, Nikhil I-81
Joshi, Puneet I-268
Juneja, Mamta II-136

Kabir, Nasrin I-338
Kamat, R. K. II-183, II-195
Kamble, Pooja II-292
Kanani, Pratik II-33
Kannan, A. I-68
Kashyap, Anwesha II-313
Katkar, S. V. II-183, II-195
Kaur, Harjinder II-136
Kaur, Paramjeet I-26
Kaur, Sanmeet II-343
Kekre, Kaustubh II-33
Kesav, R. Sai I-538, II-42
Khanna, Abhishek II-343
Kharade, K. G. II-183, II-195
Kharade, S. K. II-183, II-195
Khosla, Arun I-410
Kolhekar, Megha II-292
Kumar, Ajay I-290
Kumar, Krishanth I-538
Kumar, Parmalik I-571
Kumar, Pradeep I-290
Kumar, Ravindra II-404
Kumar, S. V. N. Santhosh I-68
Kumar, Sandeep II-116
Kumar, Sanjay I-279
Kumar, Vidit I-701
Kumari, Nandini I-648

Londhe, Alka I-677
Long, Nguyen Thanh II-412
Longe, Olumide I-317

Madhavan, R. I-638
Madkar, Sanat II-63
Mahadevan, Krish I-81
Mahendru, Mansi I-512
Maheshwari, Tanay II-63
Mangaiyarkarasi, P. I-47
Mary Dayana, A. I-401
Mathai, Titty II-292
Meena, Kunj Bihari II-364
Mellah, Youssef II-97
Menon, Vijay Krishna II-208
Merani, Mann II-63
Merchant, Rahil II-63
Mishra, Girish I-711
Mishra, Preeti I-37
Mistree, Kinjal II-107
Mohite, Nilima I-193
Moon, Ayaz Hussain I-410
More, Bhushan Sanjay I-124
Muddana, Akkalakshmi I-214
Murthy, Jamuna S. II-252
Musa, Md. I-244
Musku, Ujwala I-488

Naik, Shankar B. II-22
Nain, Neeta I-627
Nair, Jayashree I-112
Namratha, R. II-252
Nand, Parma I-26
Nisat, Nahid Kawsar I-244
Norta, Alex I-660

Padmaja, K. V. I-452
Paliwal, Deepa I-15
Pant, Bhaskar I-701
Parmar, Khinal II-172
Patel, Manthan I-224
Patel, Vibha II-12
Patil, Kavita. T. I-102
Patil, Manisha I-193
Patil, Mrinmayi I-303
Patil, N. S. II-183, II-195

Patneedi, Shakti Swaroop I-648
Paul, Josephina I-463
Pawanekar, Sameer I-420
Pawar, B. V. I-102
Pawar, Sanjay I-303
Pawar, T. S. II-195
Petrovic, Aleksandar I-604
Pokharel, Manish II-85
Poulkov, Vladimir I-528
Premjith, B. II-42
Priyadarshini, R. II-230
Priyarup, Ankit II-218

Rahman, Md. Zıa Ur I-690
Rahul Nath, R. I-151
Rajasekaran, Vidya II-230
Rajendran, V. I-47
Raju, G. I-720
Rakshit, Sandip I-317
Ramaguru, R. II-52
Ramanathan, R. I-550
Ramteke, Ramyashri B. II-333
Rao, P. V. R. D. Prasada I-677
Rao, Pamarthi Rama Koteswara II-389
Rao, S. Venkata Achuta II-389
Reddy, Bhimireddy Shanmukha Sai I-690
Resmi, K. R. I-720
Rhouati, Abdelkader II-97

Sachdeva, Mani I-204
Sachin Kumar, S. I-475
Saheed, Yakub Kayode I-317
Saini, Rakesh Kumar II-324
Salimath, Shwetha II-292
Sam Emmanuel, W. R. I-401, I-443
Sam, I. Shatheesh II-262
Sanadi, Meenaxi M. I-617
Sandhya, R. II-76
Saquib, Zia I-303
Saranya, P. I-162
Sarath, S. I-151
Sarishma I-37
Sarkar, Paramita I-56
Saseendran, Nisheel I-171
Saxena, Geetansh I-711
Sayf Hussain, Z. I-500
Selvi, M. I-68, I-181
Sethia, Divyashikha II-218
Sethumadhavan, M. II-52
Sevugan, Prabu II-147
Shah, Parin Jigishu II-241
Shah, Priyam II-281
Shaji, C. II-262
Shankar, B. Uma I-463
Sharma, Arun I-15, I-739
Sharma, D. K. II-271
Sharma, Pankaj I-729
Sharma, Sarika II-125, II-353
Shekar, Amulya C. II-252
Shenai, Hrishikesh II-33
Shete, Prasanna II-281
Shete, Shambhavi I-234
Shivakumar, Vanaja I-361
Shrotriya, Noopur I-711
Shukla, Sharvari Rahul I-90
Shyamala, K. II-76
Siddik, Md. Abu Bakkar II-160
Sil, Arpan I-90
Singh, Mayank II-324
Singh, Meher I-81
Singh, Satbir I-410
Singhal, Niraj I-290
Sipani, Ridhi I-162
Smmarwar, Santosh K. I-279
Sneha, C. I-171
Solanki, Devansh II-281
Soman, K. P. I-475, I-538, II-42
Sonawane, V. R. II-183, II-195
Sonu, P. I-151
Sowmya, V. I-538
Srinivasareddy, Putluri I-690
Sripriya, D. II-252
Srivastava, Rohit II-116
Sruthy, P. R. I-112
Subramani, Shalini I-68
Sudhanvan, Sughosh II-241
Suganya, T. I-47
Sunil, Sharath I-151
Surekha, Sala I-690
Suryaa, V. S. I-500
Syed, Hussian I-431

Tailor, Jaishree II-12
Tajanpure, Rupali I-214
Tandon, Righa I-351
Thakor, Devendra I-327, II-107
Thool, Vijaya R. II-333
Thusar, Ashraful Hossen I-244
Timbadia, Devansh Hiren II-241
Tomar, Ravi II-116

Tripathi, Vikas I-701
Tyagi, Vipin II-353, II-364

Udgirkar, Geetanjali I-420
Utkarsh I-638

Vaidya, Anagha II-125, II-353
Vajjhala, Narasimha Rao I-317
Varkey, Mariam I-181
Verma, Mitresh Kumar I-268
Vidhya, J. V. I-638
Vijaya, J. I-431
Vins, Ajil D. S. I-443
Viswan, Vivek I-151

Waghmare, Laxman I-193

Yadav, Lokesh Kumar I-268
Yadav, Prakhar I-488
Yadav, Sumit Kumar I-739

Zivkovic, Miodrag I-604
Zope, Vidya I-81
Zulfiker, Md. Sabab I-338

Printed in the United States
by Baker & Taylor Publisher Services

Printed in the United States
by Baker & Taylor Publisher Services